PARTICLE PHYSICS

PARTICLE PHYSICS
A Comprehensive Introduction

Abraham Seiden
University of California, Santa Cruz

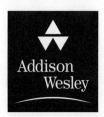

San Francisco   Boston   New York
Capetown   Hong Kong   London   Madrid   Mexico City
Montreal   Munich   Paris   Singapore   Sydney   Tokyo   Toronto

Senior Executive Editor: Adam Black
Assistant Editor: Stacie Kent
Senior Production Supervisor: Corinne Benson
Senior Marketing Manager: Christy Lawrence
Managing Editor: Erin Gregg
Manufacturing Manager: Pam Augspurger
Composition: Integre Technical Publishing Co., Inc.
Cover Designer: Blakeley Kim

Library of Congress Cataloging-in-Publication Data
Seiden, A. (Abraham)   Particle physics : a comprehensive introduction / Abraham Seiden.
    p. cm.
Includes bibliographical references and index.
ISBN 0-8053-8736-6
1. Particles (Nuclear physics) I. Title.

QC793.2.S42 2004
539.7'2–dc22

2004009922

(ISBN) 0-8053-8736-6

Copyright © 2005 Pearson Education, Inc., publishing as Addison Wesley, 1301 Sansome St., San Francisco, CA 94111. All rights reserved. Manufactured in the United States of America. This publication is protected by Copyright and permission should be obtained from the publisher prior to any prohibited reproduction, storage in a retrieval system, or transmission in any form or by any means, electronic, mechanical, photocopying, recording, or likewise. To obtain permission(s) to use material from this work, please submit a written request to Pearson Education, Inc., Permissions Department, 1900 E. Lake Ave., Glenview, IL 60025. For information regarding permissions, call (847) 486-2635.

1 2 3 4 5 6 7 8 9 10 —DOC— 07 06 05 04
www.aw-bc.com/physics

Preface

This text covers the fundamental concepts and theories that make up the Standard Model of particle interactions. The effort to develop this model continues the tradition of looking inside the atom and the nucleus to see how these systems are constructed and the principles that govern their behavior, providing an understanding of matter at ever-deeper levels.

The Standard Model describes these interactions through a number of fundamental constituents, three families of quarks and leptons and the gauge field quanta with which they interact. These constituents have been revealed through the study of natural objects, as well as controlled interactions at several generations of accelerator facilities. The Standard Model has provided the quantitative basis for understanding the very large body of data accumulated at particle accelerators of ever-increasing energy and luminosity. In this text the various types of interactions are presented in sequence, allowing each part of the Standard Model to be examined in some detail. The various concepts are always introduced in conjunction with many detailed phenomena explained by the given idea; general principles, stemming for example from quantum mechanics, are also emphasized.

Particle physics, at the level of this text, is usually taught in the second year of graduate study, following advanced quantum mechanics, and often during the same year as a field theory class. Since the sequence and numbers of classes taken varies from one institution to another, I have attempted to make the material presented in the text self-contained. Thus the Dirac equation is presented, as well as some material regarding field theory. This inclusion is not meant to replace a class covering these topics more fully. I have, however, assumed that the student has an understanding of quantum mechanics, including angular momentum and related topics, such as Clebsch–Gordan coefficients. In the case of field theory, I have tried to show what the mathematics is "doing," rather than attempting rigorous derivations of the material. The material is presented using a multi-particle theory, rather than the single particle equation that is often used, and hopefully this will contribute to the appreciation of the content of field theory. The approach taken allows a uniform treatment of fermions and bosons, as well as particles and antiparticles. Following the introductory material, which is contained in the first three chapters, the full phenomenology of particle physics is presented, based on the fundamental constituents. This includes the spectrum of states of the strong interactions, decays and mixing phenomena due to the various interactions, and the behavior of scattering cross sections. In addition, a major goal is to work through a large number of calculations, particularly for the benefit of experimentalists who may not have an opportunity to see these calculations in any other class. A number of homework problems accompany each chapter, allowing students to test their understanding of the material and in some cases extend it.

In presenting particle physics we can write down a number of equations and try to deduce their consequences, or work up through the phenomena toward a theoretical synthesis. The material in the text is presented mostly using the latter approach. This allows the student the opportunity to work through elements of the theory more slowly and to understand some of the evidence that led to the components of the Standard Model. In addition, the nonperturbative parts of the strong interaction are mostly not rigorously calculable, so we have to develop a separate understanding of these phenomena, which show their own regularities. The lightest hadrons have been used for this purpose. The spectrum of these states and their decays illustrate a number of interesting phenomena. The text covers the strong (Chapter 5), electromagnetic (Chapter 6), and weak (Chapter 8) decays of these particles in some detail.

This text provides material for a one-year long class. It can, however, also be used for a semester-long special topics class, for example, on the weak interactions. The fraction of the material covered in the first three chapters can be adjusted based on which other classes are taught at a given institution. Topics covered in some detail, such as the constituent quark model and weak decays, can be covered more briefly, with additional reading left to the student. One topic not covered is the connection of particle physics to astrophysics and cosmology. This is an exciting connection and I would expect that at some institutions this material will be taught as a semester-long special topics class following the particle physics class.

The emphasis here is pedagogical. No attempt has been made to present topics in their historical setting. In addition, a number of important topics, such as techniques used in particle detectors and particle acceleration, are not covered due to space limitations. A number of books now cover these topics; some of these are listed in a bibliography. They would provide interesting supplementary reading. This text generally does not contain the references to the original scientific literature; however, many useful references are provided in the bibliography. These are chosen for pedagogical reasons and no credit for original discovery is implied by the selection.

The text should allow the instructor the scope for some interesting additions. For example, although specific experiments are not discussed, the instructor can include material relevant to the experiments at a given institution. Material can also be added about particle detection techniques relevant to these experiments. The calculations in Chapter 11 can form the basis for such a discussion. These additions can help students make the transition to research at the given institution. In addition, the instructor can add graphic material to discussions that are mostly algebraic or descriptive. Examples include additional Feynman diagrams and various unitarity triangles for quark mixing.

The material presented illustrates the beautiful picture we have of particle physics at the start of the 21st century. Important missing pieces in this picture should be found in the next decade. The last chapter presents some ideas of where nature may lead us. The discussion is very brief, since it is not clear which ideas will prove to be a key to the future. The student can expect to learn more about these topics over the coming years.

ACKNOWLEDGMENTS

Being a participant in a vigorous scientific community is a special privilege. The training one receives provides the common language needed to appreciate and enjoy the scientific progress made by the many practitioners in the field. As a truly international activity, particle physics also has the additional advantage that one can participate with a mix of interesting singular individuals from many countries. Thus the approach that one takes to the subject reflects many fruitful interactions with others in the community. The writing of a textbook offers the opportunity to share the ideas and language of the field with a new generation of scientists.

The author wishes to thank Nicolo Cartiglia for a critical reading of many of the chapters. The work on word processing by Nora Rogers, Rachel Cunningham, and Emily Williams, over a period of several years, has made it possible to produce the text. The author also wishes to thank the people at Addison-Wesley for their help.

A. Seiden
Santa Cruz, California, 2004

Contents

Preface v

1 ■ The Particle Physics Program 1
1.1 Introduction 1
1.2 What Do We Measure? 3
1.3 Fundamental Constituents and Interactions 6
Chapter 1 Homework 13

2 ■ How to Calculate Amplitudes 15
2.1 Free Particles 15
2.2 Spin Zero Particles 16
2.3 Lagrangian Density 18
2.4 Symmetries and the Lagrangian 20
2.5 Spin 1 Particles 23
2.6 Spin 1 Photons 24
2.7 Spin $\frac{1}{2}$ Particles 27
    2.7.1 The Dirac Equation 28
    2.7.2 Lagrangian and Symmetries for Spin $\frac{1}{2}$ 30
    2.7.3 Explicit Plane Wave Solutions 32
    2.7.4 Bilinear Covariants 35
2.8 Interpretation of Extra Solutions 37
2.9 Time History of States 38
2.10 Time Evolution and Particle Exchange 43
2.11 Momentum Space Propagator 50
2.12 Calculation of Decay Rates and Cross Sections 52
    2.12.1 Particle Decay Rates 52
    2.12.2 Cross Sections 54
2.13 Particle Exchange and the Yukawa Potential 55
2.14 Quantum Field Theory 58
    2.14.1 Charged Scalar Field 63

2.14.2 Fermion Field   64
Chapter 2 Homework   68

3 ■ Scattering of Leptons and Photons   71

3.1 Interaction Hamiltonian and Lepton Currents   71
3.2 Electron–Muon Scattering   74
    3.2.1 The Photon Propagator   75
    3.2.2 Some Implications of the Photon Propagator   78
    3.2.3 Electron–Muon Scattering; Methods for Facilitating Calculations for Spin $\frac{1}{2}$ 79
    3.2.4 Two-Body Phase Space Factor   83
    3.2.5 Cross Section in the Relativistic Limit   85
3.3 e^+e^- Annihilation to $\mu^+\mu^-$ 86
    3.3.1 s, t, and u Variables   86
    3.3.2 Calculation of Annihilation Cross Section   87
3.4 e^-e^- Scattering   89
3.5 e^+e^- Production of e^+e^- 90
    3.5.1 Helicity Conservation in the Relativistic Limit   92
3.6 Processes with Two Leptons and Two Photons   94
    3.6.1 Matrix Element for γe^- Scattering   97
    3.6.2 Compton Scattering Cross Section   100
3.7 Higher-Order Terms in the Perturbation Expansion, the Feynman Rules   103
3.8 Lorentz Covariance and Field Theory   106
3.9 Special Symmetries of the S Matrix   108
Chapter 3 Homework   109

4 ■ Hadrons   112

4.1 The Charge Structure of the Strong Interactions   112
4.2 Quantum Numbers   113
    4.2.1 Additive Quantum Numbers   114
    4.2.2 Vectorially Additive Quantum Numbers   115
    4.2.3 Multiplicative Quantum Numbers   116
4.3 Internal Symmetries   117
4.4 Generators for $SU(2)$ and $SU(3)$ 122
4.5 The Color Interaction   126
4.6 The Color Potential   130
4.7 Size of Bound States   136
4.8 Baryons   141
Chapter 4 Homework   142

Contents xi

5 ■ Isospin and Flavor $SU(3)$, Accidental Symmetries   146
    5.1   The Light Quarks   146
    5.2   Light Mesons   149
    5.3   Breit–Wigner Propagators and Meson Life Histories   155
            5.3.1   Narrow Resonances and Independent Events   159
            5.3.2   Propagators and Mass Eigenstates   161
    5.4   Vector Meson Decays   163
            5.4.1   Form of Meson Decay Amplitudes   168
    5.5   Physical Picture of Decay Process   169
            5.5.1 G Parity   177
            5.5.2   Large Invariant Mass Processes and Hadronic Jets   179
    5.6   Baryon States of Three Quarks   180
    Chapter 5 Homework   186

6 ■ The Constituent Quark Model   193
    6.1   Constituent Quarks   193
    6.2   Baryon Magnetic Moments   193
            6.2.1 Σ^0 Decay to $\Lambda^0 + \gamma$, Magnetic Dipole Transition   197
            6.2.2 ϕ Decay to $\eta' + \gamma$ 198
    6.3   Meson and Baryon Masses   200
            6.3.1   Meson Masses   203
            6.3.2   Baryon Masses   204
            6.3.3   Meson Isospin Violating Mass Splittings   205
            6.3.4   Baryon Isospin Violating Mass Splittings   207
            6.3.5   Decays of the η Meson   209
    6.4   Photon Coupling to the Vector Mesons   211
            6.4.1   Vector Meson Dominance   212
            6.4.2 ρ Dominance in the π^+, π^- Channel with $J^P = 1^-$ 214
            6.4.3 ρ, ω Mixing   217
    6.5   Radiative Transitions between Pseudoscalar and Vector Mesons   220
    6.6   Pseudoscalar Meson Decays to Two Photons   224
            6.6.1   Vector Meson Dominance and Radiative Decays   225
    6.7   Conclusion   226
    Chapter 6 Homework   227

7 ■ The Full Color Gauge Theory   231
    7.1   Local Gauge Symmetry   231
    7.2   Paradox of No Scales   234
            7.2.1   The Running Coupling Constant in Electrodynamics   234
            7.2.2   Expression for $\alpha(q^2)$ in Electrodynamics   236

- 7.2.3 Running Coupling Constant for QCD 238
- 7.3 Approximate Chiral Symmetry of the Strong Interactions 240
  - 7.3.1 Goldberger–Treiman Relation 241
- 7.4 Spontaneously Broken Symmetry 244
  - 7.4.1 The Role of the Vacuum 245
- 7.5 Quark Masses in the Lagrangian 246
- 7.6 Other Issues 247
- Chapter 7 Homework 249

8 ■ Weak Interactions of Fermions 253

- 8.1 Weak Gauge Group 253
- 8.2 Muon Decay 258
- 8.3 Decays of the Tau Lepton 262
- 8.4 Charged Weak Currents for Quarks 267
  - 8.4.1 Cabibbo–Kobayashi–Maskawa Matrix 269
  - 8.4.2 CP Violation 272
- 8.5 Charged Pion Decay 274
  - 8.5.1 Conserved Vector Current 277
  - 8.5.2 Charge Operators 278
  - 8.5.3 Rate for π^- Semileptonic Decay 280
- 8.6 Strangeness Changing Current Operator and Kaon Decay 280
  - 8.6.1 Vector Dominance Model for Kaon to Pion Current Matrix Element 283
  - 8.6.2 Operator for K Decay to All Hadronic Final States 286
- 8.7 General Framework for Weak Decay of Pseudoscalar Mesons 292
- 8.8 Amplitudes for Kaon Decay to Two Pions 295
- 8.9 Amplitudes for K Decay to 3π 297
- 8.10 Rare K^0 Decays 298
- 8.11 Weak Decays Involving the Heavy Quarks c, b, t 300
  - 8.11.1 Weak Decay of Charm 302
  - 8.11.2 Weak Decay of b Quark Systems 305
- 8.12 Heavy Quark Effective Theory 307
- 8.13 Conclusion 309
- Chapter 8 Homework 310

9 ■ Weak Mixing Phenomena 313

- 9.1 Interplay of Production, Propagation, and Detection 313
- 9.2 Mixing for Weakly Decaying Pseudoscalar Mesons 315
- 9.3 $K^0, \bar{K}^0$ System 318
  - 9.3.1 Kaon Oscillations 320

          9.3.2 CP Violation in the Kaon System   321
   9.4 $D^0, \bar{D}^0$ System   327
   9.5 $B^0, \bar{B}^0$ and $B_s^0, \bar{B}_s^0$ Systems   329
          9.5.1 CP Violation in the $B^0, \bar{B}^0$ System   331
   9.6   Neutrino Oscillations   333
          9.6.1   Three Neutrino Generations   336
          9.6.2   Oscillations of Neutrinos from the Sun   338
          9.6.3   Matter-Induced Oscillations in the Sun   341
          9.6.4 CP Violation in Neutrino Mixing   344
          Chapter 9 Homework   346

10 ■ The Electroweak Gauge Theory and Symmetry Breaking   348

   10.1   Weak Neutral Current   348
   10.2   Neutral Current Mixing   348
   10.3 Z^0 Phenomenology   352
   10.4   Interactions among the Gauge Bosons Themselves   358
   10.5   Higgs Mechanism   360
   10.6   The Theory of Weinberg and Salam   365
   10.7   Corrections to the W and Z Masses   370
   10.8   Generation of Fermion Masses   373
   10.9   Majorana Neutrinos   375
          Chapter 10 Homework   376

11 ■ Large Cross Section Processes   378

   11.1   Types of Processes   378
   11.2   Multiple Coulomb Scattering   380
          11.2.1   Multiple Scattering Angle   382
          11.2.2   Radiation Length   383
          11.2.3   Energy Loss   385
   11.3   Radiative Processes for Electrons and Photons   389
          11.3.1   Rate Calculation for Pair Production   391
   11.4   Inclusive Distributions in Hadronic Scattering   395
          Chapter 11 Homework   401

12 ■ Scattering with Large Momentum Transfer   404

   12.1   Types of Processes   404
   12.2 e^+e^- Annihilation to Hadrons   405
          12.2.1   Energies below Z^0, the Electromagnetic Regime   405
          12.2.2   Quark Fragmentation Functions   406
          12.2.3 e^+e^- Annihilation in the Electroweak Regime   410

           12.2.4 Energy Scale for Production of the Higgs Particle   412
   12.3 Hadron Structure and Short Distance Scattering   412
           12.3.1 Momentum Spectrum for Constituents   415
   12.4 Deep Inelastic Lepton-Proton Scattering   417
           12.4.1 Cross Section for Deep Inelastic Scattering   419
           12.4.2 Structure Functions and Constituents   420
           12.4.3 The Quark Picture for the Structure Function   423
   12.5 Scaling Violations   429
           12.5.1 Evolution Equations for Structure Functions   434
   12.6 Results for Large Transverse Momentum Scattering
           in $\bar{p}p$ Reactions   436
   12.7 The Next Frontier   437
           Chapter 12 Homework   438

13 ■ Physics at Higher Energies   441

A ■ Conventions   445
   A   Units   445
   B   Use of Lorentz Indices   446

Bibliography   449

Index   455

CHAPTER 1

The Particle Physics Program

1.1 ■ INTRODUCTION

Particle physics aims at the discovery of the basic forms in which matter can occur and the interactions among these forms. The search for the simplest, most basic objects has led to the study of matter at very small distances. We call the forms of matter on these short distance scales particles. Interactions of the particles must be described by quantum mechanics and the present state of experimentation has allowed the study of physics at distance scales down to about 10^{-16} cm. These studies typically require high energies; in fact, the energies in such particle processes are frequently large compared to the masses of the particles involved, implying relativistic motion. As a consequence, the production of new particles through the interactions is typical, since the large kinetic energy carried by the interacting particles can provide the energy for the creation of additional particles.

To incorporate relativity into quantum mechanics will require some new ideas. In particular, the requirement of relativistic invariance is a strong constraint on how we can picture the interactions between the particles. Basic interactions must be local in character. There are no instantaneous interactions at a distance. We thus picture a particle process or quantum event as an interaction at a point in space-time of objects of very small size (perhaps vanishingly small for the "fundamental" particles). This is shown in Figure 1.1. We, of course, need to define the meaning of such a picture, including the possible participants and the strength of the interaction.

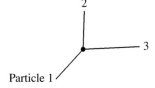

FIGURE 1.1 Representation of a local interaction involving three quantum particles at a given space-time point. This could represent the decay of particle 1 into particles 2 and 3.

In a given quantum event, all the particles obey the same quantum rules, which replace the classical physical picture of a relativistically covariant local interaction of a particle with a different entity, the classical field. In fact, we will use the word particle or field rather interchangeably to describe the particles we discuss. We can think of a single particle as the simplest field configuration after the vacuum.

The basic quantum event allows for the creation and absorption of particles, as first seen most clearly in the photo-electric effect. The particles involved carry energy and momentum, as well as other attributes such as spin and charge. These are absorbed or created fully in the event and the possible values are governed by the allowed quantum states for each particle type. Although the interactions are local, the quantum particles are not fully localized, but rather their "presence" is spread over space-time (e.g., described by a wave function). Thus, calculating the effect of the local interactions in real situations always involves adding contributions spread over space-time.

The space-time evolution and detection of the particle systems obey the basic principles of quantum mechanics. Primarily, these are:

1. For any system there exists a complete set of base states, which forms a basis that we can use to describe the system when measuring its characteristics. The values found for physical observables correspond to the values that characterize the appropriate base states. The allowed values may be either discrete or continuous. Various types of measurements may or may not be mutually compatible, leading to the limitations quantified in the uncertainty principle in the case of incompatible measurements.

Examples of base states are the quantized energy states of atoms, molecules, and nuclei. These begin at lowest energy with discrete bound states, followed by a continuum representing the breakup of the system. These states are also characterized by internal angular momenta, which are discrete.

A physical system we will often talk about is a beam of particles to be used in a scattering experiment. The particles in such beams are usually described in terms of a continuum of energy and momentum base states. To achieve a complete description also requires specification of the particle spin orientation, which is a discrete variable.

2. The evolution of a system from an initial state to a final base state is described by a complex amplitude whose square gives the probability of finding the system in the final base state. An example of an amplitude is the nonrelativistic wave function, for which the particle state has been projected onto position states. After squaring, it gives the probability density for finding the particle at a given position as a function of time.

3. Quantum mechanics is linear. This implies that the amplitude for the evolution of the system from a linear combination of initial base states to a given final state is a linear combination of the amplitudes describing the evolution of the individual base states, had they alone been the initial state.

For a single particle that goes from a spatially localized initial state to a well-separated localized final state, we can calculate an amplitude for each trajectory or path in space-time that the particle could have taken. The linearity condition applies to these amplitudes. For indistinguishable alternative paths connecting the initial and final state, the complex amplitudes for the alternatives add when constructing the full amplitude. This has long been known for light, which exhibits various diffraction and interference phenomena, where the quantum events between which propagation occurs are the initial emission and final absorption of the light quantum. The alternative paths are the available trajectories the light quantum can take. Analogous interference phenomena are seen for other particles such as neutrons or electrons.

The use of amplitudes, and the interference of alternatives, gives a wave-like propagation between quantum events. Combined with the particle-like local interaction, this gives rise to the wave-particle duality so characteristic of quantum mechanics.

4. The amplitudes must have a Lorentz covariant description, since we insist that the physical picture of the system and its evolution must allow an equivalent interpretation in all inertial reference frames.

The principles above have been found to be universal. For example, there can be no strictly classical objects; all objects are quantum in nature. In addition, there are no arbitrary potentials; rather, a potential is a consequence of the local particle interactions and propagation. This is remarkable and exciting progress in the development of a unified view of the rules by which nature works.

Typical of physics, the variety of phenomena allowed by the same basic rules is truly surprising. For example, the quantum rules, when combined with the different particle masses and interaction strengths, provide a multiplicity of structures or systems with different characteristic dimensions and energy scales. Each such system has its base states and amplitudes, which satisfy equations usually solved using approximations emphasizing the relevant degrees of freedom of the system and therefore valid over some range of distances and energies. Some examples of increasing characteristic energy are the states within a crystal, the behavior of the atom, the structure of the nucleus, and the nucleons within the nucleus.

1.2 ■ WHAT DO WE MEASURE?

The study of the forms and interactions of matter is based on observations of natural systems and on experiments arranged to create controlled interactions. The latter, using large accelerators, have become the primary research tool for looking at matter at very short distance scales. The present state of experimentation has probed energy scales up to approximately 200 GeV. This corresponds to distances on the order of 10^{-16} cm. For some types of measurements we understand the physics down to 10^{-17} cm. The types of observations typically made in experiments are characterized below.

The Particle Spectrum This involves the determination of the base states consisting of a single stable isolated particle, along with any quantum numbers needed for a complete description of these energy eigenstates. Such a particle may be a composite object. These states are eigenstates of the Hamiltonian operator that determine the time evolution of the system; they provide information on the objects and interactions the Hamiltonian describes.

Scattering of Particles In addition to seeking the spectrum of individual particles, we can create collisions between particles. This leads to measuring the results of a scattering experiment. Here beams of particles are directed onto each other or onto stationary targets. The resulting collisions allow us to discover new final state particles, as well as the characteristics of the interactions of the initial particles.

Beams for scattering experiments typically originate with either low energy electrons stripped from materials, or protons from ionized hydrogen. These particles are accelerated and can be used directly in experiments, or they can be used to make beams of secondary particles created in collisions with intermediate targets. These secondary beams can then be used in scattering experiments.

The final collisions of interest involve two particles at a time, one from the beam and one from the final target. Prior to collision, these were two widely separated, noninteracting particles. Following the collision, we measure the final particles at a distance far removed from the collision region. Here the state can again be described in terms of separate noninteracting isolated particles. To achieve maximum total energy, the target is often chosen to be a beam of high momentum particles.

The states we will use to describe the motion of the initial and final particles will usually be momentum eigenstates. Beams are usually constructed to have reasonably well-defined momenta, and measurements on final state particles typically aim at a momentum determination. We will occasionally use states of given angular momentum, but this is cumbersome for final states with many particles.

To complete the space-time description of a momentum eigenstate requires specification of the spin projection along the direction of motion, called the *helicity*. Final states of different helicity are distinguishable and do not interfere. For the initial colliding particles, the state can be specially prepared to have unequal populations of the various helicities providing a polarized initial state. In the common case, where nothing is done to preferentially populate the spin states, the initial particles contain an incoherent statistical mixture of all helicities, often reflecting the state from the initial particle source.

Since the evolution of the system is determined by the quantum principles, the description of the scattering process is contained in an amplitude that is a function of the initial and final momenta and helicities

$$A\left(\vec{P}_1, \lambda_1^i; \vec{P}_2, \lambda_2^i; \vec{k}_1, \lambda_1^f; \ldots; \vec{k}_n, \lambda_n^f\right), \tag{1.1}$$

where $\vec{P}_1, \lambda_1^i; \vec{P}_2, \lambda_2^i$ are the initial momenta and helicities and $\vec{k}_1, \lambda_1^f; \ldots; \vec{k}_n, \lambda_n^f$ are the analogous quantities for the final particles, for the case of n final particles. Rates are determined by the square of the amplitude yielding a density of events in the momentum space describing the n final particles. The calculation must yield an answer that is Lorentz covariant, which will provide a constraint on how the momenta appear in the amplitude. Finally, the amplitude must be linear in the spin degrees of freedom by the linearity requirement.

In the case of identical particles in the final state, the amplitude must reflect the spin-statistics relation obeyed by all systems, including identical complex composite systems emerging as separate isolated particles. If particles i and j are identical, the amplitude must satisfy

$$A\left(\ldots; \vec{k}_i, \lambda_i^f; \ldots; \vec{k}_j, \lambda_j^f \ldots\right) = \pm A\left(\ldots; \vec{k}_j, \lambda_j^f; \ldots; \vec{k}_i, \lambda_i^f \ldots\right),$$

1.2 What Do We Measure?

with the $+$ sign for particles of integral spin (called bosons) and the $-$ sign for particles of half-integral spin (called fermions). Care must also be taken that physically indistinguishable configurations are not counted more than once when calculating rates by integrating over the momentum space.

Production of Resonances A third type of measurement involves a determination of the properties of resonances. These are isolated particles that live long enough to leave a scattering process, but subsequently decay spontaneously. Resonances are described by a mass m, and decay lifetime τ. We will generally use the reciprocal of the lifetime, $\Gamma = 1/\tau$, called the decay width, which has units of energy.[1] For these states the mass is usually determined as an eigenstate of a part of the full Hamiltonian. Thus the mass spectrum and pattern for these states is often similar to the spectrum for fully stable particles and we will seldom distinguish between the two. The width is generated by the remaining interactions, which may have little to do with the mass, but cause transitions to final states of several particles. In general, several decay modes are possible, each contributing a fixed fraction of the decay width. Resonances provide a nice system in which to study the interactions responsible for their decay, since they are the simplest initial state we can prepare whose probability is time dependent. We give some examples below.

A familiar and simple example of a stable state and a set of related resonances is the ground state and excited states of hydrogen. Here, all the bound state energies can be calculated to high accuracy in terms of the Coulomb potential and electron kinetic energy. The decay of the excited states comes from the emission of photons leading to a transition to a state of lower energy. Both binding energies and widths result from electromagnetic interactions; however, these can be split into a static potential (with no real photons present) and the emission of a separate real photon from the resonance. Note that in thermal equilibrium, as opposed to in isolation, the relative populations of the ground state and the resonances are determined by the temperature. This is not the situation of accelerator experiments where particles are detected as isolated objects and transitions occur from heavier to lighter states.

Another interesting example is provided by the three π mesons π^+, π^0, and π^-. All three have nearly the same mass, determined by the strong interactions. The π^+ and π^- decay via weak interactions and live a sufficiently long time to allow production of beams of these particles. The π^0 decays into two photons, a process forbidden by charge conservation for the π^+ and π^-, and has a lifetime about 10^8 times shorter than that of the charged states. Here the masses and widths come from entirely different interactions.

As a final example, we can look at the ρ^0 and ω^0 resonances that decay into $\pi^+\pi^-$ and $\pi^+\pi^-\pi^0$, respectively. The masses of these are again determined by the strong interactions and the two have nearly the same mass and are made of the same constituents in the same spatial state. If we think in terms of a poten-

---

[1] We use units $\hbar = c = 1$. See the Appendix for discussion of this.

tial, we might guess that it is the same for both states, given the near equality of the masses. Both particles decay via the strong interactions, so that they live a very short time. The decay comes again from the creation of new quanta, which combine with the initial constituent quanta to make the final state pions. We will discuss later the principles that dictate why one state decays into two and the other into three pions. The net result is that Γ_ρ is about 20 times Γ_ω. We see again that widths can differ significantly for resonant states that have much in common.

In all the examples above the widths are determined by the interaction dynamics. Particles have no "intrinsic" widths. Masses are also at least partially dynamically determined, certainly for bound systems, and one could ask whether there are any "intrinsic" masses. One of the surprising discoveries is that there appear to be no "intrinsic" masses for the stable states of the spectrum, including our familiar particle, the electron! These masses result from dynamics, although the dynamics are not fully understood at present. We will discuss the mass-generating interaction in some detail in Chapter 10.

Jets at High Energies In very high-energy scattering processes producing strongly interacting particles, the number of particles produced can be very large. Focusing on the distribution function that keeps track of all of the many particles can obscure the global characteristics of the final state. We find that the final state particles can often be grouped together into individual "jets" that reflect the underlying physics more clearly. Each jet contains a number of particles close together in momentum space and separated from the other particles in the scattering event. Unlike resonances, which escape the collision volume and decay through an independent process yielding a well-defined average number of final particles independent of the momentum of the resonance or how it originated, the jets are not independent of each other in an event. Thus there is no well-defined average number of particles characterizing the decay of a single jet.

The simplest process illustrating the production of jets is the annihilation of a positron with an electron into strongly interacting particles emerging in two jets. An example of how such a two-jet event might look is shown in Figure 1.2. The average number of particles in the final state is determined by the pair of jets and grows with the invariant mass (or total center of mass energy) of the jet pair. As this energy grows, the final particles are more and more collimated into the two jet structures, making it progressively easier to define their directions and energies. The question then becomes, what determines the jet energy and direction? Whose direction is it giving us? We will discuss this in the next section, where we turn from what we measure to the forms of matter underlying the measurements.

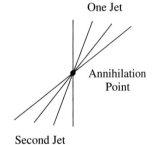

FIGURE 1.2 Two-Jet final state that might be produced in e^+e^- annihilation. Each jet has four particles.

1.3 ■ FUNDAMENTAL CONSTITUENTS AND INTERACTIONS

The set of all known particles and their interactions up to an energy scale on the order of 200 GeV can be described in terms of a small set of constituents of matter. We exclude here, however, gravity, which is an extremely weak interaction for particle pairs at presently available energies. The particles and interactions have a certain simplicity and unity and form what is called the **Standard Model**. We will spend most of our time developing this picture and relating what is seen experimentally to the postulated constituents. This section serves to introduce these ideas.

We begin by describing the simplest set of particles, called *leptons*, and their interactions, starting with interactions that are large at a scale of a few GeV.

There are a total of six leptons, e^-, μ^-, τ^-, ν_e, ν_μ, and ν_τ (we do not list the antiparticles as separate particles). Three of the leptons are electrically charged, having the same charge as the electron, and three, the neutrinos, are neutral. The masses of the neutrinos are small, and presently under very active investigation. The charged leptons have masses:

$$m_e = 0.5110 \text{ MeV},$$

$$m_\mu = 105.7 \text{ MeV},$$

$$m_\tau = 1777 \text{ MeV}.$$

There is no fundamental understanding of the mass pattern. All leptons have spin $\frac{1}{2}$ and are therefore fermions.

To first approximation, at the few GeV scale, leptons interact only with the electromagnetic field, which is described by a gauge theory that we discuss in Chapter 2. This gauge interaction results in only one field particle, the photon, transmitting the force. The gauge field theory guarantees several crucial features of electromagnetism: The photon is massless, it has spin 1, and it couples to a conserved current, determined by charged particles. The latter property is already familiar from classical electrodynamics, as is gauge invariance. The basic quantum event describing the electromagnetic interaction of the leptons is the absorption or the emission of individual photons, shown in Figure 1.3. Such an event is called an interaction vertex. It is generally part of some larger process, for example the scattering processes discussed in Chapter 3.

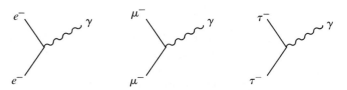

FIGURE 1.3 Basic vertices for the electromagnetic interaction of the charged leptons.

Neutrinos have no charge and thus have no interactions at all to this approximation. The photon also carries no charge, so it does not interact with itself. In this approximation there are no events that couple different leptons. Thus, all six would be stable; none would decay into any others.

However, already at low energy, we find an additional interaction, which appears to be very weak compared to electromagnetism. It is visible mainly because it has different interaction vertices, allowing new types of reactions that do not occur at all for vertices involving photons. These reactions treat the six leptons as three families (called doublets), which we group as follows:

$$\begin{pmatrix} \nu_e \\ e^- \end{pmatrix}, \begin{pmatrix} \nu_\mu \\ \mu^- \end{pmatrix}, \begin{pmatrix} \nu_\tau \\ \tau^- \end{pmatrix};$$

with transitions occurring between pairs within a family via the emission of new particles called the W^+, or its antiparticle, the W^-. These have spin 1, like the γ, and are called gauge bosons. They are heavy and have finite lifetimes when produced as real final state particles.

With this interaction, the μ^- and τ^- decay via the processes shown in Figure 1.4. In these diagrams time flows in the direction that takes us from the initial to the final state, which is upward in the figure. Such pictures are called Feynman diagrams. These are second order, involving two interaction vertices and the propagation (or exchange) of a W between. The initial energy is $\ll M_W$; the propagating W is called a virtual particle. All decays shown are energetically allowed and are observed. What is now conserved by these diagrams is the sum of the number of leptons of each family type, for example, μ^- plus ν_μ. Note that the final e^- particle plus $\bar{\nu}_e$ (an antiparticle) in μ^- or τ^- decay produce no net electron lepton number, which is conserved separately.

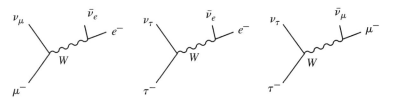

FIGURE 1.4 Weak decay processes involving only leptons. Processes include two vertices.

The stable leptons, considering the electromagnetic and weak interactions, are e^-, ν_e, ν_μ, and ν_τ. These are the lightest members of each family, plus one charged lepton whose stability is required by electric charge conservation. These stable particles are the leptons we find around us. The others have a fleeting existence on earth at accelerator laboratories or in cosmic ray showers.

The foregoing discussion could have focused equally on the various antiparticles, for example, e^+, and did not tell us what to expect for the net value of the various lepton numbers in the cosmos. We believe that this was determined during the very early evolution of the universe, although the details are not well understood. The net result, however, was the generation of an asymmetry between the numbers of particles and antiparticles, resulting in an excess of particles.

1.3 Fundamental Constituents and Interactions

In addition to causing μ^- and τ^- decays, the weak interactions provide the mechanism for the neutrinos to have very weak scattering processes with other particles at low energies. In these processes the neutrino can turn into its charged lepton partner via W exchange, or it can remain unchanged. The latter process requires a new, electrically neutral intermediate particle called the Z^0, which is closely related to the W particles. Examples of such scattering processes are shown in Figure 1.5.

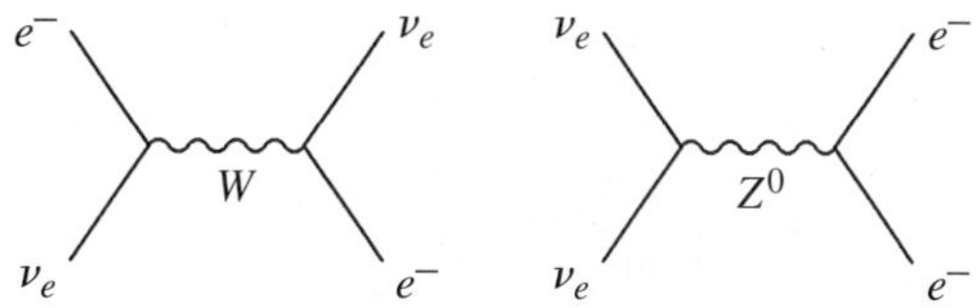

FIGURE 1.5 Two Feynman diagrams leading to the same final state and involving different virtual intermediate states.

The transitions between the members of a doublet by three gauge bosons is described by a non-Abelian gauge theory denoted $SU(2)$, which we will discuss extensively in Chapter 8. The theory has an $SU(2)$ symmetry, that is, a symmetry under unitary transformations among the doublet members and gauge bosons. An expected consequence of the symmetry is that particles that transform into each other via the unitary transformations have the same mass. Such a theory would be expected to yield three massless spin 1 gauge bosons, analogous to the massless photon, and requires degenerate fermions within a doublet family. Neither of these is true! The deviation of what is seen from what is expected is called symmetry breaking. For example, $M_W = 80.4$ GeV and $M_Z = 91.2$ GeV. The large value of the gauge boson masses is responsible for the apparent weakness of the interaction at low energies and the fact that there are no bound states of this interaction. The interactions are not weak compared to electromagnetism for energies much greater than the boson mass.

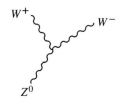

FIGURE 1.6 Vertex involving $Z^0W^+W^-$ at a point in space-time.

The breaking of the $SU(2)$ symmetry is attributed to a new interaction; we leave to later discussion how the details of this theory are emerging experimentally. This is the mass generating interaction alluded to earlier. Before continuing to the next set of constituents, we mention one other interesting point. The photon does not interact with itself, which we attribute to it not carrying electric charge. In contrast, in a non-Abelian theory the gauge bosons carry the "charge" that creates the interaction and interact with each other. This is true of all particles that are not singlets under the unitary transformations of the gauge group, whereas singlets do not interact. Thus there are vertices such as the one shown in Figure 1.6, a process that is visible at sufficiently high energy. Since the W^+ and W^- carry electric charge, there must also be a vertex of the type shown in Figure 1.7. Apparently, the photon is not a singlet under the weak $SU(2)$ symmetry. This process is a hint that the photon is somehow linked to the three vector bosons. This will

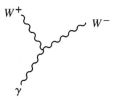

FIGURE 1.7 Vertex involving γW^+W^- at a point in space-time.

be seen to be true and the set of reactions involving all four bosons, which is the full set of known reactions for the leptons, are collectively called the electroweak interaction.

We turn next to the remaining fermions, called *quarks*. These again come in three doublet families that interact via the electroweak interaction. We denote the quark types (called flavors) u (up), d (down), c (charm), s (strange), t (top), and b (bottom) with doublet families:

$$\begin{pmatrix} u \\ d \end{pmatrix}, \begin{pmatrix} c \\ s \end{pmatrix}, \begin{pmatrix} t \\ b \end{pmatrix}.$$

All upper members of a doublet have electric charge $+\frac{2}{3}$ (we use units of the proton or positron charge, hence the electron has charge -1) and the lower members a charge of $-\frac{1}{3}$. The masses are:

$$m_u \sim \text{few MeV} \qquad m_c \cong 1.5 \text{ GeV}$$
$$m_d \sim \text{few MeV} \qquad m_b \cong 4.8 \text{ GeV}$$
$$m_s \cong 170 \text{ MeV} \qquad m_t \cong 175 \text{ GeV}.$$

These particles, however, are never freely found, making it difficult to determine the masses. The u and d are very light, with $m_d > m_u$; the large variation in masses is an unsolved puzzle.

Again, the electromagnetic interactions alone would have provided six separate stable quark types. The weak interactions are visible at low energies by creating transitions between quark types, reducing the number of stable objects. Considering only transitions involving quarks, we would expect to find the stable types (choosing the lightest doublet members for the second and third generations): u, d, s, b. However, two effects further reduce the number of stable quark types. The first is the mixing of members of the three quark families, again believed to be due to the $SU(2)$ symmetry breaking interaction that generates the gauge boson masses. It also generates the fermion masses and the family mixing. The mixing allows transitions of the type shown in Figure 1.8 in addition to those to the partner in the doublet, and is specified by a unitary matrix that we will discuss later. The net effect is to allow the s and b quarks to weakly decay to lighter quarks.

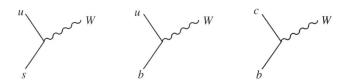

FIGURE 1.8 Some of the basic vertices for the weak interaction of the quarks leading to transitions outside a doublet pair.

1.3 Fundamental Constituents and Interactions

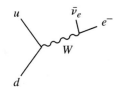

FIGURE 1.9 Weak decay involving quarks and leptons.

The second effect is that the weak interactions couple the quarks to the leptons. Thus, the process shown in Figure 1.9 allows d quark decay because the total mass $m_u + m_e + m_{\nu_e}$ is $< m_d$ by a small amount. Thus we would expect to find one quark type, u, to be stable and one conserved quantity, the total number of quarks of all types.

The above would be the whole story if quarks were just like leptons. Instead they have an additional interaction, the strong interaction, which dominates their behavior. In fact, the quark families we have written down each come in three copies, which have identical masses and electroweak properties. These types are called colors; each quark flavor comes in three colors.

Quarks can make transitions between color types, emitting a new type of spin 1 gauge boson. Since the transitions involve triplets now, the gauge group is $SU(3)$. The symmetry is unbroken, so the gauge bosons are massless and coupled to conserve color. To describe all types of transitions, the number of gauge bosons, now called gluons, has to be 8, as discussed in Chapter 4. They are the glue binding together quarks in the proton and neutron. The gluons carry "color" charges and interact with each other as well as with the quarks. To explain the much greater strength of the strong interaction, as compared to the electromagnetic interaction, the amplitude for a quark to emit a gluon must be much greater than the amplitude for an electron to emit a photon. Choosing one family of quarks, the possible transitions that occur are shown in Figure 1.10.

$$\begin{pmatrix} u \\ d \end{pmatrix} \xleftrightarrow{\text{Gluon Transitions}} \begin{pmatrix} u \\ d \end{pmatrix} \xleftrightarrow{\text{Change Colors}} \begin{pmatrix} u \\ d \end{pmatrix} \updownarrow \begin{array}{l} \text{Weak Transitions} \\ \text{Change Flavor} \end{array}$$

Color # 1   Color # 2   Color # 3

FIGURE 1.10 Types of transitions for u and d quarks considering both strong and weak interactions.

According to this picture, we would expect the quark types that are stable under the weak interactions to come in three versions, the three colors. These are not found in nature; in fact, no quarks or gluons are found at all among the spectrum of isolated states we see. This is the surprising feature we have to deal with for color theory, just as symmetry breaking is the surprise we have to deal with for electroweak theory. These features most clearly distinguish these interactions from electromagnetism. However, while the electroweak theory has missing pieces, namely the dynamics responsible for the symmetry breaking and mass generation, the unusual features of the color theory follow just from the strength of the interaction and the properties of the gauge theory.

The fact that free quarks and gluons are not seen is called confinement and is a consequence of the behavior of the gluon exchange force with distance. The net effect is that quarks are confined in bound states consisting of color-correlated quark clusters, whose net color charge vanishes. Attempts to knock the quarks out using high-energy collisions always result in showers of colorless bound-state particles, which form the strong interaction spectrum. These jets of final particles, discussed in the previous section, are the result of the particle showers following the direction of the parent quarks or gluons and carrying their large energy and momentum. The existence of jets allows an approximate determination of the parent quark direction and energy and a test of our understanding of the interactions at the quark level. This was critical for gaining acceptance of a theory whose basic objects cannot be isolated individually.

The strongly interacting particles made of quarks are called hadrons. We will look at the spectrum and properties of hadrons in Chapters 4, 5, and 6. They come in states with half-integral spin (made of three quarks) called baryons, and integral spin, called mesons. The one stable quark type with respect to the electroweak transitions results in one stable hadron, the proton. However, the neutron–proton mass difference is sufficiently small that the binding energy of a neutron in a stable nucleus forbids (by energy conservation) neutron decay, which occurs in free space. Note that the free neutron decay, $n \rightarrow p + e^- + \bar{\nu}_e$, is just the decay of the d quark discussed earlier but now in the context of a bound quark system.

The neutron and proton differ by the exchange of a d quark for a u quark within the bound state, resulting in a mass difference of about 1.29 MeV compared to a typical binding energy per nucleon of about 8 MeV, for a wide range of nuclei. The stability of the neutron in nuclei is thus a fortunate accident of the quark mass spectrum, as all other quarks are more than 100 MeV heavier than the u or d. We note that neutrinos and the weak interactions are extremely important for our understanding of the processes in stars responsible for building heavy nuclei, since these are the only processes that lead to nuclear transmutations via the basic transition of an n and a p.

We conclude by noting that the stable spin $\frac{1}{2}$ constituents of the low energy world, as presently known, turn out to be: e^- and p, which are stable; n, which is stable in many nuclei; and the three neutrino types, which are nearly invisible since they interact only via the weak interactions. This simple set of objects then build up the various structures around us using the interactions we have discussed. For these interactions, the photon is the only gauge particle that appears as a stable single particle state.

The Standard Model, as sketched above, provides an understanding of atomic phenomena including chemical behavior, the principles underlying nuclear behavior, and, after including the gravitational potential, an understanding of why the sun shines, and why some stars explode. Not well understood are the questions of why the universe is made of particles and not antiparticles, the cosmic ratio of massive particles to photons, as well as various phenomena visible mainly through gravitational effects. This includes the quantum nature of gravity itself and dark matter, which may well consist of the most abundant of the massive particles in nature, but whose nongravitational effects are very small. An understanding of gravity is an important motivation for much of the recent work on theories such as string theory. The search for the nature of dark matter, using both experiments that look for this material as it passes through the earth, and at future accelerators, is an exciting undertaking.

We might ask about the role of the unstable particles, the heavier leptons and quarks. Has nature just been generous with the number of particle families, or have they been crucial in the evolution of the universe? They are like dinosaurs, remnants of an earlier age—in this case the very early universe where temperatures were large enough to produce these objects. It may well be that we need the several families to produce the difference between particle and antiparticle processes required for a world of only particles. The mass and mixing parameters that distinguish the family members arise through the mass-generating dynamics. How does the mass generation mechanism work? The very heavy top quark may have some special role in mass generation; in any case its very heavy mass may carry a message we do not yet understand. These kinds of questions, along with whether there is a deeper relationship of the several interactions to each other (and finally gravity) are some of the exciting forefront questions in particle physics.

CHAPTER 1 HOMEWORK

1.1. Given a particle's lifetime τ we can define the width Γ and a decay distance ℓ by multiplying by the velocity of light. The mean decay length then equals $\beta\gamma\ell$, where β and γ are the usual relativistic kinematic parameters.

(a) The charged pion lifetime = 2.60×10^{-8} sec. What are Γ and ℓ for this particle?
(b) The ω^0 decay width $\Gamma = 8.4$ MeV. What are τ and ℓ for this particle?

For the units used, see the Appendix.

1.2. In Chapter 2 we will show that, in the low-energy limit, the exchange of a virtual particle of mass m corresponds to a potential with a range $\sim 1/m$. What is the range for Z^0 gauge boson exchange? Compare this to the size of a hydrogen atom. Do you expect Z^0 exchange to have much of an effect on atomic binding energies?

1.3. The uncertainty relation for position and momentum is

$$\Delta x \Delta p \sim 1.$$

For an uncertainty in x of 10^{-8} cm (a typical atomic dimension), what is the typical spread in the momentum?

1.4. The momentum transfer, q, in a scattering process tells us approximately the distance scale probed for "structure" in the target by the reaction. This is quantified by the relation $q\Delta x \sim 1$, where Δx is the distance scale probed. What must q be to probe distance scales $\sim 10^{-16}$ cm?

1.5. Draw the simplest Feynman diagrams for the scattering processes contributing to the following:

(a) $\nu_\mu e^- \to \nu_\mu e^-$
(b) $\bar{\nu}_\mu e^- \to \bar{\nu}_\mu e^-$
(c) $\bar{\nu}_e e^- \to \bar{\nu}_e e^-$

(d) $\nu_e e^- \to \nu_e e^-$
(e) $\nu_\mu e^- \to \mu^- \nu_e$.

1.6. Suppose the electron mass were 5 MeV instead of approximately 0.5 MeV, with all other particle masses unchanged. What particles would you expect to populate the universe?

1.7. Suppose the mass difference between the d and u quarks was 50 MeV, with all other particle masses unchanged. What major difference would this make in the universe?

CHAPTER 2

How to Calculate Amplitudes

2.1 ■ FREE PARTICLES

Particle physics incorporates and builds on the ideas of quantum mechanics and classical field theory as developed in electrodynamics. We will describe particle propagation using field equations that generalize electrodynamics. The results in this chapter are quite far-reaching, and the interpretation of how to use the field equations will develop as we explore their solutions.

As discussed in Chapter 1, we will often be looking at states for asymptotically noninteracting single particles that we detect individually. The quantities needed to describe such single particle states can be specified without actually having to identify explicitly the detailed quantum theory. The possible choices for such states follow from looking at the group of transformations that result from the space-time symmetries of the universe: translations, rotations, and Lorentz transformations. This is analogous to the situation where we can figure out the various spin choices by considering the representations of the rotation group, without having to introduce an explicit Hamiltonian.

We will not present this group theory analysis, but the result has been mentioned already. The free single particle base states can be chosen to have a given energy, momentum, and helicity for a given reference frame. The amplitude for the field associated with the particle is a plane wave given by

$$\phi(\vec{x}, t) = \chi(p_\mu) e^{-ip\cdot x}, \qquad (2.1)$$

where $p \cdot x = p_\mu x_\mu = Et - \vec{p} \cdot \vec{x}$, with E = energy and $\vec{p}$ = particle momentum. We use the summation convention for repeated indices, but will not distinguish between upper and lower indices. Only special relativity will be relevant for the topics covered here, so that the four-dimensional inner product is the same everywhere in space-time. The conventions used in the text are discussed in the Appendix. For Lorentz indices the convention will include the minus sign for the term involving the dot product of the space components.

The factor $\chi(p_\mu)$ in Eq. 2.1 carries the spin information for the particle and is independent of $\vec{x}$ and t. E and $\vec{p}$ transform under Lorentz transformations in the familiar way, so only the expression for $\chi(p_\mu)$ is likely to be unfamiliar. The amplitude satisfies a space-time relativistic wave equation, called the Klein–Gordon equation, but we will require more than this to get the spin dependence right. In the rest frame of the particle, the choices for χ are just those functions (spinors, vectors, etc.) that provide representations of the rotation group, which are transformations that leave the particle at rest. In this chapter we will first look at the field equations and amplitudes for several spin choices. We then discuss how the amplitudes are used in quantum calculations. These lead to the quantities we measure in experiments, which are transition rates between states.

The phase variation in space-time of the amplitude in Eq. 2.1 is the basis for many interference phenomena, for example, in the case of photon detection following a diffraction grating, where several sources contribute to the full amplitude. In all cases, however, particles are confined to some finite region in space. This means that the particle momentum is not perfectly defined and the amplitude is not a perfect plane wave. Thus Eq. 2.1, which we will typically use in calculations, is an idealization of the realistic situation. For example, an electron constrained to an atomic dimension $\sim 10^{-8}$ cm will have a momentum smearing that we can estimate from the uncertainty principle to be a few thousand eV. If accelerated to a much larger energy, such an electron will have a very small uncertainty in both momentum and position, allowing us to describe it by a state of well-defined momentum even when making a rather accurate simultaneous position determination in an experiment.

The state corresponding to a single particle will transform under the group transformations when changing reference frames. However, several quantities are invariant, that is, the same in all frames. These can be shown to be the mass of the particle, $m = \sqrt{E^2 - p^2}$, and its total spin. Thus we can speak about these quantities as characteristic of the particle itself.

2.2 ■ SPIN ZERO PARTICLES

The simplest amplitude is that for a spin zero particle. This will not describe any of the constituents discussed in Section 1.3, all of which have nonzero spin, but will serve in this chapter as the mathematically simplest example of a field amplitude. In this case, except for a normalization constant, Eq. 2.1 becomes

$$\phi(\vec{x}, t) = e^{-ip \cdot x}. \tag{2.2}$$

The Klein–Gordon equation, satisfied by ϕ, is:

$$\left[\nabla^2 - \frac{\partial^2}{\partial t^2}\right]\phi = m^2\phi, \tag{2.3}$$

where m is the particle mass. This can be verified by direct substitution. Alternatively, using the energy and momentum operators $E = i(\partial/\partial t)$ and $\vec{p} = -i\nabla$, the Klein–Gordon equation is the operator equivalent of $E^2 - p^2 = m^2$, acting on the amplitude ϕ. Note that the covariant differential operator corresponding to p_μ, $i(\partial/\partial x_\mu)$, has the components $i(\partial/\partial t, -\nabla)$. The normalization of ϕ is not constrained for a classical theory. In the transition to the quantum case, the normalization is fixed by the requirement that the quanta come in discrete units.

2.2 Spin Zero Particles

Interactions of the field with other particles requires the introduction of a source term into the field equation. The simplest modification, which parallels electromagnetism and the role of the electromagnetic current, is to introduce a source ρ that modifies the field equation as follows:

$$\left[\nabla^2 - \frac{\partial^2}{\partial t^2}\right]\phi - m^2\phi = \rho. \tag{2.4}$$

Since ϕ is a Lorentz scalar, ρ must be as well. It should be calculable in terms of properties and space-time trajectories of the other particles that are present. A simple example is a static source; this is a source that (to good approximation) is localized and whose change in energy due to interactions can be ignored, such as a very heavy particle at rest. In this case, choosing the origin as the source location,

$$\rho = g\,\delta^3(\vec{x}).$$

The time independent solution for ϕ from Eq. 2.4 is

$$\phi(r) = \frac{-ge^{-mr}}{4\pi r}, \tag{2.5}$$

which is called the Yukawa potential. We see that the strength of the potential at a given point is determined by g (called the coupling constant) and m. A large value of m gives a very short range interaction. This is the reason, in fact, for the weakness of the weak interactions. It is an example of a general result, that high mass physics is very hard to see at low energies, since it corresponds to very short distance phenomena only. The Yukawa potential occurs as a nonrelativistic limit using a classical source. We will make the connection to the quantum particle picture in Section 2.13.

In nonrelativistic quantum mechanics, we can define a probability density and current in terms of the particle wave function. The density and current satisfy the continuity equation but do not form a four-vector in the nonrelativistic theory. The current describes the local flow of the probability for finding the particle, with the total probability remaining constant. In the relativistic case, the creation of new particles in interactions implies that we cannot expect a constant probability for finding every individual particle. However, for free particles we might expect to find a fixed probability. This is true for the solutions of the free field (Eq. 2.3). We define a four-vector current density for complex solutions:

$$J_\mu = i\left(\phi^* \frac{\partial \phi}{\partial x_\mu} - \phi \frac{\partial \phi^*}{\partial x_\mu}\right),$$

which has components

$$J_0 = i\left(\phi^* \frac{\partial \phi}{\partial t} - \phi \frac{\partial \phi^*}{\partial t}\right) \quad \text{and} \quad \vec{J} = -i(\phi^* \vec{\nabla}\phi - \phi \vec{\nabla}\phi^*).$$

The current density satisfies the continuity equation:

$$\frac{\partial J_0}{\partial t} + \nabla \cdot \vec{J} = 0,$$

which follows from the Klein–Gordon equation. Substituting the solution $\phi = e^{-ip\cdot x}$ gives for J_μ, $J_0 = 2E$, $\vec{J} = 2\vec{p}$, or $J_\mu = 2p_\mu$. If we have a situation where the state we have prepared involves one free particle in a volume V, then the expression for J_0 implies that the normalization we have to choose for ϕ is

$$\phi = \frac{1}{\sqrt{2EV}} e^{-ip\cdot x}. \tag{2.6}$$

We will typically do calculations using covariant fields like $e^{-ip\cdot x}$. To then translate these calculations into results for one asymptotically free particle in a volume V, we have to multiply the derived expressions for probabilities by $1/2EV$ for each particle involved.

In this discussion, the plane wave for ϕ is assumed to be complex, as is the wave function in nonrelativistic quantum mechanics. However, classically, fields such as the electromagnetic field are real, so we might not expect complex solutions in this case. We will discuss the meaning of the complex solutions later in the chapter, starting in Section 2.8, for the quantum theory.

The equation for ϕ with a source present will not automatically allow for current conservation, since Eq. 2.4 does not imply that the current satisfies the continuity equation. Thus, even if we start with individual free particles, the source can, in general, modify the number of particles. This is also true for electromagnetism, where the number of photons is not fixed. However, we expect to find some quantities conserved, for example, electric charge, as we explore more general equations.

2.3 ■ LAGRANGIAN DENSITY

A technique for arriving at the equations of motion in classical mechanics is the use of a Lagrangian and a variational principle. It is possible to derive the field equations using an analogous approach based on a Lagrangian. Since the field is defined over all space, the Lagrangian is specified by a density, also called the Lagrangian for simplicity, which is a Lorentz invariant space-time function of all of the fields and their first derivatives. For the case of the free scalar field, the function is $\mathcal{L}(\phi, \partial\phi/\partial x_\mu)$.

We define the action, S, as:

$$S = \int_{t_1}^{t_2} dt \int d^3x \, \mathcal{L}. \tag{2.7}$$

Requiring the action to be stationary, that is, $\delta S = 0$ for ϕ varying in space-time arbitrarily, generates the field equations from $\mathcal{L}$. The variation in ϕ is, however, taken to vanish at t_1, t_2 and the extremities of the space integration. For $\mathcal{L}$

2.3 Lagrangian Density

depending on more fields, this procedure applied to each field will generate equations for each of the independent fields.

For the scalar field,

$$\delta S = \int_{t_1}^{t_2} dt \int d^3x \left[\frac{\partial \mathcal{L}}{\partial \phi} \delta\phi + \frac{\partial \mathcal{L}}{\partial \left(\frac{\partial \phi}{\partial x_\mu} \right)} \delta \left(\frac{\partial \phi}{\partial x_\mu} \right) \right].$$

Integrating the second term by parts, with the total derivative yielding a term that vanishes at the extremities of the integration region, gives

$$\delta S = \int_{t_1}^{t_2} dt \int d^3x \left[\frac{\partial \mathcal{L}}{\partial \phi} - \frac{\partial}{\partial x_\mu} \frac{\partial \mathcal{L}}{\partial \left(\frac{\partial \phi}{\partial x_\mu} \right)} \right] \delta\phi.$$

Requiring this term to vanish for $\delta\phi$ an arbitrary function of $\vec{x}$ and t yields the Euler–Lagrange equation:

$$\frac{\partial}{\partial x_\mu} \left(\frac{\partial \mathcal{L}}{\partial \left(\frac{\partial \phi}{\partial x_\mu} \right)} \right) - \frac{\partial \mathcal{L}}{\partial \phi} = 0. \tag{2.8}$$

To derive the free field Klein–Gordon equation using the Euler–Lagrange equation, a choice for $\mathcal{L}_{\text{Free}}$ is

$$\mathcal{L}_{\text{Free}} = \frac{1}{2} \left[\frac{\partial \phi}{\partial x_\mu} \frac{\partial \phi}{\partial x_\mu} - m^2 \phi^2 \right]. \tag{2.9}$$

To include a source-term interaction, as in Eq. 2.4, we can add to $\mathcal{L}_{\text{Free}}$

$$\mathcal{L}_{\text{Int}} = -\phi\rho, \tag{2.10}$$

which then yields the correctly modified field equation. Since we are particularly interested in the interactions, the usual focus of attention is the form and consequences of $\mathcal{L}_{\text{Int}}$, which must be generalized to apply to the correct set of particles and interactions present.

We can define a Hamiltonian density, by analogy with the mechanical case, as

$$\mathcal{H} = \frac{\partial \phi}{\partial t} \frac{\partial \mathcal{L}}{\partial \left(\frac{\partial \phi}{\partial t} \right)} - \mathcal{L}, \tag{2.11}$$

with the Hamiltonian $H = \int \mathcal{H} d^3x$. We will make the connection to quantum mechanics in Section 2.9, through the Hamiltonian. $\mathcal{H}$ must then be Hermitian, which implies that $\mathcal{L}$ is as well. The contribution of the interaction term in $\mathcal{L}$ provides an interaction term in H:

$$H_{\text{Int}} = \int \mathcal{H}_{\text{Int}} d^3x, \tag{2.12}$$

where

$$\mathcal{H}_{\text{Int}} = \frac{\partial \phi}{\partial t} \frac{\partial \mathcal{L}_{\text{Int}}}{\partial (\frac{\partial \phi}{\partial t})} - \mathcal{L}_{\text{Int}}.$$

For the scalar field just discussed, $\mathcal{H}_{\text{Int}} = -\mathcal{L}_{\text{Int}} = \rho\phi$. As an example, for a static source at $\vec{x}_2$ in the field of another static source at $\vec{x}_1$, we can calculate the interaction energy by integrating $\rho\phi$ over all space. Using the Yukawa potential given in Eq. 2.5 for $\phi(\vec{x})$:

$$\int \phi(\vec{x}) g\, \delta^3(\vec{x} - \vec{x}_2)\, d^3x = \frac{-g^2}{4\pi} \frac{e^{-m|\vec{x}_1 - \vec{x}_2|}}{|\vec{x}_1 - \vec{x}_2|}. \tag{2.13}$$

The interaction potential energy is attractive, with a range determined by $1/m$. This is the scalar theory analog of the Coulomb interaction energy in electrostatics.

2.4 ■ SYMMETRIES AND THE LAGRANGIAN

If the Lagrangian is invariant under a continuous group of transformations, then there exist locally conserved quantities constructed from the fields and their derivatives. These quantities, like the electric charge, flow in space-time and are described in terms of currents. This is called Noether's theorem. The conserved quantities in nature tell us which symmetries to build into the Lagrangian. In addition, an internal symmetry will result in a spectrum of quantum states that are related by symmetry transformations. The mechanism for evasion of this expectation, that is, a symmetry of the Lagrangian not seen in the spectrum of states, is called spontaneous symmetry breaking. This intriguing phenomenon, believed to be responsible for mass generation, will be discussed in Chapter 10. We look below at symmetries for the case of a scalar field, but the result is directly generalizable to other spins.

A symmetry familiar from mechanics is the invariance of the equations of motion under translation of the origin in space-time. For the fields this symmetry corresponds to the Lagrangian having the same form if the x_μ are used as coordinates, or if $x'_\mu = x_\mu + \varepsilon_\mu$ are used. This requires that x_μ does not appear explicitly, only the fields and their derivatives can appear. We calculate the change in $\mathcal{L}$ due to an infinitesimal displacement:

$$\delta\mathcal{L} = \mathcal{L}' - \mathcal{L} = \varepsilon_\mu \frac{\partial \mathcal{L}}{\partial x_\mu}.$$

We can, however, write this in terms of changes of ϕ and $\partial\phi/\partial x_\mu$, since $\mathcal{L}$ depends only on these; therefore, using the chain rule for differentiation,

$$\varepsilon_\mu \frac{\partial \mathcal{L}}{\partial x_\mu} = \frac{\partial \mathcal{L}}{\partial \phi} \varepsilon_\nu \frac{\partial \phi}{\partial x_\nu} + \frac{\partial \mathcal{L}}{\partial(\frac{\partial \phi}{\partial x_\mu})} \varepsilon_\nu \frac{\partial}{\partial x_\nu}\left(\frac{\partial \phi}{\partial x_\mu}\right).$$

2.4 Symmetries and the Lagrangian

Using the Euler–Lagrange equation to replace

$$\frac{\partial \mathcal{L}}{\partial \phi} \quad \text{by} \quad \frac{\partial}{\partial x_\mu}\left(\frac{\partial \mathcal{L}}{\partial \left(\frac{\partial \phi}{\partial x_\mu}\right)}\right)$$

gives

$$\varepsilon_\mu \frac{\partial \mathcal{L}}{\partial x_\mu} = \varepsilon_\nu \frac{\partial}{\partial x_\mu}\left[\frac{\partial \mathcal{L}}{\partial \left(\frac{\partial \phi}{\partial x_\mu}\right)}\left(\frac{\partial \phi}{\partial x_\nu}\right)\right].$$

We can write this as

$$\varepsilon_\nu \frac{\partial}{\partial x_\mu}\left[\frac{\partial \mathcal{L}}{\partial \left(\frac{\partial \phi}{\partial x_\mu}\right)}\left(\frac{\partial \phi}{\partial x_\nu}\right) - g_{\mu\nu}\mathcal{L}\right] = 0.$$

Since the ε_ν are arbitrary, the tensor

$$T_{\mu\nu} = \frac{\partial \mathcal{L}}{\partial \left(\frac{\partial \phi}{\partial x_\mu}\right)}\left(\frac{\partial \phi}{\partial x_\nu}\right) - g_{\mu\nu}\mathcal{L} \quad \text{satisfies} \quad \frac{\partial T_{\mu\nu}}{\partial x_\mu} = 0. \quad (2.14)$$

Using the divergence theorem, it is then easy to show that the quantities $P_\nu = \int d^3x T_{0\nu}$ are conserved over time, that is $dP_\nu/dt = 0$. T_{00} is the Hamiltonian density, $\mathcal{H}$, and the corresponding conserved quantity H is the total energy. Its conservation follows from time translational invariance.

We turn next to a symmetry more specific to quantum mechanics. For the non-relativistic case, we know that the absolute phase of the wave function is not measurable. We look at the analog of this for the scalar field. Stated as a symmetry, we want the replacement of ϕ by $\phi' = e^{i\theta}\phi$ to yield the same form of the Lagrangian. To achieve this we have to introduce ϕ as a complex field explicitly, with ϕ and ϕ^* appearing symmetrically. The Lagrangian,

$$\mathcal{L} = \frac{\partial \phi^*}{\partial x_\mu}\frac{\partial \phi}{\partial x_\mu} - m^2\phi^*\phi, \quad (2.15)$$

is clearly invariant with regard to the overall phase change for ϕ. Such a change in phase over all space is called a global gauge transformation.

By varying ϕ and ϕ^* separately, it is seen from Eq. 2.8 that both fields satisfy the free field Klein–Gordon equation with the same mass. We next show that the symmetry leads to a conserved current J_μ. The associated time independent quantity $Q = \int J_0 d^3x$ is called a conserved charge.

Under an infinitesimal phase change:

$$\phi \to (1+i\theta)\phi, \quad \phi^* \to (1-i\theta)\phi^* \quad \text{and} \quad \mathcal{L} \to \mathcal{L}.$$

Thus

$$0 = \delta\mathcal{L} = \frac{\partial\mathcal{L}}{\partial\phi}\delta\phi + \frac{\partial\mathcal{L}}{\partial\phi^*}\delta\phi^* + \frac{\partial\mathcal{L}}{\partial\left(\frac{\partial\phi}{\partial x_\mu}\right)}\delta\frac{\partial\phi}{\partial x_\mu} + \frac{\partial\mathcal{L}}{\partial\left(\frac{\partial\phi^*}{\partial x_\mu}\right)}\delta\frac{\partial\phi^*}{\partial x_\mu}.$$

Here

$$\delta\phi = i\theta\phi, \quad \delta\left(\frac{\partial\phi}{\partial x_\mu}\right) = i\theta\frac{\partial\phi}{\partial x_\mu}.$$

Replacing

$$\frac{\partial\mathcal{L}}{\partial\left(\frac{\partial\phi}{\partial x_\mu}\right)}\frac{\partial\phi}{\partial x_\mu} \quad \text{by} \quad \frac{\partial}{\partial x_\mu}\left[\frac{\partial\mathcal{L}}{\partial\left(\frac{\partial\phi}{\partial x_\mu}\right)}\phi\right] - \left[\frac{\partial}{\partial x_\mu}\left(\frac{\partial\mathcal{L}}{\partial\left(\frac{\partial\phi}{\partial x_\mu}\right)}\right)\right]\phi$$

gives:

$$0 = i\theta\left[\frac{\partial\mathcal{L}}{\partial\phi} - \frac{\partial}{\partial x_\mu}\left(\frac{\partial\mathcal{L}}{\partial\left(\frac{\partial\phi}{\partial x_\mu}\right)}\right)\right] - i\theta\left[\frac{\partial\mathcal{L}}{\partial\phi^*} - \frac{\partial}{\partial x_\mu}\left(\frac{\partial\mathcal{L}}{\partial\left(\frac{\partial\phi^*}{\partial x_\mu}\right)}\right)\right]$$
$$+ i\theta\left[\frac{\partial}{\partial x_\mu}\left(\frac{\partial\mathcal{L}}{\partial\left(\frac{\partial\phi}{\partial x_\mu}\right)}\right)\phi - \frac{\partial}{\partial x_\mu}\left(\frac{\partial\mathcal{L}}{\partial\left(\frac{\partial\phi^*}{\partial x_\mu}\right)}\right)\phi^*\right].$$

The first two terms vanish by the Euler–Lagrange equations. Since θ is arbitrary, the last term, using the explicit formula for $\mathcal{L}$ above, gives

$$\frac{\partial}{\partial x_\mu}J_\mu = 0, \quad J_\mu = i\left(\phi^*\frac{\partial\phi}{\partial x_\mu} - \phi\frac{\partial\phi^*}{\partial x_\mu}\right), \tag{2.16}$$

where the sign convention has been chosen to be the same as in Section 2.2.

To arrive at a conserved current we had to double explicitly the number of fields being considered. This can be seen in another way. Taking

$$\phi = \frac{\phi_1 + i\phi_2}{\sqrt{2}}, \quad \phi^* = \frac{\phi_1 - i\phi_2}{\sqrt{2}},$$

we can write $\mathcal{L}$ in terms of the real fields ϕ_1 and ϕ_2. The result is $\mathcal{L} = \mathcal{L}(\phi_1) + \mathcal{L}(\phi_2)$, that is, two independent scalar field Lagrangians, but with both fields having the same mass. The phase transformation for ϕ and ϕ^* is equivalent to an orthogonal transformation (a real rotation) in the ϕ_1 and ϕ_2 space. From the point of view of ϕ_1 and ϕ_2, the symmetry arises because the ϕ_1 and ϕ_2 particles are degenerate in mass. The doubling of the degrees of freedom for a field describing a particle that carries charge will be seen later in this chapter to correspond to the existence of antiparticles of opposite charge, but identical mass, for each particle type. Thus the total charge Q can be either positive or negative, depending on how many particles or antiparticles are present.

Maintaining charge conservation even in the presence of interactions constrains the choices available for $\mathcal{L}_{\text{Int}}$; $\mathcal{L}_{\text{Int}}$ must maintain the global gauge symmetry. For example, a term of the type $-\lambda(\phi^*\phi)^2$ maintains the symmetry. This represents an interaction between the scalar particles themselves, which conserves the charge and leaves the expression for J_μ unchanged.

For the simplest fields and their interactions described in this chapter, we focus mostly on the field equations themselves with the Lagrangian playing a secondary role. Since the Lagrangian formalism is the same for the classical and quantum case, we can use the familiar classical field equations for electromagnetism directly. When we come to the more complex equations and symmetries of the strong and weak interactions, the Lagrangian will provide the most direct approach for formulating the complete theory.

The construction of an appropriate $\mathcal{L}$ and the corresponding field equations follow closely what experiment tells us. That is, we've learned how to build the symmetries and interactions into our theories and the use of a Lagrangian is typically the most direct way to accomplish this. In subsequent chapters we return repeatedly to symmetry questions. In addition, $\mathcal{L}$ provides the Hamiltonian that we will use for construction of the quantum theory. Except for bound states of spin $\frac{1}{2}$ particles, no spin 0 particle has yet been seen, so much of our subsequent discussion will focus on particles of spin $\frac{1}{2}$ and spin 1.

2.5 ■ SPIN 1 PARTICLES

We turn next to the case of a free massive particle with spin 1, called a vector field. The spin dependence $\chi(p_\mu)$ is specified by a 4-vector e_μ called the polarization vector. In the particle rest frame, e_μ is a space vector, which transforms as spin 1 under spatial rotations. There are thus three independent spin choices in the rest frame, with space components $\hat{e}_x$, $\hat{e}_y$, or $\hat{e}_z$ providing the simplest set. These satisfy the Lorentz covariant conditions $e \cdot p = 0, e \cdot e = -1$, which then hold in any frame. We write for the free particle vector field in a general frame: $A_\mu = e_\mu e^{-ip \cdot x}$. This is the amplitude in Eq. 2.1 for spin 1.

Choosing $\vec{p}$ in the z-direction, we can Lorentz transform the individual spin vectors given above to get three choices for e_μ: $(0, 1, 0, 0)$, $(0, 0, 1, 0)$, or $1/m(p, 0, 0, E)$, where the first component in parentheses is the time-component and $p \equiv |\vec{p}|$. The first two choices are called transverse polarization, the third is called longitudinal.

A convenient, related basis is given by the helicity states, which are states with fixed angular momentum component along $\vec{p}$. To calculate these we can take states of given angular momentum in the rest frame along the axis given by $\vec{p}$ and then boost these so the momentum is $\vec{p}$. Taking as an example $\vec{p}$ along z, the states in the rest frame have spatial components:

$$J_z = \pm 1, \ \mp\left(\frac{\hat{e}_x \pm i\hat{e}_y}{\sqrt{2}}\right)$$

$$J_z = 0, \ \hat{e}_z.$$

Under a rotation of the vectors by an angle θ about the z-axis, these change by a factor $e^{-iJ_z\theta}$, indicating the correct J_z. Boosting, the value of J_z becomes the helicity, which we denote by λ. Thus (keeping the unit vector notation for the spatial part of the transverse spin states):

$$\lambda = \pm 1, \ e_\mu(\lambda) = \mp\left(0, \ \frac{\hat{e}_x \pm i\hat{e}_y}{\sqrt{2}}\right)$$

$$\lambda = 0, \ e_\mu(\lambda) = \frac{1}{m}(p, 0, 0, E).$$

To specify fully a given free particle state, we have to specify the momentum $\vec{p}$ and the helicity λ.

We can generalize the procedure above to arrive at fields for massive particles with other integral values of spin. For example, for spin 2 there are $2J + 1 = 5$ base states. These are specified in the rest frame by five choices for J_z. Based on the representations of the rotation group, the spin states are given by traceless symmetric tensors. The five linearly independent such tensors correspond to five choices we can use to construct the field amplitudes.

2.6 ■ SPIN 1 PHOTONS

An example of a spin 1 field is the electromagnetic field, whose associated particle, the photon, is massless. Massless particles differ in some important respects from massive ones. For the case of massive particles of spin J discussed above, the $2J + 1$ base states in the rest frame must have the same mass as a consequence of rotational invariance. When looking at interactions of moving massive particles, those with opposite helicity will have related interactions as a consequence of Lorentz invariance, since a Lorentz transformation along $\vec{p}$ with velocity greater than the particle velocity changes λ into $-\lambda$. Thus the same event involves a particle with helicity λ or $-\lambda$ depending on the reference frame. Furthermore, all internal properties (for example, electric charge) that do not depend on the reference frame must be the same for all helicities. These arguments cannot be invoked for massless particles, which always travel at the speed of light (so there is no rest frame and λ is the same in all frames) and we will see that the consequences don't hold in this case. For the photon, only states of $\lambda = \pm 1$ exist, which correspond in the classical limit to transverse fields. There is no state of $\lambda = 0$ at all. When we get to the weak interactions, we will find that the massless fermions of the Lagrangian with $\lambda = +\frac{1}{2}$ (called right-handed) and $\lambda = -\frac{1}{2}$ (called left-handed) behave very differently, and have different internal properties.

2.6 Spin 1 Photons

An alternative way to relate states with opposite sign λ is through a parity transformation, which changes the sign of $\vec{p}$ but leaves the sense of rotations unchanged. Therefore λ also changes sign under a parity transformation. This transformation works for both massive and massless particles and, if parity is a good symmetry, we can relate interactions and internal properties for massless particles with opposite helicities. The very different behavior of left-handed and right-handed fermions for the weak interactions will therefore violate parity. The electromagnetic interaction conserves parity.

We turn now to the photon field, beginning with the classical equations. These are Maxwell's equations, which we can write in a 4-dimensional notation. For the basic field A_μ, we introduce the electric and magnetic fields through

$$F_{\mu\nu} = \frac{\partial}{\partial x_\mu} A_\nu - \frac{\partial}{\partial x_\nu} A_\mu,$$

$$F_{\mu\nu} = \begin{pmatrix} 0 & -E_x & -E_y & -E_z \\ E_x & 0 & -B_z & B_y \\ E_y & B_z & 0 & -B_x \\ E_z & -B_y & B_x & 0 \end{pmatrix}. \quad (2.17)$$

The Maxwell's equations that relate the fields to their sources are

$$\frac{\partial}{\partial x_\mu} F_{\mu\nu} = J_\nu, \quad (2.18)$$

where J_ν is the 4-dimensional electromagnetic current density. Since $F_{\mu\nu}$ is antisymmetric, by differentiating again we get the current conservation equation:

$$\frac{\partial}{\partial x_\nu} J_\nu = 0. \quad (2.19)$$

The sign conventions have been chosen to correspond to the classical case where an electron has a negative charge (of value $-e$, where e will be a positive number everywhere it occurs). The units for charge are discussed in the Appendix.

The relation between $F_{\mu\nu}$ and A_μ allows various A_μ to correspond to the same $F_{\mu\nu}$. In particular, if θ is any scalar function, then

$$A'_\mu = A_\mu + \frac{\partial \theta}{\partial x_\mu}$$

gives the same $F_{\mu\nu}$. Such a transformation is called a local gauge transformation (local, since θ may vary in space and time and is not the same everywhere). The field equations for the photon are gauge invariant, requiring the source to be a conserved current.

We can arrive at the field equations from a Lagrangian. In particular,

$$\mathcal{L} = -\tfrac{1}{4} F_{\mu\nu} F_{\mu\nu} - J_\mu A_\mu \quad (2.20)$$

yields Maxwell's equations by variation of the vector potential field components. $\mathcal{L}_{\text{Free}} = -\frac{1}{4}F_{\mu\nu}F_{\mu\nu}$ is the Lagrangian for a free photon field. $\mathcal{L}_{\text{Int}}$ is given by $-J_\mu A_\mu$.

We could introduce a photon mass term by adding to $\mathcal{L}_{\text{Free}}$ a term of the form

$$\tfrac{1}{2}m^2 A_\mu A_\mu.$$

This explicitly violates the gauge symmetry. Thus, the gauge symmetry forbids a photon mass. An interaction term is not automatically gauge invariant but requires that the source terms are introduced into $\mathcal{L}$ in a very specific way, described in Section 2.7.2.

Turning to the free particle field, we have:

$$A_\mu = e_\mu e^{-ip\cdot x}. \tag{2.21}$$

The relation found earlier for the massive vector field $e \cdot p = 0$, can be shown to be equivalent to

$$\frac{\partial A_\mu}{\partial x_\mu} = 0$$

by explicitly differentiating the expression for A_μ in Eq. 2.21. This equation is called the Lorentz condition and corresponds to a specific (but still not unique) choice of gauge for A_μ. With this choice of gauge, Maxwell's equations can also be written as:

$$\frac{\partial}{\partial x_\mu}\frac{\partial}{\partial x_\mu} A_\nu = \frac{\partial^2}{\partial t^2} A_\nu - \nabla^2 A_\nu = J_\nu. \tag{2.22}$$

Each component of A_μ then satisfies the massless Klein–Gordon equation in the free field case where $J_\mu = 0$.

Turning to the polarization vectors, the longitudinal polarization of the massive vector field $e_\mu = 1/m(p, 0, 0, E)$ cannot be taken over to the photon case, since $m \to 0$. The two transverse polarizations are independent of mass and translate into $\lambda = \pm 1$ states for the photon. In fact, we can use gauge invariance to show that the two transverse choices for the free field are sufficient.

Consider a gauge transformation

$$A'_\mu = A_\mu + \frac{\partial \theta}{\partial x_\mu},$$

where both A_μ and A'_μ satisfy the Lorentz condition. This restricts θ to functions that satisfy the massless Klein–Gordon equation. An example of such a function is $\theta = \text{constant} \times e^{-ip\cdot x}$. Substituting this into the gauge relation with $A_\mu = e_\mu e^{-ip\cdot x}$ and $A'_\mu = e'_\mu e^{-ip\cdot x}$ gives the relation:

$$e'_\mu = e_\mu + \text{constant} \times p_\mu.$$

Both e'_μ and e_μ must describe the same physics. We can choose the constant to eliminate the time-like component of e_μ. The new polarization vector will then satisfy the Lorentz condition in the simplified form

$$0 = e'_\mu p_\mu = \hat{e}' \cdot \vec{p}.$$

This means that $\hat{e}'$ can be chosen in the transverse plane. Since there are only two linearly independent transverse vectors, there are only two polarization choices.

Note, the above means that if you wish, you can use for the two polarization choices:

$$e^{(1)}_\mu = e_\mu(\lambda = +1) + a p_\mu$$
$$e^{(2)}_\mu = e_\mu(\lambda = -1) + b p_\mu$$

in all calculations and the terms proportional to p_μ must drop out. This is a useful way of checking that a calculation is gauge invariant. Stated another way, if we replace e_μ by p_μ in a quantum calculation involving a real photon, the resulting amplitude must vanish.

2.7 ■ SPIN $\frac{1}{2}$ PARTICLES

The electron is a spin $\frac{1}{2}$ particle, which implies that each momentum state has two possible helicities, $\lambda = +\frac{1}{2}$ or $\lambda = -\frac{1}{2}$. The states in the particle rest frame can be determined by looking at the spin $\frac{1}{2}$ representation of the rotation group.

We can describe the two spin choices in terms of base states:

$$\chi^+ = \begin{pmatrix} 1 \\ 0 \end{pmatrix} \quad \text{and} \quad \chi^- = \begin{pmatrix} 0 \\ 1 \end{pmatrix}. \tag{2.23}$$

These states, called spinors, correspond to spin $+\frac{1}{2}$ and $-\frac{1}{2}$ along a chosen axis, which we take to be the z-axis.

The spin operator in the fermion rest frame is $\vec{S}$, which is given in the basis above by

$$\vec{S} = \frac{\vec{\sigma}}{2} \tag{2.24}$$

with

$$\sigma_x = \begin{pmatrix} 0 & 1 \\ 1 & 0 \end{pmatrix}, \quad \sigma_y = \begin{pmatrix} 0 & -i \\ i & 0 \end{pmatrix}, \quad \sigma_z = \begin{pmatrix} 1 & 0 \\ 0 & -1 \end{pmatrix},$$

the Pauli spin matrices. Why can $\vec{S}$ be a legitimate spin operator? The answer is that it satisfies the angular momentum commutation relations:

$$[S_i, S_j] = i \varepsilon_{ijk} S_k, \tag{2.25}$$

where ε_{ijk} is the fully antisymmetric symbol. As a result, the three numbers $\chi^\dagger \vec{S} \chi$, for a general state χ, transform as a vector and thus are appropriate for coupling to other vectors, for example the magnetic field, to make an operator that transforms correctly as a scalar under rotations. The spin state is measurable; for example, in a magnetic field the two spin states of an electron split in energy.

For integral spin, our knowledge of how to Lorentz transform vectors allows us to take the rest-frame solutions and boost them, generating the general free particle solution. For the spinor case this procedure is not so obvious. In fact, what was done by Dirac was to invent an equation that gave the general solution in any frame. We turn to this next, expecting it to give the rest-frame solution as a special case.

2.7.1 ■ The Dirac Equation

The Dirac equation is a relativistically covariant equation describing spin $\frac{1}{2}$ particles. It is a first order equation in the various space and time derivatives. The free field equation can be written:

$$i\gamma_\mu \frac{\partial \psi}{\partial x_\mu} - m\psi = 0. \qquad (2.26)$$

Our field is now ψ, containing both the momentum and spin dependence, and m is the particle mass. The γ_μ are four matrices, one for each index μ. It will turn out that each γ_μ is a 4×4 matrix; therefore ψ is a 4-component column vector called a Dirac spinor. Since we expect two helicity choices for the electron, the number of degrees of freedom is twice as large as expected. The doubling will correspond to the existence of positrons—particles of opposite charge and the same mass and spin as the electron!

The dependence on spin and the space-time coordinates should factorize for a free field, as given by Eq. 2.1. To see how it can work, we operate on both sides of the equation with $i\gamma_\mu(\partial/\partial x_\mu)$, which gives:

$$-\gamma_\mu \gamma_\nu \frac{\partial}{\partial x_\mu} \frac{\partial \psi}{\partial x_\nu} = im\gamma_\mu \frac{\partial \psi}{\partial x_\mu} = m^2 \psi.$$

For this to give us the Klein–Gordon equation for the common space-time dependence of each component of ψ, requires that the first term reduce to the correct operator. Rewriting the first term in a symmetric form as

$$-\frac{1}{2}(\gamma_\mu \gamma_\nu + \gamma_\nu \gamma_\mu) \frac{\partial}{\partial x_\mu} \frac{\partial \psi}{\partial x_\nu}$$

we will get

$$\frac{-\partial}{\partial x_\mu} \frac{\partial \psi}{\partial x_\mu}$$

2.7 Spin $\frac{1}{2}$ Particles

provided the γ matrices satisfy $\frac{1}{2}\{\gamma_\mu, \gamma_\nu\} = g_{\mu\nu} I$. $\{\gamma_\mu, \gamma_\nu\}$ is defined to be $(\gamma_\mu \gamma_\nu + \gamma_\nu \gamma_\mu)$ and I is the identity matrix. A given set of matrices that satisfy the above anti-commutation relations provides a representation for the Dirac matrices. The lowest dimension for which these matrices exist is 4×4.

Assuming the relations above, we get:

$$\frac{\partial}{\partial x_\mu} \frac{\partial \psi}{\partial x_\mu} + m^2 \psi = 0, \tag{2.27}$$

which will have solutions of the form $\psi \sim$ column vector $\times\, e^{-ip\cdot x}$ with $p \cdot p = m^2$.

We will use, whenever looking at explicit solutions, the Dirac–Pauli representation for γ_μ. This representation is most convenient for looking at the nonrelativistic limit of the Dirac equation. In this representation:

$$\gamma_0 = \begin{pmatrix} I & 0 \\ 0 & -I \end{pmatrix},$$

where $I = \begin{pmatrix} 1 & 0 \\ 0 & 1 \end{pmatrix}$ is a 2×2 submatrix, as is 0, and

$$\vec{\gamma} = \begin{pmatrix} 0 & \vec{\sigma} \\ -\vec{\sigma} & 0 \end{pmatrix},$$

where again σ_i is a 2×2 submatrix given by the appropriate Pauli matrix. For example, in this notation:

$$\gamma_1 = \gamma_x = \begin{pmatrix} 0 & \sigma_1 \\ -\sigma_1 & 0 \end{pmatrix} = \begin{pmatrix} 0 & 0 & 0 & 1 \\ 0 & 0 & 1 & 0 \\ 0 & -1 & 0 & 0 \\ -1 & 0 & 0 & 0 \end{pmatrix}.$$

You can show that these γ_μ satisfy the correct anti-commutation relations.

Since measurables involve Hermitian operators, it is useful to check the properties of these matrices. The individual σ_i are Hermitian, so that $\gamma_0^\dagger = \gamma_0$ and $\vec{\gamma}^\dagger = -\vec{\gamma}$. Using the anti-commutation relations, a useful way of writing these is $\gamma_\mu^\dagger = \gamma_0 \gamma_\mu \gamma_0$. We can also look at the Hermitian conjugate of the Dirac equation, which is

$$-i \frac{\partial \psi^\dagger}{\partial x_\mu} \gamma_\mu^\dagger - m \psi^\dagger = 0.$$

Here $\psi^\dagger$ is a row vector. Multiplying from the right by γ_0 and inserting a $\gamma_0^2 = 1$ between $\partial \psi^\dagger / \partial x_\mu$ and $\gamma_\mu^\dagger$ gives the equation:

$$i \frac{\partial}{\partial x_\mu} \bar{\psi} \gamma_\mu + m \bar{\psi} = 0. \tag{2.28}$$

The spinor $\bar{\psi} = \psi^\dagger \gamma_0$, called the adjoint spinor, will be the one that appears most often with ψ in various operators.

Multiplying Eq. 2.26 for ψ by $\bar{\psi}$ and Eq. 2.28 for $\bar{\psi}$ by ψ and adding the two gives:

$$i\bar{\psi}\left(\gamma_\mu \frac{\partial}{\partial x_\mu}\psi\right) + i\left(\frac{\partial}{\partial x_\mu}\bar{\psi}\gamma_\mu\right)\psi = 0 \quad \text{or} \quad i\frac{\partial}{\partial x_\mu}(\bar{\psi}\gamma_\mu\psi) = 0.$$

Hence the operator $\bar{\psi}\gamma_\mu\psi$ is a conserved current!

Returning to the Dirac equation, we can multiply by γ_0 to give, after using $\gamma_0^2 = 1$,

$$i\frac{\partial \psi}{\partial t} = \gamma_0 \vec{\gamma} \cdot (-i\vec{\nabla}\psi) + \gamma_0 m \psi.$$

This has the form of a Schrödinger equation where the Hamiltonian operator is, in terms of the momentum operator,

$$H = \gamma_0 \vec{\gamma} \cdot \vec{p} + \gamma_0 m. \tag{2.29}$$

For this to be Hermitian requires again the relation $\gamma_\mu^\dagger = \gamma_0 \gamma_\mu \gamma_0$.

2.7.2 ■ Lagrangian and Symmetries for Spin $\frac{1}{2}$

The Lagrangian for the free field is

$$\mathcal{L} = \bar{\psi}\left(i\gamma_\mu \frac{\partial}{\partial x_\mu} - m\right)\psi.$$

This expression depends on the fields and their derivatives, and through the Euler–Lagrange equation, results in Eqs. 2.26 and 2.28 for ψ and $\bar{\psi}$. The Hamiltonian density corresponding to this $\mathcal{L}$ is $\mathcal{H} = \psi^\dagger(\gamma_0 \vec{\gamma} \cdot \vec{p} + \gamma_0 m)\psi = \bar{\psi}(\vec{\gamma} \cdot \vec{p} + m)\psi$. The Lagrangian has a global gauge symmetry, since taking $\psi \to e^{i\theta}\psi$ and $\bar{\psi} \to e^{-i\theta}\bar{\psi}$ leaves the form of $\mathcal{L}$ unchanged. Thus we expect a conserved current, which is $J_\mu = \bar{\psi}\gamma_\mu\psi$, derived in the previous section directly from the Dirac equation.

We would like to look next at the question of angular momentum for the free spin $\frac{1}{2}$ field. Consider an expression for the expectation value of an operator O:

$$\langle O \rangle = \int \psi^\dagger O \psi \, d^3 x.$$

We assume O has no explicit time dependence, and differentiate this expression with respect to time. Replacing the time derivative of ψ by the Hamiltonian, $\langle O \rangle$ is conserved in time provided $[H, O] = 0$. For a spin zero particle the orbital angular momentum operator $\vec{x} \times \vec{p}$ is conserved. In the nonrelativistic limit for a spin $\frac{1}{2}$ particle the analogous quantity is $\vec{J} = \vec{x} \times \vec{p} + \vec{S}$, where $\vec{S} = \vec{\sigma}/2$ and the 2-component spinor formalism is adequate.

2.7 Spin ½ Particles

What is the relativistic extension for the angular momentum? By commuting $\vec{x} \times \vec{p}$ with $H = \gamma_0 \vec{\gamma} \cdot \vec{p} + \gamma_0 m$, we can try to determine what must be added to the orbital angular momentum to make a conserved quantity. It turns out that the simplest extension of the nonrelativistic expression works. Defining the 4×4 matrices:

$$\vec{\Sigma} = \begin{pmatrix} \vec{\sigma} & 0 \\ 0 & \vec{\sigma} \end{pmatrix},$$

$\vec{J} = \vec{x} \times \vec{p} + (\vec{\Sigma}/2)$ satisfies $[H, \vec{J}] = 0$.

The plane wave fields we are using are not eigenstates of $\vec{J}$. These fields are eigenstates of $\vec{p}$, and $\vec{J}$ and $\vec{p}$ don't commute! This happens because the $\vec{x}$ in the $\vec{x} \times \vec{p}$ term doesn't commute with the momentum operator. If we can isolate the spin operator alone in an expression that commutes with H, we can get a quantity that does commute with the momentum. Taking $\vec{J} \cdot \vec{p}$, the term $(\vec{x} \times \vec{p}) \cdot \vec{p}$ vanishes, leaving $\vec{\Sigma} \cdot \vec{p}/2$. This helicity operator now commutes with H and $\vec{p}$ and can therefore be simultaneously diagonalized. The resulting states are the helicity states. Note that this works because the internal spin operator is independent of position, as is the resulting spin dependent factor in the amplitude in Eq. 2.1. Both of these features are true for particles with other values of the spin.

As a final issue we look at examples of what interactions can be added to $\mathcal{L}$. For electromagnetism, the classical interaction of an electron of charge $-e$ can be introduced by the substitution $p_\mu \to p_\mu + eA_\mu$ into the basic equations for the electron. This is called the minimal substitution and maintains gauge invariance for the classical case. In the quantum case, the replacement

$$p_\mu \to i \frac{\partial}{\partial x_\mu}$$

gives the prescription:

$$i \frac{\partial}{\partial x_\mu} \to i \frac{\partial}{\partial x_\mu} + eA_\mu.$$

Making this substitution in the Dirac $\mathcal{L}$ gives:

$$\mathcal{L} = \bar{\psi} \gamma_\mu \left(i \frac{\partial}{\partial x_\mu} + eA_\mu \right) \psi - m \bar{\psi} \psi. \tag{2.30}$$

To include the free photon field in $\mathcal{L}$ we add the term $-\frac{1}{4} F_{\mu\nu} F_{\mu\nu}$. The local gauge transformation now is

$$\psi \to e^{ie\theta} \psi \quad \text{and} \quad A_\mu \to A_\mu + \frac{\partial \theta}{\partial x_\mu},$$

leaving $\mathcal{L}$ unchanged. The function θ is an arbitrary space-time function and the fields must transform in a carefully coordinated fashion. The interaction term arises directly out of the modifications to the free field Lagrangian if we add the local gauge symmetry requirement. The Dirac equation, Eq. 2.26, does not have a local gauge symmetry. The main degree of freedom left in Eq. 2.30 is the value of the coupling constant e. For particles with different charge, we replace e with $-Q$, where Q is the particle charge. For several kinds of fermions (for example, electrons and muons) the Lagrangian is the sum of terms, one for each fermion.

The interaction term resulting from Eq. 2.30 is:

$$\mathcal{L}_{\text{Int}} = eA_\mu \bar{\psi}\gamma_\mu\psi. \tag{2.31}$$

Comparing this expression to the general expression for $\mathcal{L}_{\text{Int}}$ given in Eq. 2.20, we see that $J_\mu = -e\bar{\psi}\gamma_\mu\psi$ is the conserved electromagnetic current. Thus, the expression for the conserved current is not modified in the presence of an electromagnetic field (that is, when going from a global to a local gauge symmetry) for the Dirac field. This is a consequence of the derivative not explicitly appearing in the expression for the free particle current for spin $\frac{1}{2}$ (all derivative terms are modified because of the minimal substitution used to introduce the electromagnetic interaction). For other spins, such as spin zero, the expression for the current is modified. This will be important in the discussion of symmetry breaking in Chapter 10.

Other interactions can be introduced in an analogous way. Since $\mathcal{L}$ is a Lorentz scalar, the term multiplying m, $\bar{\psi}\psi$ is also a scalar. Thus an interaction with a scalar particle, if one is found, can be introduced via:

$$\mathcal{L}_{\text{Int}} = -g\phi\bar{\psi}\psi \tag{2.32}$$

where ϕ is the scalar field. This interaction also leaves the expression for the conserved current unchanged provided ϕ is real and unchanged under the gauge transformations. $\phi\bar{\psi}\psi$ will go into $\phi\bar{\psi}\psi$ after a phase transformation of ψ provided $\phi \to \phi$ under this change. Physically the case where $\phi \to \phi$ under the gauge transformation corresponds to the ϕ particles carrying no charge; only the particles of the ψ field are charged. Finally, note that g is a real number since $\mathcal{L}_{\text{Int}}$ has to be Hermitian.

2.7.3 ■ Explicit Plane Wave Solutions

We have not yet looked explicitly at the spin dependent parts of the solution of the Dirac equation. To look at the solutions for the free field, we start first with the simplest case, which is a particle at rest. In this case the Dirac equation reduces to

$$i\gamma_0 \frac{\partial \psi}{\partial t} - m\psi = 0.$$

Since

$$\gamma_0 = \begin{pmatrix} I & 0 \\ 0 & -I \end{pmatrix},$$

2.7 Spin $\frac{1}{2}$ Particles

this is, taking

$$\psi = \begin{pmatrix} \psi_1 \\ \psi_2 \\ \psi_3 \\ \psi_4 \end{pmatrix} : \quad \begin{aligned} i\frac{\partial \psi_1}{\partial t} - m\psi_1 &= 0, & i\frac{\partial \psi_2}{\partial t} - m\psi_2 &= 0 \\ i\frac{\partial \psi_3}{\partial t} + m\psi_3 &= 0, & i\frac{\partial \psi_4}{\partial t} + m\psi_4 &= 0. \end{aligned}$$

The equations for the four components of ψ separate in this case. As mentioned earlier, this representation for the γ_μ gives a simple nonrelativistic limit. The separation shown above is an example.

Taking a solution of the form e^{-iEt}, we get $E = m$ for ψ_1 and ψ_2. This solution for ψ_3 and ψ_4, however, would give $E = -m$. Since all particle masses are positive, this is clearly a problem. To get a positive mass we have to take for the solution e^{iEt}. If we assume the energy operator is $i(\partial/\partial t)$, then this state has a negative energy. We have a dilemma, since all energies are also positive. This problem actually exists for all of the fields. The Klein–Gordon equation has solutions of the form $e^{-ip \cdot x}$ and also $e^{ip \cdot x}$. So far we have just ignored the latter. Classically A_μ is real: We expect to have $e^{ip \cdot x}$ appear in A_μ if $e^{-ip \cdot x}$ appears. We will resolve this dilemma in Section 2.8. It points to the fact that the relativistic theory requires some extensions beyond nonrelativistic quantum mechanics. These extensions will lead to the existence of antiparticles as well as the connection of spin and statistics. In the nonrelativistic case these are ad hoc additions; in the relativistic case they emerge as a consequence of physically interpreting the extra solutions.

For the particle at rest only two solutions of the form e^{-iEt} exist with $E = m$. Thus the four-component spinor has only two degrees of freedom that behave in the expected way—just what we need for the two spin degrees of freedom of an electron.

We now turn to the solutions for particles in motion. We take the space-time dependence to be $e^{-ip \cdot x}$ for the two solutions which reduce to e^{-imt} as $\vec{p} \to 0$. After differentiating,

$$i\gamma_\mu \frac{\partial}{\partial x_\mu} e^{-ip \cdot x} = p_\mu \gamma_\mu e^{-ip \cdot x}.$$

To simplify the notation we introduce the definition $\not{p} = p_\mu \gamma_\mu$, which is a 4×4 matrix.

$$\not{p} = \begin{pmatrix} E & -\vec{\sigma} \cdot \vec{p} \\ \vec{\sigma} \cdot \vec{p} & -E \end{pmatrix}, \quad \text{where} \quad \vec{\sigma} \cdot \vec{p} = \begin{pmatrix} p_z & p_x - ip_y \\ p_x + ip_y & -p_z \end{pmatrix}.$$

For the other two solutions we take the dependence to be $e^{ip \cdot x}$, so these also have $E = m$ as $\vec{p} \to 0$. We write the four solutions as:

$$\psi^{(1)} = u^{(1)} e^{-ip \cdot x}$$
$$\psi^{(2)} = u^{(2)} e^{-ip \cdot x}$$

$$\psi^{(3)} = v^{(1)} e^{ip \cdot x}$$
$$\psi^{(4)} = v^{(2)} e^{ip \cdot x}.$$

For $\vec{p} = 0$, we've found the four solutions, which can be written as $u(0) = \begin{pmatrix} \chi \\ 0 \end{pmatrix}$, $v(0) = \begin{pmatrix} 0 \\ \chi \end{pmatrix}$, where $\chi = \begin{pmatrix} 1 \\ 0 \end{pmatrix}$ or $\begin{pmatrix} 0 \\ 1 \end{pmatrix}$. Substituting into the Dirac equation the more general solutions satisfy:

$$(\not{p} - m)u = 0$$
$$(\not{p} + m)v = 0.$$

Noting that $(\not{p} - m)(\not{p} + m) = p^2 - m^2 = 0$, we can find a general solution by taking

$$u(p) = (\not{p} + m)u(0),$$
$$v(p) = (\not{p} - m)v(0).$$

We can now choose $u(0) = \begin{pmatrix} \chi \\ 0 \end{pmatrix}$, with $\chi = \begin{pmatrix} 1 \\ 0 \end{pmatrix}$ or $\begin{pmatrix} 0 \\ 1 \end{pmatrix}$, to generate the two choices for $u(p)$, with analogous solutions for $v(p)$.

To pick a normalization for the spinors, we will use the same convention as for the scalar field. Using the current $\bar{\psi} \gamma_\mu \psi$, we shall choose the normalization so that the number of particles per unit volume is $2E$. When calculating using covariant spinors we must multiply probabilities by $1/2EV$ when the physical situation involves one particle in a volume V. Using the density $\bar{\psi} \gamma_0 \psi = \psi^\dagger \psi$, the normalization choice for the particle at rest is then

$$u(0) = \sqrt{2m} \begin{pmatrix} \chi \\ 0 \end{pmatrix}, v(0) = \sqrt{2m} \begin{pmatrix} 0 \\ \chi \end{pmatrix}. \tag{2.33}$$

Using these rest frame solutions and including a constant to get the correct normalization, we now have our general result by applying $(\not{p} + m)/2m$ to $u(0)$ or $(-\not{p} + m)/2m$ to $v(0)$:

$$u = \sqrt{E+m} \begin{pmatrix} \chi \\ \frac{\vec{\sigma} \cdot \vec{p} \chi}{E+m} \end{pmatrix}, v = \sqrt{E+m} \begin{pmatrix} \frac{\vec{\sigma} \cdot \vec{p} \chi}{E+m} \\ \chi \end{pmatrix}. \tag{2.34}$$

$\vec{\sigma} \cdot \vec{p}$ is the operator proportional to the helicity for the two component spinor χ. For $\vec{p}$ along z, taking $\chi = \begin{pmatrix} 1 \\ 0 \end{pmatrix}$ or $\begin{pmatrix} 0 \\ 1 \end{pmatrix}$ will yield two helicity eigenstates for u or v. If $\vec{p}$ is along an arbitrary direction, we can rotate $\begin{pmatrix} 1 \\ 0 \end{pmatrix}$ and $\begin{pmatrix} 0 \\ 1 \end{pmatrix}$ using 2×2 rotation operators and the Pauli spin formalism to generate helicity eigenstates along $\vec{p}$. For $\vec{p}$ in an arbitrary direction, we can still use $\chi = \begin{pmatrix} 1 \\ 0 \end{pmatrix}$ and $\begin{pmatrix} 0 \\ 1 \end{pmatrix}$ to give us two base states for u (or v); these are then each a mix of the two helicity states.

2.7 Spin $\frac{1}{2}$ Particles

2.7.4 ■ Bilinear Covariants

Interaction terms in $\mathcal{L}$ involve bilinear quantities such as $\bar{\psi}\gamma_\mu\psi$ or $\bar{\psi}\psi$. These must transform appropriately as a 4-vector or scalar, respectively, to maintain the Lorentz invariance of the theory. Had we specified how ψ behaves under a Lorentz transformation we could check this explicitly. Since the Dirac equation is not tied to any frame, we are able to get a general result and assume that an appropriate Lorentz transformation can be defined. This is true, although we will not take the time to work through such a transformation. You can check by explicit substitution that, for example, the quantities $\bar{u}\gamma_\mu u = 2p_\mu$ and $\bar{u}u = 2m$ transform correctly.

A question we can ask is whether we can find other operators O that are 4×4 matrices, for which $\bar{\psi} O \psi$ is both Hermitian and has a well-defined transformation property under Lorentz transformations. Such operators, called bilinear covariants, are legitimate candidates to appear in $\mathcal{L}$ for terms involving only the fields and no derivatives. So far, the choices for O are 1(scalar) and γ_μ(vector).

We can approach this question by looking at operators involving products of γ_μ just as we generate tensors by taking products of vectors. The choices are:

$$\gamma_\mu\gamma_\nu - \gamma_\nu\gamma_\mu \quad \text{for } \mu, \nu \text{ different}$$

$$\gamma_\mu\gamma_\nu + \gamma_\mu\gamma_\nu \quad \text{reduces to } 2g_{\mu\nu}$$

$$\gamma_\mu\gamma_\nu\gamma_\sigma \quad \text{for } \mu, \nu, \sigma \text{ different}$$

$$\gamma_\mu\gamma_\nu\gamma_\sigma\gamma_\rho \quad \text{for } \mu, \nu, \sigma, \rho \text{ different.}$$

Five or more γ's reduces to at most a product of four since some γ's are repeated and $\gamma_0\gamma_0 = 1$, $\gamma_i\gamma_i = -1$ for $i = 1, 2, 3$.

The product of four different γ matrices is an important operator. We define the Hermitian matrix:

$$\gamma_5 = i\gamma_0\gamma_1\gamma_2\gamma_3. \tag{2.35}$$

In the Dirac–Pauli representation:

$$\gamma_5 = \begin{pmatrix} 0 & I \\ I & 0 \end{pmatrix},$$

while in general,

$$\{\gamma_5, \gamma_\mu\} = 0, \quad \text{and} \quad \gamma_5^2 = \begin{pmatrix} I & 0 \\ 0 & I \end{pmatrix}.$$

Using γ_5 we can form the Hermitian bilinear operator $\bar{\psi}i\gamma_5\psi$. By explicit substitution we can calculate that, in the nonrelativistic limit, $\bar{u}(\vec{p}_f)i\gamma_5 u(\vec{p}_i) = i\chi_f^\dagger \vec{\sigma} \cdot (\vec{p}_i - \vec{p}_f)\chi_i$. Since $\chi^\dagger \vec{\sigma} \chi$ transforms under rotations as an axial or pseudovector, just like the angular momentum, the bilinear operator transforms as a pseudoscalar in the nonrelativistic limit. It turns out to be a pseudoscalar in general, although demonstrating this requires figuring out the form of a parity transformation for the Dirac theory, which we will do in Chapter 4. Making use of γ_5, we can also write the product of three different γ matrices as $\gamma_5\gamma_\mu$ by suitable choice of μ. This leaves then the choices for bilinear covariants in Table 2.1, where we have made each quantity Hermitian. Here $\sigma_{\mu\nu}$ is defined as $\frac{i}{2}(\gamma_\mu\gamma_\nu - \gamma_\nu\gamma_\mu)$. Adding together the number of possible operators gives a total of 16, which is the number of independent terms in an arbitrary Hermitian 4×4 matrix. Since a general operator O is a 4×4 matrix, the bilinear covariants provide a complete set for expanding an arbitrary Hermitian operator that has no derivatives.

TABLE 2.1   Possible Spin $\frac{1}{2}$ Hermitian Operators.

Bilinear covariant	Transformation property	Number of operators
$\bar{\psi}\psi$	Scalar	1
$\bar{\psi}i\gamma_5\psi$	Pseudoscalar	1
$\bar{\psi}\gamma_\mu\psi$	Vector	4
$\bar{\psi}\gamma_5\gamma_\mu\psi$	Pseudovector	4
$\bar{\psi}\sigma_{\mu\nu}\psi$	Antisymmetric Tensor	6

In the relativistic limit the helicity states have simple representations. For $\chi^{(+)}$ along $\vec{p}$ and $\chi^{(-)}$ opposite $\vec{p}$:

$$u^{(\pm)} = \sqrt{E}\begin{pmatrix} \chi^{(\pm)} \\ \pm\chi^{(\pm)} \end{pmatrix} \quad \text{and} \quad v^{(\pm)} = \sqrt{E}\begin{pmatrix} \pm\chi^{(\pm)} \\ \chi^{(\pm)} \end{pmatrix}.$$

In this limit the operators

$$O^{(+)} = \frac{1+\gamma_5}{2} \quad \text{and} \quad O^{(-)} = \left(\frac{1-\gamma_5}{2}\right)$$

project out states of given helicity. By explicitly multiplying:

$$O^{(\pm)}u^{(\pm)} = u^{(\pm)}, \quad O^{(\pm)}v^{(\pm)} = v^{(\pm)},$$
$$O^{(\pm)}u^{(\mp)} = 0, \quad O^{(\pm)}v^{(\mp)} = 0.$$

For massless fermions the left- and right-handed states behave differently under the weak interactions. The above projection operators become very useful as a way of keeping the Dirac spinor notation when writing a $\mathcal{L}$ that differentiates between right- and left-handedness.

2.8 ■ INTERPRETATION OF EXTRA SOLUTIONS

The free field equations we looked at had solutions of the form $e^{-ip\cdot x}$ and $e^{ip\cdot x}$, and now we want to interpret the two types of solutions. We want the solutions to provide amplitudes for particles participating in given quantum events where particles are absorbed and created at points in space-time. We will use two observations below to motivate the interpretation.

In nonrelativistic quantum mechanics we calculate transitions via matrix elements of the form $\int \psi_f^* H \psi_i \, d^3x$. For momentum states, $\psi_i \sim e^{-ip_i \cdot x}$ and $\psi_f^* \sim e^{ip_f \cdot x}$. Thus incoming particles in an event have a dependence $e^{-ip\cdot x}$ and outgoing particles have a dependence $e^{ip\cdot x}$ in these calculations. We will maintain this interpretation for the amplitudes in the relativistic case. This is the simplest extension of the nonrelativistic result. The amplitudes in some sense replace the wave functions of the nonrelativistic theory.

What is new in the relativistic case is that a given field has both incoming and outgoing solutions, as does its conjugate. Thus a given field contains as solutions both ψ_i and ψ_f^* of the nonrelativistic theory. For example, the Lagrangian and interaction Hamiltonian for the electromagnetic field contain only A_μ; there is no term of the type A_μ^* (we will make this more precise in Section 2.14, where A_μ becomes a Hermitian operator). Since photons can be both absorbed and emitted, A_μ must be sufficient to describe both processes. Based on the discussion above, the solutions where $A_\mu \sim e^{-ip\cdot x}$ correspond to absorption of (that is, incoming) photons and the solution $e^{ip\cdot x}$ corresponds to emission. The two solutions are needed to allow the one field to describe both processes.

With this interpretation we can state the general result for the fields:

1. A field, for example ϕ or ψ, will have solutions of the form $e^{-ip\cdot x}$ and $e^{ip\cdot x}$. In quantum processes, we take the former to correspond to incoming particles of type A and the latter to correspond to outgoing particles of type B. We will use these amplitudes when these particles participate in processes.

2. The conjugate field, for example ϕ^* or ψ^*, has the exponential complex conjugated, so the solutions for this field correspond to outgoing particles of type A and incoming particles of type B. We use these amplitudes when the conjugate field appears.

3. For a real (Hermitian) field, the conjugate field is the same as the field itself. For this case the type A and type B particles are the same. Examples are the photon field and the real (Hermitian) version of ϕ.

4. If the conjugate field is different from the initial field, for example ψ or the complex (non-Hermitian) ϕ field, particles of type B are called the antiparticles of type A. Stemming from the same initial equation, particles and antiparticles have the same mass and spin. This interpretation accommodates the four independent solutions of the Dirac equation.

Although we are using the nonrelativistic expression for a transition as guidance, it is important to appreciate some of the differences in the relativistic case. For the nonrelativistic theory, H is a fixed operator typically involving given potential functions. The information about the states is contained entirely in ψ_i and ψ_f^* and these represent two states of the same particles. In the relativistic case, the Hamiltonian density, and the resulting H, are dynamically determined in terms of the fields themselves, which does not necessarily result in the same particles in the initial and final states. As an example, the Hamiltonian density for the interacting scalar field in Section 2.3 was $\rho\phi$. Using this expression to calculate transitions, the field ϕ is dynamically determined by the physics process under consideration. The field amplitude will take on the value for a single incoming (or outgoing) particle when we calculate the overall transition amplitude for the absorption (or emission) of such a scalar particle.

In a more abstract notation, the states in nonrelativistic quantum mechanics are indicated by $|i\rangle$ and $\langle f|$. We want to extend this notation to the relativistic theory. Taking the scalar field as an example, and $|\vec{p}\rangle$ as a single scalar particle state of momentum $\vec{p}$, our interpretation for the fields corresponds to the matrix elements for quantum transitions:

$$\langle 0|\phi|\vec{p}\rangle \sim e^{-ip\cdot x} \quad \text{and} \quad \langle \vec{p}|\phi|0\rangle \sim e^{ip\cdot x}. \quad (2.36)$$

The vacuum state $|0\rangle$ has no scalar particle, so these amplitudes correspond to the absorption and emission, respectively, of one scalar particle in a quantum process. We have not distinguished very carefully until now between the states, for example $|\vec{p}\rangle$ or $|0\rangle$, and the corresponding amplitudes. We will, in the next few sections, clarify the distinctions as we develop the quantum theory.

2.9 ■ TIME HISTORY OF STATES

We will use the above interpretation to begin calculating probabilities for processes. We use $|i\rangle$ to denote an initial state that we prepare experimentally. It consists of free particles as $t \to -\infty$. We use $|f\rangle$ to denote a final state whose probability we wish to determine. $|f\rangle$ also consists of free particles as $t \to \infty$. The state into which $|i\rangle$ evolves over time we denote by $|I, t\rangle$. Thus we want to calculate $|\langle f | I, t \to \infty\rangle|^2$ as the experimentally determinable information. Note that we do not follow the system while it is interacting; the real dimensions and times for measurements (taken as infinite) are large enough so that uncertainties in momentum and energy created by restricting the system are insignificant.

The operator that determines the time evolution is the Hamiltonian. However, we want to separate out the effect on the state of the free field Hamiltonian, which creates no transitions, from the effect of the interaction Hamiltonian, which has the fields acting as sources for each other. To accomplish this we introduce the interaction representation. We thus follow the method used to develop time-dependent perturbation theory in nonrelativistic quantum mechanics.

2.9 Time History of States

Consider the time evolution equation

$$i\frac{\partial}{\partial t}|I,t\rangle_S = H|I,t\rangle_S,$$

where H is the full Hamiltonian. The subscript S indicates that this state is in what is called the Schrödinger representation. We will omit the subscript for the state in the interaction representation below. We now separate H into $H_F + H_I$, where H_F describes the free field evolution and H_I the rest of H, which corresponds to the interactions. We now make the following transformation:

$$|I,t\rangle = e^{iH_Ft}|I,t\rangle_S$$

for the states and

$$O = e^{iH_Ft}O_Se^{-iH_Ft}$$

for all operators. Note that O is an explicit function of time, whereas O_S typically is not—the latter characterizes the Schrödinger representation for operators. Using the above we can explicitly differentiate, giving the following equations:

$$i\frac{\partial}{\partial t}|I,t\rangle = H_I|I,t\rangle \tag{2.37}$$

and

$$\frac{d}{dt}O = i[H_F, O]. \tag{2.38}$$

These tell us the following:

1. In the interaction representation the time-dependence of the state is determined entirely by H_I. $|I,t\rangle$ would be just a constant $|i\rangle$ if there were no interactions.

2. All the operators in this representation have a time-dependence determined entirely by the free field Hamiltonian. In particular, this will mean that the field operators we eventually define in Section 2.14, for example the A_μ we will use in calculations, can be determined in the free field case. Since we have already found the free field solutions, we already know much about the amplitudes embedded in these operators for various particle types.

We proceed to solve formally the Hamiltonian Eq. 2.37. We introduce the unitary operator $U(t, t_0)$, called the time evolution operator, which satisfies

$$|I,t\rangle = U(t, t_0)|I, t_0\rangle.$$

This implies that $U(t_0, t_0) = 1$. The Hamiltonian equation, then, is equivalent to

$$i\frac{\partial}{\partial t}U(t, t_0) = H_I(t)U(t, t_0). \tag{2.39}$$

Chapter 2 How to Calculate Amplitudes

This differential equation can be recast into an integral equation:

$$U(t, t_0) = 1 - i \int_{t_0}^{t} H_I(t_1) U(t_1, t_0) \, dt_1.$$

We can solve the equation for $U(t, t_0)$ by the procedure of iteration.

$$U(t, t_0) = 1 - i \int_{t_0}^{t} dt_1 \, H_I(t_1) \left[1 - i \int_{t_0}^{t_1} dt_2 \, H_I(t_2) U(t_2, t_0) \right]$$

$$= 1 - i \int_{t_0}^{t} dt_1 \, H_I(t_1) + (-i)^2 \int_{t_0}^{t} dt_1 \int_{t_0}^{t_1} dt_2 \, H_I(t_1) H_I(t_2) + \cdots$$

$$+ (-i)^n \int_{t_0}^{t} dt_1 \int_{t_0}^{t_1} dt_2 \cdots \int_{t_0}^{t_{n-1}} dt_n \, H_I(t_1) H_I(t_2) \cdots H_I(t_n) + \cdots$$

Each $H_I(t_i)$ can be written in terms of the Hamiltonian density:

$$H_I(t_i) = \int \mathcal{H}_{\text{Int}}(\vec{x}, t_i) \, d^3x.$$

To simplify the notation we will write $\mathcal{H}_{\text{Int}}$ as $\mathcal{H}$, that is, we take $\mathcal{H}$ to refer to the interaction term only.

A few important observations:

1. In the equation $|I, t\rangle = U(t, t_0)|I, t_0\rangle$, the operator $U(t, t_0)$ is independent of $|I, t_0\rangle$. This is a statement of the linearity of quantum mechanics for time evolution. $U(t, t_0)$ in fact must contain within it the information for the evolution of any valid initial system.

2. We will evaluate the interaction Hamiltonian, whose repeated action appears in the iteration expansion, in terms of the fields we have been discussing. This evaluation will provide the mathematical expression corresponding to the point-like local interaction described in Section 1.1. The basic local quantum event will contribute to the time evolution calculation through the matrix element of $\mathcal{H}(\vec{x}, t)$ at a given point $\vec{x}, t$. Quantities conserved by $\mathcal{H}(\vec{x}, t)$, for example, electric charge, will be conserved over time, since the evolution is determined by repeated actions of $\mathcal{H}$.

3. The iteration solution for the operator will yield a coherent sum of amplitudes for all quantum events that can be experienced by the system as it goes from t_0 to t.

4. We can find an approximate solution by cutting off the sum in the iteration solution after n terms, which requires that the next order term be sufficiently small. Such an approximation is known as a perturbation solution. Since the total sum is not known for the cases of interest, most of our results will be perturbation results or results that follow from symmetry arguments.

2.9 Time History of States

Taking the limit as $t_0 \to -\infty, t \to \infty$ for matrix elements between initial and final states gives us the quantity we want, called the S matrix:

$$S_{fi} = \lim_{\substack{t_0 \to -\infty, \\ t \to \infty}} \langle f | U(t, t_0) | i \rangle. \tag{2.40}$$

For the states of given momentum that we are using for $|i\rangle$ and $|f\rangle$, we show that we can write S_{fi} as

$$S_{fi} = \delta_{fi} - i(2\pi)^4 \delta^4(p_f - p_i) M_{fi}, \tag{2.41}$$

where p_i = total initial momentum, p_f = total final momentum, and the 4-dimensional δ function expresses energy and momentum conservation. The quantity M_{fi}, called the matrix element or amplitude for the process, contains the nontrivial physics of the problem. We will try to determine the quantity, often in the approximation of perturbation theory, in order to predict the time evolution of the system. In a perturbation expansion we can write $M_{fi} = M_{fi}^{(1)} + M_{fi}^{(2)} + \cdots$, where the terms correspond to terms in the iteration expansion.

We look at some examples of the perturbation solution, starting with the simplest example. The lowest order term for the time evolution operator is:

$$U^{(1)} = 1 - i \int_{t_0}^{t} H_I(t) \, dt.$$

Substituting for $H_I(t)$ and taking the appropriate limits gives, for the S matrix,

$$S^{(1)} = 1 - i \int_{-\infty}^{\infty} \mathcal{H}(x) \, d^4x.$$

Since $\mathcal{H}$ is a Lorentz scalar and the integral is over all space-time, $S^{(1)}$ is Lorentz invariant. This holds to all orders. We want to take the matrix element of this operator between the initial and final state.

Some examples for $\mathcal{H}$ we have examined earlier are $\rho\phi$ for a scalar interaction, or $J_\mu A_\mu$ for electromagnetism. In terms of the spin $\frac{1}{2}$ particles, we have $\rho = g\bar{\psi}\psi$ and $J_\mu = -e\bar{\psi}\gamma_\mu\psi$, derived from Eqs. 2.32 and 2.31, respectively. For such interactions each factor of $\mathcal{H}$ contains a g for the scalar case or an e for the electromagnetic case. These are called the coupling constants for the given interaction. The applicability of the perturbation expansion will therefore depend on the size of the coupling constant.

A matrix element of $\mathcal{H}$ will correspond to an interaction vertex. For the first order expansion, the one vertex directly involves the initial and final state, so that what we want is $\langle f | \mathcal{H} | i \rangle$. For the vector or scalar interaction, the processes corresponding to particle emission

$$\psi(\text{initial}) \to \psi(\text{final}) + \gamma(\text{final}) \quad \text{or} \quad \phi(\text{final}),$$

are indicated (along with the momenta) in Figure 2.1.

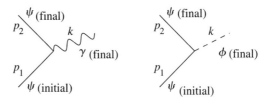

FIGURE 2.1 First order interaction vertex.

We get the matrix element of $\mathcal{H}$ by substituting into $\mathcal{H}$ the appropriate single particle amplitudes for ψ, $\bar{\psi}$, ϕ, or A_μ. We assume for these amplitudes the expressions from Section 2.8, but now also include the appropriate spin dependent factors when the particles are not scalars. Note that ψ has solutions corresponding to incoming particles, the adjoint $\bar{\psi}$ has solutions corresponding to outgoing particles, and ϕ or A_μ has solutions corresponding to either incoming or outgoing particles, where the outgoing solutions are relevant here. Looking at interactions involving spin $\frac{1}{2}$ particles (rather than antiparticles), the two interaction types yield the following matrix elements of $\mathcal{H}$:

$$g\bar{u}(p_2)u(p_1)e^{i(p_2+k-p_1)\cdot x} \quad \text{or} \quad -e\bar{u}(p_2)\gamma_\mu u(p_1)e_\mu^* e^{i(p_2+k-p_1)\cdot x}.$$

Terms from the various exponentials have been combined and the helicity choices for the spin $\frac{1}{2}$ particles and the photon are not explicitly indicated to simplify the notation. For the outgoing photon, the amplitude is the complex conjugate of that for an incoming photon, resulting in e_μ^* appearing rather than e_μ. Note our expression for the matrix element quantifies the amplitudes for the vertices shown in Figure 1.3.

We now look at the structure of this result. Since the spin functions are independent of space-time, the space-time dependence occurs entirely in the phase factors $e^{\pm ip\cdot x}$, making the space-time dependence very simple. For given momenta and helicities the spin part will give a Lorentz invariant function of the momenta that factorizes from the space-time calculation. We are using a density, so we can view the final amplitude for finding the particles as a sum over emissions, which can take place spread over space-time. For a given point in space-time, taking $p_f = p_2 + k$ and $p_i = p_1$ to be the total final and initial 4-momenta, respectively, the amplitude varies as $e^{i(p_f - p_i)\cdot x}$ and momentum conservation is not required. Momentum conservation is therefore not a local property of the Hamiltonian density, which allows in higher-order diagrams for the propagation of virtual particles over restricted regions in space-time. However, adding amplitudes from different locations, the relative phase will vary in the contributions because of the factor $e^{i(p_f - p_i)\cdot x}$. We get a large overall amplitude when the phases are similar over the full space that contributes; this requires $p_f - p_i = 0$ or $p_f = p_i$ when we integrate over all of space-time. Thus, we only get momentum conservation as a result of adding amplitudes over a large space-time region. Clearly, all incoming particles must give a contribution $e^{-ip\cdot x}$ and all outgoing particles $e^{ip\cdot x}$, so that we can separately add together all incoming and all outgoing momenta to get the correct p_i and p_f, respectively.

Writing $\langle f|\mathcal{H}|i\rangle$ as $\mathcal{H}_{fi}(0)e^{i(p_f-p_i)\cdot x}$ and integrating, we have

$$S_{fi}^{(1)} = \delta_{fi} - i(2\pi)^4 \mathcal{H}_{fi}(0)\, \delta^4(p_f - p_i), \qquad (2.42)$$

where the relation

$$\int_{-\infty}^{\infty} e^{i(p_f-p_i)\cdot x}\, d^4x = (2\pi)^4\, \delta^4(p_f - p_i)$$

has been used.

Thus $M_{fi}^{(1)} = \mathcal{H}_{fi}(0)$. Note that $M_{fi}^{(1)}$ is Hermitian since $\mathcal{H}_{fi}$ is; this will not be true for higher orders.

The vertices we've looked at are simple, involving a small number of particles like $e^- \to e^- + \gamma$. This will remain true as we look at other interactions. The S matrix element for $e^- \to e^- + \gamma$, calculated to lowest order, vanishes, since this process would violate the momentum conservation enforced by the $\delta^4(p_f - p_i)$ factor. The first order process, however, is responsible for excited atomic states making transitions to lower energy states with the emission of a photon. Also, the heavy gauge particles such as W^- decay into lighter final states such as $e^- \bar{\nu}_e$ via a first order matrix element. We will look at this process when we discuss the weak interactions.

2.10 ■ TIME EVOLUTION AND PARTICLE EXCHANGE

We turn next to the second order term for the S matrix. This term is extremely important; it will provide an accurate result for particle scattering in cases where the perturbation expansion is adequate—for example, for electrodynamics. Specifically, we will look at the case of muon-electron scattering via the electromagnetic interaction, or a hypothetical scalar interaction. We set up the calculation using electromagnetism, but complete it taking instead the scalar interaction, which is the simplest. This illustrates the common features in such a calculation, which will be completed for electromagnetism in Chapter 3.

The interaction Hamiltonian density for this case is obtained by adding terms for each fermion interacting separately:

$$\mathcal{H} = J_\alpha^{e^-} A_\alpha + J_\alpha^{\mu^-} A_\alpha$$

for the electromagnetic interaction, and

$$\mathcal{H} = \rho^{e^-} \phi + \rho^{\mu^-} \phi$$

for the scalar interaction.

Using

$$U^{(2)} = (-i)^2 \int_{t_0}^{t} dt_1 \int_{t_0}^{t_1} dt_2 \, H_I(t_1) H_I(t_2),$$

then gives for the second order term:

$$S_{fi}^{(2)} = (-i)^2 \int_{-\infty}^{\infty} \int_{-\infty}^{\infty} \theta(t_1 - t_2) \, d^4x_1 \, d^4x_2 \, \langle f | \mathcal{H}(x_1) \mathcal{H}(x_2) | i \rangle. \qquad (2.43)$$

The function $\theta(t_1 - t_2)$, which is 1 for $t_1 > t_2$ and 0 for $t_1 < t_2$, restricts the time integrals to the region where t_2 is less than t_1, as in the expression for $U^{(2)}$.

We next calculate $\mathcal{H}(x_1)\mathcal{H}(x_2)$ using the expression for $\mathcal{H}$ above. It has four terms, of which one gives contributions for e^-e^- scattering, one for $\mu^-\mu^-$ scattering, and two that are relevant to $e^-\mu^-$ scattering. Since U contains the terms to describe all processes, it should not be surprising that only a subset contribute in a given process. The appropriate terms are:

$$J_\alpha^{e^-}(x_1) A_\alpha(x_1) J_\beta^{\mu^-}(x_2) A_\beta(x_2) + J_\beta^{\mu^-}(x_1) A_\beta(x_1) J_\alpha^{e^-}(x_2) A_\alpha(x_2).$$

We indicate the initial and final states, which are composed of separate free particle states, as:

$$|i\rangle = |\mu_{\text{in}}\rangle \, |e_{\text{in}}\rangle \, |0\gamma\rangle$$

and

$$\langle f | = \langle \mu_{\text{out}} | \, \langle e_{\text{out}} | \, \langle 0\gamma |.$$

We have included the information for the photon, indicating that no initial or final photons are present. This gives:

$$\langle f | \mathcal{H}(x_1) \mathcal{H}(x_2) | i \rangle$$
$$= \langle e_{\text{out}} | J_\alpha^{e^-}(x_1) | e_{\text{in}} \rangle \langle \mu_{\text{out}} | J_\beta^{\mu^-}(x_2) | \mu_{\text{in}} \rangle \langle 0 | A_\alpha(x_1) A_\beta(x_2) | 0 \rangle$$
$$+ \langle e_{\text{out}} | J_\alpha^{e^-}(x_2) | e_{\text{in}} \rangle \langle \mu_{\text{out}} | J_\beta^{\mu^-}(x_1) | \mu_{\text{in}} \rangle \langle 0 | A_\beta(x_1) A_\alpha(x_2) | 0 \rangle.$$

In this expression, we exchanged the order in which the currents are written in the second term, since after taking matrix elements the terms are just 4-vectors. Note, as an example, that the expression

$$\langle e_{\text{out}} | J_\alpha^{e^-}(x_2) | e_{\text{in}} \rangle = -e \bar{u}_{\text{out}}^{e^-} \gamma_\alpha u_{\text{in}}^{e^-} e^{i(p_{\text{out}}^{e^-} - p_{\text{in}}^{e^-}) \cdot x_2}$$

is the same as the corresponding expression for the first order calculation in the previous section. Again, the space-time dependence is entirely in the exponential factor, and the spin-dependent part is a function of the momenta and helicities only.

2.10 Time Evolution and Particle Exchange

We next make a useful observation. The space-time points x_2, x_1 label the locations of quantum events in the time sequence in which events occur, irrespective of the event. We would like now to change variables so that the coordinates correspond to specific types of vertices, irrespective of the time-sequence. If we exchange the labels $x_1 \longleftrightarrow x_2$ in the second term, then we can interpret x_1 as the space-time point where the electron scatters and x_2 as the space-time point where the muon scatters. The coordinates then correspond to well-defined physical events. We can do this, provided we also exchange t_1 and t_2 in the θ function from Eq. 2.43 accompanying the second term. This gives us now the expression:

$$(-i)^2 \int_{-\infty}^{\infty} \int_{-\infty}^{\infty} d^4x_1 \, d^4x_2 \left\langle e_{\text{out}} \right| J_\alpha^{e^-}(x_1) \left| e_{\text{in}} \right\rangle \left\langle \mu_{\text{out}} \right| J_\beta^{\mu^-}(x_2) \left| \mu_{\text{in}} \right\rangle T_{\alpha\beta}(x_1, x_2),$$

$$T_{\alpha\beta}(x_1, x_2) = \left\langle 0 \right| A_\alpha(x_1) A_\beta(x_2) \theta(t_1 - t_2) + A_\beta(x_2) A_\alpha(x_1) \theta(t_2 - t_1) \left| 0 \right\rangle.$$

The currents are each evaluated at arbitrary space-time points, with the tensor $T_{\alpha\beta}$ depending on the locations of the two points.

You might imagine that, since no photons are in the initial or final state that the term involving $A_\alpha A_\beta$ vanishes; but this isn't true. Since the field appears twice, the first field can correspond to the emission of a photon and the second its absorption, so that no photon ends up in the final state. Such a particle is called a virtual particle. Thus the scattering occurs because of the exchange of virtual particles. This idea is crucial in understanding how the particle picture explains scattering, which in the nonrelativistic limit is pictured as due to a potential.

We now develop the same calculation for scalar exchange and work through the effect of the virtual particle. In this case,

$$J_\alpha, J_\beta \to \rho \quad \text{and} \quad A_\alpha, A_\beta \to \phi.$$

Thus the analogous result is:

$$S_{fi}^{(2)} = (-i)^2 \int d^4x_1 \, d^4x_2 \left\langle e_{\text{out}} \right| \rho^{e^-}(x_1) \left| e_{\text{in}} \right\rangle$$

$$\cdot \left\langle \mu_{\text{out}} \right| \rho^{\mu^-}(x_2) \left| \mu_{\text{in}} \right\rangle D_F(x_1, x_2),$$

$$D_F(x_1, x_2) = \left\langle 0 \right| \phi(x_1)\phi(x_2)\theta(t_1 - t_2) + \phi(x_2)\phi(x_1)\theta(t_2 - t_1) \left| 0 \right\rangle,$$

$$\left\langle e_{\text{out}} \right| \rho^{e^-}(x_1) \left| e_{\text{in}} \right\rangle = g \bar{u}_{\text{out}}^{e^-} u_{\text{in}}^{e^-} e^{i(p_{\text{out}}^{e^-} - p_{\text{in}}^{e^-}) \cdot x_1},$$

$$\left\langle \mu_{\text{out}} \right| \rho^{\mu^-}(x_2) \left| \mu_{\text{in}} \right\rangle = g \bar{u}_{\text{out}}^{\mu^-} u_{\text{in}}^{\mu^-} e^{i(p_{\text{out}}^{\mu^-} - p_{\text{in}}^{\mu^-}) \cdot x_2}.$$

We next focus on D_F, which involves the scalar fields. We insert a complete set of states involving one ϕ particle between the two ϕ fields:

$$1 = \sum_{\vec{p}} |p\rangle \langle p|,$$

Chapter 2 How to Calculate Amplitudes

so that this expression becomes:

$$\sum_{\vec{p}} \langle 0|\phi(x_1)|p\rangle \langle p|\phi(x_2)|0\rangle \theta(t_1-t_2) + \sum_{\vec{p}} \langle 0|\phi(x_2)|p\rangle \langle p|\phi(x_1)|0\rangle \theta(t_2-t_1).$$

Here $|p\rangle$ represents a scalar particle of 4-momentum p. The term of the form $\langle p|\phi(x_2)|0\rangle$ represents an outgoing scalar particle of momentum p created at x_2. The field representing an outgoing scalar is $e^{ip\cdot x_2}$. Similarly, $\langle 0|\phi(x_2)|p\rangle$ represents the absorption of an incoming particle. The field corresponding to an incoming particle absorbed at x_2 is $e^{-ip\cdot x_2}$. We can thus picture the two terms by Figure 2.2.

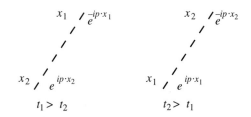

FIGURE 2.2 Amplitudes for virtual scalar exchange.

Defining $D_F(x_2 - x_1)$ as the contribution to the amplitude from the scalars, we have

$$D_F(x_2 - x_1) = \sum_{\vec{p}} \left[e^{-ip\cdot(x_1-x_2)}\theta(t_1 - t_2) + e^{-ip\cdot(x_2-x_1)}\theta(t_2 - t_1) \right]. \quad (2.44)$$

D_F is called the space-time propagator for the exchange of a scalar particle of arbitrary momentum.

Because of the integrals over phase factors, the simplest way to complete the calculation is to introduce the Fourier transform of D_F. Defining the 4-dimensional transform $f(p)$ via

$$D_F(x) = \int f(p) e^{-ip\cdot x} \frac{d^4 p}{(2\pi)^4}, \quad (2.45)$$

we can now write the expression for $S_{fi}^{(2)}$:

$$S_{fi}^{(2)} = \left[-i \langle e_{\text{out}} | \rho^{e^-}(0) | e_{\text{in}} \rangle \right] \left[-i \langle \mu_{\text{out}} | \rho^{\mu^-}(0) | \mu_{\text{in}} \rangle \right] I.$$

The space-time integral

$$I = \int d^4x_1 \, d^4x_2 \, e^{i(p_{\text{out}}^{e^-} - p_{\text{in}}^{e^-})\cdot x_1} e^{i(p_{\text{out}}^{\mu^-} - p_{\text{in}}^{\mu^-})\cdot x_2} f(p) e^{-ip\cdot(x_2-x_1)} \frac{d^4 p}{(2\pi)^4}.$$

2.10 Time Evolution and Particle Exchange

Defining

$$x = x_2 - x_1, \quad p_i = p_{in}^{e^-} + p_{in}^{\mu^-}, \quad p_f = p_{out}^{e^-} + p_{out}^{\mu^-},$$

$$I = \int d^4x_1 \, d^4x \, \frac{d^4p}{(2\pi)^4} e^{i(p_f - p_i) \cdot x_1} e^{i(p_{out}^{\mu^-} - p_{in}^{\mu^-} - p) \cdot x} f(p)$$

$$= (2\pi)^4 \delta^4(p_f - p_i) \int d^4x \frac{d^4p}{(2\pi)^4} e^{i(p_{out}^{\mu^-} - p_{in}^{\mu^-} - p) \cdot x} f(p).$$

For any given $x_2 - x_1$, momentum conservation again comes from adding amplitudes spread over space-time. The rest of the amplitude depends on the physics for relative separations $x_2 - x_1$. We can complete the integration simply. The integral over x gives a delta function, which then gets rid of the integral over d^4p. The end result is:

$$(2\pi)^4 \delta^4(p_f - p_i) f(p_{out}^{\mu^-} - p_{in}^{\mu^-}).$$

Defining $q = p_{in}^{e^-} - p_{out}^{e^-}$, which also $= p_{out}^{\mu^-} - p_{in}^{\mu^-}$; it is the 4-momentum transferred in the reaction. The function f, which is called the momentum-space propagator, is a function of q. We show in the next section that

$$f(q) = \frac{i}{q^2 - m^2 + i\varepsilon}, \tag{2.46}$$

where m = scalar particle mass, and the $i\varepsilon$ tells us how to handle the poles from the denominator (which are not important for the present calculation, since q^2 cannot equal m^2). The coherent effect of the space-time sum involving all single particle scalar states, each of which satisfy $p^2 = m^2$, is given by the simple momentum-space propagator where $q^2 \neq m^2$.

We can now put together all of the terms and write the matrix element (leaving out the $i\varepsilon$):

$$-iM_{fi}^{(2)} = \left[-i \left\langle e_{out} | \rho^{e^-}(0) | e_{in} \right\rangle \right] \frac{i}{q^2 - m^2} \left[-i \left\langle \mu_{out} | \rho^{\mu^-}(0) | \mu_{in} \right\rangle \right].$$

Cancelling the factors of i and putting in the spin dependent terms:

$$M_{fi}^{(2)} = g^2 \bar{u}_{out}^{e^-} u_{in}^{e^-} \frac{1}{q^2 - m^2} \bar{u}_{out}^{\mu^-} u_{in}^{\mu^-}.$$

We can summarize the elements in the calculation by a Feynman diagram, which shows the correct linkages of the particles. For the lines in the picture we can associate the terms in the matrix element shown in Figure 2.3. Such a diagram represents an explicit calculable contribution to M_{fi}. The propagator in the picture represents the sum over both time-orderings for the scalar exchange in space-time. In calculating q, the 4-momentum is conserved at each vertex.

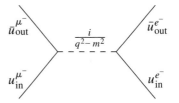

FIGURE 2.3 Feynman diagram for $e^-\mu^-$ scattering via exchange of uncharged scalar particles. Terms shown are ingredients needed to calculate $-iM_{fi}$.

Before turning to an explicit calculation of the crucial propagator function, we look at some points related to the fact that the propagator includes two different time sequences. First, we would like to see how our calculation might work for $\nu_\mu e^-$ scattering via W exchange. The Feynman diagram analogous to the diagram in Figure 2.3 is shown in Figure 2.4.

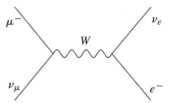

FIGURE 2.4 Feynman diagram for $\nu_\mu e^- \to \mu^- \nu_e$.

We redraw this diagram in Figure 2.5, indicating the two time orderings in space-time and the associated flow of charge, which is a conserved quantity.

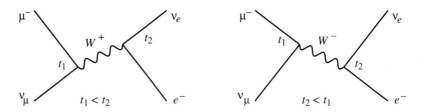

FIGURE 2.5 Two time sequences represented by one Feynman diagram.

The virtual W that contributes is sometimes a W^- and sometimes a W^+—both are needed to conserve charge and allow two time sequences. The propagator in general includes both particles and antiparticles! The coexistence of charges for the particles and relativistic covariance, which requires both time ordered processes, implies the existence of antiparticles. Note that the difference between the

2.10 Time Evolution and Particle Exchange

two time sequences for the virtual particle at each vertex is that we go from incoming to outgoing (or vice versa) solutions for the field. The exchange of particle and antiparticle required by the diagram when we go from incoming to outgoing solutions of the field equation fits perfectly with the interpretation of these solutions given in Section 2.8. For the neutral scalar field the particles and antiparticles are identical, so we did not encounter this in our calculation involving ϕ.

The space-time behavior of scattering is determined by the integrals over the product of the exponential phase factors. Writing the net phase factor from the incoming and outgoing particles (that is, excluding the virtual particle) at vertex 1 as $e^{-iq \cdot x_1}$, we can write the analogous factor for vertex 2 as $e^{iq \cdot x_2}$ if we constrain the momenta to a configuration which will satisfy overall 4-momentum conservation. The phase factor terms in the calculation of the amplitude, now including the virtual particle, are:

$$\sum_{\vec{p}} e^{-i(q-p) \cdot (x_1 - x_2)} \theta(t_2 - t_1) + e^{-i(q+p) \cdot (x_1 - x_2)} \theta(t_1 - t_2).$$

Returning to the $e^- \mu^-$ scattering case, $q = (0, \vec{q})$ in the center of mass. For this q the two time orderings contribute equal amplitudes after integration over momenta, since all values of $\vec{p}$ occur. In this case q is a space-like 4-vector (that is, $q^2 < 0$).

We will have frequent situations, beginning in Chapter 3, where q is time-like (that is, $q^2 > 0$). We assume that we have chosen vertex 1 to be the one for which $q_0 > 0$. The situation is illustrated in Figure 2.6. If $q^2 = m^2$, where m = mass of the virtual particle, there is a pole in the momentum-space propagator reflecting the fact that a real intermediate particle can be produced. Looking at the two time ordered terms in the center of mass, with $E(\vec{p}) = \sqrt{|\vec{p}|^2 + m^2}$ always > 0 and $q_0 > 0$, the term

$$e^{-i(q+p) \cdot (x_1 - x_2)} \theta(t_1 - t_2)$$

is rapidly oscillating over time and contributes little to the overall amplitude when integrated over time. The other term,

$$e^{-i(q-p) \cdot (x_1 - x_2)} \theta(t_2 - t_1),$$

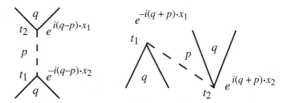

FIGURE 2.6 Two time orderings for a process where a virtual particle can be real. In this case, one of the two terms dominates the final full amplitude after summing over momenta of the virtual particle and the space-time points for the quantum events.

will give in phase contributions over a large space-time region if $p \approx q$, providing most of the amplitude. In fact, the contributions can exist over a macroscopic region where we can detect the virtual particle as a physically present object. This term has the characteristics with which we are familiar from a causal chain of events:

1. Event 1 happens first in time at t_1 followed later by event 2 at time t_2. The final state exists after the initial state.
2. The intermediate physical particle carries away the missing 4-momentum and imparts it to the second system.
3. The charge of the intermediate state is fixed; it carries away the charge lost at vertex 1 and brings it to vertex 2.

2.11 ■ MOMENTUM SPACE PROPAGATOR

We want to calculate

$$\sum_{\vec{p}} \left[e^{-ip\cdot(x_1-x_2)}\theta(t_1-t_2) + e^{-ip\cdot(x_2-x_1)}\theta(t_2-t_1) \right] = \int f(p) e^{-ip\cdot(x_2-x_1)} \frac{d^4p}{(2\pi)^4}$$

in order to show that

$$f(p) = \frac{i}{p^2 - m^2 + i\varepsilon}.$$

The first step is to make sense of the $\sum_{\vec{p}}$, since there are an infinite number of momentum states. In addition, we have to correctly calculate the field normalization. We are exchanging one quantum but the field used in the calculation, $e^{-ip\cdot x}$, corresponds to $2(EV)$ quanta in a given volume V.

To keep track conceptually of quantities that go to ∞ or 0, or products of such quantities, we assume that the process we are observing is constrained to occur inside a large volume V. The momentum states now form a very finely spaced denumerable set. We can approximate the sum over states by an integral, which is familiar from momentum states in a box:

$$\sum_{\vec{p}} \to \frac{d^3p}{(2\pi)^3} V.$$

Putting in the correct normalization for the fields will require changing, at both $x = x_1$ and x_2, $e^{\pm ip\cdot x}$ to $e^{\pm ip\cdot x}/\sqrt{(2E)V}$, which is the amplitude for one quantum particle in the volume V. Including this normalization correction gives an extra factor of $1/2EV$ in the sum. Thus the correct space-time propagator, assuming one quantum to be exchanged, is:

2.11 Momentum Space Propagator

$$D_F(x) = \int \left[e^{-ip\cdot x}\theta(t) + e^{ip\cdot x}\theta(-t)\right] \frac{d^3p}{2E(2\pi)^3}.$$

We write here $x = x_2 - x_1$. The volume V has disappeared. The factor $d^3p/2E(2\pi)^3$ is called the Lorentz invariant momentum space factor. It is a Lorentz scalar, that is, under a Lorentz transformation it has the same form in any frame.

The expression for $D_F(x)$ is already in the form of a 3-dimensional Fourier transform. We want to show that the 4-dimensional form satisfies

$$\int \frac{i}{p^2 - m^2 + i\varepsilon} e^{-ip\cdot x} \frac{d^4p}{(2\pi)^4} = \int \left[e^{-ip\cdot x}\theta(t) + e^{ip\cdot x}\theta(-t)\right] \frac{d^3p}{2E(2\pi)^3}.$$

Note that the time-component of p in the 4-dimensional integral, which we write as p_0, is a parameter that varies from $-\infty$ to $+\infty$, whereas E in the 3-dimensional integral (both in the exponential and in the denominator) is the number $E = \sqrt{|\vec{p}|^2 + m^2}$ for a given $\vec{p}$. The space-time point x is an arbitrary point, which is the same in both integrals.

To turn the 4-dimensional integral into a 3-dimensional one, we integrate over p_0. We use the method of complex variables to figure out the integral, with p_0 considered to be a complex number. The denominator $p^2 - m^2 + i\varepsilon$ produces poles near $p_0 = \pm\sqrt{|\vec{p}|^2 + m^2}$ for a given $|\vec{p}|$, with the $i\varepsilon$ telling us whether they are just above or below the real p_0 axis. Defining $E = \sqrt{|\vec{p}|^2 + m^2}$, we can write $p^2 - m^2$ as $(p_0 - E)(p_0 + E)$, from which it is clear that the pole at $p_0 = -E$ lies just above the real axis and that at $p_0 = E$ lies just below the real axis.

We now do the integral along dp_0 by closing the contour with a large semicircle and using the residue theorem. We have to close the contour with a semicircle over which the integrand vanishes. Looking at the dependence on p_0, taken to be $p_0^R + ip_0^I$, $e^{-ip_0 t} = e^{-ip_0^R t}e^{p_0^I t}$. Thus, the integrand vanishes for large positive p_0^I for $t < 0$ and large negative p_0^I for $t > 0$. Depending on the sign of t, we have to use the upper or lower semicircle. The situation is shown in Figure 2.7.

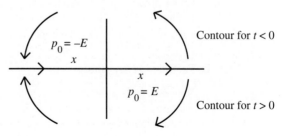

FIGURE 2.7 Contours for evaluating the integral appearing in the propagator. Cross indicates pole location. Integral is in complex p_0 plane.

Including a -1 for $t > 0$, since the contour runs clockwise instead of counter-clockwise, gives the following factors after using the residue theorem:
For $t > 0$, pole at $p_0 = E$:

$$(-1)(2\pi i) \frac{i}{2E} e^{-iEt} e^{i\vec{p}\cdot\vec{x}} \frac{d^3 p}{(2\pi)^4},$$

for $t < 0$, pole at $p_0 = -E$:

$$(2\pi i) \frac{i}{-2E} e^{iEt} e^{i\vec{p}\cdot\vec{x}} \frac{d^3 p}{(2\pi)^4}.$$

To complete the proof, we note that E is independent of the sign of $\vec{p}$ and the remaining 3-dimensional integral will be over all $\vec{p}$. We can thus make a change of variables, replacing $\vec{p}$ with $-\vec{p}$ in the $t < 0$ result, leaving the integral unchanged. Putting all of this together we can write the integral as

$$\int \left[e^{-ip\cdot x} \theta(t) + e^{ip\cdot x} \theta(-t) \right] \frac{d^3 p}{2E(2\pi)^3},$$

which is the desired result.

2.12 ■ CALCULATION OF DECAY RATES AND CROSS SECTIONS

Having calculated a matrix element, we next want to get a number that can be compared to what might be seen in an experiment. So far we have an expression for the S matrix involving a singular quantity, $\delta^4(p_f - p_i)$, with S evaluated using fields normalized to $(2E)V$ particles for each incoming and outgoing particle, instead of the individual single particles actually detected. To deal with these issues we again assume that we are calculating a process constrained to occur in a large volume V. The formulas we derive will in the end be independent of V; they will be the foundation for the calculations of rates in various processes!

Generally, we assume that we are calculating either a decay or a 2-body scattering process. The S matrix calculation is similar in the two cases; we just have to choose the appropriate $|i\rangle$ and $\langle f|$. We will return to the $e^-\mu^-$ scattering calculation after deriving the general expression for a scattering cross section.

2.12.1 ■ Particle Decay Rates

We start with a decay where we assume an initial particle, mass m, decays to particles $1, \ldots, n$ in the final state. The transition probability to a given final state is $|S_{fi}|^2$. However, to get a total probability we want to sum over final states, so that we are interested in $\sum_f |S_{fi}|^2$. In this expression we sum over the final state particle types, their helicities, and momenta. To understand the physics, we

2.12 Calculation of Decay Rates and Cross Sections

normally look at terms in the sum, for example, breaking it into pieces with given particle types. We shall do this below.

For an individual particle k, the sum over momentum states, for a large volume V, is given by integrating over $d^3 p_k V/(2\pi)^3$. Similarly we must put in a factor $1/2E_k V$ to normalize the field for k to one particle in V. Putting these two together, the sum over final states for $1, \ldots, n$ particles, after correctly normalizing the outgoing fields, will be given by

$$\frac{d^3 p_1 V}{2E_1 V (2\pi)^3} \cdots \frac{d^3 p_n V}{2E_n V (2\pi)^3} = \prod_{k=1}^{n} \frac{d^3 p_k}{2E_k (2\pi)^3}.$$

This is the Lorentz invariant phase space for the final state. By sticking to the covariant fields when calculating S_{fi}, we actually arrive at an elegant, manifestly Lorentz invariant result, since S_{fi} and the phase space factor are separately invariant functions!

Since we are looking at particle decay, the quantity of interest will be the decays per second or transition rate. To get this, we have to divide the number of transitions by the time elapsed, T. Finally, putting in the normalization factor $1/2E_i V$ for the initial state, so that we are dealing with the transition rate for an individual particle, and multiplying by the true number of particles we start with, N_i, gives:

Number of transitions/sec $(i \to f)$

$$= N_i \frac{1}{2E_i} \int \prod_{k=1}^{n} \frac{d^3 p_k}{2E_k (2\pi)^3} |M_{fi}|^2 \frac{|(2\pi)^4 \delta^4(p_f - p_i)|^2}{VT}.$$

To make sense of the rate expression, we have to understand the meaning of the δ-function term. To do this, we return to the expression from which it originated, namely,

$$(2\pi)^4 \delta^4(p_f - p_i) = \int e^{i(p_f - p_i) \cdot x} d^4 x.$$

Thus what we really want is the limit, for $T = t - t_0$, V both large, of:

$$\frac{\left| \int_V \int_{t_0}^{t} e^{i(p_f - p_i) \cdot x} dt \, d^3 x \right|^2}{TV}.$$

To do this properly, one should actually perform the integral and then take the limit. The following shortcut gives the correct result. Writing the square as two terms in which we separately take limits gives:

$$(2\pi)^4 \delta^4(p_f - p_i) \left[\frac{\int_{T,V} e^{i(p_f - p_i) \cdot x} d^4 x}{TV} \right].$$

The $\delta^4(p_f - p_i)$ term tells us that in the second integral we should set $p_f = p_i$, in which case the integral is just TV, cancelling the denominator.

This gives finally the crucial result:

$$\text{Number of transitions/sec } (i \to f) = N_i \Gamma_{i \to f},$$

where

$$\Gamma_{i \to f} = \frac{1}{2E_i} \int |M_{fi}|^2 (2\pi)^4 \delta^4(p_f - p_i) \prod_{k=1}^{n} \frac{d^3 p_k}{2E_k (2\pi)^3}. \tag{2.47}$$

Γ is called the decay width.

In defining the state f, each set of helicity choices $\lambda_1, \ldots, \lambda_n$ is a separate decay choice. Usually, we want to consider just the separate particle types. Thus what is usually used for $\Gamma_{i \to f}$ is

$$\Gamma_{i \to f} = \sum_{\text{All } \lambda} \Gamma_{i \to f}(\lambda_1, \ldots, \lambda_n).$$

If final states $f_1, f_2, \ldots, f_m$ are possible, then we define $\Gamma_{\text{tot}} = \sum_{j=1}^{m} \Gamma_{f_j}$. We also define the branching ratio for f_j as

$$B_{f_j} = \frac{\Gamma_{f_j}}{\Gamma_{\text{tot}}}.$$

The appearance of any f_j corresponds to a decay of i; thus Γ_{tot} determines the actual full decay rate of i. B_{f_j} gives the fraction of the decays of i in which the final state is f_j.

To figure out the population of i we write down the rate of change, which decreases N_i:

$$\frac{dN_i}{dt} = -\Gamma_{\text{tot}} N_i,$$

which gives

$$N_i = N_0 e^{-\Gamma_{\text{tot}} t}.$$

Considering the particle at rest, we can write this as $N_0 e^{-t/\tau}$, where τ is called the lifetime, and defined as $\tau = (1/\Gamma_{\text{tot}})$, where Γ_{tot} is calculated for i at rest, which involves setting $E_i = m$ in Eq. 2.47. For a moving particle, Eq. 2.47 gives $N_i = N_0 e^{-mt/(E_i \tau)}$. The apparent lifetime is now longer by a factor E_i/m. This is exactly the time dilation factor given by special relativity.

2.12.2 ■ Cross Sections

The rate for two-body scattering, $1 + 2 \to f$, is defined in terms of a cross section. It keeps track of rates in a manner dependent on the basic physics, with fluxes of particles divided out. We define the cross section σ by

$$\sigma = \left[\frac{\text{number of final states detected}}{\text{unit time}} \right] \text{per}$$

[target particle $\times$ flux of projectile particles relative to the target particles].

Again taking the events to be restricted to a large volume V and normalizing to one projectile particle in V, the flux factor is $v_{\text{rel}} \times$ density of projectile particles $= v_{\text{rel}}/V$. v_{rel} is the relative velocity $= |\vec{v}_1 - \vec{v}_2|$. The numerator is, prior to normalizing the particle numbers and summing over final states, $|S_{fi}|^2/T$. Putting in the same normalization factors for the sum over final states as in the decay rate calculation, gives the cross section for transitions to a given final state:

$$\sigma = \frac{1}{2E_1 2E_2 v_{\text{rel}}} \int |M_{fi}|^2 (2\pi)^4 \delta(p_f - p_i) \prod_{k=1}^{n} \frac{d^3 p_k}{2E_k (2\pi)^3}. \qquad (2.48)$$

The two factors of $1/V$ coming from normalizing the particle density for each initial state particle to one particle in V, are used to cancel the factor of V in v_{rel}/V and, when combined with the time T, to eliminate the extra $(2\pi)^4 \delta^4(p_f - p_i)$ factor, as was done in calculating the width.

In general, for several possible final states f_j, we define partial cross sections σ_{f_j}. The total cross section is $\sigma_{\text{tot}} = \sum_{j=1}^{m} \sigma_{f_j}$. The total cross section is related to the disappearance of beam particles as they travel through a target. For example, if we start with a beam of N_0 particles passing through a medium of target particles with a density ρ of target particles per unit volume, then as a function of $z = $ distance through the medium, the number of particles remaining in the beam, $N(z)$, is

$$N(z) = N_0 e^{-\rho \sigma_{\text{tot}} z}. \qquad (2.49)$$

The cross section has units of area.

2.13 ■ PARTICLE EXCHANGE AND THE YUKAWA POTENTIAL

We look next at the nonrelativistic limit of the $e^- \mu^-$ scattering cross section for scalar exchange. In Chapter 3 we will look at cross sections for the electromagnetic interactions systematically. To calculate, we choose a reference frame in which μ^- is at rest and the incident electron has a kinetic energy much less than its mass. We indicate the various masses as $m_e = $ electron mass, $m_\mu = $ muon mass, and $m = $ mass of the exchanged hypothetical scalar particle. In this calculation the muon acts essentially as a static source and, to good approximation, no energy is transferred, so $q^2 = -|\vec{q}|^2$. We can calculate $|\vec{q}|^2 = |\vec{p}_{\text{in}}^{e^-} - \vec{p}_{\text{out}}^{e^-}|^2$. Taking the incident direction as the z-axis and the scattering angle of the electron as θ, $|\vec{q}|^2 = 4|\vec{p}_{e^-}|^2 \sin^2(\theta/2)$. We assume that the coupling constant g^2 is small enough so that $M_{fi} \approx M_{fi}^{(2)}$, that is, we take the lowest order term that contributes to this scattering process.

From Section 2.10, we have:

$$M_{fi} = g^2 \bar{u}_{\text{out}}^{e^-} u_{\text{in}}^{e^-} \frac{1}{q^2 - m^2} \bar{u}_{\text{out}}^{\mu^-} u_{\text{in}}^{\mu^-}.$$

In the nonrelativistic limit we can use Eq. 2.33 for the spinors. So in this case, $\bar{u}_{\text{out}}^{e^-} u_{\text{in}}^{e^-} \approx 2m_e$ for the outgoing spinor = incoming spinor and ≈ 0 for the outgoing spinor orthogonal to the incoming spinor. (Note that the final helicity state is, however, not the same as the initial). A similar result holds for the muon. Thus we can simplify:

$$M_{fi} = \frac{-g^2 (2m_e)(2m_\mu)}{|\vec{q}|^2 + m^2}$$

for the spinors that give a finite contribution. M_{fi} is now a function of the electron scattering angle only.

We next have to integrate this over the phase space. Writing

$$p = |\vec{p}_{\text{out}}^{e^-}|, \quad p' = |\vec{p}_{\text{out}}^{\mu^-}|$$

and the solid angle elements for the outgoing e^- and μ^- as $d\Omega$ and $d\Omega'$, respectively,

$$\prod_{k=1}^{2} \frac{d^3 p_k}{2E_k (2\pi)^3} = \frac{p^2 \, dp \, d\Omega}{2E(2\pi)^3} \frac{p'^2 \, dp' \, d\Omega'}{2E'(2\pi)^3}.$$

Writing σ as differential in Ω gives

$$\frac{d\sigma}{d\Omega} = \frac{1}{(2E_1)(2E_2) v_{\text{rel}}} \int |M_{fi}|^2 \frac{p^2 \, dp}{2E(2\pi)^3} \frac{p'^2 \, dp' \, d\Omega'}{2E'(2\pi)^3} (2\pi)^4 \delta^4(p_f - p_i).$$

We can integrate over the p' variables, which eliminates the three momentum δ-functions and fixes the relationship of the various momenta. In the nonrelativistic limit, for the frame chosen:

$$E_1 = E = m_e, \quad E_2 = E' = m_\mu, \quad v_{\text{rel}} = \frac{p}{m_e}.$$

Making these substitutions and multiplying out the various factors of 2 and 2π gives

$$\frac{d\sigma}{d\Omega} = \frac{1}{16 m_\mu^2 m_e (2\pi)^2} \int |M_{fi}|^2 p \, dp \, \delta(E_f - E_i).$$

Using the nonrelativistic expression for $E_f - E_i$ and the fact that no energy is transferred to the muon,

2.13 Particle Exchange and the Yukawa Potential

$$E_f - E_i = \frac{p^2}{2m_e} - \frac{|\vec{p}_{\text{in}}^{e^-}|^2}{2m_e}.$$

We next need to integrate $\delta(E_f - E_i)$ over dp. We use the general result:

$$\int dx\, \delta(f(x)) = \frac{1}{|\frac{df}{dx}|_{f(x)=0}}.$$

Thus

$$\int p\, dp\, \delta(E_f - E_i) = \frac{p}{\frac{dE_f}{dp}} = m_e$$

for the function above for E_f. This gives, finally:

$$\frac{d\sigma}{d\Omega} = \left(\frac{1}{2m_\mu}\right)^2 \frac{|M_{fi}|^2}{(4\pi)^2},$$

where

$$|M_{fi}|^2 = \left| \frac{-g^2(2m_e)(2m_\mu)}{|\vec{q}|^2 + m^2} \right|^2.$$

Thus:

$$\frac{d\sigma}{d\Omega} = \left(\frac{m_e}{2\pi}\right)^2 \frac{g^4}{(|\vec{q}|^2 + m^2)^2}.$$

For $m^2 \gg |\vec{q}|^2$ the scattering is isotropic. We expect in the nonrelativistic limit to be able to calculate the same quantity using a potential and the Born approximation. We can thus ask, what potential must we choose to get this result?

Using formulas from nonrelativistic quantum mechanics,

$$\frac{d\sigma}{d\Omega} = |f(\theta, \phi)|^2,$$

where

$$f(\theta, \phi) = \frac{-m_e}{2\pi} \int e^{i\vec{q}\cdot\vec{x}} V(\vec{x})\, d^3x.$$

The latter expression is the Born approximation. Comparing the two expressions for $d\sigma/d\Omega$, we have within a $\pm$ sign:

$$\int e^{i\vec{q}\cdot\vec{x}} V(\vec{x})\, d^3x = \frac{\pm g^2}{|\vec{q}|^2 + m^2}.$$

To get the sign correct we have to decide whether M_{fi} is related to $+f(\theta, \phi)$ or $-f(\theta, \phi)$. For this we need something related to $f(\theta, \phi)$ rather than $|f(\theta, \phi)|^2$.

The relation that involves $f(\theta, \phi)$ is the unitarity relation, also called the optical theorem, which relates Imf(0) to σ_{tot}. The analogous relation for M_{fi} will be discussed in Chapter 5. We conclude that M_{fi} is related to $-f(\theta, \phi)$. Thus, we have the relation:

$$\int e^{i\vec{q}\cdot\vec{x}} V(\vec{x}) \, d^3x = \frac{-g^2}{|\vec{q}|^2 + m^2}.$$

Inverting this Fourier transform gives:

$$V(\vec{x}) = \frac{-g^2 e^{-mr}}{4\pi r}.$$

This is just the static source Yukawa potential energy derived in Eq. 2.13. Thus the particle exchange picture has given us the expected nonrelativistic potential, which includes both the coupling constants and the Fourier transform of the propagator.

2.14 ■ QUANTUM FIELD THEORY

We have included, when evaluating $\langle f| \mathcal{H} |i \rangle$, the field amplitudes corresponding to the various particles participating in the process. We have done this by analogy with nonrelativistic quantum mechanics and by interpreting the solutions for particle amplitudes that arise from the various field equations. We want to put this on a more solid mathematical basis, where the results will follow more clearly from the mathematical structure. If we can write $\mathcal{H}$ in terms of operators that create and destroy particles at points in space-time, leaving in their place the field amplitudes, then these operators will yield the same formulas we have been using. This is called Quantum Field Theory. It has some major advantages; it allows the extension of the theory to the case of many particles present in the same state (for example, photons in a laser), and a clear approach to the exchange question for bosons and fermions.

We have specified our free particle states $|i\rangle$ by the various particles and their momentum and helicity. The number of particles were 1 or 0. Thus we had the simple situation where $|i\rangle = |0\rangle$ or $|\vec{p}, \lambda\rangle$, the latter a single particle state. This space of states, extended to an arbitrary number of particles, is called a Fock space. We have taken for the matrix elements of the fields, taking A_μ as an example:

$$\langle 0| A_\mu |\vec{p}, \lambda\rangle = e_\mu(\lambda) e^{-ip\cdot x}$$
$$\langle \vec{p}, \lambda| A_\mu |0\rangle = e_\mu^*(\lambda) e^{ip\cdot x}.$$

We would like to consider A_μ as an operator in the Fock space. We expand it as follows:

2.14 Quantum Field Theory

$$A_\mu = \sum_{\vec{p},\lambda} a_{\vec{p},\lambda} e_\mu(\lambda) e^{-ip\cdot x} + a^\dagger_{\vec{p},\lambda} e^*_\mu(\lambda) e^{ip\cdot x}. \tag{2.50}$$

Provided $a_{\vec{p},\lambda}$ and $a^\dagger_{\vec{p},\lambda}$ are independent of $\vec{x}$ and t, A_μ will satisfy the free field equation; $a_{\vec{p},\lambda}$ is called an annihilation operator, and $a^\dagger_{\vec{p},\lambda}$ is called a creation operator. To get the desired results for the matrix elements of A_μ, we require:

$$\langle 0| a_{\vec{p}',\lambda'} |\vec{p},\lambda\rangle = \delta_{\vec{p}',\vec{p}} \delta_{\lambda',\lambda}$$

$$\langle 0| a^\dagger_{\vec{p}',\lambda'} |\vec{p},\lambda\rangle = 0$$

$$\langle \vec{p},\lambda| a^\dagger_{\vec{p}',\lambda'} |0\rangle = \delta_{\vec{p}',\vec{p}} \delta_{\lambda',\lambda}$$

$$\langle \vec{p},\lambda| a_{\vec{p},\lambda} |0\rangle = 0.$$

For the operators above, $a^\dagger$ is the Hermitian conjugate of a and therefore $A^\dagger_\mu = A_\mu$. This is the generalization of the classical relation $A^*_\mu = A_\mu$ to the case of an operator relation.

To make the field an operator in the entire Fock space, we have to specify the behavior of the creation and annihilation operators beyond just the single particle matrix elements. We assume that the matrix element relations follow from the operator relations:

$$a_{\vec{p},\lambda} |\vec{p},\lambda\rangle = |0\rangle \,, \quad a_{\vec{p},\lambda} |0\rangle = 0, \quad a^\dagger_{\vec{p},\lambda} |0\rangle = |\vec{p},\lambda\rangle \,,$$

hence the name creation and annihilation operators. From these we see that

$$\left[a_{\vec{p},\lambda} a^\dagger_{\vec{p},\lambda} - a^\dagger_{\vec{p},\lambda} a_{\vec{p},\lambda} \right] |0\rangle = |0\rangle \,.$$

The a and $a^\dagger$ operators do not commute. We assume the above commutator relation reflects a general operator relation, namely:

$$\left[a_{\vec{p},\lambda}, a^\dagger_{\vec{p}',\lambda'} \right] = \delta_{\vec{p},\vec{p}'} \delta_{\lambda,\lambda'}. \tag{2.51}$$

In addition,

$$\left[a_{\vec{p},\lambda}, a_{\vec{p}',\lambda'} \right] = \left[a^\dagger_{\vec{p},\lambda}, a^\dagger_{\vec{p}',\lambda'} \right] = 0. \tag{2.52}$$

These relations are sufficient to define the behavior of the operators in the entire space of states.

We can make a Hermitian operator using a and $a^\dagger$ by taking a product. We choose the product

$$N_{\vec{p},\lambda} = a^\dagger_{\vec{p},\lambda} a_{\vec{p},\lambda}. \tag{2.53}$$

With this order of the operators $N_{\vec{p},\lambda}|0\rangle = 0$. We can consider eigenstates of N, $|n\rangle$, with eigenvalue n. Thus $N_{\vec{p},\lambda}|n\rangle = n|n\rangle$. n is real and positive definite since $\langle n|N_{\vec{p},\lambda}|n\rangle = n$ can be interpreted as the norm of the state $a_{\vec{p},\lambda}|n\rangle$. We will show that n is an integer shortly.

From the relations for a and $a^\dagger$ applied to $|0\rangle$ and $|\vec{p},\lambda\rangle$, $N_{\vec{p},\lambda}|\vec{p},\lambda\rangle = |\vec{p},\lambda\rangle$. Therefore $|1\rangle = |\vec{p},\lambda\rangle$. Generalizing the above, we will interpret the state $|n\rangle$ as a state of n photons of momentum $\vec{p}$ and helicity λ!

Consider next the relation that follows from the commutator Eq. 2.51:

$$N_{\vec{p},\lambda} a^\dagger_{\vec{p},\lambda}|n\rangle = \left(a^\dagger_{\vec{p},\lambda} N_{\vec{p},\lambda} + a^\dagger_{\vec{p},\lambda}\right)|n\rangle = (n+1) a^\dagger_{\vec{p},\lambda}|n\rangle.$$

Therefore $a^\dagger_{\vec{p},\lambda}|n\rangle$ is an eigenstate of N, a state with one more photon. We can show similarly that $N_{\vec{p},\lambda} a_{\vec{p},\lambda}|n\rangle = (n-1) a_{\vec{p},\lambda}|n\rangle$ and therefore $a_{\vec{p},\lambda}|n\rangle$ is a state with one less photon. We can repeat the latter, arriving at states with coefficient $n-m$ in front. Since n is positive definite and m is an integer, we require n to be an integer or else $n-m$ would become negative. This justifies our assertion that n is an integer. We can also figure out the normalization of the states above. Writing $a^\dagger_{\vec{p},\lambda}|n\rangle$ as $|a^\dagger n\rangle$, gives

$$\langle a^\dagger n | a^\dagger n\rangle = \langle n|aa^\dagger|n\rangle = \langle n|N + [a,a^\dagger]|n\rangle = n+1.$$

Similarly $\langle an|an\rangle = n$. Thus, using normalized states,

$$a^\dagger_{\vec{p},\lambda}|n\rangle = \sqrt{n+1}\,|n+1\rangle$$

$$a_{\vec{p},\lambda}|n\rangle = \sqrt{n}\,|n-1\rangle.$$

The operators $a^\dagger$ and a have only matrix elements with states shifted up or down by one quantum. This generalizes the results we started with, which applied to the vacuum and the single particle state only. We note an important result: In the expansion

$$\langle 0|A_\alpha A_\beta|0\rangle = \sum_n \langle 0|A_\alpha|n\rangle\langle n|A_\beta|0\rangle,$$

where a complete set $\sum_n |n\rangle\langle n|$ has been inserted, only states with one quantum $|n\rangle = |\vec{p},\lambda\rangle$ contribute. We used this result when calculating the propagator for spin 0 (for which we can drop the helicity label), which involved only single particle exchange.

Consider next a state with two photons of different momenta that can be obtained by applying two creation operators to the vacuum state. Since the $a^\dagger$ commute

$$a^\dagger_{\vec{p},\lambda} a^\dagger_{\vec{p}',\lambda'} = a^\dagger_{\vec{p}',\lambda'} a^\dagger_{\vec{p},\lambda},$$

so that $|\vec{p},\lambda;\vec{p}',\lambda'\rangle = |\vec{p}',\lambda';\vec{p},\lambda\rangle$. The state is automatically symmetric under particle exchange. The theory we are writing down describes particles that are bosons!

2.14 Quantum Field Theory

We next want to look at the Hamiltonian. We know experimentally that a photon carries energy $E(\vec{p}) = |\vec{p}|$; thus we expect to arrive at the operator

$$H = \sum_{\vec{p},\lambda} E(\vec{p}) a^\dagger_{\vec{p},\lambda} a_{\vec{p},\lambda} = \sum_{\vec{p},\lambda} E(\vec{p}) N_{\vec{p},\lambda}.$$

For the free field, the expression for the energy is given by the classical expression

$$H_F = \int \frac{\vec{E}^2 + \vec{B}^2}{2} d^3 x. \tag{2.54}$$

Using transverse vectors for the polarization, we can write

$$\vec{B} = \nabla \times \vec{A}$$

and

$$\vec{E} = \frac{-\partial \vec{A}}{\partial t},$$

and calculate H_F in terms of $\vec{A}$.

In calculating H_F, we again have to consider integrals over all space and momenta, and will have to get correctly the field normalization and sum over states. As before, we imagine the system constrained to a large volume V. We then have to normalize the fields correctly so that we are talking about one quantum in V for the single particle state. With these normalization requirements we need to choose

$$\vec{A} = \sum_\lambda \int \left[\frac{a_{\vec{p},\lambda} \hat{e}(\lambda) e^{-ip\cdot x}}{\sqrt{2E(\vec{p})V}} + \frac{a^\dagger_{\vec{p},\lambda} \hat{e}^*(\lambda) e^{ip\cdot x}}{\sqrt{2E(\vec{p})V}} \right] \frac{d^3 p\, V}{(2\pi)^3}$$

in the calculation of H_F. In doing the integrals in Eq. 2.54, we will get quadratic terms in the creation and annihilation operators. We get two types of terms, $aa^\dagger$ or $a^\dagger a$, multiplying an integral of the type

$$\int e^{i(\vec{p}-\vec{p}')\cdot\vec{x}} d^3 x \frac{d^3 p}{(2\pi)^3} \frac{d^3 p'}{(2\pi)^3} = \int \delta^3(\vec{p}-\vec{p}') d^3 p \frac{d^3 p'}{(2\pi)^3} = \frac{d^3 p}{(2\pi)^3} \delta_{\vec{p},\vec{p}'},$$

for which $\vec{p} = \vec{p}'$; and aa or $a^\dagger a^\dagger$, multiplying an integral of the type

$$\int e^{i(\vec{p}+\vec{p}')\cdot\vec{x}} d^3 x \frac{d^3 p}{(2\pi)^3} \frac{d^3 p'}{(2\pi)^3} = \frac{d^3 p}{(2\pi)^3} \delta_{\vec{p},-\vec{p}'},$$

for which $\vec{p} = -\vec{p}'$.

We now calculate H_F given by Eq. 2.54. Using the above integrals, the terms of the type aa and $a^\dagger a^\dagger$ cancel between $\vec{B}^2$ and $\vec{E}^2$ and the terms of the type $a^\dagger a$ and $aa^\dagger$ add. Keeping only the latter terms:

$$H_F = \int \sum_{\lambda,\lambda'} E^2(\vec{p}) \left[\hat{e}(\lambda) \cdot \hat{e}^*(\lambda') a_{\vec{p},\lambda} a^\dagger_{\vec{p}',\lambda'} \frac{e^{-i(p-p')\cdot x}}{2\sqrt{E(\vec{p})E(\vec{p}')}V} \right.$$

$$\left. + \hat{e}^*(\lambda) \cdot \hat{e}(\lambda') a^\dagger_{\vec{p},\lambda} a_{\vec{p}',\lambda'} \frac{e^{i(p-p')\cdot x}}{2\sqrt{E(\vec{p})E(\vec{p}')}V} \right] d^3x \frac{d^3p\,V}{(2\pi)^3} \frac{d^3p'\,V}{(2\pi)^3}.$$

Integrating over d^3x gives $(2\pi)^3 \delta^3(\vec{p}-\vec{p}')$, which then eliminates the integral over d^3p setting $\vec{p} = \vec{p}'$. Also, we use $\hat{e}(\lambda) \cdot \hat{e}^*(\lambda') = \delta_{\lambda\lambda'}$. Thus

$$H_F = \int E(\vec{p}) \sum_\lambda \left[\frac{a_{\vec{p},\lambda} a^\dagger_{\vec{p},\lambda} + a^\dagger_{\vec{p},\lambda} a_{\vec{p},\lambda}}{2} \right] \frac{d^3p\,V}{(2\pi)^3}$$

$$= \sum_{\vec{p},\lambda} E(\vec{p}) \left[\frac{a_{\vec{p},\lambda} a^\dagger_{\vec{p},\lambda} + a^\dagger_{\vec{p},\lambda} a_{\vec{p},\lambda}}{2} \right].$$

Using the commutation relation for $a, a^\dagger$ gives:

$$H_F = \sum_{\vec{p},\lambda} E(\vec{p}) N_{\vec{p},\lambda} + \sum_{\vec{p},\lambda} \frac{1}{2} E(\vec{p}). \tag{2.55}$$

We indeed get the expected Hamiltonian except for an added overall constant. The constant represents the zero-point or vacuum energy of the field. This will not affect commutators of operators with H_F, which determine the time-evolution of such quantities, or energy differences between states and the vacuum. Its presence is, however, troubling, since it is infinite for an infinite number of possible states.

Using the commutation relations for $a, a^\dagger$ the operator $\vec{A}$ satisfies:

$$\frac{d\vec{A}}{dt} = i[H_F, \vec{A}].$$

In the discussion of the interaction representation in Section 2.9, it was pointed out that the field operators in this representation satisfy this relation, as given by Eq. 2.38. This indicates that $\vec{A}$, with the expansion chosen in terms of $a, a^\dagger$, is a consistent choice for use in the time evolution equation. Equation 2.38 is where quantum mechanics supplements the expressions for $\vec{A}$ and H_F, which look like the classical expressions. Using Eq. 2.55 for H_F results in a nonlinear relation between the creation and annihilation operators, which is solved by the commutation relations, Eqs. 2.51 and 2.52. This results in photons coming in fixed quanta of energy. Finally, note that H_F is a time independent operator, unlike the vector potential, as required by Eq. 2.38.

2.14.1 ■ Charged Scalar Field

For a charged particle we have seen that the field operator is not Hermitian. We have interpreted the solutions for such a field in terms of particles and antiparticles, following the discussion in Section 2.8. The simplest example is the complex scalar field. We write this as:

$$\phi = \sum_{\vec{p}} a_{\vec{p}}\, e^{-ip\cdot x} + c_{\vec{p}}^{\dagger}\, e^{i\vec{p}\cdot\vec{x}}.$$

$\phi^{\dagger}$ contains analogous terms that involve $a_{\vec{p}}^{\dagger}$ and $c_{\vec{p}}$. The Fock space now has particles of the a and c type and we can have single particle states $|\vec{p}_a\rangle$ or $|\vec{p}_c\rangle$. We assume that the $a_{\vec{p}}$ and $c_{\vec{p}}$ operators obey among themselves the same commutation relations as for $a_{\vec{p},\lambda}$ in the previous section. In addition, all $a_{\vec{p}}$ and $c_{\vec{p}}$ operators commute. With this interpretation, $a_{\vec{p}}^{\dagger}$ creates particles and $c_{\vec{p}}^{\dagger}$ creates antiparticles. The Hamiltonian, leaving out constants, can be shown to be

$$H_F = \sum_{\vec{p}} E(\vec{p}) N_{\vec{p}}^{a} + E(\vec{p}) N_{\vec{p}}^{c},$$

where

$$N_{\vec{p}}^{a} = a_{\vec{p}}^{\dagger} a_{\vec{p}}, \quad N_{\vec{p}}^{c} = c_{\vec{p}}^{\dagger} c_{\vec{p}}$$

and

$$E(\vec{p}) = \sqrt{|\vec{p}|^2 + m^2}.$$

We have yet to show, however, the most characteristic feature of antiparticles, which is that $|\vec{p}_c\rangle$ and $|\vec{p}_a\rangle$ have opposite charge. To do this we now look at the scalar particle charge operator:

$$Q = \int d^3x\, J_0 = \int i\left(\phi^{\dagger}\frac{\partial\phi}{\partial t} - \phi\frac{\partial\phi^{\dagger}}{\partial t}\right) d^3x,$$

where we use $\phi^{\dagger}$ instead of ϕ^*, since ϕ is now an operator in the Fock space. The calculation of the charge operator is similar to that for H_F in the previous section, except that one fewer power of $E(\vec{p})$ appears in the integrand. The net result is

$$Q = \sum_{\vec{p}} \left[\frac{a_{\vec{p}}^{\dagger} a_{\vec{p}} + a_{\vec{p}} a_{\vec{p}}^{\dagger}}{2}\right] - \left[\frac{c_{\vec{p}}^{\dagger} c_{\vec{p}} + c_{\vec{p}} c_{\vec{p}}^{\dagger}}{2}\right].$$

Using the commutation relations we rewrite this as:

$$Q = \sum_{\vec{p}} a_{\vec{p}}^{\dagger} a_{\vec{p}} - c_{\vec{p}}^{\dagger} c_{\vec{p}} = \sum_{\vec{p}} N_{\vec{p}}^{a} - N_{\vec{p}}^{c}.$$

This shows explicitly that for the scalar field the antiparticles have the opposite sign of the charge from the particles!

2.14.2 ■ Fermion Field

We now look at the fermion field, for example the electron field. We again write the fields in terms of creation and annihilation operators:

$$\psi = \sum_{\vec{p},\lambda} b_{\vec{p},\lambda} u(\vec{p},\lambda) e^{-ip\cdot x} + d^\dagger_{\vec{p},\lambda} v(\vec{p},\lambda) e^{ip\cdot x} \qquad (2.56)$$

$$\psi^\dagger = \sum_{\vec{p},\lambda} b^\dagger_{\vec{p},\lambda} u^\dagger(\vec{p},\lambda) e^{ip\cdot x} + d_{\vec{p},\lambda} v^\dagger(\vec{p},\lambda) e^{-ip\cdot x}. \qquad (2.57)$$

These expressions would look more symmetrical if $u^*(\vec{p},\lambda)$ appeared instead of $v(\vec{p},\lambda)$. This actually can be done in a different representation for the γ_μ, called the Majorana representation. It is the simplest representation for looking at the particle-antiparticle symmetry; however, all results are, in the end, independent of representation.

The creation and annihilation operators must provide the correct single particle matrix elements. Thus, as in the photon case:

$$b^\dagger_{\vec{p},\lambda} |0\rangle = |\vec{p},\lambda\rangle, \; b_{\vec{p},\lambda} |\vec{p},\lambda\rangle = |0\rangle \quad \text{and} \quad b_{\vec{p},\lambda} |0\rangle = 0.$$

Analogous results hold for the $d^\dagger_{\vec{p},\lambda}$ operators and the particles they create.

For the electron we know that no two particles can be in the same state and that states are antisymmetric under exchange of electrons. We can arrange this by requiring for the field operators:

$$b^\dagger_{\vec{p},\lambda} b^\dagger_{\vec{p}\,',\lambda'} = -b^\dagger_{\vec{p}\,',\lambda'} b^\dagger_{\vec{p},\lambda}.$$

Applying the above to the vacuum we get:

$$|\vec{p}\,',\lambda';\vec{p},\lambda\rangle = -|\vec{p},\lambda;\vec{p}\,',\lambda'\rangle.$$

For $\vec{p}\,' = \vec{p}$, $\lambda' = \lambda$, the state cannot exist and the operator product has to be chosen to vanish. This also means, applying the product of raising operators to the vacuum, that $b^\dagger_{\vec{p},\lambda} |\vec{p},\lambda\rangle = 0$. Thus the creation operator cannot add a second particle to a state already occupied.

For a fermion field we therefore need to use anti-commutation relations

$$\{b^\dagger_{\vec{p},\lambda}, b^\dagger_{\vec{p}\,',\lambda'}\} = 0, \; \{b_{\vec{p},\lambda}, b_{\vec{p}\,',\lambda'}\} = 0. \qquad (2.58)$$

What about the anti-commutator formed from b and $b^\dagger$? The result is

$$\{b_{\vec{p},\lambda}, b^\dagger_{\vec{p}\,',\lambda'}\} = \delta_{\vec{p},\vec{p}\,'} \delta_{\lambda,\lambda'}. \qquad (2.59)$$

2.14 Quantum Field Theory

We can check this by applying it to the vacuum:

$$\left[b_{\vec{p},\lambda}b^{\dagger}_{\vec{p},\lambda} + b^{\dagger}_{\vec{p},\lambda}b_{\vec{p},\lambda}\right]|0\rangle = b_{\vec{p},\lambda}|\vec{p},\lambda\rangle = |0\rangle.$$

Applying the operator to $|p,\lambda\rangle$ and using $b^{\dagger}_{\vec{p},\lambda}|\vec{p},\lambda\rangle = 0$, we get the consistent result:

$$\left[b_{\vec{p},\lambda}b^{\dagger}_{\vec{p},\lambda} + b^{\dagger}_{\vec{p},\lambda}b_{\vec{p},\lambda}\right]|\vec{p},\lambda\rangle = |\vec{p},\lambda\rangle.$$

Since the only states are the vacuum and $|\vec{p},\lambda\rangle$ for momentum and helicity $\vec{p},\lambda$, the relation is true in general for the operators. We assume the analogous anti-commutation relations hold for the $d, d^{\dagger}$ operators and that all b and d operators anti-commute.

We can again define the Hermitian number operator

$$N_{\vec{p},\lambda} = b^{\dagger}_{\vec{p},\lambda}b_{\vec{p},\lambda}, \tag{2.60}$$

which now has eigenvalue 0 for $|0\rangle$ and 1 for the single particle state. An analogous operator can be defined for the antiparticles.

We would like to look now at both the free field Hamiltonian and the charge operator:

$$H_F = \int \bar{\psi}(\vec{\gamma}\cdot\vec{p} + m)\psi\, d^3x = \int \psi^{\dagger}i\frac{\partial}{\partial t}\psi\, d^3x$$

by the Dirac equation,

$$Q = -e\int \psi^{\dagger}\psi\, d^3x.$$

The calculations for H_F and Q are rather similar. The Hamiltonian calculation has an extra $E(\vec{p})$ for the terms involving u, and $-E(\vec{p})$ for the terms involving v. We follow the calculation for H_F, using again a quantization volume V and a normalization factor $1/\sqrt{2E(\vec{p})V}$. We keep only the terms of the type $e^{ip'\cdot x}e^{-ip\cdot x}$, which are the only ones that survive since the u and v spinors are orthogonal.

$$H_F = \sum_{\lambda,\lambda'}\int \left[\frac{E(\vec{p})u^{\dagger}(\vec{p}',\lambda')u(\vec{p},\lambda)b^{\dagger}_{\vec{p}',\lambda'}b_{\vec{p},\lambda}e^{i(p'-p)\cdot x}}{2\sqrt{E(\vec{p})E(\vec{p}')}V}\right.$$
$$\left. - \frac{E(\vec{p})v^{\dagger}(\vec{p}',\lambda')v(\vec{p},\lambda)d_{\vec{p}',\lambda'}d^{\dagger}_{p,\lambda}e^{-i(p-p')\cdot x}}{2\sqrt{E(\vec{p})E(\vec{p}')}V}\right]d^3x\frac{d^3pV}{(2\pi)^3}\frac{d^3p'V}{(2\pi)^3}.$$
$$\tag{2.61}$$

We first integrate over d^3x, followed by an integral over d^3p to get rid of the resulting $(2\pi)^3\delta^3(\vec{p}-\vec{p}')$. This gives:

$$H_F =$$

$$\sum_{\vec{p}} \sum_{\lambda,\lambda'} \left[E(\vec{p}) \frac{u^\dagger(\vec{p},\lambda') u(\vec{p},\lambda)}{2E(\vec{p})} b^\dagger_{\vec{p},\lambda'} b_{\vec{p},\lambda} - E(\vec{p}) \frac{v^\dagger(\vec{p},\lambda') v(\vec{p},\lambda)}{2E(\vec{p})} d_{\vec{p},\lambda'} d^\dagger_{\vec{p},\lambda} \right].$$

Using $u^\dagger(\vec{p},\lambda') u(\vec{p},\lambda) = 2E(\vec{p}) \delta_{\lambda',\lambda}$ and $v^\dagger(\vec{p},\lambda') v(\vec{p},\lambda) = 2E(\vec{p}) \delta_{\lambda',\lambda}$, we get, finally:

$$H_F = \sum_{\vec{p},\lambda} \left[E(\vec{p}) b^\dagger_{\vec{p},\lambda} b_{\vec{p},\lambda} - E(\vec{p}) d_{\vec{p},\lambda} d^\dagger_{\vec{p},\lambda} \right].$$

The analogous relation for Q is

$$Q = -e \sum_{\vec{p},\lambda} \left[b^\dagger_{\vec{p},\lambda} b_{\vec{p},\lambda} + d_{\vec{p},\lambda} d^\dagger_{\vec{p},\lambda} \right].$$

We see above in H_F, the potentially negative energy corresponding to the antiparticle solutions. Since we require the energy to be positive, we need $d_{\vec{p},\lambda} d^\dagger_{\vec{p},\lambda}$ to be replaceable by $-d^\dagger_{\vec{p},\lambda} d_{\vec{p},\lambda}$; this is exactly the result we have by using the anti-commutation relations. In this sense, the spin-statistics relation is required by Quantum Field Theory and the positivity of energies. Anti-commuting the $d\, d^\dagger$ terms and including the resulting constant:

$$H_F = \sum_{\vec{p},\lambda} \left(E(\vec{p}) \left[b^\dagger_{\vec{p},\lambda} b_{\vec{p},\lambda} + d^\dagger_{\vec{p},\lambda} d_{\vec{p},\lambda} \right] - E(\vec{p}) \right)$$

$$Q = -e \sum_{\vec{p},\lambda} \left(\left[b^\dagger_{\vec{p},\lambda} b_{\vec{p},\lambda} - d^\dagger_{\vec{p},\lambda} d_{\vec{p},\lambda} \right] + 1 \right).$$

We see that indeed the antiparticles, the positrons, have the opposite charge of the electrons. Both H_F and Q have constants, which are formally infinite but don't have any physical significance, assuming we measure quantities relative to the values in the vacuum.

If we look back at the expressions for the energy and charge carefully, then the sign difference between particle and antiparticle terms arises partly from the number of time derivatives and partly from the choice of commutation relations. This is also true for the analogous expressions for the boson fields discussed earlier. For fermions, the extra minus sign from the anti-commutation relations compensates for one less time derivative in the various expressions for the energy and charge when compared to the boson fields.

Finally, we can look at the commutator of H_F and ψ. The result is

$$\frac{d\psi}{dt} = i[H_F, \psi].$$

2.14 Quantum Field Theory

Thus ψ is a consistent choice for the correct field operator to be used in the time evolution equation in the interaction representation. This equation requires that the number operator satisfies the same commutation relations with the creation and annihilation operators for both bosons and fermions. This is perhaps a surprise, given that the fermion creation and annihilation operators satisfy anti-commutation relations, but can be derived from these relations.

The expressions for H_F and Q indicate that we have to be careful about the signs of quantities because of the anti-commutation relations of the field operators. A measurable we looked at earlier was the spin operator. This depended on the quantity $\int \psi^\dagger \vec{\Sigma} \psi \, d^3x$. The important terms in this operator, for each momentum and helicity choice, are of the form:

$$u^\dagger \sum \vec{u} \, b^\dagger b \quad \text{and} \quad v^\dagger \sum \vec{v} \, d \, d^\dagger.$$

For simplicity, we will look at the terms in the spin operator Σ_z with the particles at rest. In this case the spin operator can be written in terms of the two component spinors, defined in Section 2.7.3. These give, for the expressions above:

$$\chi_u^\dagger \sigma_z \chi_u b^\dagger b \quad \text{and} \quad \chi_v^\dagger \sigma_z \chi_v d \, d^\dagger.$$

The term involving $d \, d^\dagger$ can be turned into a term of the form $d^\dagger d$ by using the anti-commutation relations. Thus the relevant terms will be:

$$\chi_u^\dagger \sigma_z \chi_u b^\dagger b - \chi_v^\dagger \sigma_z \chi_v d^\dagger d,$$

which express the z-component of spin in terms of the number of particles and antiparticles. This implies that, using χ in the 4-component notation:

$$\begin{pmatrix} 1 \\ 0 \\ 0 \\ 0 \end{pmatrix} \text{ represents a spin up particle,}$$

$$\begin{pmatrix} 0 \\ 1 \\ 0 \\ 0 \end{pmatrix} \text{ represents a spin down particle,}$$

$$\begin{pmatrix} 0 \\ 0 \\ 0 \\ 1 \end{pmatrix} \text{ represents a spin up antiparticle,}$$

$$\begin{pmatrix} 0 \\ 0 \\ 1 \\ 0 \end{pmatrix} \text{ represents a spin down antiparticle.}$$

Boosting these then gives the general free particle states of given spin. As a consequence of this, if we look at highly relativistic states, the projection operator $(1 + \gamma_5)/2$ discussed in Section 2.7.4 projects out right-handed particles and left-handed antiparticles, whereas $(1 - \gamma_5)/2$ projects out left-handed particles and right-handed antiparticles.

CHAPTER 2 HOMEWORK

2.1. The minimal substitution, discussed in the text for spin $\frac{1}{2}$, can be used more generally to arrive at the field equations for charged particles coupled to the electromagnetic field. What field equation do you expect for a spin 0 particle coupled to the electromagnetic field A_μ?

2.2. Consider the three operators $(S_i)_{jk} = -i\varepsilon_{ijk}$. For example,

$$S_3 = -i \begin{pmatrix} 0 & 1 & 0 \\ -1 & 0 & 0 \\ 0 & 0 & 0 \end{pmatrix}$$

in matrix notation. Show that

(a) $[S_i, S_j] = i\varepsilon_{ijk} S_k$, $\quad S^2 = S_1^2 + S_2^2 + S_3^2 = 2I$.

Therefore, the S_i provide a representation for the angular momentum operators for spin 1.

(b) Consider the orthonormal polarization vectors:

$$\hat{e}^{(1)} = -\frac{(\hat{e}_x + i\hat{e}_y)}{\sqrt{2}}, \quad \hat{e}^{(0)} = \hat{e}_z, \quad \hat{e}^{(-1)} = \frac{(\hat{e}_x - i\hat{e}_y)}{\sqrt{2}}.$$

Show that $S_3 \hat{e}^{(m)} = m\hat{e}^{(m)}$.

(c) We can choose an alternative basis to represent the polarization states in an abstract vector space by the correspondence:

$$\hat{e}^{(1)} \leftrightarrow \begin{pmatrix} 1 \\ 0 \\ 0 \end{pmatrix}, \quad \hat{e}^{(0)} \leftrightarrow \begin{pmatrix} 0 \\ 1 \\ 0 \end{pmatrix}, \quad \hat{e}^{(-1)} \leftrightarrow \begin{pmatrix} 0 \\ 0 \\ 1 \end{pmatrix}.$$

This will generate an alternative representation for the S_i, which we denote $\tilde{S}_i$. Show that:

$$\tilde{S}_1 = \frac{1}{\sqrt{2}} \begin{pmatrix} 0 & 1 & 0 \\ 1 & 0 & 1 \\ 0 & 1 & 0 \end{pmatrix}, \tilde{S}_2 = \frac{1}{\sqrt{2}} \begin{pmatrix} 0 & -i & 0 \\ i & 0 & -i \\ 0 & i & 0 \end{pmatrix}, \tilde{S}_3 = \begin{pmatrix} 1 & 0 & 0 \\ 0 & 0 & 0 \\ 0 & 0 & -1 \end{pmatrix}.$$

We get these matrices when considering angular momentum in a general way, using raising and lowering operators.

2.3. Consider a photon with spatial momentum $\vec{k} = k(\sin\theta \cos\phi, \sin\theta \sin\phi, \cos\theta)$. What are two transverse polarization vectors we can use to form a basis?

2.4. For the Dirac–Pauli representation of the γ matrices, calculate explicitly the matrices for $\sigma_{\mu\nu}$ and $\gamma_5 \gamma_\mu$. What are the values of the 16 bilinear covariants if we take for ψ

the four-component $u(p)$ of a particle at rest specified by a two-component spinor χ_i, and for $\bar{\psi}$ an analogous expression given in terms of a two-component spinor χ_f?

2.5. The existence of five anti-commuting matrices allows a straightforward way to create other representations of the four γ_μ matrices. The Weyl or Chiral representation is obtained by replacing γ_0 with γ_5 in the Dirac–Pauli representation. In this representation:

$$\gamma_0 = \begin{pmatrix} 0 & I \\ I & 0 \end{pmatrix}, \quad \vec{\gamma} = \begin{pmatrix} 0 & \vec{\sigma} \\ -\vec{\sigma} & 0 \end{pmatrix}, \quad \gamma_5 = -\begin{pmatrix} I & 0 \\ 0 & -I \end{pmatrix}.$$

It leads to solutions that are particularly simple in the relativistic limit. Show that in this representation:

(a) $\not{p}$ is $\begin{pmatrix} 0 & E - \vec{p} \cdot \vec{\sigma} \\ E + \vec{p} \cdot \vec{\sigma} & 0 \end{pmatrix}$.

(b) If we try a solution for the spinor

$$\begin{pmatrix} \chi_L \\ \chi_R \end{pmatrix},$$

then in the massless limit, χ_R and χ_L satisfy separate equations:

$$(E + \vec{p} \cdot \vec{\sigma}) \chi_L = 0, \quad (E - \vec{p} \cdot \vec{\sigma}) \chi_R = 0.$$

(c) Since the particle solutions satisfy $E = |\vec{p}|$ in this limit, the particle solution for χ_L is a left-handed two-component spinor; χ_R is a right-handed two-component spinor.

2.6. The Majorana representation is given by the choices:

$$\gamma_0 = \begin{pmatrix} 0 & i\sigma_1 \\ -i\sigma_1 & 0 \end{pmatrix}, \quad \gamma_1 = \begin{pmatrix} iI & 0 \\ 0 & -iI \end{pmatrix}, \quad \gamma_2 = \begin{pmatrix} 0 & \sigma_2 \\ -\sigma_2 & 0 \end{pmatrix},$$

$$\gamma_3 = \begin{pmatrix} 0 & iI \\ iI & 0 \end{pmatrix}, \quad \gamma_5 = \begin{pmatrix} 0 & i\sigma_3 \\ -i\sigma_3 & 0 \end{pmatrix}.$$

(a) Show that this is a valid representation, that is,

$$\{\gamma_\mu, \gamma_\nu\} = 2g_{\mu\nu} I, \quad \gamma_\mu^\dagger = \gamma_0 \gamma_\mu \gamma_0, \quad \{\gamma_5, \gamma_\mu\} = 0, \quad \gamma_5^2 = I.$$

(b) Show that $\gamma_\mu^* = -\gamma_\mu$.

(c) Show that if ψ satisfies

$$i\gamma_\mu \frac{\partial \psi}{\partial x_\mu} = m\psi,$$

then by complex conjugating,

$$-i\gamma_\mu^* \frac{\partial \psi^*}{\partial x_\mu} = m\psi^*.$$

Based on $\gamma_\mu^* = -\gamma_\mu$, this means that in this representation ψ^* is also a solution of the original Dirac equation. Thus any solution $u(p)e^{-ip\cdot x}$ has a corresponding solution $u(p)^* e^{ip\cdot x}$.

(d) Find the four solutions for $\vec{p} = 0$ in this representation.

2.7. Consider an electromagnetic current density that can be described adequately as a classical source:
$$\vec{J} = \vec{J}(\vec{x})\cos(\omega t).$$

(a) In first order perturbation theory calculate the rate to emit a photon of given polarization into a solid angle element $d\Omega$ in terms of $\vec{J}$ and the photon momentum. Write the rate as:
$$\frac{dN}{dt\, d\Omega}$$

(b) What is the average power, $dP/d\Omega$, radiated into $d\Omega$?

(c) Compare to the classical expectation for the time-averaged power radiated.

2.8. Consider an interaction Hamiltonian density $\mathcal{H} = g\phi_1\phi_2\phi_3$ involving three different, uncharged, scalar fields ϕ_1, ϕ_2, ϕ_3 (g is a constant). Assume the scalar masses are m_1, m_2, m_3, respectively, with $m_3 \gg m_2 \gg m_1$.

(a) What is the first order matrix element, M_{fi}, for the decay $3 \to 1+2$?

(b) What is the second order matrix element for the scattering process $1+2 \to 1+2$? The initial momenta are p_1, p_2; the final momenta are p_1', p_2'. (*Hint:* Use creation and annihilation operators to keep track of the alternatives contributing. The final result can be described in terms of the sum of two Feynman diagrams).

(c) Calculate the cross section $d\sigma/d\Omega$ in the nonrelativistic limit for scalar 1 incident on 2 at rest.

2.9. Show that
$$\frac{d\vec{A}}{dt} = i\left[H_F, \vec{A}\right]$$
for the free field vector potential.

2.10. For the free complex scalar field, find H_F in terms of an integral over fields and their derivatives. Expressing the fields in terms of creation and annihilation operators, rewrite H_F.

CHAPTER 3

Scattering of Leptons and Photons

3.1 ■ INTERACTION HAMILTONIAN AND LEPTON CURRENTS

In this chapter we will look at the scattering of two initial particles to two final particles. We restrict the particles to be charged leptons or photons with the interaction always electromagnetic. We take the particle energies to be small enough that the weak interactions can be ignored. The calculations presented involve Quantum Electrodynamics, the first successful particle theory. The calculations will be second-order in the perturbation expansion, that is, we use the approximation $M_{fi} = M_{fi}^{(2)}$. The accuracy of this choice can be checked by looking at the next order term; the fact that it is a good approximation is justified by the smallness of

$$\alpha = \frac{e^2}{4\pi} = \frac{1}{137}.$$

This leads to corrections to $M_{fi}^{(2)}$ being $\sim 1\%$.

The interaction Hamiltonian is:

$$\mathcal{H} = J_\alpha^{e^-} A_\alpha + J_\alpha^{\mu^-} A_\alpha + J_\alpha^{\tau^-} A_\alpha, \tag{3.1}$$

corresponding to the three charged leptons. An interaction vertex contributes through a nonvanishing matrix element of $\mathcal{H}$ at a point in space-time, and a complete process requires two such vertices. Since a given vertex for this $\mathcal{H}$ involves three particles, one particle will be virtual. It is exchanged between the two vertices.

To understand the kinds of vertices, we will look at the lepton current in more detail. Specifically, we take the electron current. Using the expression for ψ from Section 2.14.2 and multiplying out the terms, we get four types of terms:

$$\bar{\psi}\gamma_\mu\psi = \sum_{\vec{p},\vec{p}',\lambda,\lambda'} \sum_{i=1}^{4} J_\mu^{(i)}(\vec{p}, \lambda, \vec{p}', \lambda', x).$$

The four terms are:

1. $\bar{u}(\vec{p}', \lambda')\gamma_\mu u(\vec{p}, \lambda) b^\dagger_{\vec{p}',\lambda'} b_{\vec{p},\lambda} e^{i(p'-p)\cdot x}$
2. $\bar{v}(\vec{p}', \lambda')\gamma_\mu v(\vec{p}, \lambda) d_{\vec{p}',\lambda'} d^\dagger_{\vec{p},\lambda} e^{i(p-p')\cdot x}$

3. $\bar{v}(\vec{p}\,',\lambda')\gamma_\mu u(\vec{p},\lambda) d_{\vec{p}\,',\lambda'} b_{\vec{p},\lambda} e^{-i(p'+p)\cdot x}$

4. $\bar{u}(\vec{p}\,',\lambda')\gamma_\mu v(\vec{p},\lambda) b^\dagger_{\vec{p}\,',\lambda'} d^\dagger_{\vec{p},\lambda} e^{i(p'+p)\cdot x}$.

The creation and annihilation operators keep track of the process to which each term corresponds (and go along with the correct exponential phase factors for incoming and outgoing particles). For term 1, using the properties of the annihilation and creation operator, we get a nonzero result for an initial state with an electron of momentum $\vec{p}$ and helicity λ, and a final state of momentum $\vec{p}\,'$, λ'. Thus,

$$\langle e^-, \vec{p}\,', \lambda' | J_\mu^{(1)} | e^-, \vec{p}, \lambda \rangle = \bar{u}(\vec{p}\,',\lambda')\gamma_\mu u(\vec{p},\lambda) e^{i(p'-p)\cdot x}.$$

We used this expression to set up the calculation $e^-\mu^- \to e^-\mu^-$ in Chapter 2.

Terms 3 and 4 correspond to an initial e^+ and $e^- \to$ vacuum and vacuum $\to e^+ e^-$, respectively, since these terms have all annihilation or all creation operators. For term 2 we might try the analog of term 1, that is, for the initial state $|e^+, \vec{p}, \lambda\rangle$ and the final $|e^+, \vec{p}\,', \lambda'\rangle$. However, the creation operator satisfies $d^\dagger_{\vec{p},\lambda}|e^+, \vec{p}, \lambda\rangle = 0$. The state that matches the creations and annihilations is $|e^+, \vec{p}\,', \lambda'\rangle$ for the initial state and $|e^+, \vec{p}, \lambda\rangle$ for the final state. To make this clear, we can use the anti-commutation relation to reverse the order of the operators. Leaving out the constant, which has vanishing matrix element for $p' \neq p$, we can rewrite the second term as

$$-\bar{v}(\vec{p}\,',\lambda')\gamma_\mu v(\vec{p},\lambda) d^\dagger_{\vec{p},\lambda} d_{\vec{p}\,',\lambda'} e^{i(p-p')\cdot x}.$$

Thus each term gives a finite matrix element for a different specific initial and final state, always involving two particles. We tabulate these in Table 3.1 and pictorially show the vertices in Figure 3.1. Note the order in which the spinors occur: $\bar{v}$ for an incoming e^+ and v for an outgoing e^+; $\bar{u}$ for an outgoing e^- and u for an incoming e^-. The extra minus sign for term 2 corresponds to the opposite sign of the charge for the positron, compared to the electron.

TABLE 3.1 Matrix elements of $J_\mu^{e^-}$, excluding a factor of $-e$.

Term number	$|i\rangle$	$|f\rangle$	Matrix element
1	$|e^-, \vec{p}, \lambda\rangle$	$|e^-, \vec{p}\,', \lambda'\rangle$	$\bar{u}(\vec{p}\,',\lambda')\gamma_\mu u(\vec{p},\lambda) e^{i(p'-p)\cdot x}$
2	$|e^+, \vec{p}\,', \lambda'\rangle$	$|e^+, \vec{p}, \lambda\rangle$	$-\bar{v}(\vec{p}\,',\lambda')\gamma_\mu v(\vec{p},\lambda) e^{i(p-p')\cdot x}$
3	$|e^-, \vec{p}, \lambda; e^+, \vec{p}\,', \lambda'\rangle$	$|0\rangle$	$\bar{v}(\vec{p}\,',\lambda')\gamma_\mu u(\vec{p},\lambda) e^{-i(p+p')\cdot x}$
4	$|0\rangle$	$|e^-, \vec{p}\,', \lambda'; e^+, \vec{p}, \lambda\rangle$	$\bar{u}(\vec{p}\,',\lambda')\gamma_\mu v(\vec{p},\lambda) e^{i(p+p')\cdot x}$

3.1 Interaction Hamiltonian and Lepton Currents

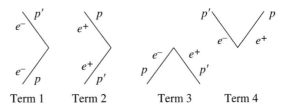

FIGURE 3.1 Processes contained in $J_\mu^{e^-}$.

Another important point concerns the order of the individual states in the two-particle (or more general multiparticle) state. Because we are dealing with fermions, if we exchange the order of a pair, for example, and we use

$$|e^+, \vec{p}, \lambda; e^-, \vec{p}\,', \lambda'\rangle \quad \text{instead of} \quad |e^-, \vec{p},', \lambda'; e^+, \vec{p}, \lambda\rangle,$$

then we must change the sign of the state, thereby changing the sign of the matrix element. In general we will order the particles so that the state created by the action of several creation operators has the first created particle furthest to the right. If we apply an annihilation operator to a state, we must move the particle annihilated all the way to the left, with minus signs for every interchange, to get the correct sign. The two particle states in Table 3.1, for terms 3 and 4, have been ordered in this way, giving a positive sign for the matrix element shown. The minus sign due to particle interchange in the state will show up in calculations later and is something to watch out for. With this warning, the basic matrix elements in Table 3.1 are the ingredients needed for any electromagnetic lepton vertex, with a given process selecting out one term from the table.

Before returning to real calculations, we look at a few other useful relations. After taking matrix elements we see that we can write

$$\langle f| J_\mu^{e^-} |i\rangle = -e J_\mu(0) e^{iq \cdot x},$$

where $J_\mu(0)$ is one of the four terms evaluated at $x = 0$, and q is the appropriate sum or difference of p and p'. Examining each of the four terms, q is the momentum carried away or brought in by the photon in each process. Current conservation tells us that

$$\frac{\partial}{\partial x_\mu} J_\mu^{e^-} = 0,$$

which for the matrix element implies that $q_\mu J_\mu(0) = 0$. This provides a useful check in a calculation. This has been discussed from a different point of view in Section 2.6.

We can also use the above to simplify relations involving photons. For example, for an incoming photon, we will end up with matrix elements involving (using $A_\mu J_\mu^{e^-}$) $e_\mu(\lambda) J_\mu(0)$. To calculate rates we square this, giving

$$|e_\mu(\lambda)J_\mu(0)|^2 = e_\mu^*(\lambda)e_\nu(\lambda)J_\mu^*(0)J_\nu(0).$$

If the photon state is unpolarized we must average this expression over polarizations. For an outgoing, instead of incoming, photon, we have the same expression except that now we sum over final polarizations to get a total rate. These will lead to the expression

$$\left[\sum_\lambda e_\mu^*(\lambda)e_\nu(\lambda)\right][J_\mu^*(0)J_\nu(0)].$$

In doing this calculation, we take the direction of $\vec{q}$ in 3-space to be the z-direction. A basis for $e_\mu(\lambda)$ will then be $\hat{e}_x$ and $\hat{e}_y$, using unit vector notation. This then gives $[\hat{e}_x\hat{e}_x + \hat{e}_y\hat{e}_y]$ for $\sum_\lambda e_\mu^*(\lambda)e_\nu(\lambda)$. We can, however, put the full expression into a more manifestly covariant form. The relation $q_\mu J_\mu(0) = 0$ implies that $q_0 J_0(0) - q_z J_z(0) = 0$, for $\vec{q}$ in the z-direction. For a real photon, this gives $J_0(0) = J_z(0)$, which then implies that we can write the above as

$$[\hat{e}_x\hat{e}_x + \hat{e}_y\hat{e}_y + \hat{e}_z\hat{e}_z - e_t e_t] \cdot [J_\mu^*(0)J_\nu(0)] = -g_{\mu\nu}J_\mu^*(0)J_\nu(0).$$

This expression is now manifestly covariant, which shows explicitly that the use of transverse photons combined with current conservation (equivalently gauge invariance) gives a covariant result. The above establishes the useful relation:

$$\sum_\lambda e_\mu^*(\lambda)e_\nu(\lambda) = -g_{\mu\nu},$$

when dotted with conserved currents.

3.2 ■ ELECTRON–MUON SCATTERING

We now return to the first complete calculation of a two-body electromagnetic scattering process, $e^-\mu^-$ scattering. To complete this we will work through a number of very important details that are used repeatedly in doing calculations and need to be fully mastered. These are:

1. evaluation and use of the propagator for a particle with spin (in this case the photon).
2. formalism for dealing with spin $\frac{1}{2}$ particles,
3. phase space integral needed to complete the calculation; in this case, it will be the simplest example of a phase space calculation, since there are only two particles in the final state.

We start where we left off in Section 2.10 with the second-order calculation for the electromagnetic interaction:

3.2 Electron–Muon Scattering

$$S^{(2)}_{fi} = (-i)^2 \int_{-\infty}^{\infty}\int_{-\infty}^{\infty} d^4x_1 d^4x_2 \, \langle e_{\text{out}}| J^{e^-}_\alpha(x_1) |e_{\text{in}}\rangle \langle \mu_{\text{out}}| J^{\mu^-}_\beta(x_2) |\mu_{\text{in}}\rangle$$
$$\cdot \left(\langle 0| A_\alpha(x_1) A_\beta(x_2) \theta(t_1 - t_2) + A_\beta(x_2) A_\alpha(x_1) \theta(t_2 - t_1) |0\rangle \right).$$

Both electron and muon currents are of the Term 1 type in Table 3.1. They contain spinor products that depend on the momenta, with the position dependence contained entirely in the exponential phase factors. We will soon see that the same is true of the term coming from the exchange of photons of all momenta, when written as a 4-dimensional Fourier transformation. This will mean that the calculation for $S^{(2)}_{fi}$ involves a product of spin factors that depend only on the momenta, and a separate space-time integral that does not depend on the spins. Since the space-time integral doesn't depend on spin, the result for it has already been calculated in Section 2.10 using spinless particle exchange. In this way, we can always represent the second-order calculation by a Feynman diagram—we just have to be careful to associate the appropriate spin factors at the vertices and in the propagator for the particles involved. For the $e^- \mu^-$ scattering case, the currents are:

for the electron, $-e\bar{u}(p^{e^-}_{\text{out}})\gamma_\alpha u(p^{e^-}_{\text{in}}) e^{i(p^{e^-}_{\text{out}} - p^{e^-}_{\text{in}})\cdot x_1}$,

for the muon, $-e\bar{u}(p^{\mu^-}_{\text{out}})\gamma_\beta u(p^{\mu^-}_{\text{in}}) e^{i(p^{\mu^-}_{\text{out}} - p^{\mu^-}_{\text{in}})\cdot x_2}$.

Thus, in the Feynman diagram, after the space-time integration, we can represent the process as:

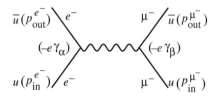

Note that in the diagram the $-e\gamma_\alpha$ is associated with the vertex itself, since it is determined by the interaction type (vector in this case). The spinors tell us which spin factors to include in the calculation and would be there for any other covariant interaction involving spin $\frac{1}{2}$ particles. The last missing ingredient is the photon propagator, which we turn to next.

3.2.1 ∎ The Photon Propagator

In order to have all the terms needed to complete the scattering calculation, we need to figure out the space-time propagator:

$$\langle 0| A_\alpha(x_1) A_\beta(x_2) \theta(t_1 - t_2) + A_\beta(x_2) A_\alpha(x_1) \theta(t_2 - t_1) |0\rangle,$$

and then its Fourier transform. To evaluate the propagator, we expand in terms of virtual particles by inserting a complete set of states between the two field operators in the propagator. However, for a particle with spin, we will have to include the sum over helicity choices for the virtual particle, since these interfere in the overall amplitude. In general, for arbitrary spins, after summing over helicities and taking $x = x_1 - x_2$, we are led to a space-time propagator expression of the type:

$$\int [g(p)e^{-ip\cdot x}\theta(t) + g(-p)e^{ip\cdot x}\theta(-t)]\frac{d^3p}{2E(2\pi)^3}. \tag{3.2}$$

We will explicitly show this for a massive spin 1 particle. To arrive at a momentum-space propagator, we repeat the calculation done in Section 2.11, where $g(p) = 1$ for spin zero. Provided $g(p)$ doesn't have poles and doesn't spoil the convergence as $p_0 \to \infty$ in the complex plane, the calculation from Section 2.11 can be generalized easily. This allows the representation of the space-time propagator in terms of the 4-dimensional integral:

$$\int g(p)\frac{i}{p^2 - m^2 + i\varepsilon}e^{-ip\cdot x}\frac{d^4p}{(2\pi)^4}.$$

The term

$$\frac{ig(p)}{p^2 - m^2 + i\varepsilon}$$

is called the momentum-space propagator for the exchange of the given particle type. It generalizes the analogous spin 0 propagator. For bosons we will find that $g(p) = g(-p)$, so we do not have to be careful of signs; for fermions we will have to keep track of the signs.

Before looking at photons, we look at the case of a massive spin 1 boson for which all three polarizations contribute. Inserting a complete set of states between the two factors $A_\alpha(x_1)A_\beta(x_2)$ and $A_\beta(x_2)A_\alpha(x_1)$ in the space-time propagator, we can then follow the procedure in Section 2.10 to arrive at Eq. 3.2. The terms multiplying the exponential phase factors (the exponential factors are the same as for the spin zero case) are represented pictorially below for the two time orderings, where the spin vectors at each vertex are indicated. The product of these spin vectors, summed over helicities, yield the tensors $g(p)$ and $g(-p)$.

$$g(p) = \sum_\lambda e_\beta^*(\lambda) e_\alpha(\lambda) \qquad g(-p) = \sum_\lambda e_\alpha^*(\lambda) e_\beta(\lambda)$$

3.2 Electron–Muon Scattering

For the massive spin 1 particle there are three choices for λ. In the rest frame of the particle, the three choices for the spin vectors are $\hat{e}_x, \hat{e}_y, \hat{e}_z$. This gives, in this case, $g(p) = \hat{e}_x\hat{e}_x + \hat{e}_y\hat{e}_y + \hat{e}_z\hat{e}_z$. We can write this as a covariant tensor that reduces in the rest frame to the above:

$$g(p) = -\left[g_{\alpha\beta} - \frac{p_\alpha p_\beta}{m^2}\right], \quad (3.3)$$

where m is the particle mass. Since it is covariant, it provides the expression for the tensor in an arbitrary frame and describes $g(p)$ for an exchanged particle with a general value of momentum. The expression for $g(-p)$ above also gives

$$-\left[g_{\alpha\beta} - \frac{p_\alpha p_\beta}{m^2}\right],$$

so $g(p) = g(-p)$ in this case.

Changing to Lorentz indices μ and ν and using q as the 4-momentum transfer, we have found the momentum-space propagator for a massive spin 1 particle to be:

$$\frac{-i\left[g_{\mu\nu} - \frac{q_\mu q_\nu}{m^2}\right]}{q^2 - m^2 + i\varepsilon}.$$

In this expression, all four momentum components of q_μ are variables, so typically $q^2 \neq m^2$, whereas each term in the 3-dimensional time-ordered integral of Eq. 3.2 had $p^2 = m^2$. For a massive spin 1 particle, the propagator, which represents the sum of contributions from intermediate particles with many momenta, will not correspond to the exchange of one unit of angular momentum in any given frame except when $q^2 = m^2$.

In the calculation of a Feynman diagram, the two vertices linked by the virtual spin 1 particle will each contain a vector that is dotted into one of the two Lorentz indices of the propagator. For vectors corresponding to conserved currents, the term proportional to $q_\mu q_\nu$ in the propagator will drop out, since it gives a vanishing result. In this case, the propagator is effectively:

$$-i\frac{g_{\mu\nu}}{q^2 - m^2 + i\varepsilon}.$$

This has a well-defined limit as $m \to 0$, which we can guess is then the photon propagator:

$$-i\left[\frac{g_{\mu\nu}}{q^2 + i\varepsilon}\right]. \quad (3.4)$$

This is correct; but it is not easy to arrive at this expression if we start with real photons with two transverse polarizations. We will not attempt a rigorous derivation of this result.

We demand of the photon propagator:

1. It must be covariant to give a Lorentz invariant theory.
2. For the propagation of real photons, that is, $q^2 \to 0$, the propagator must correspond to the exchange of the two transverse polarizations only.

Because of current conservation, terms proportional to q_μ give a vanishing contribution; thus the only covariant choice with a nonvanishing result is the function above. We will see that this choice, which satisfies (1), will imply (2). The choice of any other function of the type

$$-i\left[\frac{g_{\mu\nu} - c\frac{q_\mu q_\nu}{q^2}}{q^2 + i\varepsilon}\right],$$

where c is a constant, also works and gives the same result as taking $c = 0$. This is an example of the freedom due to gauge invariance. We will typically choose $c = 0$.

3.2.2 ■ Some Implications of the Photon Propagator

After integrating over all of space-time to calculate the amplitude, as was done in Section 2.10, we get for the matrix element for one photon exchange:

$$-iM_{fi}^{(2)} = \left(-iJ_\alpha^A(0)\right)\left[\frac{-ig_{\alpha\beta}}{q^2 + i\varepsilon}\right]\left(-iJ_\beta^B(0)\right).$$

We will drop the explicit indication that we are working to second order and just write $M_{fi}^{(2)}$ as M_{fi}; in addition, we omit the explicit indication that $x = 0$ in the currents. J_α^A and J_β^B, the current matrix elements at each of the vertices for $x = 0$, are momentum-space functions only. For the case of $e^-\mu^-$ scattering:

$$J_\alpha^A = -e\bar{u}(p_{\text{out}}^{e^-})\gamma_\alpha u(p_{\text{in}}^{e^-}), \quad J_\beta^B = -e\bar{u}(p_{\text{out}}^{\mu^-})\gamma_\beta u(p_{\text{in}}^{\mu^-}).$$

The result for single photon exchange,

$$M_{fi} = \frac{-J_\alpha^A g_{\alpha\beta} J_\beta^B}{q^2 + i\varepsilon} = \frac{-J^A \cdot J^B}{q^2 + i\varepsilon}, \quad (3.5)$$

is rather general, however, and we look at some of the implications. Choosing axes so that $\vec{q}$ defines the z-direction, we have $q_\alpha = (q_0, 0, 0, |\vec{q}|)$, and for the matrix element,

$$M_{fi} = \frac{-[J_0^A J_0^B - J_z^A J_z^B]}{q^2 + i\varepsilon} + \frac{\vec{J}_T^A \cdot \vec{J}_T^B}{q^2 + i\varepsilon}. \quad (3.6)$$

$\vec{J}_T^A$ and $\vec{J}_T^B$ are the transverse components of the current. For a real photon $q^2 = 0$ and, therefore, current conservation $q_\alpha J_\alpha = 0$ implies that $J_0^A = J_z^A$, $J_0^B = J_z^B$.

3.2 Electron–Muon Scattering

In this case, the first term in M_{fi} vanishes compared to the second term and only the transverse currents contribute. This is exactly what we would get from using only the two transverse polarizations in the photon exchange expression.

For virtual photon exchange, $q^2 \ne 0$, we can use the relation $q_0 J_0 = |\vec{q}| J_z$ to eliminate $J_z^A J_z^B$. The resulting expression is:

$$M_{fi} = \frac{J_0^A J_0^B}{|\vec{q}|^2} + \frac{\vec{J}_T^A \cdot \vec{J}_T^B}{q^2}.$$

Note that the first term, called the Coulomb interaction term, involves only the 3-momentum squared for the virtual photon and depends on the matrix element of the charge density only. For situations where $q^2 < 0$ and at least one vertex contains only nonrelativistic charged particles, the current operator $|\vec{J}| \sim |\vec{v} J_0| \ll |J_0|$ so that the Coulomb term alone gives a good approximation. A good example is provided by the interactions in the hydrogen atom. Here the electron is nonrelativistic and the proton is very nonrelativistic, so that the Coulomb interaction provides a very good approximation to the electromagnetic effects. First-order corrections to the fully nonrelativistic spinless approximation for hydrogen arise from relativistic corrections to the electron's kinetic energy and spin dependent corrections arising from the electron charge density.

For $e^- \mu^-$ scattering in the nonrelativistic limit we need only keep the Coulomb term. Using $\bar{u} \gamma_0 u = 2m$ for initial and final states with the same spin direction, we can write

$$M_{fi} = \frac{e^2 (2m_e)(2m_\mu)}{|\vec{q}|^2}.$$

This is the same result, except for the overall sign, as for particle exchange of a spinless and massless particle, which was discussed in detail in Section 2.13. Following that discussion we can associate a nonrelativistic potential acting between the electron and muon, which in this case is the repulsive Coulomb potential

$$V(\vec{x}) = \frac{e^2}{4\pi r}.$$

3.2.3 ■ Electron–Muon Scattering; Methods for Facilitating Calculations for Spin $\frac{1}{2}$

We return now to the $e^- \mu^-$ scattering matrix element, without the nonrelativistic assumption. Equation 3.5 gives:

$$M_{fi} = \frac{-e^2 \bar{u}(p_{\text{out}}^{e^-}) \gamma_\alpha u(p_{\text{in}}^{e^-}) \bar{u}(p_{\text{out}}^{\mu^-}) \gamma_\alpha u(p_{\text{in}}^{\mu^-})}{q^2}. \quad (3.7)$$

Squaring gives

$$|M_{fi}|^2 = \frac{e^4}{q^4} L_{\alpha\beta}^{e^-} L_{\alpha\beta}^{\mu^-},$$

where the tensor $e^2 L^{e^-}_{\alpha\beta}$ is $J^{e^-}_\alpha J^{e^-}_\beta{}^*$ and $L^{\mu^-}_{\alpha\beta}$ is given by the analogous expression for the μ^-. To complete the calculation requires evaluation of the two tensors. One could substitute specific spinors and do the calculation of the tensors; however, a rather general method has been devised to facilitate such calculations. We want to work out the details of this method, which allows us to avoid introducing any specific representation of the γ matrices or spinors.

Consider an expression in a matrix element $\bar{u}(p_2) O_\alpha u(p_1)$, where O_α is a 4×4 matrix made from momenta and γ matrices. Thus O_α can have Lorentz indices that we will indicate collectively as α. Here, no indices indicates a scalar, one index a 4-vector, etc., for the overall expression. For each Lorentz index the expression is a complex number. Squaring the matrix element we get a tensor, which will make up a part of $|M_{fi}|^2$. Since the complex conjugate and Hermitian conjugate are the same for a complex number, the tensor is:

$$(\bar{u}(p_2) O_\alpha u(p_1))\ (\bar{u}(p_2) O_\beta u(p_1))^* = (\bar{u}(p_2) O_\alpha u(p_1))\ (u^\dagger(p_1) O^\dagger_\beta \gamma_0 u(p_2))$$

$$= (\bar{u}(p_2) O_\alpha u(p_1))\ (\bar{u}(p_1) \gamma_0 O^\dagger_\beta \gamma_0 u(p_2)).$$

We define $\bar{O}_\beta = \gamma_0 O^\dagger_\beta \gamma_0$. For example, for $O_\beta = \gamma_\mu, \gamma_\mu \gamma_5, i\gamma_5, \sigma_{\mu\nu}$, $\bar{O}_\beta = O_\beta$. Writing the product expression in terms of indices:

$$(\bar{u}(p_2) O_\alpha u(p_1))(\bar{u}(p_2) O_\beta u(p_1))^*$$
$$= \bar{u}(p_2)_i (O_\alpha)_{ij} u(p_1)_j \bar{u}(p_1)_k (\bar{O}_\beta)_{km} u(p_2)_m$$
$$= [u(p_2)_m \bar{u}(p_2)_i][O_\alpha]_{ij} [u(p_1)_j \bar{u}(p_1)_k][\bar{O}_\beta]_{km}.$$

Defining a matrix whose m, nth component is $u(p)_m \bar{u}(p)_n$, the above expression is just the trace of a matrix resulting from multiplying four matrices. We can write this, without explicitly indicating the indices, as:

$$\text{Tr}[u(p_2)\bar{u}(p_2) O_\alpha u(p_1)\bar{u}(p_1) \bar{O}_\beta]. \tag{3.8}$$

If we can find a simple expression for the 4×4 matrix $u(p)\bar{u}(p)$, then our calculation will involve just a trace calculation. Simplifying the calculation thus involves figuring out the 4×4 matrix made of the spinors, as well as developing a table of results for taking traces.

For many calculations, the states involved are initially unpolarized, or in the case of final states, we are often interested in a total rate summed over final polarizations. Summing $|M_{fi}|^2$ over polarizations for a given spin $\frac{1}{2}$ particle will lead in Eq. 3.8 to an additional sum over helicities for the relevant particle. Note the matrices $O_\alpha, \bar{O}_\beta$ are not helicity dependent, so the helicity appears only in the 4×4 matrix constructed from the spinors. This sum over helicities leads to the following simple result (which can be derived from the spinors in Eq. 2.34):

$$\sum_{\lambda=1,2} u(p,\lambda)\bar{u}(p,\lambda) = \not{p} + m \tag{3.9}$$

where the helicity λ is now indicated explicitly in the spinor. In our calculation we have considered particles only. For antiparticles we get completely analogous expressions in matrix elements with the resulting trace, after squaring, containing a 4×4 matrix of the form $v(p)\bar{v}(p)$. For this case, the sum over helicities is just

$$\sum_{\lambda=1,2} v(p,\lambda)\bar{v}(p,\lambda) = \slashed{p} - m. \tag{3.10}$$

After summing over helicities for all spin $\frac{1}{2}$ particles, Eq. 3.8 becomes $\text{Tr}[\slashed{p}_2 + m_2) O_\alpha (\slashed{p}_1 + m_1) \bar{O}_\beta]$. For antiparticles, each $\slashed{p} + m$ is replaced by $\slashed{p} - m$. These expressions occur repeatedly in calculations of processes involving incoming and outgoing spin $\frac{1}{2}$ particles. They embody the linearity requirement of quantum mechanics, where M_{fi} must contain a single power of each spinor leading to a product of the form $u(p)\bar{u}(p)$ in the square of the matrix element.

If the helicity of the spin $\frac{1}{2}$ particle is fixed initially or determined after scattering, the expression we need to use for $u(p,\lambda)\bar{u}(p,\lambda)$ will be a function of λ. One could use explicit spinors to calculate this but an alternative form has been derived that can be used simply in trace calculations. Consider the spin $\frac{1}{2}$ particle in its rest-frame. We define a spin vector in this frame as $\vec{s} = \chi^\dagger \vec{\sigma} \chi$, where χ is normalized so that $\chi^\dagger \chi = 1$. We can make a 4-vector from $\vec{s}$; the 4-vector is $s = (0, \vec{s})$. This satisfies $p \cdot s = 0$. We now boost the particle so it has 4-momentum p, resulting in a boosted 4-vector s. In terms of $s(\lambda)$ we can show (for example, using the spinors in Eq. 2.34) that:

$$u(p,\lambda)\bar{u}(p,\lambda) = (\slashed{p} + m)\frac{(1 + \gamma_5 \slashed{s}(\lambda))}{2}. \tag{3.11}$$

Note that summing this expression over λ gives us $\slashed{p} + m$, the result in Eq. 3.9.

Defining $\hat{e}(\lambda)$ as a unit vector along $\vec{p}$ for positive helicity and as a unit vector opposite $\vec{p}$ for negative helicity

$$s = \left(\frac{\hat{e}(\lambda) \cdot \vec{p}}{m}, \frac{\hat{e}(\lambda) E}{m}\right) \tag{3.12}$$

provides $s(\lambda)$ for the two helicity choices. Note that $s \cdot p = 0$, as in the rest frame.

An analogous calculation for antiparticles gives the result:

$$v(p,\lambda)\bar{v}(p,\lambda) = (\slashed{p} - m)\left(\frac{1 + \gamma_5 \slashed{s}(\lambda)}{2}\right). \tag{3.13}$$

We will use these spin-dependent expressions mainly in Chapter 8, where we study the weak interactions. For these interactions, parity violation makes the dependence on the helicity a particularly interesting phenomenon.

With the above substitutions, the trace we have to evaluate involves a number of vectors (p, s, spin vectors for photons in later calculations) and γ matrices. We tabulate below the main results for various traces involving γ matrices. These follow from the basic representation independent relation:

$$\gamma_\mu \gamma_\nu + \gamma_\nu \gamma_\mu = 2g_{\mu\nu} I.$$

The first result concerns the trace of an odd number of γ matrices. We write, using $\gamma_5 \gamma_5 = I$:

$$\text{Tr}[\gamma_{\mu_1} \cdots \gamma_{\mu_n}] = \text{Tr}[\gamma_{\mu_1} \cdots \gamma_{\mu_n} \gamma_5 \gamma_5].$$

Moving one γ_5 factor all the way to the left and using the fact that it anti-commutes with each γ_{μ_i} gives the relation:

$$\text{Tr}[\gamma_{\mu_1} \cdots \gamma_{\mu_n}] = (-1)^n \text{Tr}[\gamma_5 \gamma_{\mu_1} \cdots \gamma_{\mu_n} \gamma_5].$$

Using next $\text{Tr}[AB] = \text{Tr}[BA]$ for any two matrices, we get that $\text{Tr}[\gamma_{\mu_1} \cdots \gamma_{\mu_n}] = (-1)^n \text{Tr}[\gamma_{\mu_1} \cdots \gamma_{\mu_n}]$, so that the trace vanishes for n an odd integer.

We consider next the case of an even number of γ matrices. Since only four are independent, if we have more than four γ matrices the expression can be reduced to the case of at most four. In practice it will be convenient to have independent expressions for traces with and without a γ_5 factor included. For no γ matrices we have

$$\text{Tr}[I] = 4.$$

For two γ matrices we have

$$\text{Tr}[\gamma_{\mu_1} \gamma_{\mu_2}] = \tfrac{1}{2} \text{Tr}[\gamma_{\mu_1} \gamma_{\mu_2} + \gamma_{\mu_2} \gamma_{\mu_1}]$$
$$= \text{Tr}[g_{\mu_1 \mu_2} I] = g_{\mu_1 \mu_2} \text{Tr}[I] = 4 g_{\mu_1 \mu_2}.$$

Our expression for the γ matrices can be turned easily into expressions involving $\not{p}$ instead. For example:

$$\text{Tr}[\not{p}_1 \not{p}_2] = p_{1\mu} p_{2\nu} \text{Tr}[\gamma_\mu \gamma_\nu] = 4 p_1 \cdot p_2.$$

We give in Table 3.2 the remaining interesting traces as well as a useful formula that arises when summing over products of traces.

TABLE 3.2 Useful Trace Formulas.

$\text{Tr}[I] = 4$
$\text{Tr}[\gamma_{\mu_1} \gamma_{\mu_2}] = 4 g_{\mu_1 \mu_2}$
$\text{Tr}[\gamma_{\mu_1} \gamma_{\mu_2} \gamma_{\mu_3} \gamma_{\mu_4}] = 4[g_{\mu_1 \mu_2} g_{\mu_3 \mu_4} - g_{\mu_1 \mu_3} g_{\mu_2 \mu_4} + g_{\mu_1 \mu_4} g_{\mu_2 \mu_3}]$
$\text{Tr}[\gamma_{\mu_1} \cdots \gamma_{\mu_n}] = 0$ for n odd
$\text{Tr}[\gamma_5 \gamma_{\mu_1} \cdots \gamma_{\mu_n}] = 0$ for $n \leq 3$
$\text{Tr}[\gamma_5 \gamma_{\mu_1} \gamma_{\mu_2} \gamma_{\mu_3} \gamma_{\mu_4}] = 4i \varepsilon_{\mu_1 \mu_2 \mu_3 \mu_4}$
$\text{Tr}[\gamma_{\mu_1} \gamma_\alpha \gamma_{\mu_2} \gamma_\beta (1 - \gamma_5)] \text{Tr}[\gamma_{\nu_1} \gamma_\alpha \gamma_{\nu_2} \gamma_\beta (1 - \gamma_5)] = 64 g_{\mu_1 \nu_1} g_{\mu_2 \nu_2}$

3.2 Electron–Muon Scattering

The expression for traces involving γ_5 could be anticipated since γ_5 is used to make pseudoscalar or pseudovector bilinear covariants. The resulting $\varepsilon_{\mu_1\mu_2\mu_3\mu_4}$ from the trace makes a Lorentz invariant pseudoscalar when dotted with four independent 4-vectors. We cannot make such a pseudoscalar with less than four such 4-vectors.

Returning finally to the $e^- \mu^-$ scattering calculation, $e^2 L_{\alpha\beta}^{e^-} = J_\alpha^{e^-} J_\beta^{e^-*}$, where $J_\alpha^{e^-} = -e\bar{u}(p_{\text{out}}^{e^-})\gamma_\alpha u(p_{\text{in}}^{e^-})$. Summing over final e^- spins and averaging over initial spins (which is achieved by summing over initial spins and dividing by 2), we get:

$$L_{\alpha\beta}^{e^-} = \tfrac{1}{2}\text{Tr}[(\not{p}_{\text{out}}^{e^-} + m_e)\gamma_\alpha(\not{p}_{\text{in}}^{e^-} + m_e)\gamma_\beta].$$

This expression gives two terms with an even number of γ matrices, which can be evaluated from the trace table:

$$L_{\alpha\beta}^{e^-} = \frac{1}{2}\text{Tr}[\not{p}_{\text{out}}^{e^-}\gamma_\alpha \not{p}_{\text{in}}^{e^-}\gamma_\beta] + \frac{m_e^2}{2}\text{Tr}[\gamma_\alpha\gamma_\beta]$$

$$= 2[(p_{\text{out}}^{e^-})_\alpha(p_{\text{in}}^{e^-})_\beta + (p_{\text{out}}^{e^-})_\beta(p_{\text{in}}^{e^-})_\alpha - (p_{\text{out}}^{e^-}\cdot p_{\text{in}}^{e^-} - m_e^2)g_{\alpha\beta}].$$

The tensor $L_{\alpha\beta}^{\mu^-}$ is given by an analogous formula. Multiplying these we get finally the unpolarized result for $|M_{fi}|^2$, summed over final spins:

$$\frac{1}{4}\sum_{\text{spins}}|M_{fi}|^2 = \frac{8e^4}{q^4}[(p_{\text{out}}^{e^-}\cdot p_{\text{out}}^{\mu^-})(p_{\text{in}}^{e^-}\cdot p_{\text{in}}^{\mu^-}) + (p_{\text{out}}^{e^-}\cdot p_{\text{in}}^{\mu^-})(p_{\text{out}}^{\mu^-}\cdot p_{\text{in}}^{e^-})$$

$$- m_e^2(p_{\text{out}}^{\mu^-}\cdot p_{\text{in}}^{\mu^-}) - m_\mu^2(p_{\text{out}}^{e^-}\cdot p_{\text{in}}^{e^-}) + 2m_e^2 m_\mu^2].$$

The final result is a Lorentz invariant quantity symmetric under the exchange of $e^- \leftrightarrow \mu^-$. The theory provides the dot products we should use in constructing this invariant and the various constants that appear.

What remains is to calculate the cross section by integrating over the phase space. We can leave the cross section differential in the independent variables that describe the final state, after imposing energy and momentum conservation, or integrate over all variables. For two final particles, the only spatial variable that describes the final state is the scattering angle in the center of mass. We look at this phase space factor next. It will occur for both scattering and resonance decays to two final state particles.

3.2.4 ■ Two-Body Phase Space Factor

For two-body final states, the phase space factor for a decay or a scattering process is:

$$\int \frac{d^3 p_1}{2E_1(2\pi)^3}\frac{d^3 p_2}{2E_2(2\pi)^3}(2\pi)^4\delta^4(p_f - p_i). \qquad (3.14)$$

Chapter 3 Scattering of Leptons and Photons

The integral is over four variables in order to eliminate the δ functions, leaving two independent variables to describe the final state configuration. The labels 1 and 2 specify the two final particles. Integrating over p_2, we get:

$$\frac{1}{(4\pi)^2} \int \frac{d^3 p_1}{E_1 E_2} \delta(E_1 + E_2 - E),$$

where in this integral p_2 and E_2 are constrained by $\vec{p}_2 = \vec{p} - \vec{p}_1$, where $\vec{p}$ is the initial state momentum and E is the initial state energy. Using

$$d^3 p_1 = p_1^2 dp_1 d\Omega_1,$$

where $p_1 = |\vec{p}_1|$, and integrating over dp_1 gives for Eq. 3.14:

$$\frac{1}{(4\pi)^2 E_1 E_2} \frac{p_1^2 d\Omega_1}{\frac{d}{dp_1}(E_1 + E_2)}.$$

Taking the derivatives

$$\frac{dE_1}{dp_1} = \frac{p_1}{E_1}$$

and

$$\frac{dE_2}{dp_1} = \frac{d}{dp_1}\sqrt{(\vec{p} - \vec{p}_1)^2 + m_2^2} = \frac{-(\vec{p} - \vec{p}_1) \cdot \vec{p}_1}{E_2 p_1},$$

these finally give the expression:

$$\frac{1}{(4\pi)^2} \frac{p_1^3 d\Omega_1}{E p_1^2 - E_1(\vec{p} \cdot \vec{p}_1)}. \tag{3.15}$$

The simplest frame in which to use this expression is the center of mass, where $\vec{p} = 0$ and p_1 is a constant. In this case the two-body phase space factor is

$$\frac{1}{(4\pi)^2} \frac{p_1 d\Omega_1}{E}. \tag{3.16}$$

As an example, the rate for the decay of a particle of mass m at rest is

$$\Gamma = \left(\frac{1}{2m}\right)\left(\frac{1}{4\pi}\right)^2 \frac{p_1}{m} \int |M_{fi}|^2 d\Omega_1.$$

Writing the integral as $4\pi |\overline{M_{fi}}|^2$,

$$\Gamma = \frac{p_1}{8\pi m^2} |\overline{M_{fi}}|^2. \tag{3.17}$$

3.2 Electron–Muon Scattering

For a two-body scattering process in the center of mass, the analogous expression is:

$$\frac{d\sigma}{d\Omega_f} = \frac{1}{v_{\text{rel}}}\left(\frac{1}{2E_1}\right)\left(\frac{1}{2E_2}\right)\left(\frac{1}{4\pi}\right)^2 \frac{p_f}{E}|M_{fi}|^2. \tag{3.18}$$

We have, in this case, indicated the two initial particles as 1 and 2, and the final momenta as p_f. In terms of the initial momenta p_i, we have

$$v_{\text{rel}} = \frac{p_i}{E_1} + \frac{p_i}{E_2},$$

so that

$$\frac{d\sigma}{d\Omega_f} = \left(\frac{1}{8\pi}\right)^2 \left(\frac{p_f}{p_i}\right)\frac{|M_{fi}|^2}{E^2}. \tag{3.19}$$

3.2.5 ■ Cross Section in the Relativistic Limit

We can now use the various terms calculated above to get the differential cross section. We will do this in the relativistic limit in the center of mass frame.

In the center of mass each particle has an energy E_{cm} and momentum $|\vec{p}_{cm}| = E_{cm}$ in the relativistic limit. Ignoring the particle masses in this limit:

$$\frac{1}{4}\sum_{\text{spins}}|M_{fi}|^2 = \frac{8e^4}{q^4}[(p_{\text{out}}^{e^-}\cdot p_{\text{out}}^{\mu^-})(p_{\text{in}}^{e^-}\cdot p_{\text{in}}^{\mu^-}) + (p_{\text{out}}^{e^-}\cdot p_{\text{in}}^{\mu^-})(p_{\text{out}}^{\mu^-}\cdot p_{\text{in}}^{e^-})].$$

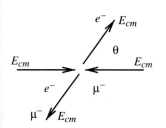

FIGURE 3.2
Scattering arrangement for $e^-\mu^- \to e^-\mu^-$.

The scattering configuration is shown in Figure 3.2.

Squaring $p_{\text{out}}^{e^-} + p_{\text{out}}^{\mu^-} = p_{\text{in}}^{e^-} + p_{\text{in}}^{\mu^-}$ and dropping mass terms gives

$$p_{\text{out}}^{e^-}\cdot p_{\text{out}}^{\mu^-} = p_{\text{in}}^{e^-}\cdot p_{\text{in}}^{\mu^-} = 2E_{cm}^2.$$

Squaring $p_{\text{out}}^{e^-} - p_{\text{in}}^{\mu^-} = p_{\text{in}}^{e^-} - p_{\text{out}}^{\mu^-}$, we get

$$p_{\text{out}}^{e^-}\cdot p_{\text{in}}^{\mu^-} = p_{\text{out}}^{\mu^-}\cdot p_{\text{in}}^{e^-} = E_{cm}^2(1+\cos\theta).$$

We can also calculate $|q^2| = 2E_{cm}^2(1-\cos\theta)$. Using these expressions in Eq. 3.19 we get finally:

$$\frac{d\sigma}{d\Omega} = \left(\frac{1}{8\pi}\right)^2 \frac{1}{4E_{cm}^2}\frac{8e^4}{16E_{cm}^4\sin^4\frac{\theta}{2}}\left[4E_{cm}^4\left(1+\cos^4\frac{\theta}{2}\right)\right]$$

$$= \frac{\alpha^2}{8E_{cm}^2\sin^4\frac{\theta}{2}}\left(1+\cos^4\frac{\theta}{2}\right).$$

The cross section peaks strongly in the forward direction and falls like $1/E_{cm}^2$ with energy at a fixed angle. The latter dependence can be anticipated from dimensional analysis.

3.3 ■ e^+e^- ANNIHILATION TO $\mu^+\mu^-$

We next look at the case of $e^+e^- \rightarrow \mu^+\mu^-$. This is an annihilation process from an initial state having no net lepton number. The final leptonic states that are allowed, all with no net lepton number, are $e^+e^-, \mu^+\mu^-, \tau^+\tau^-$. We will look specifically at $\mu^+\mu^-$ here and e^+e^- later in this chapter. We introduce first a clearer notation to keep track of the Lorentz invariant quantities for two-body scattering.

3.3.1 ■ s, t, and u Variables

We specify the four-momenta of the initial particles as p_1 and p_2 and the final particles as p_3 and p_4. We take p_3 to correspond to the particle closest in character (for example, the same lepton number) to the particle with momentum p_1. Thus for $e^-\mu^-$ scattering, if p_1 is the momentum of the initial electron, then p_3 is the momentum of the final e^-. If there is no special relationship between the particles, either particle choice for p_3 is appropriate.

We define Lorentz invariant quantities from the momenta in the problem $s = (p_1 + p_2)^2, t = (p_1 - p_3)^2$, and $u = (p_1 - p_4)^2$. These are called Mandelstam variables. Since $p_1 + p_2 = p_3 + p_4$, u also is $(p_2 - p_3)^2$. In the previous section we used the alternative common notation q^2 instead of t. We will often, based on common practice, use q^2 interchangeably with one of the $s, t,$ or u variables.

The variable s is the square of the total energy in the center of mass; t is called the momentum transfer squared; u is introduced in a symmetric way to t. The invariant u will appear symmetrically with t if nothing distinguishes the various particles, as happens, for example, in $e^- e^-$ scattering. In the center of mass, both t and u depend on the scattering angle as well as the energies. However, $s, t,$ and u are not all independent. Squaring the various momenta in the expressions for $s, t,$ and u, and adding, results in:

$$s + t + u = m_1^2 + m_2^2 + m_3^2 + m_4^2.$$

Although we will, in the expression for cross sections, often keep the dependence on $s, t,$ and u for a symmetrical notation, one of the three could be eliminated.

We have seen that the matrix element is a Lorentz invariant; thus it must be a function of invariant variables only. For specific helicities, a scattering process involving two initial and two final particles is therefore specified completely by a function of two independent invariants, for example, s and t (and the particle masses) and the invariants we can form using the spin directions. If we square and sum over final helicities and average over initial helicities, we get a function $f(s, t) = |\overline{M_{fi}}|^2$ that describes the rate for the unpolarized scattering process.

3.3 e^+e^- Annihilation to $\mu^+\mu^-$

For example, for $e^-\mu^-$ scattering this function can be written, in the relativistic limit, as

$$2e^4 \left[\frac{s^2 + u^2}{t^2} \right],$$

using the result of the previous section.

In general, the perturbation expansion to lowest order gives simple ratios of polynomials involving s, t, and u for the functions describing the rates. A function of the type $1/t^2$ (as results from the $e^-\mu^-$ matrix element) will lead to a large cross section, since t can go to zero even at high energies. A function of the type $1/s^2$ (as we will find for e^+e^- annihilation to $\mu^+\mu^-$) will lead to a cross section that falls with increasing center of mass energy. These factors result from the various propagators and determine which processes are large and which small at high energies. Including higher-order terms in the perturbation expansion, or more complex interactions, will lead to more complicated functions, but still of the same invariants s, t, and u. Finally, we note that it is often convenient to write the differential cross section in terms of the invariant momentum transfer. For example, the $e^-\mu^-$ differential elastic scattering cross section at high energies can be written:

$$\frac{d\sigma}{dt} = \frac{2\pi\alpha^2}{t^2} \left[\frac{s^2 + u^2}{s^2} \right].$$

This is most easily derived in the center of mass. However, since the expression is invariant, it can be used directly in any frame.

3.3.2 ■ Calculation of Annihilation Cross Section

The $e^+e^- \to \mu^+\mu^-$ matrix element involves one pair annihilation and one pair creation vertex with a photon propagator between. Using the appropriate terms in Table 3.1:

$$M_{fi} = \frac{-e^2 \bar{v}(p_2)\gamma_\alpha u(p_1)\bar{u}(p_3)\gamma_\alpha v(p_4)}{s}. \tag{3.20}$$

Thus,

$$|M_{fi}|^2 = \frac{e^4}{s^2} L^e_{\alpha\beta} L^\mu_{\alpha\beta},$$

where now

$$L^e_{\alpha\beta} = \text{Tr}[v(p_2)\bar{v}(p_2)\gamma_\alpha u(p_1)\bar{u}(p_1)\gamma_\beta]$$

and $L^\mu_{\alpha\beta}$ is the analogous expression for $\mu^+\mu^-$. Averaging over all initial spins and summing over final spins gives for the traces:

$\frac{1}{4}\text{Tr}[(\not{p}_2 - m_e)\gamma_\alpha(\not{p}_1 + m_e)\gamma_\beta]$ and $\text{Tr}[(\not{p}_3 + m_\mu)\gamma_\alpha(\not{p}_4 - m_\mu)\gamma_\beta]$.

We complete the calculation in the relativistic limit, in which case we can ignore the various particle masses. In this case we have for the traces:

$$\tfrac{1}{4}\text{Tr}[\not{p}_2\gamma_\alpha\not{p}_1\gamma_\beta] = (p_{2\alpha}p_{1\beta} + p_{2\beta}p_{1\alpha} - p_1 \cdot p_2 g_{\alpha\beta})$$

$$\text{Tr}[\not{p}_3\gamma_\alpha\not{p}_4\gamma_\beta] = 4(p_{4\alpha}p_{3\beta} + p_{4\beta}p_{3\alpha} - p_3 \cdot p_4 g_{\alpha\beta}).$$

Using these expressions, we get

$$\frac{1}{4}\sum_{\text{spins}}|M_{fi}|^2 = \frac{8e^4}{s^2}[(p_1 \cdot p_3)(p_2 \cdot p_4) + (p_1 \cdot p_4)(p_2 \cdot p_3)].$$

In the relativistic limit:

$$t = -2(p_1 \cdot p_3) = -2(p_2 \cdot p_4) \quad \text{and} \quad u = -2(p_1 \cdot p_4) = -2(p_2 \cdot p_3).$$

Therefore,

$$\frac{1}{4}\sum_{\text{spins}}|M_{fi}|^2 = \frac{2e^4}{s^2}(t^2 + u^2).$$

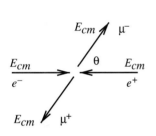

FIGURE 3.3 Scattering arrangement for $e^+e^- \to \mu^+\mu^-$ in the center of mass.

The scattering arrangement in the center of mass is shown in Figure 3.3.

We can calculate the invariants s, t, and u in terms of the energy of each beam E_{cm} and the scattering angle θ. These are

$$s = 4E_{cm}^2$$
$$t = -2E_{cm}^2(1 - \cos\theta)$$
$$u = -2E_{cm}^2(1 + \cos\theta).$$

Therefore,

$$\frac{1}{4}\sum_{\text{spins}}|M_{fi}|^2 = e^4(1 + \cos^2\theta).$$

Multiplying by the phase space factor gives a differential cross section:

$$\frac{d\sigma}{d\Omega} = \frac{e^4}{(8\pi)^2}\frac{(1 + \cos^2\theta)}{4E_{cm}^2} = \frac{\alpha^2(1 + \cos^2\theta)}{16E_{cm}^2}.$$

Unlike the $e^-\mu^-$ differential cross section, this cross section is not peaked sharply in the forward direction. Integrating this over all angles gives a total cross section:

$$\sigma = \frac{\pi\alpha^2}{3E_{cm}^2} = \frac{4\pi\alpha^2}{3s}.$$

3.4 ■ e^-e^- SCATTERING

We specify the initially prepared and finally measured momenta as p_1, p_2, p_3, p_4 as discussed in Section 3.3.1. The labeling of which final momentum is p_3 or p_4 does not matter, since the particles are identical. We will calculate the matrix element for the transition $|p_1, p_2\rangle \longrightarrow |p_3, p_4\rangle$. The possible leptonic transitions and time-orderings are listed below, including sign changes due to permutations in the state in order to have the annihilated particles at the extreme left in the state. We do not indicate the virtual photons that link the two vertices, but will include them when we write down the full amplitude. The calculation of all possible transitions involves finding all choices of annihilation and creation operators appearing in the product of current operators acting at two space-time points that give transitions from the initial to the final state. The transitions and times are, taking $t_1 < t_2$ in the space-time description of the processes (the action of the current operator is indicated; it can only scatter one electron at a vertex):

1. $|p_1, p_2\rangle \xrightarrow{J_\mu(t_1)} |p_3, p_2\rangle = -|p_2, p_3\rangle \xrightarrow{J_\mu(t_2)} -|p_4, p_3\rangle = |p_3, p_4\rangle$
2. $|p_1, p_2\rangle = -|p_2, p_1\rangle \xrightarrow{J_\mu(t_1)} -|p_4, p_1\rangle = |p_1, p_4\rangle \xrightarrow{J_\mu(t_2)} |p_3, p_4\rangle$
3. $|p_1, p_2\rangle \xrightarrow{J_\mu(t_1)} |p_4, p_2\rangle = -|p_2, p_4\rangle \xrightarrow{J_\mu(t_2)} -|p_3, p_4\rangle$
4. $|p_1, p_2\rangle = |p_2, p_1\rangle \xrightarrow{J_\mu(t_1)} -|p_3, p_1\rangle = |p_1, p_3\rangle \xrightarrow{J_\mu(t_2)} |p_4, p_3\rangle = -|p_3, p_4\rangle$

(1) and (2) correspond to the time-ordered diagrams:

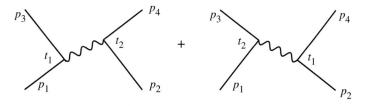

We have calculated these already in the case of $e^-\mu^-$ scattering, where we had exactly the same diagrams. (3) and (4) correspond to the diagrams:

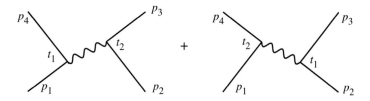

except for an overall minus sign from the use of $|p_3, p_4\rangle$ to specify the final state. These are again the same diagrams as for $e^-\mu^-$ scattering except for the exchange of the labels for $p_3 \leftrightarrow p_4$ (which exchanges the invariants $t \leftrightarrow u$). We can therefore write the full matrix element immediately using the $e^-\mu^-$ formula:

$$M_{fi} = -e^2 \left[\frac{\bar{u}(p_4)\gamma_\mu u(p_2)\bar{u}(p_3)\gamma_\mu u(p_1)}{(p_1 - p_3)^2} - \frac{\bar{u}(p_3)\gamma_\mu u(p_2)\bar{u}(p_4)\gamma_\mu u(p_1)}{(p_1 - p_4)^2} \right].$$

(3.21)

As usual, we have not explicitly indicated helicities, which have to be specified. For identical helicities and momenta in the final state (that is, $p_3 = p_4$), the two terms cancel as required for identical fermions. Had the particles been bosons (for example, spin 0 particles), the analogous terms in M_{fi} would have added, since the minus signs obtained from exchanging the order of particles in the states would have instead been plus signs. The rate does not depend on which e^- we assign the label p_3 or p_4; the behavior of the angular distribution will still tell us that we are dealing with fermions.

The result for the second-order matrix element in terms of Feynman diagrams corresponds to terms for each topologically distinct and connected figure linking the initial and final state via one propagator. The vertices are given by the local Hamiltonian density. The relative signs between diagrams reflect the exchange property for fermions or bosons. All calculations in this chapter can be organized this way.

We will not go through the trace calculation, but rather just state the result. For relativistic unpolarized initial states, summed over final spins:

$$\frac{1}{4} \sum_{\text{spins}} |M_{fi}|^2 = 2e^4 \left[\frac{s^2 + u^2}{t^2} + \frac{s^2 + t^2}{u^2} + \frac{2s^2}{tu} \right].$$

The first term is identical to the $e^-\mu^-$ result, and the second two terms result from the fact that we have identical particles in the e^-e^- case. This matrix element squared gives a differential cross section in the center of mass:

$$\frac{d\sigma}{d\Omega} = \frac{\alpha^2}{8E_{cm}^2} \left[\frac{(1 + \cos^4 \frac{\theta}{2})}{\sin^4 \frac{\theta}{2}} + \frac{(1 + \sin^4 \frac{\theta}{2})}{\cos^4 \frac{\theta}{2}} + \frac{2}{\sin^2 \frac{\theta}{2} \cos^2 \frac{\theta}{2}} \right].$$

Note that, to calculate a total cross section, only the integral over $d\Omega$ in the forward direction should be calculated (or alternatively the integral can be taken for both forward and backward angles and the result divided by 2) since we should sum only over distinguishable configurations.

3.5 ■ e^+e^- PRODUCTION OF e^+e^-

We now specify the momenta of the initial and final e^- as p_1 and p_3, respectively. The analogous quantities for e^+ are specified by p_2 and p_4. We again have annihilation diagrams with two time orderings, corresponding to the situation where the initial pair annihilation or final pair creation occur first or last. These combine

3.5 e^+e^- Production of e^+e^-

to give the same matrix element as in the case of $e^+e^- \to \mu^+\mu^-$. However, additional terms occur in the current operators that can be paired to give a transition to the correct final state. These involve scattering, rather than annihilation. At one vertex we can have the initial $e^- \to$ final e^- and at the other vertex e^+ can scatter to e^+. The Feynman diagrams contributing to the matrix element are shown below.

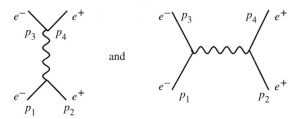

Using the current matrix elements from Table 3.1 and the photon propagator gives

$$M_{fi} = -e^2 \left[\frac{\bar{u}(p_3)\gamma_\mu v(p_4)\bar{v}(p_2)\gamma_\mu u(p_1)}{s} - \frac{\bar{v}(p_2)\gamma_\mu v(p_4)\bar{u}(p_3)\gamma_\mu u(p_1)}{t} \right].$$
(3.22)

The minus sign in the second term comes from the minus sign for the $e^+ \to e^+$ vertex compared to the $e^- \to e^-$ vertex, as appropriate for the opposite sign of charge.

The electron-positron system allows bound states that are analogous to those of the hydrogen atom. These nonrelativistic states, called positronium, have a finite lifetime since the e^-e^+ can annihilate into photons. The major binding term for positronium comes from the Coulomb potential, which we could calculate by taking the Fourier transform of the nonrelativistic limit of the scattering term (the $1/t$ term in Eq. 3.22) in M_{fi}, following the procedure in Section 2.13. This would give the expected attractive potential.

The ground state of positronium has zero spatial angular momentum. The e^-e^+ spins, however, will couple, to give states with total angular momentum 0 or 1. These will be split in energy due to small corrections from the spin-spin interaction as well as a contribution from the annihilation diagram (the $1/s$ term in Eq. 3.22) in M_{fi}, which contributes a repulsive term to the spin 1 state. We would like to show that, generally, the annihilation term contributes only to the $J = 1$ initial and final states in the center of mass. We consider the numerator of the photon propagator given by Eq. 3.4, which is proportional to $g_{\mu\nu}$. Because of gauge invariance we could equally well use in the propagator

$$g_{\mu\nu} - \frac{q_\mu q_\nu}{s}$$

instead of $g_{\mu\nu}$, where q_μ is the virtual photon 4-momentum. This expression is just the sum over spins for a real massive spin 1 particle (of mass-squared s), as given by Eq. 3.3. By angular momentum conservation, a real particle at rest will couple only to final states of the same total angular momentum, $J = 1$ in the present case. The expected effect of the annihilation diagram is indeed seen through its contribution to the energies of the positronium bound states with total spin 1. In Chapter 4 we will look at the production of hadronic states through e^+e^- annihilation. These states must have total angular momentum 1, based on the foregoing discussion.

Returning to e^+e^- scattering, the calculation of $|M_{fi}|^2$ will yield three terms; two come from squaring individual terms in M_{fi}, with the third an interference term between the two terms in Eq. 3.22. The differential cross section in the center of mass, resulting from Eq. 3.22 (in the relativistic limit for unpolarized particles) is:

$$\frac{d\sigma}{d\Omega} = \frac{\alpha^2}{8E_{cm}^2}\left[\frac{s^2+u^2}{t^2} + \frac{t^2+u^2}{s^2} + \frac{2u^2}{st}\right].$$

The first (scattering) term is the same as the result for $e^-\mu^- \to e^-\mu^-$, while the second (annihilation) term is the same as the result for $e^+e^- \to \mu^+\mu^-$. The third term is the interference contribution. Rewriting this in terms of the electron scattering angle in the center of mass gives

$$\frac{d\sigma}{d\Omega} = \frac{\alpha^2}{16E_{cm}^2}\left(\frac{3+\cos^2\theta}{1-\cos\theta}\right)^2.$$

This process is called Bhabha scattering. A comparison of data with this prediction is shown in Figure 3.4.

3.5.1 ■ Helicity Conservation in the Relativistic Limit

In this chapter we have summed over all final helicities and averaged over initial helicities in rate calculations. This does not, however, imply that all helicity combinations contribute equally to the various cross sections calculated. In fact we saw that in the nonrelativistic limit the final and initial spin states of the fermions were the same in the Coulomb scattering process. The relativistic current matrix element also has a simple spin structure, but now for helicity states rather than states with fixed spin directions. For relativistic helicity states, the scattering current operators $\bar{u}\gamma_\mu u$ and $\bar{v}\gamma_\mu v$ are nonzero only for the initial and final states having the same helicity, while the annihilation current operators $\bar{v}\gamma_\mu u$ and $\bar{u}\gamma_\mu v$ are nonzero only for the two fermions having opposite helicities. These relations can be checked by using the explicit helicity states. We can, however, demonstrate these helicity relations using the following method. We take $\bar{u}\gamma_\mu u$ as our example and show that transitions between a right-handed and left-handed state vanish. For

3.5 e^+e^- Production of e^+e^-

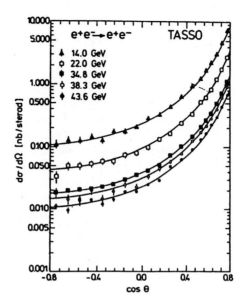

FIGURE 3.4 Data showing the $e^+e^- \to e^+e^-$ cross section and the theoretical prediction for a number of center of mass energies. [From W. Braunschweig et al., *Z. Phys.* C 37, 171 (1988).]

u a right-handed state, we have from Section 2.7.4 that

$$u_R = \left(\frac{1+\gamma_5}{2}\right) u_R.$$

For $\bar{u}_L$ the adjoint of a left-handed state,

$$\bar{u}_L = \bar{u}_L \left(\frac{1+\gamma_5}{2}\right).$$

Therefore,

$$\bar{u}_L \gamma_\mu u_R = \bar{u}_L \left(\frac{1+\gamma_5}{2}\right) \gamma_\mu \left(\frac{1+\gamma_5}{2}\right) u_R$$

$$= \bar{u}_L \gamma_\mu \left(\frac{1-\gamma_5}{2}\right)\left(\frac{1+\gamma_5}{2}\right) u_R = 0, \text{ since } \gamma_5^2 = 1.$$

The helicity relations imply, for example, that the amplitudes for the two Feynman diagrams for Bhabha scattering have different mixes of helicity states contributing (as well as different orbital angular momenta). The annihilation diagram requires that the initial e^+ and e^- have opposite helicity, while the scattering diagram has contributions from all initial helicity states.

In Section 3.5, it was shown that the annihilation diagram only populates final states of total angular momentum 1. For unpolarized relativistic initial e^+ and e^- particles colliding in the center of mass, the helicity relations provide additional constraints. They imply that the angular momentum component along the collision axis (z-axis) is an incoherent mix of $J_z = +1$ and -1. For two-body final states, this well-defined initial state will often uniquely determine the final state angular distribution based on angular momentum conservation. This is true for two final spinless particles or a fermion pair, given the helicity relation between the pair. A general method has been devised for expressing the two-body decay of an initial spin state (for example, the virtual photon or a narrow resonance decaying at rest) in terms of contributing helicity amplitudes. This method uses angular momentum conservation in the decay, but allows for helicity states of the various particles. We will not present this method, but provide a reference in the bibliography where it is described.

3.6 ■ PROCESSES WITH TWO LEPTONS AND TWO PHOTONS

Four processes involve two leptons of a given type and two photons. Taking the electron as our example, these are:

1. $\gamma + e^- \longrightarrow \gamma + e^-$
2. $\gamma + e^+ \longrightarrow \gamma + e^+$
3. $e^- + e^+ \longrightarrow \gamma + \gamma$
4. $\gamma + \gamma \longrightarrow e^- + e^+$.

These processes all share common virtual particle types that link the two vertices. These virtual particles are electrons and positrons. Diagrams that contribute, for example, to (1) and (3) above are shown below, where both time orderings are shown separately. Note that the incoming and outgoing photons of given momentum attach to the same initial or final lepton in both time-ordered terms that make up one Feynman diagram. Attaching given photons to a different set of initial and final particles changes the topology and leads to a separate Feynman diagram. In fact, each of the four processes have two Feynman diagrams contributing to the matrix element. A complete calculation for process (1) will be shown in the next section.

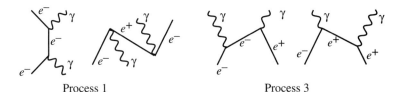

Process 1   Process 3

3.6 Processes with Two Leptons and Two Photons

As in the case of the exchange of spin 0 or spin 1 particles, the two time-ordered terms can be combined into a single covariant expression in momentum space. Thus the sum is written as one diagram, indicated as shown in the figure below.

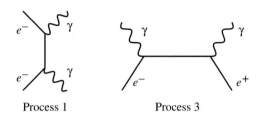

Process 1   Process 3

We do not specify the charge of the intermediate fermion; both charges contribute. The space-time propagator, which results from a coherent sum over all momenta for the virtual particle, can be written for each process as:

$$\int \left[g(p) e^{-ip \cdot x} \theta(t) + g(-p) e^{ip \cdot x} \theta(-t) \right] \frac{d^3 p}{2E(2\pi)^3}.$$

Here $g(p)$ is a 4×4 matrix, constructed from a product of spinor terms for the virtual particle (with a factor in the product from each vertex) and summed over all helicities of the virtual particle of momentum p. The second term in the integral is expected to contain $g(-p)$, as discussed earlier, in order for the two terms to combine into a covariant expression. The signs between the two terms are determined partly by the various anti-commutation properties, which are important for maintaining Lorentz covariance.

Using the matrix elements in Table 3.1, we will determine $g(p)$ and $g(-p)$ for all four processes, including the overall sign coming from particle exchanges in the states involved. To see how the sign works, we take as an example the two time-ordered terms contributing to process (1). We label

e_i^- the initial electron,

e_f^- the final electron,

e_v^- the virtual electron for the first time ordering,

e_v^+ the virtual positron for the second time ordering.

Following the particles in time:

First time ordering: $|e_i^-\rangle \to |e_v^-\rangle \to |e_f^-\rangle$

Second time ordering: $|e_i^-\rangle \to |e_f^-, e_v^+, e_i^-\rangle = -|e_i^-, e_v^+, e_f^-\rangle \to -|e_f^-\rangle.$

96   Chapter 3   Scattering of Leptons and Photons

Thus, the second time ordering requires an extra minus sign. This occurs for the processes (1) and (2) that have three particles propagating in one time ordering, which combine with a diagram having one particle at all times, to form the covariant propagator. The processes (3) and (4), with two particles propagating at all times, have no net extra signs, which can be checked by following the sequence of creations and annihilations as we did for process (1) above. They do, however, contain a sign from the vertex itself, in that positron scattering and electron scattering have an opposite sign in Table 3.1.

Including the various minus signs, the spinor expressions for $g(p)$ and $g(-p)$, corresponding to the two time orderings, respectively, are given in Table 3.3. The helicity sums needed to complete the calculation were calculated earlier; for the electron spinors we get $\not{p} + m$, for the positron spinors

$$\sum_\lambda v(p, \lambda) \bar{v}(p, \lambda) = \not{p} - m.$$

TABLE 3.3   Factors for Fermion Propagator.

Two time-ordered processes	$g(p)$	$g(-p)$
(1)	$\sum_\lambda u(p, \lambda) \bar{u}(p, \lambda)$	$-\sum_\lambda v(p, \lambda) \bar{v}(p, \lambda)$
(2)	$\sum_\lambda v(p, \lambda) \bar{v}(p, \lambda)$	$-\sum_\lambda u(p, \lambda) \bar{u}(p, \lambda)$
(3)	$\sum_\lambda u(p, \lambda) \bar{u}(p, \lambda)$	$-\sum_\lambda v(p, \lambda) \bar{v}(p, \lambda)$
(4)	$\sum_\lambda u(p, \lambda) \bar{u}(p, \lambda)$	$-\sum_\lambda v(p, \lambda) \bar{v}(p, \lambda)$

With these substitutions, we always get for the second term in each row of Table 3.3 an expression equal to the first term after the exchange $p \to -p$. We

3.6 Processes with Two Leptons and Two Photons

can put $g(p)$ for all four processes into a common form by the convention:

$$g(p) = \slashed{p} + m,$$

to within an overall sign, where p is the momentum of the electron traveling forward in time. The overall sign for the terms in Table 3.3 is $+1$ for (1), (3), and (4), and -1 for (2). The minus sign for (2), like the minus sign for an individual positron scattering vertex in Table 3.1, has real consequences. An example occurs in a higher order (loop) correction to the photon propagator shown in Figure 3.5, which we will look at quantitatively in Chapter 7. In this loop correction, we indicate the positron and electron momenta as p_{e^+} and p_{e^-}. The sum over spins for each particle will give a factor of $(\slashed{p}_{e^+} - m)$ and $(\slashed{p}_{e^-} + m)$ when calculating the propagation of the pair. Instead of the positron term $\slashed{p}_{e^+} - m$, we can use the rule that we should always calculate the propagator by taking the momentum of the electron traveling forward in time, but then we must include the overall minus sign, to get the same (correct) result.

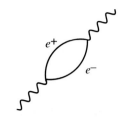

FIGURE 3.5 Loop correction to photon propagation.

3.6.1 ■ Matrix Element for γe^- Scattering

The process $\gamma + e^- \rightarrow \gamma + e^-$ is called Compton scattering. We will look at this in detail as an example of a process involving two leptons and two photons. We start with the second-order expression for the S matrix and again end up with Feynman diagrams in momentum space from which we can write down rather directly the expression for M_{fi}. For Compton scattering:

$$S^{(2)}_{fi} = (-i)^2 \int_{-\infty}^{\infty}\int_{-\infty}^{\infty} dx_1^4 dx_2^4 \, \langle f| J_\mu^{e^-}(x_1) A_\mu(x_1) J_\nu^{e^-}(x_2) A_\nu(x_2) |i\rangle \, \theta(t_1 - t_2)$$

The initial and final state momenta are given by p_1 and p_3 for the e^-, and p_2 and p_4 for the γ, respectively. By working through $S^{(2)}_{fi}$ carefully, we will arrive at a derivation of the electron propagator discussed in Section 3.6.

Taking product states for the initial and final states we have:

$$S^{(2)}_{fi} = (-i)^2 \int_{-\infty}^{\infty}\int_{-\infty}^{\infty} dx_1^4 dx_2^4 \, \langle p_3| J_\mu^{e^-}(x_1) J_\nu^{e^-}(x_2) |p_1\rangle$$
$$\times \langle p_4| A_\mu(x_1) A_\nu(x_2) |p_2\rangle \, \theta(t_1 - t_2).$$

For notational simplicity, none of the helicities are explicitly indicated for the initial and final states. The term involving the photon field has two field operators and two photons in the matrix element and will give a tensor function of x_2 and x_1, to which we return later. We focus first on the term involving the electrons, which will have virtual propagating particles when evaluating the matrix element. The two currents allow two processes. In the first process the current $J_\nu^{e^-}(x_2)$ annihilates the initial electron and creates a virtual electron. In order to end up in the correct final state, $J_\mu^{e^-}(x_1)$ must annihilate the virtual electron and create the final state electron. In the second process, $J_\nu^{e^-}(x_2)$ leaves the initial electron unchanged and creates a pair. For this to give us the correct final state, the electron in the pair must match the final state electron while the positron is virtual, and can have any momentum. Finally, to arrive in the correct final state, the current $J_\mu^{e^-}(x_1)$ must annihilate the initial electron with the virtual positron. The current matrix elements corresponding to these processes are listed in Table 3.1. They give (including the correct signs discussed in the previous section):

$$\langle p_3 | J_\mu^{e^-}(x_1) J_\nu^{e^-}(x_2) | p_1 \rangle$$
$$= \sum_{p,\lambda} e^2 \bar{u}(p_3) \gamma_\mu [u(p,\lambda)\bar{u}(p,\lambda)] \gamma_\nu u(p_1) e^{ip\cdot(x_2-x_1)} e^{i(p_3\cdot x_1 - p_1\cdot x_2)}$$
$$+ \sum_{p,\lambda} e^2 \bar{u}(p_3) \gamma_\nu [-v(p,\lambda)\bar{v}(p,\lambda)] \gamma_\mu u(p_1) e^{ip\cdot(x_2-x_1)} e^{i(p_3\cdot x_2 - p_1\cdot x_1)}$$

We will now redefine coordinates to correspond to physical events rather than the time order in which events occur. We define x_2 as the coordinate at which the initial electron in annihilated, and x_1 as the coordinate at which the final state electron is created, irrespective of which is earlier. This requires that in the second term we exchange x_1 and x_2. Including $\theta(t_1-t_2)$, which becomes $\theta(t_2-t_1)$ for the second term, and using the fact that the photon-field matrix element is symmetric under the exchange $\mu \leftrightarrow \nu$ and $x_1 \leftrightarrow x_2$, we get from the electron currents:

$$e^2 \bar{u}(p_3) \gamma_\mu G_F(x_1 - x_2) \gamma_\nu u(p_1) e^{i(p_3\cdot x_1 - p_1\cdot x_2)}$$

where

$$G_F(x) = \sum_{p,\lambda} u(p,\lambda)\bar{u}(p,\lambda) e^{-ip\cdot x} \theta(t) - v(p,\lambda)\bar{v}(p,\lambda) e^{ip\cdot x} \theta(-t).$$

$G_F(x)$ is of the form given by Eq. 3.2 and is called the electron space-time propagator. It gives a momentum-space propagator:

$$\frac{i(\not{p}+m)}{p^2 - m^2 + i\varepsilon} \tag{3.23}$$

after taking a 4-dimensional Fourier transform.

We now return to the photon matrix elements. In the expression $A_\mu(x_1) A_\nu(x_2)$ either field can annihilate the incoming photon, with the second field creating the outgoing photon. This gives, therefore, two terms:

$$\langle p_4 | A_\mu(x_1) A_\nu(x_2) | p_2 \rangle = e_\mu^*(\lambda_{\text{out}}) e_\nu(\lambda_{\text{in}}) e^{i(p_4\cdot x_1 - p_2\cdot x_2)}$$
$$+ e_\mu(\lambda_{\text{in}}) e_\nu^*(\lambda_{\text{out}}) e^{i(p_4\cdot x_2 - p_2\cdot x_1)}.$$

The incoming and outgoing photon helicities are indicated as λ_{in} and λ_{out}, respectively. Dotting together the various tensors, we get for the S-matrix element a sum of two terms (arising from the sum of two Feynman diagrams):

3.6 Processes with Two Leptons and Two Photons

$$S_{fi}^{(2)} = (-i)^2 e^2 \int_{-\infty}^{\infty}\int_{-\infty}^{\infty} d^4x_1\, d^4x_2 [T_1(x_1,x_2) + T_2(x_1,x_2)]$$

where

$$T_1(x_1,x_2) = \bar{u}(p_3)\,\epsilon^*(\lambda_{\text{out}}) G_F(x_1-x_2)\,\epsilon(\lambda_{\text{in}}) u(p_1) e^{i[(p_3+p_4)\cdot x_1 - (p_1+p_2)\cdot x_2]}$$

$$T_2(x_1,x_2) = \bar{u}(p_3)\,\epsilon(\lambda_{\text{in}}) G_F(x_1-x_2)\,\epsilon^*(\lambda_{\text{out}}) u(p_1) e^{i[(p_3-p_2)\cdot x_1 - (p_1-p_4)\cdot x_2]}.$$

Writing

$$G_F(x) = \int \frac{i(\slashed{p}+m)e^{-ip\cdot x}}{p^2-m^2+i\varepsilon}\frac{d^4p}{(2\pi)^4},$$

we complete the calculation by integrating over

$$\frac{d^4p}{(2\pi)^4} d^4x_1 d^4x_2.$$

This is completely analogous to the calculation in Section 2.10 for the spinless case, or the calculation for $\mu^- e^-$ scattering, since the integrals to be evaluated merely involve integrating over phase factors. The result is:

1. A factor of $(2\pi)^4 \delta^4(p_f - p_i)$ after integrating over x_1,
2. A value for p in the momentum-space electron propagator of $p = p_1 + p_2$ for T_1 and $p = p_1 - p_4$ for T_2, after integrating over $x_1 - x_2$,
3. Remaining factors that come from various terms (spinors and polarization vectors) that are independent of the space-time coordinates.

The resulting expression for M_{fi} is:

$$M_{fi} = e^2 \bar{u}(p_3)\left[\epsilon^*(\lambda_{\text{out}})\frac{(\slashed{p}_1+\slashed{p}_2)+m}{(p_1+p_2)^2-m^2}\epsilon(\lambda_{\text{in}}) \right.$$

$$\left. +\epsilon(\lambda_{\text{in}})\frac{(\slashed{p}_1-\slashed{p}_4)+m}{(p_1-p_4)^2-m^2}\epsilon^*(\lambda_{\text{out}})\right]u(p_1). \qquad (3.24)$$

We can obtain this expression in momentum-space from two topologically different Feynman diagrams, if we consider the different quantum events that can lead to absorption of the initial photon and emission of the final photon. These are shown in Figure 3.6 along with the factors associated with the diagrams for arriving at the scattering amplitude. The other processes involving two photons and two leptons have matrix elements calculable using analogous Feynman diagrams.

To facilitate tracking the order in which the spin dependent factors appear in matrix elements, arrows are often attached to the fermion lines in the Feynman diagrams. The arrow then points from the past into the future along the elec-

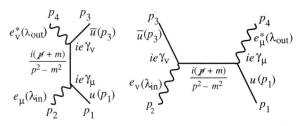

FIGURE 3.6 Feynman diagrams for Compton scattering. Factors are those needed to calculate $-iM_{fi}$. The value of the momentum p in the propagator is obtained by four-momentum conservation at the appropriate vertex.

tron lines. This is the same sequence in which the factors appear in each term in Eq. 3.24. To arrive at the correct sequence of factors for positrons, the arrow points from the future to the past.

3.6.2 ■ Compton Scattering Cross Section

Using Eq. 3.24 for the Compton scattering matrix element we can calculate the cross section for this process. For simplicity, we will do this in the relativistic limit in the center of mass frame for unpolarized initial particles. The denominators for the two terms in M_{fi} are

$$s - m^2 = (p_1 + p_2)^2 - m^2 = 2p_1 \cdot p_2 \simeq s$$

and

$$u - m^2 = (p_1 - p_4)^2 - m^2 = -2p_1 \cdot p_4 \simeq u. \quad (3.25)$$

Leaving out the mass terms, we can therefore approximate $M_{fi} = e^2[M_1 + M_2]$, where

$$M_1 = e_\nu^*(\lambda_{\text{out}})e_\mu(\lambda_{\text{in}})\bar{u}(p_3)\gamma_\nu(\not{p}_1 + \not{p}_2)\gamma_\mu u(p_1)/s$$
$$M_2 = e_\nu^*(\lambda_{\text{out}})e_\mu(\lambda_{\text{in}})\bar{u}(p_3)\gamma_\mu(\not{p}_1 - \not{p}_4)\gamma_\nu u(p_1)/u.$$

Summing over all final spins and averaging over initial spins, we calculate:

$$\frac{1}{4}\sum_{\text{all spins}} |M_{fi}|^2 = \frac{e^4}{4}\sum_{\text{all spins}}[|M_1|^2 + |M_2|^2 + M_1 M_2^* + M_1^* M_2]$$

For the photon spin sum we can use the result from Section 3.1:

$$\sum_{\lambda_{\text{in}}} e_\mu^*(\lambda_{\text{in}})e_\nu(\lambda_{\text{in}}) = \sum_{\lambda_{\text{out}}} e_\mu^*(\lambda_{\text{out}})e_\nu(\lambda_{\text{out}}) = -g_{\mu\nu}.$$

3.6 Processes with Two Leptons and Two Photons

With this simplification, and leaving out the electron mass everywhere:

$$\sum_{\text{all spins}} |M_1| = \frac{1}{s^2} \text{Tr}[\slashed{p}_3 \gamma_\nu (\slashed{p}_1 + \slashed{p}_2) \gamma_\mu \slashed{p}_1 \gamma_\mu (\slashed{p}_1 + \slashed{p}_2) \gamma_\nu].$$

The two factors of $g_{\mu_1 \mu_2}$ from the two photon spin sums leaves us with two sums over γ matrices inside the trace. Next, we use the fact that we can exchange the order of two matrices in a trace of a product of two matrices to rewrite the above as follows:

$$\sum_{\text{all spins}} |M_1|^2 = \frac{1}{s^2} \text{Tr}[(\gamma_\nu \slashed{p}_3 \gamma_\nu)(\slashed{p}_1 + \slashed{p}_2)(\gamma_\mu \slashed{p}_1 \gamma_\mu)(\slashed{p}_1 + \slashed{p}_2)].$$

Using the anti-commutation formula for the γ matrices, we can simplify the two γ matrix sums, which give

$$\gamma_\nu \slashed{p}_3 \gamma_\nu = -2\slashed{p}_3, \quad \gamma_\mu \slashed{p}_1 \gamma_\mu = -2\slashed{p}_1.$$

Therefore,

$$\sum_{\text{all spins}} |M_1|^2 = \frac{4}{s^2} \text{Tr}[\slashed{p}_3 (\slashed{p}_1 + \slashed{p}_2) \slashed{p}_1 (\slashed{p}_1 + \slashed{p}_2)].$$

This expression is now in the form of a trace of a sum of products of four γ matrices that can be evaluated from Table 3.2. Leaving out factors of m^2, we get:

$$\sum_{\text{all spins}} |M_1|^2 = \frac{32}{s^2}(p_1 \cdot p_2)(p_3 \cdot p_2) = 8\left(\frac{-u}{s}\right).$$

The calculation of $\sum_{\text{all spins}} |M_2|^2$ is identical to that for $\sum_{\text{all spins}} |M_1|^2$ with the exchange $p_2 \leftrightarrow -p_4$. This gives:

$$\sum_{\text{all spins}} |M_2|^2 = 8\left(\frac{-s}{u}\right).$$

The cross terms for $M_1 M_2^*$ and $M_1^* M_2$ summed over spins are equal and both vanish in the limit where we ignore the electron mass. We can see this by calculating:

$$\sum_{\text{all spins}} M_1 M_2^* = \text{Tr}[\slashed{p}_3 \gamma_\nu (\slashed{p}_1 + \slashed{p}_2) \gamma_\mu \slashed{p}_1 \gamma_\nu (\slashed{p}_1 - \slashed{p}_4) \gamma_\mu].$$

To evaluate this, we use two general results that follow from the γ matrix anti-commutation relations. The first is for the sum over ν:

$$\gamma_\nu (\slashed{p}_1 + \slashed{p}_2) \gamma_\mu (\slashed{p}_1) \gamma_\nu = -2 \slashed{p}_1 \gamma_\mu (\slashed{p}_1 + \slashed{p}_2),$$

which is an example of the general result $\gamma_\nu \not{p}_a \not{p}_b \not{p}_c \gamma_\nu = -2\not{p}_c \not{p}_b \not{p}_a$. For the trace, we now have:

$$-2\,\mathrm{Tr}[\not{p}_3 \not{p}_1 \gamma_\mu (\not{p}_1 + \not{p}_2)(\not{p}_1 - \not{p}_4) \gamma_\mu].$$

We can next sum over μ, by using the general formula

$$\gamma_\mu \not{p}_a \not{p}_b \gamma_\mu = 4 p_a \cdot p_b,$$

giving

$$-8[(p_1 + p_2) \cdot (p_1 - p_4)]\,\mathrm{Tr}(\not{p}_3 \not{p}_1) = -32(p_1 \cdot p_3)[(p_1 + p_2) \cdot (p_1 - p_4)].$$

The dot product $(p_1 + p_2) \cdot (p_1 - p_4) = m^2 + p_1 \cdot p_2 - p_1 \cdot p_4 - p_2 \cdot p_4$. Using $p_1 - p_3 = p_4 - p_2$, we can calculate that $p_2 \cdot p_4 = p_1 \cdot p_3 - m^2$; therefore the dot product can also be written as $2m^2 + p_1 \cdot (p_2 - p_3 - p_4) = 2m^2 - p_1 \cdot p_1 = m^2$. Thus in the relativistic limit where we take $m^2 \simeq 0$, this term will vanish.

Adding the various trace terms we have finally:

$$\frac{1}{4} \sum_{\text{all spins}} |M_{fi}|^2 = 2e^4 \left(\frac{-u}{s} + \frac{-s}{u} \right).$$

Including the phase space-factor, in the center of mass,

$$\frac{d\sigma}{d\Omega} = \frac{1}{64\pi^2} \frac{1}{E^2} 2e^4 \left(\frac{-u}{s} + \frac{-s}{u} \right).$$

Using $E = 2E_{cm}$, where E_{cm} is the center of mass energy of e or γ; the invariant $s = 4E_{cm}^2$, and $u = -2E_{cm}^2(1 + \cos\theta)$, where θ is the electron (or photon) scattering angle, the expression for the differential cross section is:

$$\frac{d\sigma}{d\Omega} = \frac{\alpha^2}{8E_{cm}^2} \left(\frac{1 + \cos\theta}{2} + \frac{2}{1 + \cos\theta} \right).$$

At high energies we can ignore the first term in $d\sigma/d\Omega$, with the cross section large only for $\cos\theta$ near -1, that is, for backward scattering, where $u \simeq 0$. The infinite value for $\cos\theta = -1$ comes about only in the limit $m = 0$. If we keep the lowest order correction due to the finite mass we should use

$$u - m^2 \simeq -2E_{cm}^2(1 + \cos\theta) - m^2.$$

Thus at high energies the correct expression is:

$$\frac{d\sigma}{d\Omega} \simeq \frac{\alpha^2}{4E_{cm}^2} \left(\frac{1}{1 + \cos\theta + \frac{m^2}{2E_{cm}^2}} \right).$$

Integrating this over all angles gives finally the total cross section in the high energy limit:

$$\sigma = \frac{\pi \alpha^2}{E_{cm}^2} \log\left(\frac{2E_{cm}}{m}\right). \qquad (3.26)$$

The total cross section, despite the logarithm, falls with increasing energy in the relativistic limit.

3.7 ■ HIGHER-ORDER TERMS IN THE PERTURBATION EXPANSION, THE FEYNMAN RULES

The calculation of higher-order terms in the S matrix expansion involves a larger number of quantum events linking the initial and final states. In the space-time description of a term in the S matrix, we can specify the absolute space-time location of one quantum event and then use relative space-time coordinates to describe the rest of the processes that occur. Integrating over the one absolute coordinate gives a four-momentum conserving delta function $(2\pi)^4 \delta^4(p_f - p_i)$. Integration over all the relative coordinates gives the dynamically determined contribution of the higher order process to $-iM_{fi}$.

Each term in the iteration expansion for the S matrix, derived in Section 2.9, involves a time-ordered sequence of events. We state without proof that we can recast each term in the expansion in terms of Feynman diagrams, each of which is topologically distinct and connected, and contains the same number of quantum events as the time-ordered form. The Feynman diagrams express the contribution to M_{fi} in terms of well-defined physical events. Each such event is given by a vertex that contributes, to the overall amplitude, a factor given by a local matrix element of the interaction Hamiltonian density. The vertices are linked by virtual particles that propagate between the local events. All such virtual particles are created and then absorbed internally to the process. By writing the space-time propagator in terms of its Fourier transform, we can express the formula for each Feynman diagram completely in momentum-space. For the second-order diagrams we have been calculating in this chapter, the four-momentum of the virtual particle is completely determined from the initial state by four-momentum conservation. For higher-order diagrams this is not the case; some of the virtual momenta are free to range over all values, and these need to be summed over in order to evaluate the diagram. This will involve an integral of the type

$$\int \frac{d^4 p}{(2\pi)^4}$$

for each undetermined four-momentum. An example of a diagram requiring such an integral is the loop diagram of Figure 3.5. After adding all diagrams up to a given order, we then have our perturbation approximation for M_{fi}, which has to be squared and integrated over phase space to calculate rates.

The perturbation expansion gives an amazingly simple physical picture, where the physics is given by the local quantum events and propagation between events via the free field amplitude of Eq. 2.1 (or its complex conjugate) for incoming, outgoing, and virtual particles. The addition of amplitudes for all alternatives then gives the full amplitude.

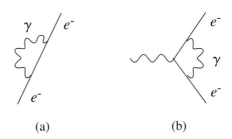

FIGURE 3.7 Higher-order corrections in Quantum Electrodynamics. (a) To electron propagation. (b) To the electron vertex.

The higher-order diagrams include terms that affect only one individual particle. These diagrams show a virtual particle being emitted and absorbed by the same particle. These will occur in any process involving the individual particle. An example is the loop correction to the photon propagator in Figure 3.5, which will be present wherever the photon propagator occurs. These events, described by the higher-order diagrams, correct or renormalize the propagator. Figure 3.7 shows analogous corrections to electron propagation and the electron vertex. The vertex corrections change the magnetic moment of the particle, which can be measured very accurately in an external static magnetic field where the higher-order corrections occur only in the electron vertex. In the nonrelativistic limit the electron interaction with the magnetic field is given by a Hamiltonian of the form $-\vec{\mu} \cdot \vec{B}$, where

$$\vec{\mu} = \frac{-ge}{2m}\vec{S}$$

and $\vec{S}$ is the expectation value of the spin operator equal to $\vec{\sigma}/2$. For the nonrelativistic limit of the Dirac equation, $g = 2$. Including the vertex corrections into an effective or renormalized g, its value is changed slightly from the value 2. The lowest-order vertex correction of Figure 3.7(b) gives for both the electron and muon:

$$g = 2\left(1 + \frac{\alpha}{2\pi}\right) = 2(1.00116).$$

The magnetic moments have been evaluated to order α^5, which involves calculating many diagrams. Comparing the calculations to the measured values provides the most exacting low-energy test of the quantum field theoretical approach.

The measured values are:

$$\frac{g_e - 2}{2} = 1159.652188 \pm 0.000004 \times 10^{-6},$$

$$\frac{g_\mu - 2}{2} = 1165.9203 \pm 0.0008 \times 10^{-6}.$$

The difference between the electron and muon come from higher-order diagrams involving particles other than the photon (for example, loops in the photon propagator) where the muon mass, being larger than the electron mass, results in a larger value. The measured values for both electrons and positrons are in excellent agreement with each other. The measured electron value is in excellent agreement with the theoretical prediction, but the comparison cannot be made at the full precision of the measurement because α is not independently measured with sufficient precision. The experimentally determined value for the muon and the theoretical expectation differ at about one standard deviation. The theoretical prediction includes measureable contributions from virtual hadrons and the weak interactions.

In addition to the higher-order diagrams, which modify the properties of the individual particles, we have diagrams that link different particles in a process. Examples are shown in Figure 3.8 for the case of $e^- \mu^-$ scattering.

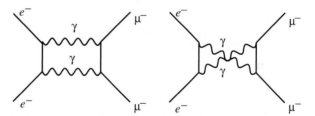

FIGURE 3.8 Higher-order diagrams where two photons are exchanged between e^- and μ^-.

The number of diagrams of a given order grows rapidly for higher orders. The expansion is therefore believed to be an asymptotic expansion rather than a convergent series. For a small coupling constant this does not preclude extremely accurate calculations, as we have seen in the case of the magnetic moments.

For completeness we list again the various factors required in electrodynamics to calculate $-iM_{fi}$ for any Feynman diagram.

1. For an incoming electron, positron, or photon, respectively, we associate a factor u, $\bar{v}$, e_μ.
2. For an outgoing electron, positron, or photon, respectively, we associate $\bar{u}$, v, e_μ^*.

3. The vertex factor is $ie\gamma_\mu$ for the electron, and $-ie\gamma_\mu$ for positrons. Some caution is needed regarding the overall sign for fermions, as discussed earlier in the chapter.

4. The propagators for electrons and photons are, respectively:

$$i\left(\frac{\not{p}+m}{p^2-m^2+i\varepsilon}\right)$$

and

$$\frac{-ig_{\mu\nu}}{q^2+i\varepsilon}.$$

Again, be careful of the overall sign for fermions.

The perturbation expansion for more complicated interactions will be similar to that for electrodynamics, with additional particles involved and factors at vertices given by the interaction Hamiltonian density for the new interaction type. Unfortunately, for the case where perturbation theory is inadequate, particularly for the scattering of relativistic bound states of the strong interactions, there is no simple physical space-time picture or simple calculation analogous to the perturbation expansion.

3.8 ■ LORENTZ COVARIANCE AND FIELD THEORY

We want to examine several constraints that arise from combining relativity with quantum field theory. The results apply to a general relativistic field theory, not just to quantum electrodynamics. In the classical case, relativity requires that no information can be transferred faster than the speed of light; in particular, regions of space-time that are space-like separated are not actively exchanging information and, in fact, the time ordering of events in the two regions changes from reference frame to reference frame (unlike events for time-like separated regions of space-time where past and future are uniquely defined).

The simplest way in which the constraints of relativity can work is if all the quantum interactions linking events at points x_1 and x_2 vanish for space-like separations. These interactions are determined by the space-time propagator, which, in fact, does not vanish anywhere. However, the combination of particles and antiparticles propagating with the various time-orderings did in fact give a final covariant result for all the examples we considered, as was evident in the covariant momentum-space representation. Thus an observer who attributes the interaction for space-like separated points $x_1 - x_2$ to particle exchange using the expression

$$\sum_{\vec{p}} e^{-ip\cdot(x_1-x_2)}\theta(t_1-t_2)$$

3.8 Lorentz Covariance and Field Theory

(assuming $t_1 > t_2$ and a scalar interaction) will get the same result as a Lorentz-transformed observer, who attributes the interaction to antiparticle exchange using the expression

$$\sum_{\vec{p}} e^{-ip\cdot(x_2'-x_1')} \theta(t_2' - t_1')$$

(where, after the Lorentz transformation, $t_2' > t_1'$). The coordinates x_1', x_2' are the result of Lorentz transforming x_1 and x_2, respectively, to the new frame. The existence of antiparticles is thus a requirement for Lorentz invariance as emphasized by Feynman. Note that the couplings must also be the same for the two versions of the calculation.

Therefore, the field operators that describe a quantum process must contain an outgoing antiparticle of momentum $\vec{p}$ for every incoming particle of momentum $\vec{p}$. Since the Hamiltonian is Hermitian, every field that appears in the Hamiltonian must also have a corresponding Hermitian conjugate, which then contains incoming antiparticles of momentum $\vec{p}$ and outgoing particles of momentum $\vec{p}$. The coupling constant that appears in a Hamiltonian density involving a field must multiply the whole field operator, that is, both the creation and annihilation part.

The operations discussed above, that is, changing incoming particles into outgoing antiparticles of the same momentum, is due to the following operations:

1. Charge conjugation, C, which exchanges particles and antiparticles;
2. Parity P, which changes the sign of the momentum $\vec{p}$;
3. Time reversal, T, which exchanges incoming and outgoing states and the direction of motion.

The combination of PT leaves the momentum unchanged after exchange of incoming particles and outgoing particles.

If we include spin, the CPT operation will also change the sign of the helicity λ. It thus relates incoming particles of momentum $\vec{p}$, helicity λ, to outgoing antiparticles of momentum $\vec{p}$, helicity $-\lambda$. In Chapter 8 we will discuss the weak interactions where the consequences of the CPT symmetry are most relevant. We will see that the sign change of the helicity is indeed necessary in identifying weak processes that have equal amplitudes.

The previous discussion involved interactions for space-like separated systems. Next, we want to look at constraints on local measurements made on such systems. Each measurement corresponds to a local Hermitian operator. We denote these operators as O_1 and O_2, for measurements at x_1 and x_2. The state on which the measurement is performed is specified for simplicity as a product state: $|1\rangle |2\rangle = |1, 2\rangle$. For $t_1 < t_2$ the result of our measurement is $O_2 O_1 |1, 2\rangle$.

We now look at the same measurements in another reference frame. The change of frames is specified by a Lorentz transformation, to which we can associate a unitary operator, U, generating the transformation. In the new frame, the result of the measurement is:

$$UO_2O_1|1,2\rangle = UO_2U^{-1}UO_1U^{-1}U(|1,2\rangle)$$

Denoting the operators and locations in the new frame by primes, we have

$$UO_2O_1|1,2\rangle = O'_2O'_1|1',2'\rangle.$$

However, for a space-like separation of x_1 and x_2, we can choose a new frame for which we have $t'_2 < t'_1$. Therefore, in this frame the measurement is given by the opposite time order:

$$O'_1O'_2|1',2'\rangle.$$

Equating the two expressions for the measurement gives

$$UO_2O_1|1,2\rangle = O'_2O'_1|1',2'\rangle = O'_1O'_2|1',2'\rangle.$$

Hence we conclude that local Hermitian operators corresponding to measurable quantities must commute for space-like separations for a relativistically covariant theory. Examples of such operators are the Hamiltonian density, the current operators, and Hermitian fields (like A_μ) that have a classical measurable limit. Note that the spin $\frac{1}{2}$ field ψ does not satisfy this commutation rule, but the bilinear operators like $\bar{\psi}\gamma_\mu\psi$ do. We will not prove these commutation relations; they can be shown to hold for the field operators we have defined.

3.9 ■ SPECIAL SYMMETRIES OF THE S MATRIX

In the previous section we saw that the fields making up the Hermitian Hamiltonian density must contain certain specific combinations of particle and antiparticle operators. These are:

Particle Annihilation(Creation)   ⟷   Antiparticle Creation(Annihilation)
    with $\vec{p}$ and λ with $\vec{p}$ and $-\lambda$

The discussion focused on particles and antiparticles propagating internally, generating an interaction. These particles and antiparticles can, however, also be real, making up part of some initial or final state. In this case, the *CPT* symmetry can be used to relate matrix elements for corresponding measurable processes.

We specify the relevant states in these processes as follows, starting from a general initial state $|n\rangle$. If we exchange particles and antiparticles, leaving $\vec{p}$ and λ unchanged, we specify the resulting state $|Cn\rangle$. We can further change the sign of the helicity and momentum with a parity operation that changes the sign of vectors and leaves axial vectors unchanged. We specify this state as $|CPn\rangle$. The final time-reversal operation reverses $\vec{p}$ as well as the sense of rotation. With both $\vec{J}$ and $\vec{p}$ reversed, the projection of $\vec{J}$ on $\vec{p}$ is unchanged, leaving the helicity unchanged. These states are specified as $|CPTn\rangle$. Finally, we have to exchange incoming and outgoing states. We can now state our symmetry relation, called the *CPT* theorem: $\langle m| S |n\rangle = \langle CPTn| S |CPTm\rangle$ for all states. We will use this theorem later; it implies, for example, that masses and lifetimes are equal for particles and antiparticles.

The *CPT* theorem follows from relativistic invariance and how we construct the *S* matrix in terms of given quantum events. It may turn out that some of the other operations discussed above also provide *S* matrix (and Hamiltonian density) symmetries, that is, the symmetry may extend beyond the minimum required. These are

1. If $\langle Cm| S |Cn\rangle = \langle m| S |n\rangle$, the system is charge conjugation invariant.
2. If $\langle CPm| S |CPn\rangle = \langle m| S |n\rangle$, the system is *CP* invariant.
3. If $\langle Pm| S |Pn\rangle = \langle m| S |n\rangle$, the system conserves parity.
4. If $\langle Tn| S |Tm\rangle = \langle m| S |n\rangle$, the system obeys time-reversal symmetry.

The electromagnetic and strong interactions satisfy all of these symmetries, leading historically to the expectation that all interactions would. However, it was subsequently found that the weak interactions only satisfy the *CPT* symmetry. As a consequence, for example, left- and right-handed particles do not have the same interactions (resulting in parity violation).

CHAPTER 3 HOMEWORK

3.1. Consider the decay of an initial particle of mass m_{in} into two final particles of mass m_1 and m_2. In the center of mass the final particles are produced with energies E_1 and E_2. Show that

$$E_1 = \frac{m_{in}^2 + m_1^2 - m_2^2}{2m_{in}}, \quad E_2 = \frac{m_{in}^2 + m_2^2 - m_1^2}{2m_{in}}$$

(*Hint:* Don't do any complicated calculations).

The two particles recoil with the same magnitude of momentum $P = |\vec{p}_1| = |\vec{p}_2|$. Show that

$$P = \frac{1}{2m_{in}} \sqrt{(m_{in})^4 + (m_1^2 - m_2^2)^2 - 2m_{in}^2(m_1^2 + m_2^2)}.$$

3.2. (a) A particle of mass m_{in} decays into n particles with energy and momenta E_i, $\vec{p}_i$, respectively, $i = 1, \ldots, n$. The initial particle is moving with velocity $\vec{v}$, which is called the velocity of the center of mass. Show that

$$m_{in}^2 = \left(\sum_{i=1}^n E_i\right)^2 - \left(\sum_{i=1}^n \vec{p}_i\right)^2$$

and that

$$\vec{v} = \frac{\sum_{i=1}^n \vec{p}_i}{\sum_{i=1}^n E_i}.$$

(b) A particle of mass m_1, four-momentum $(E_1, \vec{p}_1)$, is incident on a particle of mass m_2 at rest. What is the total center of mass energy? What is the velocity of the center of mass?

3.3. Show that $q_\mu J_\mu(0) = 0$ for the four terms in Table 3.1, where q_μ is the appropriate sum or difference of momenta.

3.4. Consider the Majorana representation for the γ_μ matrices. These satisfy the usual relations for the γ_μ matrices, and in addition are purely imaginary, that is, $\gamma_\mu^* = -\gamma_\mu$.

(a) Use the relation $\gamma_\mu^\dagger \gamma_0^\dagger = \gamma_0 \gamma_\mu$ and the imaginary condition to show that $(\gamma_0 \gamma_\mu)^T = \gamma_0 \gamma_\mu$, where T indicates the transposed matrix.

(b) In the Majorana representation the positron scattering operator is

$$\bar{v}(p_{\text{in}}) \gamma_\mu v(p_{\text{out}}) = u(p_{\text{in}})^T \gamma_0 \gamma_\mu u^*(p_{\text{out}}).$$

Using (a) above, show that this can be rewritten as

$$\bar{u}(p_{\text{out}}) \gamma_\mu u(p_{\text{in}}).$$

This establishes the basic symmetry between the particle scattering and antiparticle scattering matrix element.

(c) Show that

$$\bar{v}(p_{e^+}, \lambda') \gamma_\mu u(p_{e^-}, \lambda) = \bar{v}(p_{e^-}, \lambda) \gamma_\mu u(p_{e^+}, \lambda')$$

in the Majorana representation. This relation is important for the charge conjugation symmetry of the theory.

(d) We can change representations from the Majorana representation to another via a unitary transformation: $\gamma_\mu^{\text{new}} = M \gamma_\mu M^\dagger$. In this case, we should use $v = Mu^*$ for positrons to maintain the symmetry. The u spinors in the new representation are transformed into Mu. Show that $v e^{ip \cdot x}$ satisfies the transformed Dirac equation if the initial $u e^{-ip \cdot x}$ does. (*Note:* It can be shown that all representations for the γ_μ can be related by some unitary transformation).

3.5. Consider a beam of high energy muons passing through a material. Because of scattering, low energy electrons can be knocked out of the material (these are called δ-rays). Calculate the cross section for this process per electron for final electron energies

$$\frac{E_\mu}{m_\mu} \gg \frac{E_e}{m_e},$$

but E_e is much greater than the binding energy of the material. Show that for each electron the cross section to knock out an electron of kinetic energy $> E_{\text{min}}$ is approximately

$$\sigma = \frac{250 \text{ millibarns}}{E_{\text{min}}},$$

for E_{min} measured in MeV. Note 1 millibarn = 10^{-27} cm^2.

Chapter 3 Homework

3.6. For $\vec{p}$ along the z-direction show, using the explicit spinor solutions, that:

$$u(\vec{p}, \lambda)\bar{u}(\vec{p}, \lambda) = (\slashed{p} + m)\frac{1 + \gamma_5 \slashed{s}(\lambda)}{2}, \quad v(\vec{p}, \lambda)\bar{v}(\vec{p}, \lambda) = (\slashed{p} - m)\frac{1 + \gamma_5 \slashed{s}(\lambda)}{2}.$$

3.7. Calculate the unpolarized differential cross section, summed over final spins, for $e^+e^- \to e^+e^-$. Assume the particles are relativistic in the center of mass.

3.8. The matrix element for Compton scattering has two terms, that come from adding two Feynman diagrams. Show that both terms are needed to get a gauge invariant result.

3.9. Calculate the differential and total Compton scattering cross section in the nonrelativistic limit. Use a frame where the electron is initially at rest and the γ energy is $\ll m_e$.

3.10. What is the Compton total cross section for a photon of energy 10 GeV incident on an electron at rest? For a material with 10^{24} electrons/cm^3, what is the average distance the photon would travel before one Compton scatter, if we assume that this is the only scattering process that occurs?

3.11. Draw the Feynman diagrams for the process $\gamma + \gamma \to e^+ + e^-$. Write down the expression for M_{fi} in terms of the spinors and polarization vectors for the photons.

3.12. Write down the matrix element for the process $\gamma + e^+ \to \gamma + e^+$.

CHAPTER 4

Hadrons

4.1 ■ THE CHARGE STRUCTURE OF THE STRONG INTERACTIONS

In the introduction, we discussed that hadrons are made of quarks. We will see later that they are made of a quark-antiquark, three quarks, or three anti-quarks. We will, for the time being, borrow this observation. We will use it to see that the charge structure of the strong interactions must differ in a fundamental way from electromagnetism or the neutral scalar theory, with which we did most of our calculations in previous chapters.

To discuss the charge structure, we look first at some consequences of the charge conjugation operation that exchanges particles and antiparticles. For electromagnetism, this operation changes the sign of A_μ. If we take a configuration of charge that is negative, and change it to positive, the corresponding expectation value for A_μ changes sign. Consider now a single photon state $|1\gamma\rangle$; if we perform a charge conjugation operation, what do we have to do to $|1\gamma\rangle$? The answer is $C|1\gamma\rangle = -|1\gamma\rangle$, and C for n photons gives a factor $(-1)^n$, since we can construct the n photon state as a product state from single photon states. The photon is said to have negative charge conjugation. The single photon result follows from the relation $C^\dagger A_\mu C = -A_\mu$, which implies that $\langle 1\gamma, \vec{p}, \lambda | C^\dagger A_\mu C | 0\rangle = -\langle 1\gamma, \vec{p}, \lambda | A_\mu | 0\rangle$. Since $C|0\rangle = |0\rangle$ for the electrically neutral vacuum, we get $C|1\gamma\rangle = -|1\gamma\rangle$.

A consequence of this is that particles repel each other and particle-antiparticle attract, based on the nonrelativistic potential calculated from the single photon exchange amplitude. This follows because the amplitudes

$$e^- \to e^- + 1\gamma \quad \text{and} \quad e^+ \to e^+ + 1\gamma$$

have an opposite sign, given the negative photon charge conjugation value.

The neutral scalar field we looked at in Chapter 2 is an eigenstate of C as well, but now with positive charge conjugation. The Yukawa potential resulting from such a field was discussed in Section 2.13. The calculation done there can be extended to particle-antiparticle interactions. The interaction resulting from such a field (like gravity as well, which is spin 2) results in particles and antiparticles that all attract each other.

The question we ask is: "Could the strong interactions be due to a field analogous to our neutral photon or scalar field?" The pattern of bound states tells us that the answer is no. The strong interactions cannot be due to the exchange of a single particle of given positive or negative charge conjugation. The reason is that the pattern of states: $q\bar{q}$ bound, qqq bound, qq and $qqqq$ not bound (q is a quark of some flavor and $\bar{q}$ the antiquark) doesn't fit with a fixed charge conjugation exchange. For a photon we would expect $q\bar{q}$ bound, qq and qqq all repulsive and unbound. For the neutral scalar particle $q\bar{q}$, qq, and qqq, and in fact many quarks, would all be bound, with qq and $q\bar{q}$ having similar binding.

We thus conclude that the strong interaction is not due to the exchange of a single particle that is its own antiparticle. This interaction requires generalizing the electromagnetic interaction to the case where several types of particles can be exchanged, each type carrying a "charge." The interaction will have a more complicated symmetry than just the charge conjugation symmetry of a single neutral particle exchange. To introduce such a generalization of the charge and related issues, we first look at symmetry and conservation issues more generally.

4.2 ■ QUANTUM NUMBERS

The rules that limit the initial and final states that can transform from one to another are called the quantum conservation rules, expressed in terms of quantum numbers of the given states. Thus, for example, the Feynman diagrams we drew previously all conserved electric charge. The simplest state is one particle; the quantum numbers of this state that don't change under Lorentz transformations are called the quantum numbers of the particle. The quantum numbers for a multiparticle state are given by certain combination rules from those of the individual particles.

Thus, for example, we can define three types of lepton numbers, L_e, L_μ, and L_τ, where L_e counts the number of electrons and electron neutrinos, with analogous quantities for the μ and τ leptons. These are called additive quantum numbers, that is,

$$L_e \text{ (multiparticle state)} = \sum_{\text{particle } i} L_e^i.$$

(Note that antiparticles count as -1 here). The Feynman diagrams we have drawn conserve each lepton number individually.

Which of the quantum numbers are conserved is of course an experimental question. Some conservation rules are based on strongly held symmetry principles and others, like the conservation of the three individual lepton numbers, could fail to be true without changing our conceptual framework greatly. In fact, as mentioned in Chapter 1, the analogous conservation rule does not hold for quarks, and the most recent studies of neutrinos indicate that the conservation of individual lepton numbers is violated as well. Not all the interactions have the same invariance properties and consequently do not follow the same conservation rules. We can define quantum numbers that are partially conserved—that is, conserved by a dominant interaction and violated only by a weaker interaction.

The conservation rules are of three types: additive, like the lepton numbers, vectorially additive, and multiplicative. They can stem from space-time symmetries or internal symmetries. We illustrate the three types using space-time symmetries that should be familiar from nonrelativistic quantum mechanics. We discussed earlier the existence of conserved currents as a consequence of continuous symmetry transformations of the Lagrangian. In this chapter we look at conserved quantities from the point of view of symmetries of the Hamiltonian.

4.2.1 ■ Additive Quantum Numbers

Examples of additive quantum numbers are energy and momentum. These arise as a consequence of time and space translational invariance. The symmetry transformations are continuous (that is, the translation is a function of a real number) and yield additive quantum numbers. As an example, let us look at a spatial translation of all states by a distance x_0. Associated with the transformation is an operator $\tilde{p}_x$, called the generator of the transformation. The translation is accomplished by applying the operator $T_{x_0} = e^{-i\tilde{p}_x x_0}$ to all states. For x_0 infinitesimal, the translation operator is $T_{x_0} = 1 - i\tilde{p}_x x_0$ to the first order. We can imagine building up a full translation by repeated infinitesimal transformations. The operator $\tilde{p}_x$ is Hermitian, which implies that T_{x_0} is unitary. Thus $T_{x_0}^\dagger = T_{x_0}^{-1}$. A unitary operator preserves the normalization of the states. The statement of the symmetry is $[\tilde{p}_x, H] = 0$, where H is the Hamiltonian for the system. An eigenstate of momentum satisfies $T_{x_0}|\psi\rangle = e^{-i p_x x_0}|\psi\rangle$, where p_x is the eigenvalue.

Consider two widely separated particles, each of them momentum eigenstates. We assume that initially the complete system is specified by a product eigenstate $|\psi\rangle = |\psi_1\rangle|\psi_2\rangle$. In terms of the individual momentum operators, $T_{x_0} = e^{-i(\tilde{p}_{x_1}+\tilde{p}_{x_2})x_0}$ and $T_{x_0}|\psi\rangle = e^{-i(p_{x_1}+p_{x_2})x_0}|\psi\rangle = e^{-i p_{\text{tot}} x_0}|\psi\rangle$. The momentum for the system is $p_{\text{tot}} = p_{x_1} + p_{x_2}$, which is an additive quantum number.

We can now prove that momentum is conserved in a collision of the particles. We describe the final state as a product of momentum eigenstates, corresponding to noninteracting particles whose momenta we measure far away from the collision region. Consider $H_{fi} = \langle\psi_{\text{final}}|H|\psi_{\text{initial}}\rangle$. We shall prove that $H_{fi} = 0$, unless $p_{\text{final}} = p_{\text{initial}}$, where these are the x components of momentum. The same would work for the other components. Note that this relation is not true for the local density $\mathcal{H}_{fi}$, as seen in Chapter 2, but is true only for the full Hamiltonian.

Since $[\tilde{p}_x, H] = 0$, $[T_{x_0}, H] = 0$, which gives $H = T_{x_0}^{-1} H T_{x_0} = T_{x_0}^\dagger H T_{x_0}$. Thus $H_{fi} = \langle\psi_f|H|\psi_i\rangle = \langle\psi_f|T_{x_0}^\dagger H T_{x_0}|\psi_i\rangle = e^{i(p_{\text{final}}-p_{\text{initial}})x_0} H_{fi}$ for all x_0. Thus either $p_{\text{final}} = p_{\text{initial}}$ or $H_{fi} = 0$.

The key ideas leading to the conserved momentum are:

a. The existence of a symmetry operation specified by a unitary operator that commutes with H.

b. Infinitesimal transformations are specified in terms of a Hermitian operator. This is called the generator of the transformations.

c. Using eigenstates of the Hermitian operator, additivity of phases corresponds to the additivity of the quantum numbers given by the operator eigenvalues. Note that a key element is that the generator for a composite system is a linear combination of individual generators that mutually commute.

d. The total quantum number is then conserved by the time evolution, which is determined by H.

Note that $\tilde{p}_x, \tilde{p}_y, \tilde{p}_z$ mutually commute, so the above apply to the total $\vec{p}$. Energy conservation arises similarly by considering translations in time. Finally, to take an example of an internal symmetry, the charge for electrodynamics is an additive quantum number.

4.2.2 ∎ Vectorially Additive Quantum Numbers

In quantum mechanics we also have more complicated symmetries such as angular momentum, with associated operators J_x, J_y, J_z that are the generators for rotations about the x, y, z axes, respectively. The generators each commute with H, $[J_i, H] = 0$. This is a consequence of rotational invariance; however, $[J_i, J_j]$ do not all equal 0. For angular momentum, the generators satisfy $[J_k, J_n] = i\varepsilon_{knm} J_m$. The constants ε_{knm} are called the structure constants for the transformation group. In general, we can only find simultaneous eigenstates for operators that commute. Thus, states are not simultaneously eigenstates of J_x, J_y, J_z. What we can choose are eigenstates of J_z and $J^2 = J_x^2 + J_y^2 + J_z^2$. The operators like J^2, constructed from the generators and commuting with all of them, are called Casimir operators. For rotations there is only one Casimir operator.

For a pair of particles with spin and no orbital angular momentum: $\vec{J} = \vec{J}_1 + \vec{J}_2$, and $J^2 = J_1^2 + J_2^2 + 2\vec{J}_1 \cdot \vec{J}_2 \neq J_1^2 + J_2^2$. The Casimir operators are not linearly additive. Thus, the J^2 quantum number for a pair of particles is not just given by the individual quantum numbers, but rather by the rule for adding angular momentum: J_{tot} can be $J_1 - J_2, J_1 - J_2 + 1, \ldots, J_1 + J_2$ for $J_1 > J_2$.

When two particles with spin and no orbital angular momentum interact, the final state can be found in one of a number of total angular momentum final states. Each distinct value for J_{tot} for the initial two-body system of specified J_1 and J_2 is called a channel, in which interactions can occur. In general the strength of interaction can be different in each channel. The symmetry condition $[J^2, H] = 0$ implies that only the allowed values for J_{tot} can occur for the final states produced in the interaction.

We call the conservation rule for the Casimir operator vectorially additive, since the rule for the addition of angular momentum is called vector addition in nonrelativistic quantum mechanics. Note that J_z is, however, an additive quantum number.

The allowable states of angular momentum can all be obtained by considering composite systems of spin $\frac{1}{2}$ particles. The two basic states (up and down) of spin $\frac{1}{2}$ transform among each other under rotations of the system. They are the simplest example of a multiplet, the other important consequence of symmetries besides the conservation idea. We will define a multiplet shortly, when we discuss internal symmetries.

4.2.3 ■ Multiplicative Quantum Numbers

The transformations discussed so far are characterized by unitary operations that form a group. This means that successive transformations are also operations of the same type, and inverse operations exist. In general, successive operations might not commute. An extremely simple group is one with only two elements, an example of a discrete or finite group. In this case the elements are I, P, with $P^2 = I$, where I = identity and P is the interesting nontrivial group operation. An example of such a group element is the charge conjugation operation. Eigenstates of this operation (for example, the photon, but not e^- or e^+) must satisfy:

$$P \left|\psi\right\rangle = \pm \left|\psi\right\rangle, \quad \text{that is,} \quad e^{i\pi} \text{ or } e^{2i\pi} \left|\psi\right\rangle.$$

Rather than specify the phase, which is additive for composite systems, it is easier just to keep track of the sign that is multiplicative. Thus this type of quantum number is called multiplicative.

The most familiar example of a multiplicative quantum number is parity. The corresponding operator changes momenta and positions to $\vec{p} \to -\vec{p}, \vec{x} \to -\vec{x}$. Thus $\vec{J} = \vec{x} \times \vec{p} \to \vec{J}$. Momentum states (except for particles at rest) are not eigenstates of parity since $[P, \vec{p}] \neq 0$, but bound states at rest typically are, since these are usually specified as eigenstates of H and J^2, J_z.

One surprising discovery is that parity is not a good symmetry of all the interactions. That is, $[P, H] \neq 0$ for some interactions; in particular the weak interactions do not conserve parity. The most striking phenomenon of parity nonconservation occurs in the weak interactions of the neutrino. In the limit of a massless neutrino, its spin vector always points opposite to its momentum vector, if produced in a weak decay. The neutrino is produced in a left-handed helicity state. If we make a parity transformation on this state $\vec{p} \to -\vec{p}$, $\vec{J} \to \vec{J} \Rightarrow$ helicity changes sign—that is, we get a right-handed neutrino. If parity is conserved, for every process (for instance, unpolarized neutron decay) producing left-handed neutrinos, we should have an analogous process of equal rate, producing right-handed neutrinos. In fact, weak interactions never produce the right-handed neutrino state.

It is useful to know how fields transform under parity. Since the single particle states are obtained by applying the field operator to the vacuum, this tells us the parity transformation properties of the state. For integral spin, if the transformation of the field is the same as that of a spatial tensor of the same rank (for example, scalar or vector), the field is said to have natural spin-parity. If the field transforms with an extra minus sign (for example, a pseudoscalar or pseudovector), the field (and state) is said to have unnatural spin-parity. The photon has natural spin parity since the polarization vectors are ordinary 4-vectors. For spin $\frac{1}{2}$ particles there is no analogy to space tensors. We can, however, figure out how parity works by looking at the current, which is a 4-vector:

$$J_\mu = \bar\psi \gamma_\mu \psi.$$

We assume that under parity,

$$P(\psi) \to \text{Matrix} \cdot \psi = M\psi$$

Thus we need M to satisfy

$$\psi^\dagger M^\dagger \gamma_0 \gamma_\mu M \psi = \begin{cases} \bar\psi \gamma_0 \psi & \text{for } \mu = 0 \\ -\bar\psi \vec\gamma \psi & \text{for } \mu = 1, 2, 3. \end{cases}$$

Taking $M = \gamma_0$ works, since for $\mu = 0$ the product $\gamma_0 \gamma_0 \gamma_0 = \gamma_0$, while for $\mu \neq 0$, $\gamma_0 \vec\gamma \gamma_0 = -\vec\gamma$ by the anti-commutation property of the γ matrices. It is useful to check the other bilinear-covariants of Table 2.1 for their transformation properties under parity.

One surprise, however, comes out of the above spin $\frac{1}{2}$ transformation. Consider making a parity transformation on a state at rest. We expect this to be a parity eigenstate with the spinor corresponding to the state obtained from the transformed field via $\gamma_0 \psi$ operating on the vacuum. For particles, applying γ_0 to the spinor, we get

$$\gamma_0 u = u, \quad \text{for } \vec p = 0.$$

For antiparticles:

$$\gamma_0 v = -v, \quad \text{for } \vec p = 0.$$

Thus we must assign opposite intrinsic parity to the particle and antiparticle states for spin $\frac{1}{2}$. This has measurable consequences, for example, as seen for the positronium states of $e^- e^+$.

4.3 ■ INTERNAL SYMMETRIES

Consider a set of particle types that interact in a "similar" manner, which generates a symmetry, and which we define as follows. Suppose that exchanging the labels for these particles:

$$\text{type } i \underset{\text{exchange}}{\longleftrightarrow} \text{type } j$$

leaves the Lagrangian $\mathcal{L}$ unchanged; then we have an internal symmetry. Rather than just an exchange, we can look at a more general transformation that mixes the particle states together, preserving the normalization. Such a general complex rotation is a unitary transformation. For n particle types, the group of transformations is called $SU(n)$, and $\mathcal{L}$ is said to have an $SU(n)$ symmetry. Such a symmetry should be familiar for rotations for a spin $\frac{1}{2}$ particle in its rest frame, where the rotational symmetry translates into an $SU(2)$ symmetry. The most important symmetry groups in particle physics are $SU(2)$ and $SU(3)$, which occur for the three colored quarks.

The unitary transformations can be written in terms of Hermitian operators called generators, which generate infinitesimal transformations in the space of states. The transformations typically involve a finite number of base states, unlike the momentum operator, for which an infinite number of momentum states are possible. Thus the generators can be represented by finite dimensional matrices. The symmetry implies that we can associate quantum numbers to the particle types based on their eigenvalues for the set of mutually commuting operators. The quantum number conservation rules for composite systems will be additive for the linearly additive diagonal operators that mutually commute and vectorially additive for the Casimir operators (with the rule for addition depending on the group and operator).

We write a symmetry transformation (taking three quark types as our example below):

$$|q_i'\rangle = \sum_{j=1}^{3} U_{ij} |q_j\rangle, \qquad (4.1)$$

with U_{ij} a general unitary matrix with unit determinant. Here $|q_j\rangle$ are the particle states or types. The corresponding field operators will now have an extra particle type label and we can write the field equations for all the related particles in one matrix equation. For example, for the free field equation for the spin $\frac{1}{2}$ particles:

$$i\gamma_\mu \frac{\partial \psi}{\partial x_\mu} - m\psi = 0 \quad \text{becomes} \quad i\gamma_\mu \frac{\partial \psi_\alpha}{\partial x_\mu} - m\psi_\alpha = 0,$$

where $\alpha = 1, 2,$ or 3 for three objects. The mass must be independent of α for a symmetry. To maintain the symmetry, the anticommutation relations among the creation and annihilation operators must be extended to apply to all the particle types.

The single particle and antiparticle states can be obtained by using the creation operators $|q_\alpha\rangle = b_\alpha^\dagger |0\rangle$ and $|\bar{q}_\alpha\rangle = d_\alpha^\dagger |0\rangle$, where the antiparticle state is indicated by $|\bar{q}_\alpha\rangle$. However, since b_α and $d_\alpha^\dagger$ appear together in the field ψ_α, they must transform the same way, implying that the particle and antiparticle states do not. The antiparticle states must transform via complex conjugate matrices:

$$|\bar{q}_i'\rangle = \sum_{j=1}^{3} U_{ij}^* |\bar{q}_j\rangle. \qquad (4.2)$$

4.3 Internal Symmetries

In terms of generators $\vec{T}$, a unitary transformation can be written as

$$U = e^{i\vec{\theta}\cdot\vec{T}}.$$

$\vec{T}$ are the analogues of $\vec{J}$ and are just $\vec{J}$ for $SU(2)$. Choosing the $\vec{T}$ to be Hermitian, the $\vec{\theta}$ are real. For $n \times n$ matrices there are exactly $n^2 - 1$ Hermitian traceless matrices, $\vec{T}$, which, plus the identity, make up a complete set for expanding general $n \times n$ Hermitian matrices. The relation, determinant $U = 1$, which distinguishes "special" unitary from generic unitary matrices, requires $\text{Tr}(\vec{\theta}\cdot\vec{T}) = 0$ for a finite dimensional matrix. With $\vec{\theta}$ arbitrary, $\text{Tr}(T_i) = 0$.

The diagonal operators among the T_i correspond to additive quantum numbers and are real. Since the antiparticle states transform via matrices:

$$(e^{i\vec{\theta}\cdot\vec{T}})^* = e^{-i\vec{\theta}\cdot\vec{T}^*} = e^{i\vec{\theta}\cdot(-\vec{T}^*)}$$

(note that the matrices $-\vec{T}^*$ obey the same commutation relations as the $\vec{T}$ and are the generators for the antiparticles), each diagonal operator for the antiparticles = $-$ particle operator. Thus, taking a corresponding element of the matrix, particles and antiparticles have opposite sign eigenvalues. So the opposite charge of spin $\frac{1}{2}$ particles and antiparticles generalizes to any other linearly additive internal quantum number.

We next define the idea of a multiplet. A multiplet is a set of orthonormal states, which are eigenstates of the mutually commuting generators, and

1. share the same eigenvalues for the Casimir operators of the group,
2. are generally specified (that is, differentiated) by the eigenvalues they have for the set of generators that can be simultaneously diagonalized. Thus they typically have different additive quantum numbers.
3. provide a basis for expanding states that are related by the unitary transformations. If there are m states in the multiplet, specified by $|i\rangle$ with $i = 1, \ldots, m$, then any state of the form $\sum_{i=1}^{m} \alpha_i |i\rangle$ will transform into another state of this form under the unitary rotations.

If under a rotation, the quarks transform as

$$|q_i'\rangle = \sum_{j=1}^{3} U_{ij} |q_j\rangle,$$

then the states of a multiplet with m states will transform via

$$|i'\rangle = \sum_{j=1}^{m} \tilde{U}_{ij} |j\rangle.$$

$\tilde{U}_{ij}$ is again a unitary matrix, but now $m \times m$ instead of 3×3.

As an example, for mesons, which are quark-antiquark bound states, a state in a multiplet will have the form $|k\rangle = \sum_{i,j} \alpha_{ij}^k |q_i\rangle |\bar{q}_j\rangle$. $\tilde{U}_{ij}$ can be calculated in terms of U_{ij} and α_{ij}^k, if known. Actually, we often want to figure out α_{ij}^k. Note that in the above product state notation, we have suppressed the dependence of the states on position (or momentum) and spin and focused only on the behavior in terms of the internal quantum numbers. All states $|q_i\rangle|\bar{q}_j\rangle$ in such a product (for a perfect symmetry) have identical space-time properties; only the indices i and j change.

Why is the multiplet important? The reason is that the physical properties of all particles in a multiplet are identical for an exact symmetry! Thus particles come in well-defined families, with the family pattern indicative of the symmetry group. For example, the doublets of the weak interactions indicate an underlying $SU(2)$ symmetry. The colored quarks, however, are found only in colorless multi-quark states, requiring other evidence to deduce the underlying symmetry group.

We now look at the important result stated above for a multiplet. Let O be some operator describing a physical property (spin, energy, etc.). Let U be an arbitrary unitary operator of the symmetry group. The symmetry means that $O = U^\dagger O U$. Let's take the expectation value of O for a state $|n\rangle$ belonging to a multiplet:

$$\langle n| O |n\rangle = \langle n| U^\dagger O U |n\rangle = \langle Un| O |Un\rangle.$$

The state $|Un\rangle$ is in general a linear combination of states within the multiplet, and has the same expectation value as $|n\rangle$. This result is quite general and it can be shown that all states of the form $\sum_{i=1}^m \alpha_i |i\rangle$ share the same expectation value for O, and in particular each $|i\rangle$ does. Taking, for example, $O = H$, each particle of the multiplet has the same mass.

An interesting case of a composite system is one made of particles contained in the same multiplet, for example, multi-quark states (with no antiquarks). We can write this state as

$$|k\rangle = \sum \alpha_{ij...m}^k |q_i\rangle |q_j\rangle \cdots |q_m\rangle.$$

Suppose a given state has a permutation symmetry:

$$\alpha_{i,...,j,...n,...m}^k = +/- \alpha_{i,...,n,...j,...m}^k.$$

Because all of the $|q_i\rangle$ rotate with the same U_{ij}, this symmetry is preserved under the rotations. Stated more formally, the generators $\vec{T} = \vec{T}_1 + \cdots + \vec{T}_m$ commute with the permutation operator for any pair. Thus, all the states in a multiplet for such a composite system share the same permutation symmetry among the constituent states. The multiplets are in fact characterized by their permutation symmetries.

We have not yet discussed the constraints stemming from particles that are fermions or bosons. These constraints occur when two particles in the composite state are identical and the other attributes of the state, such as the spatial and spin state, have a well-defined symmetry. After imposing the fermion or boson

4.3 Internal Symmetries

constraints, the multiplets allowed will have a given permutation symmetry for the identical particles. However, since this permutation symmetry is shared by all the particles (not just the identical particles, that is, it is a characteristic of the entire multiplet) it is formally equivalent to extending the exchange characteristics to the full set of constituent states.

To see how this works we will take a concrete example—a system made of one neutron and one proton. Because of the way the nuclear forces work, the exchange $p \leftrightarrow n$ generates a rather good $SU(2)$ symmetry called isospin. The proton and neutron form a doublet in the abstract isospin space, where the additive quantum number I_3 is $+\frac{1}{2}$ for the proton and $-\frac{1}{2}$ for the neutron. For a composite system of p and n we have two possible multiplets. We can figure them out quickly using the rule that states in a multiplet share the same permutation symmetries. Thus the states are:

$$|p\rangle|p\rangle$$
$$\frac{|p\rangle|n\rangle + |n\rangle|p\rangle}{\sqrt{2}} \quad \text{Symmetric, isospin 1 state}$$
$$|n\rangle|n\rangle$$

and

$$\frac{|p\rangle|n\rangle - |n\rangle|p\rangle}{\sqrt{2}} \quad \text{Anti-symmetric, isospin 0 state.}$$

This notation is a shorthand for, taking isospin 0 as the example, the state

$$\frac{\psi_p(x_1, \lambda_1)\psi_n(x_2, \lambda_2) - \psi_n(x_1, \lambda_1)\psi_p(x_2, \lambda_2)}{\sqrt{2}}$$

or a sum over such states with different particle spins if we want a state of given total spin. We note several features of these multiplets that are also true for more complicated multiplets:

1. The additive quantum numbers for the composite states can be directly calculated from those of the constituents.
2. When only one state with given additive quantum numbers (for example, $|p\rangle|p\rangle$) appears among the composite systems, that state lies in one multiplet.
3. When several states appear among the composite systems with the same additive quantum numbers, linear combinations of these are the distinct multiplet states.

Let us consider the nucleon-nucleon bound state of lowest energy, where we expect the two particles to be in a symmetric state of zero spatial angular momentum. Including spins, the possible states can have $J = 0$ (anti-symmetric

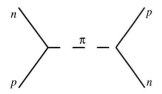

FIGURE 4.1 Pion exchange diagrams for proton and neutron states.

spin state) or $J = 1$ (symmetric spin state). To arrive at an overall state that is anti-symmetric under exchange, $J = 0$ goes with isospin 1 and $J = 1$ goes with isospin 0. Only the $J = 1$ state is bound, and therefore we find only one deuteron made of $|p\rangle$ and $|n\rangle$. Had the $J = 0$ state also been bound, we would have had three bound states of this type, since there are three states with isospin 1. The symmetry tells us that those three would be degenerate, although different in mass from the deuteron. Since interactions exist that allow transitions between p and n, for example via exchange of π^+ and π^- as shown in Figure 4.1, we should not be surprised that the multiplets have different energies.

All combinations of spin allowed by the usual rules for adding angular momentum appear for $|n\rangle$ and $|p\rangle$, as expected for nonidentical particles; they are just partitioned among the isospin states. For $|n\rangle|n\rangle$ and $|p\rangle|p\rangle$, only the spin states allowed for identical fermions occur.

4.4 ■ GENERATORS FOR $SU(2)$ AND $SU(3)$

Because of the importance of $SU(2)$ and $SU(3)$, we list below the generators that provide infinitesimal transformations for the constituents. For $SU(2)$, the 2×2 matrices are

$$T_i = \frac{\sigma_i}{2},$$

which are familiar from the case of spin $\frac{1}{2}$. These satisfy the commutation relations:

$$\left[\frac{\sigma_i}{2}, \frac{\sigma_j}{2}\right] = i\varepsilon_{ijk}\frac{\sigma_k}{2}. \qquad (4.3)$$

For a multiplet of n identical particles, the generators are:

$$T_i^{\text{multiplet}} = T_i^{(1)} + T_i^{(2)} + \cdots + T_i^{(n)}.$$

Taking matrix elements between the states of the multiplet (assumed to contain m states), $\langle k| T_i^{\text{multiplet}} |l\rangle$ gives us $m \times m$ matrices, also called generators, which satisfy the same commutation relations as the initial 2×2 matrix generators. These are called $m \times m$ irreducible matrix representations of the group. This allows us to start with the generators for the simplest multiplet and derive higher dimensional matrix representations, if the composite states forming the multiplet are known.

For $SU(3)$, the T_i for the quark states are defined to be $\lambda_i/2$ by analogy with $SU(2)$. There are eight generators in this case. A standard form for the λ_i, introduced by Gell-Mann, is:

$$\lambda_1 = \begin{pmatrix} 0 & 1 & 0 \\ 1 & 0 & 0 \\ 0 & 0 & 0 \end{pmatrix}, \quad \lambda_2 = \begin{pmatrix} 0 & -i & 0 \\ i & 0 & 0 \\ 0 & 0 & 0 \end{pmatrix}, \quad \lambda_3 = \begin{pmatrix} 1 & 0 & 0 \\ 0 & -1 & 0 \\ 0 & 0 & 0 \end{pmatrix}$$

4.4 Generators for $SU(2)$ and $SU(3)$

$$\lambda_4 = \begin{pmatrix} 0 & 0 & 1 \\ 0 & 0 & 0 \\ 1 & 0 & 0 \end{pmatrix}, \quad \lambda_5 = \begin{pmatrix} 0 & 0 & -i \\ 0 & 0 & 0 \\ i & 0 & 0 \end{pmatrix}, \quad \lambda_6 = \begin{pmatrix} 0 & 0 & 0 \\ 0 & 0 & 1 \\ 0 & 1 & 0 \end{pmatrix}$$

$$\lambda_7 = \begin{pmatrix} 0 & 0 & 0 \\ 0 & 0 & -i \\ 0 & i & 0 \end{pmatrix}, \quad \lambda_8 = \frac{1}{\sqrt{3}}\begin{pmatrix} 1 & 0 & 0 \\ 0 & 1 & 0 \\ 0 & 0 & -2 \end{pmatrix}.$$

The T_i satisfy commutation relations

$$[T_i, T_j] = if_{ijk}T_k, \tag{4.4}$$

where the f_{ijk} can be calculated explicitly using the generators. Since the T_i are Hermitian, the f_{ijk} are real. They are also fully antisymmetric.

You can check that $\text{Tr}[\lambda_i] = 0$, and that they are linearly independent. The normalization is chosen so that $\text{Tr}(\lambda_i^2) = 2$. Thus they contribute "equally" to the Casimir operator

$$\sum_{i=1}^{8} T_i^2 = \frac{1}{4}\sum_{i=1}^{8} \lambda_i^2.$$

Note that only two of these generators can be simultaneously diagonalized. The number of mutually commuting generators is called the rank of the group. In general, $SU(n)$ has $n^2 - 1$ generators and has rank $n - 1$. It can be shown that the number of Casimir operators equals the rank of the group, so there are two for $SU(3)$, although we will only use the operator given by the sum of the squares of the generators (quadratic Casimir operator) in subsequent discussions. Thus a multiplet is specified by two numbers, for example, the values of the two Casimir operators, analogous to specifying J for an $SU(2)$ multiplet. We will be working only with the few simplest multiplets that typically occur.

The behavior of the λ matrices can be understood simply as follows (they implement the particle exchange idea from which we began the internal symmetry discussion):

λ_1 exchanges quarks 1 and 2,

λ_4 exchanges quarks 1 and 3,

λ_6 exchanges quarks 2 and 3.

λ_2, λ_5, and λ_7 do the same but with a complex factor; the remaining matrices are diagonal. This pattern works in general for the case of $SU(n)$. For example, the n^2-1 generators for $SU(n)$ can be divided into $n(n-1)$ pairwise exchange operators as above, leaving $n-1$ diagonal operators. It is then also clear that the $\text{Tr}(T_i^2)$ for the exchange operators is independent of n, allowing the same normalization condition for any n by suitable normalization of the diagonal matrices.

The particle exchange property for the generators can be used to calculate additional states in a multiplet, if one state is known. Consider again the two-particle states formed from $|p\rangle$ and $|n\rangle$. There must be four such states. Consider the state $|p\rangle|p\rangle$. Only one state has the additive quantum number of this state, so it lies in a distinct multiplet. If we apply the generator $T_1^{(1)} + T_1^{(2)}$ to this state, each single particle operator will exchange $|p\rangle$ and $|n\rangle$ in one location; the net result will, however, be a state in the same multiplet. Applying it once to $|p\rangle|p\rangle$ generates the state (after normalization):

$$\frac{|p\rangle|n\rangle + |n\rangle|p\rangle}{\sqrt{2}}.$$

Applying the same generator again, we get $|n\rangle|n\rangle + |p\rangle|p\rangle$, to within a constant, allowing us to isolate the new state $|n\rangle|n\rangle$. Applying it again gives us back the previous state, so we are finished. The fourth state lies in a different multiplet and can't be reached by application of this operator; it can be found by looking for an orthogonal state to the others.

The application of a pairwise exchange operator to a state in a multiplet has the disadvantage that it needn't generate a completely new state, but only a linear combination of multiplet states. This occurs because the nondiagonal generators do not, in general, take eigenstates of the diagonal generators into other eigenstates. We can be more systematic and define raising and lowering operators, for example, $T^+ = \frac{1}{2}(\sigma_1 + i\sigma_2)$ and $T^- = \frac{1}{2}(\sigma_1 - i\sigma_2)$ for $SU(2)$, which shift between states in a multiplet with unique eigenvalues. The analogous combinations will work for $SU(n)$, where the $n(n-1)$ pairwise exchange operators can be combined into $n(n-1)$ shift operators. For $SU(3)$ the operators, and the shifts they create among the three colored quarks, are

$\frac{1}{2}(\lambda_1 \pm i\lambda_2)$ shift between $|q_1\rangle$ and $|q_2\rangle$

$\frac{1}{2}(\lambda_4 \pm i\lambda_5)$ shift between $|q_1\rangle$ and $|q_3\rangle$

$\frac{1}{2}(\lambda_6 \pm i\lambda_7)$ shift between $|q_2\rangle$ and $|q_3\rangle$.

Each operator can create one unique nonvanishing transition among the quarks. For example, the only nonvanishing transition that $\frac{1}{2}(\lambda_1 + i\lambda_2)$ can create is $|q_2\rangle \to |q_1\rangle$, while $\frac{1}{2}(\lambda_1 - i\lambda_2)$ will shift $|q_1\rangle \to |q_2\rangle$. For a general multiplet, the shift properties of these operators can be calculated from the commutation relations of the shift operators and the generators. Like the generators, the shift operators commute with the Casimir operators and the Hamiltonian and link states of the same mass within a multiplet.

For an $SU(n)$ group, we can consider a subgroup that involves rotations among a subset of the particles. The first three λ_i matrices operate in an $SU(2)$ subgroup of $SU(3)$. If we specify the three quark states as

$$\begin{pmatrix}1\\0\\0\end{pmatrix}, \begin{pmatrix}0\\1\\0\end{pmatrix}, \begin{pmatrix}0\\0\\1\end{pmatrix},$$

4.4 Generators for $SU(2)$ and $SU(3)$

then T_1, T_2, T_3 generate rotations among the first two states only. The various shift operators move us between states of each of the $SU(2)$ subgroups of $SU(3)$.

Since there are two diagonal operators for $SU(3)$, we can characterize a given state within a multiplet by its eigenvalue for each of the two operators. Thus each state can be placed in a 2-dimensional diagram whose axes are the eigenvalues of the operators. Making such a diagram for a complete multiplet will be our way of displaying the multiplet, rather than specifying the eigenvalues of the two Casimir operators. Thus, for example, for the three quark and antiquark states, we have the diagrams shown in Figure 4.2. These two multiplets are denoted 3 for the quarks and $\bar{3}$ for the antiquarks. Here the notation keeps track of the total number of states in the multiplet.

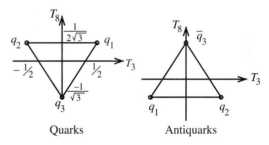

FIGURE 4.2 Quark and antiquark multiplets, 3 and $\bar{3}$, respectively.

We next look at the simplest multiplets for composite states. States composed of a colored quark and a colored antiquark can come in varieties that fall into two multiplets, one with one state (singlet) and one with eight states (octet). We write this in group theory notation for combining multiplets into composite states:

$$3 \otimes \bar{3} = 1 \oplus 8. \tag{4.5}$$

This is the $SU(3)$ vector addition, analogous to the result for composite states of particles of given angular momentum. The T_3 and T_8 quantum numbers of the nine states of the form $|q_i, \bar{q}_j\rangle$ can be calculated simply, since these are additive quantum numbers. Thus, we can easily figure out how the plot in the T_3, T_8 space looks for the multiplets all overlaid on top of each other. To then separate the states into individual multiplets requires finding those with different values for the Casimir operators.

For a state made of two quarks the multiplets are:

$$3 \otimes 3 = \bar{3} \oplus 6. \tag{4.6}$$

Since these multiplets are made of identical objects, they have a given permutation symmetry. The $\bar{3}$ is antisymmetric, the 6 symmetric. Note that the three antisymmetric states look like the antiquark multiplet when plotted in the T_3, T_8 space.

A state made of three quarks can lie in the following multiplets:

$$3 \otimes 3 \otimes 3 = 3 \otimes [\bar{3} \oplus 6] = 1 \oplus 8_a \oplus 8_s \oplus 10.$$

The 1 is fully antisymmetric, the 10 is fully symmetric, and the two 8's are symmetric or antisymmetric for some of the exchanges but not all—this is called a state of mixed symmetry. Note that both octets have the same T_3, T_8 plot and the same values for the Casimir operators, but differ by a different property, in this case the internal permutation symmetry. Under the $SU(3)$ rotations, the states in each 8 transform among themselves only.

4.5 ■ THE COLOR INTERACTION

We have already said that the hadron pattern tells us that the strong force is not due to the exchange of only one boson of fixed charge conjugation. We now look at what we would expect for the exchange of a multiplet of bosons. What we will define is an interaction with a symmetry that is called color. The symmetry is non-Abelian, defined as having generators of the symmetry transformations that do not commute. We will look at the interactions of the quarks here and the full theory in Chapter 7. The theory is called quantum chromodynamics, or QCD. To maintain the symmetry, all the particles involved must fall into well-defined multiplets with the interaction term transforming as a singlet (that is, as an invariant under the color transformations).

For an $SU(3)$ symmetric triplet of quarks, the next simplest choice of an interaction after a neutral (that is, singlet) exchange, is the exchange of eight colored gluons. This is shown in Figure 4.3. We label the gluons by an index α. The amplitude for a quark to emit such a gluon is a vertex factor depending on spins and momenta, as in electrodynamics, times a factor depending on the colors involved. This factor for a quark of color type $m \to$ type n and a gluon of type α is given by

$$g \langle q_n | T_\alpha | q_m \rangle,$$

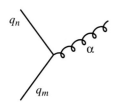

FIGURE 4.3 Quark of color type m makes a transition to gluon of type α and a quark of type n.

where g is an overall coupling constant. An interaction Lagrangian leading to a space-time dependence like electrodynamics and the correct color dependence is

$$-g A_\mu^\alpha \bar{\psi}_n \gamma_\mu T_\alpha \psi_m, \qquad (4.7)$$

where n, m can be chosen from the three colors, and α comes in eight choices. This interaction, if we add all the terms together into an interaction Lagrangian, is invariant under global $SU(3)$ transformations, provided the A_μ^α transform as a color octet. We will look at the transformation properties in Chapter 7, where we extend the symmetry to be a local gauge symmetry. Note that the number of gluons is determined by the number of generators. The Lorentz structure of each A_μ^α is the same as for electrodynamics.

4.5 The Color Interaction

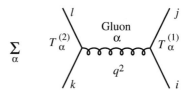

FIGURE 4.4 Quark-quark scattering via gluon exchange.

The second-order Feynman diagram for quark-quark scattering is shown in Figure 4.4. The eight exchanged gluons are summed over (interfere). The amplitude, by analogy with *QED*, is

$$g^2 \sum_{\alpha=1}^{8} \langle l| T_\alpha^{(2)} |k\rangle \langle j| T_\alpha^{(1)} |i\rangle \frac{\bar{u}_l \gamma_\mu u_k \bar{u}_j \gamma_\mu u_i}{q^2}.$$

The space-time part corresponds to Coulombic and spin-dependent forces just as in *QED*.

We expect that this interaction might be able to generate a bound state of a number of quarks and/or antiquarks, and what we want to look at first is the dependence of the interaction energy on the color multiplets involved. We will ignore the spin dependence of the energy for the time being, as would be a good approximation for a nonrelativistic system, and assume a pairwise exchange interaction. Since some of the quark types are quite heavy we expect that a nonrelativistic description should work for these. We focus on the ground state, for which we assume each constituent particle has the same s-wave wave function. Except for the color dependence, each pair of constituents, l and j, will contribute to the interaction energy by an amount $v_{lj} = V$. Absorbing the overall coupling factor g^2 into V, the total interaction energy is

$$U = V \langle f | \sum_{\text{pairs } l,j} T_\alpha^{(j)} \cdot T_\alpha^{(l)} |i\rangle . \tag{4.8}$$

The states $|i\rangle$ and $|f\rangle$ are the multiparticle states in the initial and final state and $T_\alpha^{(j)}$ are the generators for either a quark triplet or antiquark triplet depending on the constituent, with α summed over. From our analogy with *QED*, we might expect that V is the expectation value for the bound state of

$$\frac{g^2}{4\pi} \frac{1}{|\vec{x}_j - \vec{x}_l|}.$$

For a bound state, the initial and final states are identical. We will take the particle types to be the same in the initial and final state. We will see that the choice of interaction results in the color states also being the same.

We want to calculate the color factor for the various multiplets $|i\rangle$ that we can make out of the quarks and antiquarks we are considering; then we can see how the gluon exchange interaction energy depends on the multiplet choice. Note that, by the $SU(3)$ symmetry, we expect that the color state for $|f\rangle$ will be the same as that of $|i\rangle$, for $|i\rangle$ an $SU(3)$ eigenstate. We will see that the interaction operator guarantees this property, which represents the generalization of charge conservation for the non-Abelian group symmetry.

We evaluate the $SU(3)$ operator appearing in Eq. 4.8 by first rewriting

$$\sum_{\text{pairs } i,j} T_\alpha^{(i)} \cdot T_\alpha^{(j)} = \frac{1}{2}\left[\sum_i \sum_j T_\alpha^{(i)} \cdot T_\alpha^{(j)} - \sum_i T_\alpha^{(i)} \cdot T_\alpha^{(i)}\right].$$

Defining $T_\alpha^{\text{tot}} = \sum_i T_\alpha^{(i)}$, the generator for the multiparticle state, we can then write U as:

$$U = \frac{V}{2}\left[\langle f| T_\alpha^{\text{tot}} \cdot T_\alpha^{\text{tot}} |i\rangle - n T_\alpha^{(\text{quark})} \cdot T_\alpha^{(\text{quark})} \langle f|i\rangle\right].$$

Here $T_\alpha^{(\text{quark})} \cdot T_\alpha^{(\text{quark})}$ is the common value of the quadratic Casimir operator for a quark or antiquark. It is equal to $\frac{4}{3}$. We assume a total of n quarks plus antiquarks in the state. Since $T_\alpha^{\text{tot}} \cdot T_\alpha^{\text{tot}}$ is the Casimir operator for the whole multiplet in which $|i\rangle$ resides, we conclude that $|f\rangle = |i\rangle$, for $|i\rangle$ a color eigenstate, and our interaction is indeed $SU(3)$ symmetric.

The energy and stability (for example, if attractive or repulsive) will depend on the multiplet. We see immediately that if a singlet is possible, it will have the lowest energy for a given number of constituents, since the expectation value of $T_\alpha^{\text{tot}} \cdot T_\alpha^{\text{tot}}$ is positive definite and vanishes only for the singlet.

We tabulate the quadratic Casimir operator for a few $SU(3)$ multiplets in Table 4.1.

The lowest energy states will be, using the formula for U:

$$q\bar{q} \text{ in a singlet:} \quad U = -\tfrac{4}{3}V$$

$$qq \text{ in a } \bar{3}: \quad U = -\tfrac{4}{3} \times \tfrac{1}{2}V$$

$$qqq \text{ in a singlet:} \quad U = -\tfrac{4}{3} \times \tfrac{3}{2}V.$$

For $SU(3)$ (and $SU(n)$ in general), singlets have the lowest energy. If we had a bunch of quarks, we would expect them to aggregate into singlets, like electrons and protons aggregate into atoms. Finally, we note that the interaction energy per pair in the baryon (three-quark singlet) is $\frac{1}{2}$ of that in the meson (quark-antiquark singlet), and that for the meson a factor of $-\frac{4}{3}$ multiplies the expected Coulomb-like interaction energy V.

We can calculate the interaction energy between well-separated singlets in lowest order, as shown in Figure 4.5. Ignoring the size of each state relative to the overall separation $\vec{R}$, the interaction operator between state 1 and an individual constituent m of 2 is

$$V(R)\left[\sum_{\text{constituents of 1}} T_\alpha^{(j)}\right] \cdot T_\alpha^{(m)}$$

TABLE 4.1 Value of $T_\alpha \cdot T_\alpha$ for Several Multiplets.

Multiplet	$T_\alpha \cdot T_\alpha$
1	0
3	$\frac{4}{3}$
$\bar{3}$	$\frac{4}{3}$
6	$\frac{10}{3}$
8	3
10	6

4.5 The Color Interaction

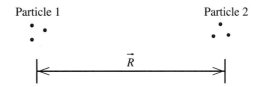

FIGURE 4.5 Two well separated states of no net color.

For a singlet,

$$\langle 1 | \sum_{\text{constituents of 1}} T_\alpha^{(j)} | 1 \rangle = 0,$$

so the interaction vanishes. The particles act as "neutral" objects.

The singlet is the lowest energy state and we can ask, based on Table 4.1, what the ionization energy would be to remove a quark from a singlet made of three quarks. We take for qq its lowest-energy color state, which is the $\bar{3}$ rather than the 6. We want the energy required for

$$(qqq)_{\text{singlet}} \to (qq)_{\bar{3}} + (q)_{\text{far away}}.$$

If the potential energy $\to 0$ for q far away, our table tells us that the ionization energy is the difference in kinetic energies plus

$$U_{\bar{3}} - U_{qqq \text{ singlet}} = \tfrac{4}{3} V.$$

However, this doesn't happen; no experiment has managed to knock a free quark out of a singlet like an individual proton! This is called confinement. We cannot ionize our quark systems; apparently the assumption that $V(|\vec{x}_i - \vec{x}_j|) \to 0$ as $|\vec{x}_i - \vec{x}_j|$ grows, as expected from the analogy to *QED*, is wrong. The singlet is not only the lowest energy state, it also seems to be the only state! We might wonder why the analogy to *QED* has failed. It must depend on how the gluons differ from the photon, that is, that they carry the $SU(3)$ charge, while a photon is electrically neutral. We will look at this feature of the color interaction as the source of the confinement in Chapter 7.

We want to return to one point. How do we know that the quark color symmetry is $SU(3)$? We can figure this out based on the fact that three quarks (for instance, the three making up a proton or neutron) make a singlet. Consider a possible state made of m quarks. For an $SU(n)$ symmetry, what value of m allows a singlet?

$$\underbrace{n \otimes \cdots \otimes n}_{m \text{ times}} = 1 \oplus \cdots$$

The answer is, for the smallest m, $m = n$. Thus the fact that baryons are made of three quarks tells us that there are three colors for each quark. The singlet state is obtained by:

$$|1\rangle = \frac{\varepsilon_{ij...k} |q_i\rangle |q_j\rangle ... |q_k\rangle}{\sqrt{n!}} \qquad (4.9)$$

for $SU(n)$. This is a singlet, just like the analogous quantity $\vec{v}_1 \cdot \vec{v}_2 \times \vec{v}_3$ is a scalar for spatial vectors. It follows from the relation of $\varepsilon_{ij...k}$, the fully antisymmetric symbol, to the calculation of the determinant, which for an $SU(n)$ transformation $= 1$. Note that each colored quark appears once in each term of this fully antisymmetric state. For mesons in $SU(n)$, $n \otimes \bar{n} = 1 \oplus (n^2 - 1)$; we always get a quark-antiquark singlet, so that the existence of mesons cannot tell us how many colors there are. The meson state is analogous to the scalar $\vec{v}_1 \cdot \vec{v}_2$ among vectors. In this singlet, each quark appears paired with its antiquark:

$$|1\rangle = \sum_{i=1}^{n} \frac{|q_i\rangle |\bar{q}_i\rangle}{\sqrt{n}}. \qquad (4.10)$$

4.6 ■ THE COLOR POTENTIAL

We said previously that the existence of only singlets implies that $V(|\vec{x}_i - \vec{x}_j|)$ does not go to zero at large distances. It would be nice to have more direct evidence for this. When looking at this issue, we are limited by the fact that interactions cannot be described by a potential except in the nonrelativistic limit. Fortunately, nature has provided us with six quarks of varying mass, allowing a variety of systems. The top quark is very heavy and decays quickly, and the u, d, s quarks are light and are relativistic within hadrons. The c (charm) and b (bottom) quarks have masses of about 1.5 GeV and 5.0 GeV, respectively, and provide systems for which potential models can be used. We can gauge which systems are nonrelativistic experimentally by the dependence of the binding energy on the quark spin correlations. Since this is a relativistic correction, a small value indicates a nonrelativistic system.

Since the heavy-quark bound states are unstable, they must be produced in interactions. The easiest reactions to study are e^+e^- annihilation to the states of interest, as shown in Figure 4.6. If the e^+e^- energy is just right, the cross

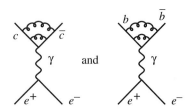

FIGURE 4.6 e^+e^- production of $c\bar{c}$ and $b\bar{b}$ resonances.

4.6 The Color Potential

section shoots up because resonant bound states are produced, as determined by the gluon interactions in the figure. These states can make transitions to other lower mass states in the spectrum (analogous to transitions in hydrogen), allowing a measurement of a number of states of the full spectrum of bound states. Based on the quark masses, these experiments are done at e^+e^- energies ~ 3 GeV (for $c\bar{c}$ states) and ~ 10 GeV (for $b\bar{b}$ states). Given a spectrum, we can then infer a potential needed to reproduce the measurements.

The quantum numbers of the states produced can be predicted on general grounds. They are the same as those of the positronium bound states of e^+e^-. The $c\bar{c}$ and $b\bar{b}$ systems are called charmonium and bottomonium. The spectrum consists of states of given energy, parity, charge conjugation (these are now C eigenstates since they contain a colorless mixture with an equal number of particles and antiparticles) and total angular momentum. To construct these states we add the two spins together to get a state of total spin $= 0$ or 1. The ground state is an s-wave, so we expect the lowest energy states to come in two varieties:

$$J = 0,$$
$$J = 1.$$

What are the parities of these? For an s-wave we expect a parity $+1$. However, a spin $\frac{1}{2}$ particle and antiparticle have opposite intrinsic parity; thus in fact the parity is -1. The two types of ground states are written as:

0^- Pseudoscalar state

1^- Vector state.

Including a spin-spin interaction, analogous to that arising from electromagnetism, we can show on general grounds that the 0^- state has a lower energy. This energy difference is expected to be small, since it is a relativistic correction due to the spin of one particle interacting with the magnetic field of the other in the case of positronium, or the color magnetic field in the case of the quarks.

What is the charge conjugation of such a state? To figure this out, we will write how such a state is defined using creation operators, in a frame where the center of mass is at rest. We take positronium as an example:

$$|e^-e^+\rangle = \left[\int \sum_{\lambda,\lambda'} b^\dagger_{-\vec{p},\lambda'} d^\dagger_{\vec{p},\lambda} \psi(\vec{p}, \lambda, \lambda') \frac{d^3p}{(2\pi)^3}\right] |0\rangle,$$

where $\psi(\vec{p}, \lambda, \lambda')$ is called the momentum space wave function and is normalized so that

$$\int \sum_{\lambda,\lambda'} |\psi(\vec{p}, \lambda, \lambda')|^2 \frac{d^3p}{(2\pi)^3} = 1.$$

Applying C:

$$C\left|e^-e^+\right\rangle = \left[\int \sum_{\lambda,\lambda'} d^\dagger_{-\vec{p},\lambda'} b^\dagger_{\vec{p},\lambda} \psi(\vec{p},\lambda,\lambda') \frac{d^3p}{(2\pi)^3}\right]|0\rangle$$

$$= -\left[\int \sum_{\lambda,\lambda'} b^\dagger_{\vec{p},\lambda} d^\dagger_{-\vec{p},\lambda'} \psi(\vec{p},\lambda,\lambda') \frac{d^3p}{(2\pi)^3}\right]|0\rangle.$$

Note that the minus sign results from the exchange of the creation operators, since we are dealing with fermions. If

$$\psi(\vec{p},\lambda,\lambda') = +/- \psi(-\vec{p},\lambda',\lambda),$$

we have a C eigenstate.

For a spatial parity eigenstate, the exchange $\vec{p} \to -\vec{p}$ gives back the same state within a sign. The exchange $\lambda \to \lambda', \lambda' \to \lambda$ is a spin exchange. For total spin = 0 or 1, this also works, that is, we get back the same state with a + or − sign. Putting together the various factors and indicating the source of these factors:

$$C\left|e^-e^+\right\rangle = \underbrace{(-1)}_{\text{fermions}} \underbrace{(-1)^L}_{\vec{p}\to-\vec{p}} \underbrace{(-1)^{S+1}}_{\text{spin exchange}} \left|e^-e^+\right\rangle = (-1)^{L+S}\left|e^-e^+\right\rangle. \quad (4.11)$$

The 1^- ground state has $C = -1$. It has the quantum numbers of the photon—thus this is the state produced by a single photon. The 0^- ground state has $C = +1$. It will decay, for example, into $\gamma\gamma$, which has $C = +1$, but is not produced by a single virtual photon. For the nonrelativistic color singlet quark-antiquark states, each quark-antiquark pair of a given color appearing in Eq. 4.10 transforms into itself under C, so the J^{PC} of the state is given by the same formulas (in terms of L and S) as for positronium.

An interesting consequence of these quantum numbers is that positronium, which comes in two very nearly degenerate states, decays via two very different lifetimes. The primary decays are:

$$0^- \text{ state} \longrightarrow 2\gamma$$

$$1^- \text{ state} \longrightarrow 3\gamma.$$

Thus the 0^- state has a much shorter lifetime!

$$\tau_{\text{singlet}} \simeq 1.25 \times 10^{-10} \text{ sec}, \quad \frac{\tau_{\text{triplet}}}{\tau_{\text{singlet}}} = 1115.$$

The Feynman diagrams for these decays are shown in Figure 4.7.

We also expect to have excited states for these bound state systems made of a particle and an antiparticle. For example, the analog of the $2s$ state in hydrogen is called a radial excitation—it has the same L and S as the ground state but has an

4.6 The Color Potential

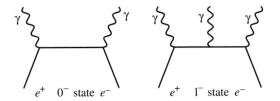

FIGURE 4.7 Positronium decays; diagrams need to be evaluated using the bound state e^+e^- wave functions. Diagrams differing in exchange of photons must also be included to get a full matrix element.

extra node in the radial wave function. The analog of the $2p$ states in hydrogen will produce four types of states by adding spin and angular momentum:

$$(S = 0) + (L = 1) \longrightarrow J = 1,$$
$$(S = 1) + (L = 1) \longrightarrow J = 0, 1, 2.$$

These states will have:

$$\text{Parity} = (-1)^{L+1}, \ C = (-1)^{L+S}.$$

Thus the two states with $J = 1$ differ in C and also are expected to have somewhat different masses.

We list in Table 4.2 a few of the states seen experimentally and the notation used for the $c\bar{c}$ and $b\bar{b}$ systems. For the $b\bar{b}$ system, states up to the $4s$ (that is, with three nodes in the wave function) have been measured. Note the smaller spin-dependent splittings for the more nonrelativistic $b\bar{b}$ states.

We can now try to fit the mass spectrum of these states with a potential. A number of choices for the potential give good fits—they have in common that they are nearly equal over distances where the wave functions are significant. We therefore believe that we have a good understanding of the potential for distances from 0.1 to 1 fermi (10^{-14} to 10^{-13} cm).

TABLE 4.2 Some of the Lightest Onium States.

State	$c\bar{c}$(mass in MeV)	J^{PC}	$b\bar{b}$(mass in MeV)	J^{PC}
$1s$	$\eta_c(2980)$	0^{-+}	still to be found	
$1s$	$J/\psi(3097)$	1^{--}	$\Upsilon(9460)$	1^{--}
$1p$	$\chi_{c0}(3415)$	0^{++}	$\chi_{b0}(9860)$	0^{++}
$1p$	$\chi_{c1}(3511)$	1^{++}	$\chi_{b1}(9892)$	1^{++}
$1p$	$\chi_{c2}(3556)$	2^{++}	$\chi_{b2}(9913)$	2^{++}
$2s$	$\psi(3686)$	1^{--}	$\Upsilon(10023)$	1^{--}

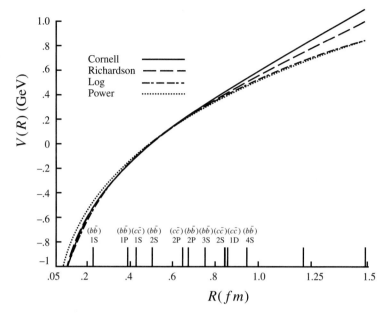

FIGURE 4.8 Form of the potential $V(R)$ for four models, each chosen to give a best fit to the spectrum of $c\bar{c}$ and $b\bar{b}$ bound states. Also indicated are the mean-square $q\bar{q}$ separations for the states of the various systems. [From E. Eichten, SLAC Report #267 (1983).]

The potentials are shown in Figure 4.8. Also shown are the mean-square $q\bar{q}$ spacing for the various $c\bar{c}$, $b\bar{b}$ states. For the analogous light quark ground state, the ρ, made of $u\bar{u}$ and $d\bar{d}$ quarks (and assumed to have the same size as the π, which is the 0^- state) the spacing is about 1.3 fermi.

The potentials are:

1. Cornell:

$$V(r) = -\frac{4}{3}\frac{\alpha_s}{r} + kr, \quad \text{where } \alpha_s = \frac{g^2}{4\pi},$$

the strong interaction analog of

$$\alpha = \frac{e^2}{4\pi}.$$

This is a Coulomb-like plus confining term. The Coulomb term is what you expect for diagrams involving single gluon exchange with the singlet color factor of $-\frac{4}{3}$. In the nonrelativistic limit, we expect this analogy to electromagnetism also to be true for the correction terms to the potential, for example, the spin dependent corrections, since the color factors and Lorentz structure factorize in the matrix elements for single gluon exchange.

4.6 The Color Potential

2. Richardson: This potential is defined via its 3-dimensional Fourier transform. From the analogy to *QED* we expect

$$V(|\vec{q}^2|) = -\frac{4}{3}\frac{\alpha_s}{|\vec{q}^2|},$$

but this does not have a confinement term. The Richardson potential involves choosing a $|\vec{q}^2|$ dependent function for $\alpha_s(|\vec{q}^2|)$, which we will have to motivate later from first principles. The choice is:

$$\alpha_s(|\vec{q}^2|) = \frac{12\pi}{33 - 2n_f} \frac{1}{\log(1 + \frac{|\vec{q}^2|}{\lambda^2})}.$$

$\alpha_s(|\vec{q}^2|)$ is called a running coupling constant; for the function above, it grows as $|\vec{q}^2|$ gets small and falls as $|\vec{q}^2|$ gets large. The growth at small $|\vec{q}^2|$ is the source of the confinement, and in fact the corresponding potential for large r is a linear potential. In the formula, n_f is the number of flavors significantly lighter than the one in question. For the states in Table 4.2, $n_f = 3$ is chosen, that is, u, d, s are the quarks taken to be the "light" quarks in the formula. Note that this potential is determined by only one free parameter.

Potentials (3) and (4) are an empirical best fitting–power law plus constant and a logarithmic potential, respectively. By comparing several forms we can see that the data very clearly constrain the potential over the distance range of the states.

The $c\bar{c}$ and $b\bar{b}$ states are typically split by several hundred MeV and all four potentials give mass splittings that are accurate to a few percent, that is, to a few MeV. To make the match to data correctly, we must take out the energy splitting due to spin interactions, for example, by taking the proper average over states with the same spatial wave function and different spin correlations, such that the spin-splitting cancels after averaging.

For an intuitive understanding, the Cornell potential is perhaps preferable, since the terms have a clear interpretation. Note that the potential has a dimensional parameter, unlike a pure Coulomb potential. We can, for example, take the parameter to be the radius where $V(r) = 0$ (that is, where the Coulombic term = the confinement term in the Cornell potential), which is about .5 fermi. Thus for much shorter distances $V(r) \sim$ Coulombic; for much larger distances it is mostly confining. We will typically choose the related mass scale, which is the reciprocal of the distance scale. This scale is 400 MeV for $r = .5$ fermi. We will call this λ_{QCD}, analogous to the λ in the Richardson potential, for which a best fit also gives 400 MeV. A definition motivated by the running coupling constant will be possible when we look at the behavior of α_s. The size of λ_{QCD}, however, allows us to decide which systems are nonrelativistic, that is, those where the quark mass is $\gg \lambda_{QCD}$.

4.7 ■ SIZE OF BOUND STATES

The potential allows a calculation of energies for the bound states. Since the quark masses are not known a priori, they must be determined as part of the calculation. The resulting masses are $m_c = 1491$ MeV, $m_b = 4883$ MeV, for example, for the Richardson potential. Note that the sign of the potential and size of the states in Figure 4.8 indicate that for Υ, the energy should be less than $2m_b$, for $J/\psi \simeq 2m_c$, while for the light quarks, the hadron masses will be substantially larger than the sum of the "free" quark masses. Also predicted from the potential are the sizes of the bound states. It would be interesting to check these predictions experimentally. For stable (or nearly stable) states, the size of the state can be measured through elastic electron scattering. Here the deviation from point-like behavior allows a determination of $\sqrt{\langle r^2 \rangle}$ for the charged constituents inside the object. This experiment has been done for pions and nucleons. The pion size is in fact about 5 times that of Υ.

For the charmonium and bottomonium states the lifetime is too short to allow a direct size measurement. We can, however, measure the wave function at the origin $\psi(0)$ because it enters into the production rate for $e^+e^- \to V$, where V is the vector meson made of $b\bar{b}$ or $c\bar{c}$. What we will calculate is the decay width $\Gamma(\Upsilon \to e^+e^-)$, that is, the rate for the reverse of the production process. The calculation proceeds in two steps: First, we calculate the rate in terms of a parameter describing the vector particle Υ; second, we calculate this parameter in terms of the bound state characteristics. The first part of the calculation illustrates how we can use general arguments of Lorentz and gauge invariance to parameterize matrix elements. We choose Υ because it is more nonrelativistic than J/ψ, since the second half of the calculation will be in the nonrelativistic limit. In addition, the potential model solution indicates that the Υ state is more confined to the Coulombic region of the potential, allowing a connection of $|\psi(0)|^2$ to the mean-square size of the meson using a hydrogen-like wave function.

The Feynman diagram for the process is shown in Figure 4.9. The vertex factors involve electromagnetic current matrix elements: $\langle e^+e^-|J_\mu|0\rangle$, which we know, and $\langle 0|J_\mu|\Upsilon\rangle$, which we don't know. We can, however, from general considerations, write the Υ current matrix element in terms of an unknown parameter. We use V to denote the vector meson (in this case Υ). This first part of the calculation works for any other vector meson bound state where the analogous parameters of the given meson must be used. Since the current matrix element must transform as a vector and only two vectors characterize the state, namely the Υ momentum and spin

$$P_\mu^V \quad \text{and} \quad e_\mu^V,$$

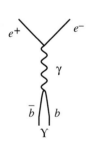

FIGURE 4.9 Vector meson decay diagram to e^+e final state.

$\langle 0|J_\mu|\Upsilon\rangle$ must be some linear combination of these. However, it must also be linear in e_μ^V by the linearity of quantum mechanics. Finally, a term proportional to P_μ^V, which also is the momentum carried by the photon, P_μ^γ, cannot exist by current conservation. Thus we are left with a matrix element proportional to e_μ^V,

4.7 Size of Bound States

$$\langle 0| J_\mu |\Upsilon\rangle = e_b g_V e_\mu^V. \quad (4.12)$$

Our expression is gauge invariant since $p_\mu^\gamma e_\mu^V = p_\mu^V e_\mu^V = 0$. Note that we have factored out the b quark charge e_b in our definition of g_V.

We can now write our Feynman diagram amplitude:

$$-iM_{fi} = ie(\bar{u}_{e^-}\gamma_\mu v_{e^+})\frac{-ig_{\mu\nu}}{q^2}(-ie_b)\left(g_V e_\nu^V\right),$$

where

$$q^2 = m_V^2, \quad \text{and} \quad e_b = \tfrac{1}{3}(-e);$$

resulting in

$$M_{fi} = \frac{-1}{3}\frac{e^2 g_V}{m_V^2}\bar{u}_{e^-}\slashed{e}_V v_{e^+}.$$

To calculate the width, we need to calculate $|M_{fi}|^2$ summed over the spins of e^- and e^+, for a given spin of the vector particle. Since the decay rate is independent of the vector particle spin direction, we can also sum over the three choices for the vector particle spin and divide by 3. Calculating and integrating over the final phase space, ignoring the electron mass, gives:

$$\Gamma_V = \frac{1}{9}\frac{e^4}{12\pi m_V^3}g_V^2 = \left(\frac{1}{9}\right)\left(\frac{4}{3}\pi\alpha^2\frac{g_V^2}{m_V^3}\right). \quad (4.13)$$

For an onium state made of quarks having charge $\tfrac{2}{3}$, the $\tfrac{1}{9}$ above would be replaced by $\tfrac{4}{9}$.

Our next job is to relate g_V to the properties of the state. The annihilation occurs because of the basic process $b + \bar{b} \to \gamma$. In the Υ rest frame the full amplitude is given by the following: amplitude to find b and $\bar{b}$ with given momenta and helicities, multiplied by the amplitude to annihilate with these momenta and helicities, summed over all choices for $\vec{p}$ and the helicities. This follows from our general quantum rules.

We thus can write:

$$\frac{\langle 0| J_\mu |\Upsilon\rangle}{\sqrt{2m_\Upsilon}} = \frac{e_b g_V e_\mu^V}{\sqrt{2m_\Upsilon}} = \int\sum_{\lambda,\lambda'}\frac{\langle 0| J_\mu |b,\vec{p},\lambda;\bar{b},-\vec{p},\lambda'\rangle}{\sqrt{2m_b 2m_{\bar{b}}}}\psi(\vec{p},\lambda,\lambda')\frac{d^3p}{(2\pi)^3}.$$

The latter expression is just the matrix element of J_μ between the vacuum and the $b\bar{b}$ analog of the positronium state $|e^-e^+\rangle$, which we wrote earlier in terms of creation and annihilation operators acting on the vacuum, and the momentum space wave function.

Note the insertion of the normalization factors $1/\sqrt{2m}$ for each field that occurs. We want to do this calculation for one Υ made of one b and one $\bar{b}$, while the calculation using covariant fields is normalized to $2E$ particles instead, necessitating the extra factors everywhere a covariant current matrix element appears. We have here approximated $E = m$ in the normalization factors, since the calculation assumes a nonrelativistic state.

The nonrelativistic momentum space wave function factorizes into a function of momentum and a spin correlation function: $\psi(\vec{p}, \lambda, \lambda') = \psi(\vec{p})\chi(\lambda, \lambda')$, where the spin correlation function depends on the Υ spin direction. To simplify the notation, we write $\langle 0| J_\mu |b, \vec{p}, \lambda; \bar{b}, -\vec{p}, \lambda'\rangle = \langle 0| J_\mu |b\bar{b}\rangle$:

$$\langle 0| J_\mu |\Upsilon\rangle = \sqrt{\frac{2m_\Upsilon}{2m_b 2m_{\bar{b}}}} \int \frac{d^3p}{(2\pi)^3} \sum_{\lambda,\lambda'} \langle 0| J_\mu |b\bar{b}\rangle \psi(\vec{p})\chi(\lambda, \lambda').$$

We next look at $\langle 0| J_\mu |b\bar{b}\rangle$, in particular at the nonrelativistic limit of this in the Υ rest frame. It will have terms $\sim m_b$ and terms proportional to $\vec{p}$. In the nonrelativistic limit, we need keep only the terms $\sim m_b$. Thus $\langle 0| J_\mu |b\bar{b}\rangle$ is approximately a constant, its $\vec{p} \to 0$ limit, and can be factored out of the integral. Thus, using

$$m_b = m_{\bar{b}} \simeq \frac{m_\Upsilon}{2}$$

in the nonrelativistic limit:

$$\langle 0| J_\mu |\Upsilon\rangle = \sqrt{\frac{2}{m_\Upsilon}} \sum_{\lambda,\lambda'} \langle 0| J_\mu |b\bar{b}\rangle \chi(\lambda, \lambda') \int \psi(\vec{p}) \frac{d^3p}{(2\pi)^3}.$$

However, the spatial wave function

$$\psi(\vec{x}) = \int \psi(\vec{p}) e^{i\vec{p}\cdot\vec{x}} \frac{d^3p}{(2\pi)^3}$$

in general. Thus:

$$\int \psi(\vec{p}) \frac{d^3p}{(2\pi)^3} = \psi(0),$$

the wave function at the origin. The annihilation in the nonrelativistic limit depends on the probability of finding the b and $\bar{b}$ on top of each other.

Before continuing we have to worry about one effect we left out. We have been calculating $b + \bar{b} \to \gamma$, ignoring color entirely. The calculation is correct for each color individually, but Υ is the color singlet state

$$\frac{1}{\sqrt{3}} \sum_{i=1}^{3} b_i \bar{b}_i,$$

where i = color index. We have been calculating the current for each $b_i \bar{b}_i$. Since the answer is independent of i, we can just add the three terms together, getting a

4.7 Size of Bound States

total amplitude for the color singlet larger by $\sqrt{3}$. Note that the inclusion of color can have clear numerical effects! So including color, the correct result is

$$\langle 0| J_\mu |\Upsilon\rangle = \sqrt{\frac{6}{m_\Upsilon}}\psi(0) \sum_{\lambda,\lambda'} \chi(\lambda,\lambda') \langle 0| J_\mu |b\bar{b}\rangle. \tag{4.14}$$

To further evaluate Eq. 4.14, we use for the annihilation amplitude, for fixed momentum and helicity states:

$$\langle 0| J_\mu |b\bar{b}\rangle = e_b \bar{v}_{\bar{b}} \gamma_\mu u_b. \tag{4.15}$$

We could calculate the current matrix element by doing the helicity sum in Eq. 4.14, which would imply averaging Eq. 4.15 over the helicity choice for a given $J = 1$ state. We will actually complete the calculation in a simple covariant way that gives us the relation we want for g_V without having to look at the three specific Υ spin states. An easy way to relate g_V to $\psi(0)$ is just to square and sum over all spins the expressions in Eq. 4.12 and Eq. 4.14. To calculate the required square of the current matrix element in Eq. 4.14, it is simplest to sum over the four possible spin states using Eq. 4.15, which corresponds to summing over one 0^- and the three 1^- states, ignoring $\chi(\lambda,\lambda')$. The 0^- state gives no contribution, since we know 1γ couples only to $J = 1$ states, so the sum over all spins gives the same result as summing over the three $J = 1$ states that we would get from including $\chi(\lambda,\lambda')$.

Looking first at the expression resulting from Eq. 4.12:

$$\sum_{\text{spins}} \langle 0| J_\mu |\Upsilon\rangle \langle 0| J_\nu |\Upsilon\rangle^* = e_b^2 g_V^2 \sum_{\Upsilon \text{ spins}} e_\mu e_\nu^* = -e_b^2 g_V^2 \left[g_{\mu\nu} - \frac{p_\mu^\Upsilon p_\nu^\Upsilon}{m_\Upsilon^2} \right].$$

The expression for $\langle 0| J_\mu |\Upsilon\rangle$ from Eq. 4.14 gives:

$$\frac{6}{m_\Upsilon} e_b^2 |\psi(0)|^2 \sum_{\text{spin}} \text{Tr}\left[v_{\bar{b}} \bar{v}_{\bar{b}} \gamma_\mu u_b \bar{u}_b \gamma_\nu \right].$$

We focus on the trace term below. Summing over spins:

$$\sum_{\text{spins}} v_{\bar{b}} \bar{v}_{\bar{b}} = (\slashed{p}_{\bar{b}} - m_{\bar{b}}), \quad \sum u_b \bar{u}_b = \slashed{p}_b + m_b.$$

Thus, the resulting trace is: $\text{Tr}(\slashed{p}_{\bar{b}} - m_{\bar{b}})\gamma_\mu (\slashed{p}_b + m_b)\gamma_\nu$. Keeping only terms with an even number of γ matrices gives

$$\text{Tr}\left[\slashed{p}_{\bar{b}} \gamma_\mu \slashed{p}_b \gamma_\nu\right] - \text{Tr}\left[\gamma_\mu \gamma_\nu\right] m_b^2 = 4\left[p_\mu^b p_\nu^{\bar{b}} + p_\mu^{\bar{b}} p_\nu^b - g_{\mu\nu}(p_b \cdot p_{\bar{b}}) \right] - 4m_b^2 g_{\mu\nu}.$$

These expressions are in an arbitrary frame and we can take (where the internal relative b momentum is ignored for a nonrelativistic bound state):

$$p_\mu^b = \frac{p_\mu^\Upsilon m_b}{m_\Upsilon}, \quad p_\mu^{\bar{b}} = \frac{p_\mu^\Upsilon m_{\bar{b}}}{m_\Upsilon},$$

which gives for the trace term:

$$-8m_b^2 \left[g_{\mu\nu} - \frac{p_\mu^\Upsilon p_\nu^\Upsilon}{m_\Upsilon^2} \right] = -2m_\Upsilon^2 \left[g_{\mu\nu} - \frac{p_\mu^\Upsilon p_\nu^\Upsilon}{m_\Upsilon^2} \right].$$

Note that we get the same Lorentz structure for both versions of the current squared, as we must, if we do things correctly.

Equating the two versions for the matrix element squared, we finally get the general expression for the vector meson decay constant:

$$g_V^2 = 12 m_V |\psi(0)|^2. \tag{4.16}$$

Substituting this into the expression for Γ_V gives

$$\Gamma_V = \frac{1}{9} \left[16\pi \alpha^2 \frac{|\psi(0)|^2}{m_V^2} \right], \tag{4.17}$$

where the factor due to the quark charge squared has been separated out. We can do the same calculation for charm; then $(\frac{1}{3})^2 \to (\frac{2}{3})^2$ due to the larger quark charge, and we use the wave function and mass for the charmonium state.

The measured $\Gamma_{\Upsilon \to e^+ e^-} = 1.32$ keV, and the measured vector meson mass is $m_\Upsilon = 9460$ MeV. Does this tell us how big the state is? To calculate this approximately, we assume that the state is reasonably well-described by a hydrogen-like ground state for which

$$|\psi(0)|^2 = \frac{1}{\pi} \left(\frac{1}{a_0} \right)^3,$$

where a_0 is the Bohr radius. This gives

$$\frac{9\Gamma}{16\alpha^2} m_V^2 = \left(\frac{1}{a_0} \right)^3.$$

Inserting the experimentally determined values gives

$$\left(\frac{1}{a_0} \right) = 1.08 \text{ GeV},$$

or $a_0 = .18$ fermi. The RMS radius for a hydrogenic ground state is $\sqrt{3} a_0 \simeq .3$ fermi for the Υ.

Note that the Coulombic assumption for the size is pretty good, as can be seen by comparing it to the more accurate potential model calculation, whose result was shown in Figure 4.8. The confinement term actually pushes the state in a little, compared to the purely Coulombic approximation.

For the hydrogen atom

$$\frac{1}{a_0} = \alpha m_e.$$

The colored version of the potential, keeping only the Coulombic term, has an analogous relation after the substitutions

$$\alpha \to \frac{4}{3}\alpha_s \quad \text{and} \quad m_e \to \frac{m_b}{2},$$

which is the reduced mass for a $b\bar{b}$ system. Thus we can estimate α_s from

$$\frac{4}{3}\alpha_s \frac{m_b}{2} = 1.08 \text{ GeV},$$

which gives $\alpha_s \simeq .3$. Note that α_s is indeed $\gg \alpha$, that is, the color interaction is much stronger than electromagnetism! Calculating the binding energy in the Coulombic approximation gives $E_{\text{binding}} = -\frac{4}{9}\alpha_s^2 m_b$, which is about -200 MeV.

We want to point out one assumption we have made when discussing a potential that works for both $c\bar{c}$ and $b\bar{b}$, that is, that the coupling due to color, the g^2 at a vertex, is independent of the quark flavor. When written in momentum space, as for the Richardson potential, this becomes more precisely the flavor independence of the coupling at given q^2. When we discuss color as a full gauge theory, we will see that this is a consequence of using an $SU(3)$ non-Abelian gauge theory. This uniformity of coupling (also for the weak doublets in the weak interactions) is thus an important consequence of the full theory.

4.8 ■ BARYONS

For baryons made of identical heavy quarks, we expect to have bound states of the type ccc and bbb, the second type of color singlet, besides the $c\bar{c}$ and $b\bar{b}$ states. There is no easy way to make these experimentally and they have not been seen. We can, however, look briefly at what we would expect. We can write a nonrelativistic Hamiltonian, including the rest masses and color factors, for the two types of color singlet states. For the meson:

$$E_M = \frac{p_1^2}{2m} + \frac{p_2^2}{2m} - \frac{4}{3}V(|\vec{r}_1 - \vec{r}_2|) + 2m.$$

For the baryon:

$$E_B = \frac{p_1^2}{2m} + \frac{p_2^2}{2m} + \frac{p_3^2}{2m} - \frac{2}{3}\left(V(|\vec{r}_1 - \vec{r}_2|) + V(|\vec{r}_1 - \vec{r}_3|) + V(|\vec{r}_2 - \vec{r}_3|)\right) + 3m.$$

The ground states have all particles in an s-wave, thus we define

$$K.E. = \left\langle \frac{p_1^2}{2m} \right\rangle, \quad \bar{V} = \langle V(|\vec{r}_1 - \vec{r}_2|) \rangle$$

$$E_M = 2(K.E._M - \tfrac{2}{3}\bar{V}_M + m) \quad E_B = 3(K.E._B - \tfrac{2}{3}\bar{V}_B + m),$$

where M indicates the meson and B the baryon. The choice of a color singlet gives an interaction term proportional to the number of constituents, rather than the number of pairs.

We can expect that the energies are minimized for approximately the same mean square spacing between particles, because of the analogous structure of the Hamiltonians. In this approximation, we can define an effective (called constituent) quark mass in the ground state, m_{eff}, by dividing the energy by the number of constituents: $m_{\text{eff}} = K.E. - \tfrac{2}{3}\bar{V} + m$. The color factor for the baryon equaling $\tfrac{1}{2}$ that for the meson results in the expectation that the constituent quark mass is approximately the same in a meson and baryon. Without the factor of $\tfrac{1}{2}$, we would expect the baryon to be more tightly bound than the meson for the heavy quark systems.

Rather remarkably, these considerations work approximately even for the states made of light quarks, to which we turn in the next chapter. For example, we can compare the sizes of the light mesons and baryons. The distances measured are shown in Figure 4.10. The mean square quantities determined in electron scattering for the pion and proton are:

$$r_\pi = .66 \times 10^{-13} \text{ cm}$$

$$r_p = .81 \times 10^{-13} \text{ cm}.$$

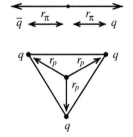

FIGURE 4.10 Arrangement of quarks and antiquarks in the meson and the baryon.

Note the $q\bar{q}$ separation in the π, $2r_\pi$, and the qq separation in the proton, $\sqrt{3}r_p$, are nearly equal and are both $\simeq 1.3$ fermi.

CHAPTER 4 HOMEWORK

4.13. Consider a unitary transformation that transforms states

$$U|i\rangle = |Ui\rangle.$$

Suppose U is a symmetry of the Hamiltonian density, which for the interaction Hamiltonian means that

$$U^\dagger \mathcal{H} U = \mathcal{H}.$$

(a) Show that for the S matrix, $\langle Uf|S|Ui\rangle = \langle f|S|i\rangle$.

(b) Charge conjugation C is such a symmetry for the electromagnetic interactions. Show that the cross sections (at the same center of mass energy) satisfy

$$\sigma\left(\mu^+ e^- \to \mu^+ e^-\right) = \sigma\left(\mu^- e^+ \to \mu^- e^+\right).$$

(c) The photon state satisfies $C\,|1\gamma\rangle = -|1\gamma\rangle$. Show that the vertices $e^+ \to e^+ + \gamma$ and $e^- \to e^- + \gamma$ have opposite sign amplitudes:

$$\langle e^- + 1\gamma|\,\mathcal{H}\,|e^-\rangle = -\langle e^+ + 1\gamma|\,\mathcal{H}\,|e^+\rangle.$$

Taking two vertices and the propagator, this line of reasoning allows us to determine the sign of the corresponding nonrelativistic potential for exchange of a particle that is a C eigenstate, without looking at the detailed field theory.

(d) Show, using the charge conjugation symmetry

$$\langle 1\gamma|\,\mathcal{H}\,|e^-, p_-, \lambda;\ e^+, p_+, \lambda'\rangle = -\langle 1\gamma|\,\mathcal{H}\,|e^+, p_-, \lambda;\ e^-, p_+, \lambda'\rangle,$$

where the momenta and helicities are explicitly indicated. Using the explicit matrix elements in the Majorana representation and the operator $\mathcal{H}$, show that this requires that we use anti-commutation relations for the particle and antiparticle annihilation operators. This was assumed, but not discussed, in the text.

(e) For bosons, for example pions, we have the same kind of relation:

$$\langle 1\gamma|\,\mathcal{H}\,|\pi^+, p_+;\ \pi^-, p_-\rangle = -\langle 1\gamma|\,\mathcal{H}\,|\pi^-, p_+;\ \pi^+, p_-\rangle.$$

The boson creation and annihilation operators satisfy commutation relations rather than anti-commutation relations. What is different about the matrix element compared to (d) so that this is not a contradiction? (*Hint:* The matrix element of the current operator for the spinless pions depends on the four-momenta in a gauge invariant way).

4.14. By using the explicit formulas for the spinors in the text, show that

$$\gamma_0 u(\vec{p}) = u(-\vec{p}),\ \gamma_0 v(\vec{p}) = -v(-\vec{p})$$

as expected for a parity transformation, where the particle and antiparticle have opposite intrinsic parity.

4.15. For $SU(3)$ the T_α matrices are $\lambda_\alpha/2$ as given in the text. By explicit calculation show that $T_\alpha T_\alpha$ is diagonal with eigenvalue $\frac{4}{3}$.

4.16. Consider the generators for unitary transformations for a composite system made of two particles:

$$T_i^{(\text{tot})} = T_i^{(1)} + T_i^{(2)}.$$

Here $T_i^{(1)}$ acts on particle 1, $T_i^{(2)}$ on particle 2 of the composite system.

(a) Show that $T_i^{(\text{tot})}$ satisfies the same commutation rules as $T_i^{(1)}$ and $T_i^{(2)}$ where we assume $[T_i^{(1)}, T_j^{(2)}] = 0$.

(b) By looking at infinitesimal transformations, show that for a product state:

$$\left(T_i^{(1)} + T_i^{(2)}\right)|1\rangle|2\rangle = \left(T_i^{(1)}|1\rangle\right)|2\rangle + |1\rangle\left(T_i^{(2)}|2\rangle\right).$$

(c) Consider an interaction of an external particle with the composite system. By choosing the interaction to behave like $T_i^{\text{ext}} \cdot (T_i^{(1)} + T_i^{(2)})$ in the group space, show that the interaction is then the sum of pairwise interactions. For nonrelativistic systems this would correspond to the total potential energy being the sum over all pairs, which generalizes this property of electromagnetic interactions.

4.17. Suppose the color interaction involved an $SU(2)$ symmetry, instead of $SU(3)$, leading to color singlet mesons (quark-antiquark states) and baryons (all quark states). Denote the two colored quark states in this model as $|q_1\rangle$ and $|q_2\rangle$.

(a) Write down the color singlet meson and baryon states in terms of $|q_1\rangle$, $|q_2\rangle$, and their antiparticles.

(b) Are these states fermions or bosons?

(c) Show that we expect the meson and baryon to be degenerate for the same spatial state in the nonrelativistic limit.

(d) How then do the ground state meson and baryon differ; that is, what quantum numbers would you expect for them?

4.18. Consider the isospin multiplets made of $|p\rangle$ and $|n\rangle$.

Triplet: $\begin{pmatrix} 1 \\ 0 \\ 0 \end{pmatrix} = |p\rangle|p\rangle, \quad \begin{pmatrix} 0 \\ 1 \\ 0 \end{pmatrix} = \frac{|p\rangle|n\rangle + |n\rangle|p\rangle}{\sqrt{2}}, \quad \begin{pmatrix} 0 \\ 0 \\ 1 \end{pmatrix} = |n\rangle|n\rangle,$

Singlet: $\frac{|p\rangle|n\rangle - |n\rangle|p\rangle}{\sqrt{2}}.$

For the triplet, calculate the 3×3 matrix representations for the $SU(2)$ generators T_1, T_2, T_3 by explicitly calculating the matrix elements of $T_i^{(1)} + T_i^{(2)}$, where $T_i^{(1)}$ and $T_i^{(2)}$ are the 2×2 operators acting separately on each of the nucleon states. Show that $T_i = 0$ for the singlet.

4.19. For states made of two colored quarks the possible multiplets would be $\bar{3}$ or 6 based on $3 \otimes 3 = \bar{3} \oplus 6$ for $SU(3)$.

(a) Denoting the three colored quarks as $|q_1\rangle$, $|q_2\rangle$, and $|q_3\rangle$, write down the 6 states in the 6.

(b) Taking the state $|q_1\rangle|q_1\rangle$ and calculating explicitly

$$\left[T_i^{(1)} + T_i^{(2)}\right]\left[T_i^{(1)} + T_i^{(2)}\right]|q_1\rangle|q_1\rangle$$

show that the value of the Casimir Operator for the 6 (given in Table 4.1 in the text) is $\frac{10}{3}$.

(c) Consider the $\bar{3}$. Denoting the states:

$\begin{pmatrix} 1 \\ 0 \\ 0 \end{pmatrix} = \frac{|q_2\rangle|q_3\rangle - |q_3\rangle|q_2\rangle}{\sqrt{2}}, \quad \begin{pmatrix} 0 \\ 1 \\ 0 \end{pmatrix} = \frac{|q_3\rangle|q_1\rangle - |q_1\rangle|q_3\rangle}{\sqrt{2}},$

$$\begin{pmatrix} 0 \\ 0 \\ 1 \end{pmatrix} = \frac{|q_1\rangle |q_2\rangle - |q_2\rangle |q_1\rangle}{\sqrt{2}},$$

we expect the generators to be $-T_i^*$, where T_i are the generators for the 3. Using $T_i^{(1)} + T_i^{(2)}$ and the states above, show that the generators are $-T_i^*$ for $i = 1, 2, 3$.

(d) For the $\bar{3}$ and 6, indicate where the states are in the T_3, T_8 space, based on the addition of these quantum numbers for the two constituents.

4.20. Consider muonium, a bound state of $\mu^+\mu^-$. For the $J^P = 1^-$ ground state, calculate the decay width into e^+e^-. For the $J^P = 0^-$ ground state, what is the Feynman diagram for decay into e^+e^-? Is this expected to be the most likely decay for this state?

4.21. The decay widths of J/ψ and $\phi \to e^+e^-$ are 5.3 keV and 1.3 keV, respectively. These are the $J^P = 1^-$ ground states of $|c\bar{c}\rangle$ and $|s\bar{s}\rangle$. Their masses are 3.1 GeV and 1.02 GeV, respectively. Assuming the nonrelativistic decay formula is adequate and that the Coulombic approximation is an adequate description of the potential, find the Bohr radii for these two systems. The assumptions above are rough approximations and give a qualitatively correct picture.

CHAPTER 5

Isospin and Flavor $SU(3)$, Accidental Symmetries

5.1 ■ THE LIGHT QUARKS

We turn now to the hadrons made of the light quark flavors. These flavors are u, d, and s. The hadrons made of these light quarks are the most common particles found in high-energy jets and also provide the longest lived hadrons, including the stable proton. We will focus mostly on the lightest (ground state) hadrons made of these flavors. These hadrons, including their decay patterns, illustrate a large number of phenomena: symmetry and symmetry breaking, the exclusion principle, spin independent and dependent forces, and mixing of states due to various types of interactions. We will also find decays coming from the weak, electromagnetic, and strong interactions.

The light quark flavors each have a mass $m_q \ll \lambda_{QCD}$; thus the states are expected to be highly relativistic, since we expect momenta $\sim \lambda_{QCD}$ for the quarks within the hadron by the uncertainty principle. As a consequence, the spin-dependent interactions are quite large. By analogy with electromagnetism, we expect these spin interactions to be magnetic, and magnetic and electric fields and forces are comparable for charged systems where the velocity $v \sim 1$ for all of the particles. This is unlike the hydrogen atom, where states that differ in spin orientation are nearly degenerate. Since the energies differ so much for different total spins, the particles are given different names, for example the spin-zero π with mass 140 MeV and the spin-one ρ with mass of 770 MeV.

The second consequence of a small mass relative to λ_{QCD} is that replacing one light flavor with another is expected to result in a state with nearly the same properties, since the energy is weakly dependent on mass for a relativistic particle. This result requires that the color force be the same for all pairs of quark flavors, as seen to be true for the $c\bar{c}$ and $b\bar{b}$ systems. This property is a consequence of QCD but is violated by the other, weaker, interactions. Therefore, the strong interactions have an approximate $SU(3)$ flavor symmetry—that is, a symmetry under unitary transformations among u, d, s. This was actually the first $SU(3)$ symmetry found, but it is accidental in the sense that it results from small ratios of quark mass to λ_{QCD}, where the ratios are not a consequence of any fundamental principles.

5.1 The Light Quarks

The flavor symmetry is not equally good for all $SU(3)$ rotations. The strange quark mass is only a few times smaller than λ_{QCD}, while the u and d masses are much smaller. Hence the $u \leftrightarrow d$ $SU(2)$ symmetry, called isospin, is a very good symmetry, while the full $SU(3)$ symmetry is more approximate. We will use I and I_3 to denote the isospin quantum numbers of states. We represent our states by

$$|u\rangle = \begin{pmatrix} 1 \\ 0 \\ 0 \end{pmatrix}, \quad |d\rangle = \begin{pmatrix} 0 \\ 1 \\ 0 \end{pmatrix}, \quad |s\rangle = \begin{pmatrix} 0 \\ 0 \\ 1 \end{pmatrix}$$

for the purpose of discussing $SU(3)$ rotations. Rotations involving only T_1, T_2, T_3 (defined analogously to the first three generators for color) generate the very good isospin subgroup symmetry.

We look first at $SU(3)$ flavor multiplets—physical states with similar properties. We will display the $SU(3)$ multiplets using 2-dimensional plots analogous to those introduced earlier in the discussion of color, but where the states are now related by flavor rather than color rotations. We will not discuss color, unless it provides an extra constraint on the available states, with an understanding that all allowed states are color singlets. Thus, for example, a quark-antiquark state such as

$$|u\rangle |\bar{d}\rangle,$$

is really

$$\sum_{i=1}^{3} \frac{|u_i\rangle |\bar{d}_i\rangle}{\sqrt{3}},$$

where i specifies the color. One more convention is typically used for writing antiparticle states. We represented the antiparticle states by the column vectors:

$$|\bar{u}\rangle = \begin{pmatrix} 1 \\ 0 \\ 0 \end{pmatrix}, \quad |\bar{d}\rangle = \begin{pmatrix} 0 \\ 1 \\ 0 \end{pmatrix}, \quad |\bar{s}\rangle = \begin{pmatrix} 0 \\ 0 \\ 1 \end{pmatrix}$$

with antiparticle generators $-T_i^*$. It turns out that for $-T_1^*, -T_2^*, -T_3^*$ there exists a unitary transformation that changes $-T_i^* \to T_i$. This is a special property of $SU(2)$. We call the transformation matrix R_I, since it acts only on the two isospin states. To accomplish the change, R_I must satisfy

$$R_I(-T_i^*)R_I^\dagger = T_i, \quad \text{for } i = 1, 2, 3.$$

Note that

$$T_2 = \frac{1}{2} \begin{pmatrix} 0 & -i & 0 \\ i & 0 & 0 \\ 0 & 0 & 0 \end{pmatrix}$$

already satisfies $-T_2^* = T_2$. Thus R_I must commute with T_2. It turns out that

$$R_I = \begin{pmatrix} 0 & -1 & 0 \\ 1 & 0 & 0 \\ 0 & 0 & 1 \end{pmatrix}$$

(that is, basically $-i\sigma_2$ for the I-spin part of R_I) works, because of the anticommutation relations of the σ matrices and the fact that σ_1 and σ_3 are real. Thus, provided we use $R_I|\bar{q}\rangle$ for the antiquark states, we can use T_1, T_2, T_3 as the generators for isospin rotations on both $|q\rangle$ and $|\bar{q}\rangle$. $R_I|\bar{q}\rangle$ gives the following column vectors representing the antiquarks:

$$|\bar{u}\rangle = \begin{pmatrix} 0 \\ 1 \\ 0 \end{pmatrix}, \quad -|\bar{d}\rangle = \begin{pmatrix} 1 \\ 0 \\ 0 \end{pmatrix}, \quad |\bar{s}\rangle = \begin{pmatrix} 0 \\ 0 \\ 1 \end{pmatrix}.$$

Applying T_3 now (instead of $-T_3^*$) to these column vectors, we see that $|\bar{u}\rangle$ has $I_3 = -\frac{1}{2}$, and $|\bar{d}\rangle$ has $I_3 = +\frac{1}{2}$, as expected.

The antiquark correspondence above quickly allows us to change known isospin states involving particles only into states involving particles and antiparticles. Thus, for example, by analogy with the composite isospin states made of $|p\rangle$ and $|n\rangle$, we can consider composite states made of the isospin doublet $|u\rangle$ and $|d\rangle$. From these, we can get corresponding states made of a quark and an antiquark by substituting, for the second particle of the composite system,

$$|u\rangle \to -|\bar{d}\rangle, \quad |d\rangle \to |\bar{u}\rangle.$$

Using the notation $|q_i, q_j\rangle$ for a composite state made of q_i and q_j, where we suppress in the notation the spatial and spin degrees of freedom for the state, the analogous quark-antiquark states are given in Table 5.1. The $I = 0$ quark-antiquark singlet is exactly what we expect the $SU(2)$ singlet to look like, based on our earlier color singlet discussion for $SU(n)$. Notice the extra minus sign for the $I = 1, I_3 = 1$ state. The three π mesons, π^+, π^0, π^-, are an example of the $I = 1$ quark-antiquark triplet. Since the isospin symmetry is rather good, the three π mesons have very nearly the same mass.

TABLE 5.1 Analog Isospin States.

$I = 1$	$I = 1$	$I = 0$	$I = 0$
$\|u, u\rangle$	$-\|u, \bar{d}\rangle$	$\dfrac{\|u, d\rangle - \|d, u\rangle}{\sqrt{2}}$	$\dfrac{\|u, \bar{u}\rangle + \|d, \bar{d}\rangle}{\sqrt{2}}$
$\dfrac{\|u, d\rangle + \|d, u\rangle}{\sqrt{2}}$	$\dfrac{\|u, \bar{u}\rangle - \|d, \bar{d}\rangle}{\sqrt{2}}$		
$\|d, d\rangle$	$\|d, \bar{u}\rangle$		

The similarity transformation R_I leaves $-T_8^*$ unchanged, so the matrix representation for this linearly additive quantum number is not altered.

5.2 ■ LIGHT MESONS

The mesons are made of a quark-antiquark in a color singlet state. With three light flavors, there are nine such states to consider for a given spatial and spin state. We can exhibit the states in a 2-dimensional space where the eigenvalues of T_3 (which we call I_3) and T_8 are displayed. To make this simpler what is often used is $\frac{2}{\sqrt{3}}T_8 = Y$, called the hypercharge. We will use a number related to Y. Since the total number of quarks is conserved we can offset the Y axis by a constant proportional to the number of quarks, which will leave our number an additive conserved quantum number still. The conservation of the number of quarks is called baryon number conservation. Here a proton is assigned one unit of baryon number. Thus a quark has $\frac{1}{3}$ unit of baryon number. We take $Y - B$, called S or strangeness, as our second axis. The u and d quarks in this scheme have zero strangeness, and the s quark -1 unit of strangeness. Thus S is -1 times what might have been the most logical definition, which would have strangeness count the number of strange quarks in a state. Strangeness is easy to calculate by just looking at the state and counting the number of strange antiquarks minus quarks. The quark and antiquark states now look as follows:

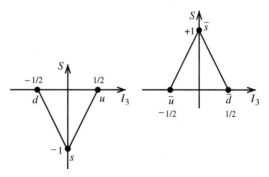

The nine meson states can be displayed in the I_3, S plane by adding the two additive quantum numbers of the constituents, as illustrated here.

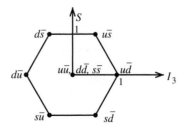

The nine states, collectively, are called a meson nonet.

We still have to figure out what multiplets the nine states form. The states on the periphery of the picture above are seen to be in the same multiplet by identifying a very simple operator, which leaves us within a multiplet and takes one state to another. Consider $|u, \bar{s}\rangle$, and the generator, made of a linear combination of quark and antiquark generators, which takes $u \to d$ and leaves s fixed. This takes $|u, \bar{s}\rangle \to |d, \bar{s}\rangle$; thus these are in the same multiplet. Continuing in this way, we can find transformations that link all the states on the periphery. Thus, they are in the same multiplet.

For the states with particle content $u\bar{u}$, $d\bar{d}$, and $s\bar{s}$, the remaining states in the multiplet can be worked out similarly using the generators. The state

$$\frac{|u, \bar{u}\rangle + |d, \bar{d}\rangle + |s, \bar{s}\rangle}{\sqrt{3}}$$

is a singlet, based on our knowledge of the analogous color singlet state. As an alternative approach, we will deduce below the multiplet assignments by using the main property of states in a multiplet—namely, that they have the same physical properties, in particular the same mass. The result will be that the states break into an octet of eight states and a singlet.

To see which states have the same properties, we make a reasonably general model for the meson masses in terms of several parameters that are physically motivated. The model provides some insight into the effects determining the masses, and we can also use it to look at the effects of the $SU(3)$ breaking. We define H_m as the meson Hamiltonian for a bound state of $q_i, \bar{q}_j$ of a given spin, parity, and spatial state at rest in the center of mass. Note that i and j are flavor indices. There are several contributions to H_m. The first is the binding due to the exchange of colored gluons. This is the same for all $q_i, \bar{q}_j$ color singlet pairs in the symmetry limit, where the q_i masses are equally small. A simple diagram of this type is as follows.

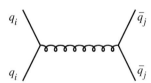

Here, $q_i, \bar{q}_j$ are in a color singlet state and all eight gluons contribute coherently. The indices i and j correspond to u, d, or s. We define the mass of the bound state to be m for the six states where $i \neq j$, in the symmetry limit.

In addition to the exchange diagrams, we can also have annihilation diagrams, but these occur only for $i = j$; that is, a particle can only annihilate an antiparticle of the same flavor, since the strong interactions do not change flavor. We show an annihilation diagram, starting from $u\bar{u}$ as an example and taking the simplest diagram.

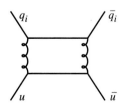

5.2 Light Mesons

Once the $u\bar{u}$ annihilate, the final state can emerge as any other $q_i, \bar{q}_i$ pair. This process produces a coupling between different pairs. In the symmetry limit, the contribution of these diagrams is the same for all three flavors. Note that the amplitude for a single gluon intermediate state vanishes for a colorless initial state. In fact, whether the simplest diagram involves two gluons or three gluons depends on whether the initial state has even or odd charge conjugation.

In the above model, the three states where $i = j$ require studying the effects of mixing in order to find the eigenstates that emerge. To diagonalize H_m for the $q_i, \bar{q}_i$ states, we define a basis of states as follows:

$$|u, \bar{u}\rangle = \begin{pmatrix} 1 \\ 0 \\ 0 \end{pmatrix}, \quad |d, \bar{d}\rangle = \begin{pmatrix} 0 \\ 1 \\ 0 \end{pmatrix}, \quad |s, \bar{s}\rangle = \begin{pmatrix} 0 \\ 0 \\ 1 \end{pmatrix}.$$

In this basis, and assuming a perfect $SU(3)$ flavor symmetry:

$$H_m = \begin{pmatrix} m + \varepsilon & \varepsilon & \varepsilon \\ \varepsilon & m + \varepsilon & \varepsilon \\ \varepsilon & \varepsilon & m + \varepsilon \end{pmatrix}. \tag{5.1}$$

Here m is due to the gluon exchange diagrams and is the same as for $i \neq j$, and ε is the contribution from annihilation. Diagonalizing this matrix gives eigenstates and masses:

$$\frac{1}{\sqrt{2}} \begin{pmatrix} 1 \\ -1 \\ 0 \end{pmatrix} = \frac{|u, \bar{u}\rangle - |d, \bar{d}\rangle}{\sqrt{2}}, \quad \text{mass} = m,$$

$$\frac{1}{\sqrt{6}} \begin{pmatrix} 1 \\ 1 \\ -2 \end{pmatrix} = \frac{|u, \bar{u}\rangle + |d, \bar{d}\rangle - 2|s\bar{s}\rangle}{\sqrt{6}}, \quad \text{mass} = m,$$

$$\frac{1}{\sqrt{3}} \begin{pmatrix} 1 \\ 1 \\ 1 \end{pmatrix} = \frac{|u, \bar{u}\rangle + |d, \bar{d}\rangle + |s\bar{s}\rangle}{\sqrt{3}}, \quad \text{mass} = m + 3\varepsilon.$$

Since we have two states that are degenerate with the other six, we expect that these eight states lie in the same multiplet—the octet. The remaining state has a different mass and hence lies in a different multiplet, the singlet. The states above are called pure octet and singlet states.

What happens in the more realistic case where the mass of the s quark is different from that of the u and d, that is, where we keep a good $SU(2)$ symmetry only? In this case, we expect different mass parameters for $q_i, \bar{q}_j$ when i or $j = 3$, than for $i, j = 1$ or 2. This breaks the degeneracy among the six states where $i \neq j$. In addition, the annihilation diagrams could vary with quark type. For $i = j$, we then have to diagonalize a more general matrix of the form

$$H_m = \begin{pmatrix} m + \varepsilon & \varepsilon & \varepsilon' \\ \varepsilon & m + \varepsilon & \varepsilon' \\ \varepsilon' & \varepsilon' & m' + \varepsilon'' \end{pmatrix}. \tag{5.2}$$

For this matrix,

$$\frac{|u, \bar{u}\rangle - |d, \bar{d}\rangle}{\sqrt{2}}$$

remains an eigenstate with mass m, while the other two states mix together with details now dependent on the detailed numbers in the matrix. The mass of

$$\frac{|u, \bar{u}\rangle - |d, \bar{d}\rangle}{\sqrt{2}}$$

is unchanged because our matrix retains the $SU(2)$ subsymmetry. Our theorem about equal masses applies now to members of each isospin multiplet and

$$\frac{|u, \bar{u}\rangle - |d, \bar{d}\rangle}{\sqrt{2}}$$

is a member of an $I = 1$ multiplet, for which the other particles ($-|u, \bar{d}\rangle$ and $|d, \bar{u}\rangle$ in Table 5.1) have mass m.

Using this general discussion as guidance, we turn now to the real ground state mesons. These have quantum numbers similar to the $b\bar{b}$ and $c\bar{c}$ systems we looked at previously, but with annihilation and spin dependent splitting much larger since the systems are now relativistic. The quantum numbers of the onium ground states were $J^{PC} = 0^{-+}$ (lowest energy) and 1^{--}. Now only the mesons made of the same quark-antiquark types are C eigenstates. We will use the generic symbol P for the pseudoscalar 0^- mesons, and V for the vector 1^- mesons.

The octet of pseudoscalar mesons is shown in Figure 5.1.

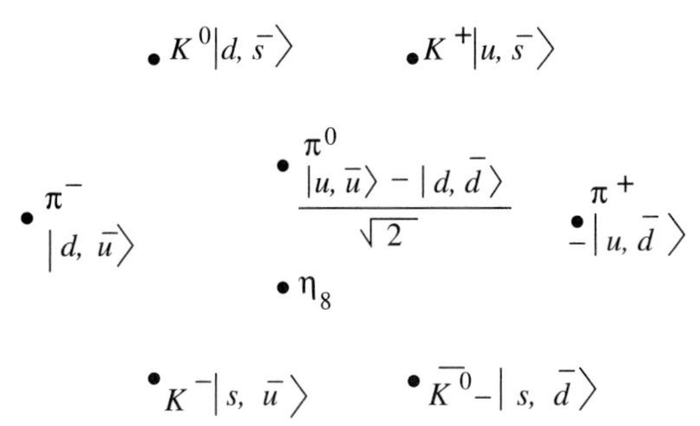

$$\bullet K^0 |d, \bar{s}\rangle \qquad \bullet K^+ |u, \bar{s}\rangle$$

$$\bullet \pi^0 \; \frac{|u, \bar{u}\rangle - |d, \bar{d}\rangle}{\sqrt{2}}$$

$$\bullet \pi^- \; |d, \bar{u}\rangle \qquad \qquad \qquad \bullet \pi^+ \; |u, \bar{d}\rangle$$

$$\bullet \eta_8$$

$$\bullet K^- |s, \bar{u}\rangle \qquad \bullet \overline{K^0} -|s, \bar{d}\rangle$$

FIGURE 5.1 Pseudoscalar octet. Note that the eighth state, η_8, is a mix of the physical states η and η'. It has the same I_3 and S as the π^0.

The octet contains two isospin doublets, with a charged and neutral kaon in each doublet, an isospin triplet of pions, and an isospin singlet state. $SU(3)$ breaking, in the limit where isospin is a perfect symmetry, will cause mixing of the isospin singlet in the octet with the isospin and $SU(3)$ singlet state. The two states

5.2 Light Mesons

of the 0^- nonet with $I = 0$, which are mixtures of the octet and singlet states, can be conveniently written as

$$\eta = X_\eta \left[\frac{|u, \bar{u}\rangle + |d, \bar{d}\rangle}{\sqrt{2}} \right] + Y_\eta |s, \bar{s}\rangle \tag{5.3}$$

$$\eta' = X_{\eta'} \left[\frac{|u, \bar{u}\rangle + |d, \bar{d}\rangle}{\sqrt{2}} \right] + Y_{\eta'} |s, \bar{s}\rangle . \tag{5.4}$$

We get the best estimates of the coefficients by studying decays of the type $P \to 2\gamma$, $V \to P + \gamma$, which are electromagnetic and therefore sensitive to both the quark charges and the amplitudes to find various quarks in the mesons. These give

$$X_\eta = Y_{\eta'} = .8, \quad X_{\eta'} = -Y_\eta = .6,$$

with uncertainties $\sim 5\%$. An alternative parameterization is in terms of the pure octet and singlet states:

$$\eta = \cos\theta_P |\eta_8\rangle - \sin\theta_P |\eta_1\rangle$$

$$\eta' = \cos\theta_P |\eta_1\rangle + \sin\theta_P |\eta_8\rangle .$$

θ_P is called the pseudoscalar mixing angle and is $\simeq -18°$, with an uncertainty of a few degrees. The particle masses are given in Table 5.2. We see from the table that

- Particle and antiparticle masses are equal.

- The $SU(2)$ breaking is ~ 5 MeV, based on the mass differences for charged and neutral kaons or pions.

- The $SU(3)$ breaking is large, based on the mass difference for kaons and pions.

- The mixing-annihilation terms are also large, based on the large η and η' masses.

The vector meson multiplet is given below in Figure 5.2. The two $I = 0$ states are now:

$$\omega^0 \simeq \frac{|u, \bar{u}\rangle + |d, \bar{d}\rangle}{\sqrt{2}} \quad \text{and} \quad \phi^0 \simeq |s, \bar{s}\rangle .$$

This mixing pattern is called ideal mixing. It follows from significant $SU(3)$ breaking due to the heavier strange quark mass but very little mixing from annihilation diagrams. A matrix that yields this pattern, where for simplicity we

TABLE 5.2
Pseudoscalar Meson Masses (MeV).

$m_{K^+} = m_{K^-} = 494$
$m_{K^0} = m_{\bar{K}^0} = 498$
$m_{\pi^+} = m_{\pi^-} = 140$
$m_{\pi^0} = 135$
$m_\eta = 547$
$m_{\eta'} = 958$

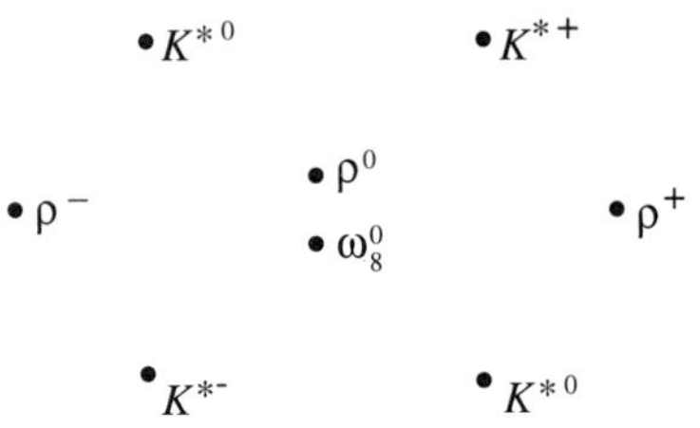

154 Chapter 5 Isospin and Flavor $SU(3)$, Accidental Symmetries

$$\begin{array}{ccc}
\bullet K^{*0} & & \bullet K^{*+} \\
& \bullet \rho^0 & \\
\bullet \rho^- & \bullet \omega_8^0 & \bullet \rho^+ \\
\bullet K^{*-} & & \bullet \bar{K}^{*0}
\end{array}$$

FIGURE 5.2 Vector meson octet. The eighth state, ω_8, is a mix of ω and ϕ.

take $SU(3)$ symmetry for the small annihilation terms, is:

$$H_m = \begin{pmatrix} m+\varepsilon & \varepsilon & \varepsilon \\ \varepsilon & m+\varepsilon & \varepsilon \\ \varepsilon & \varepsilon & m'+\varepsilon \end{pmatrix}, \quad \varepsilon \ll m'-m. \tag{5.5}$$

Using the physical masses to calculate the parameters in the matrix, $m_\omega - m_\rho \simeq 2\varepsilon$ implies $\varepsilon = 6$ MeV. The amplitude to find

$$\frac{|u,\bar{u}\rangle + |d,\bar{d}\rangle}{\sqrt{2}}$$

TABLE 5.3 Vector Meson Masses (MeV).

$m_{K^{*+}} = m_{K^{*-}} = 892$

$m_{K^{*0}} = m_{\bar{K}^{*0}} = 896$

$m_{\rho^0} \simeq m_{\rho^\pm} = 771$

$m_{\omega^0} = 783$

$m_{\phi^0} = 1019$

in the ϕ^0 (the deviation from perfect ideal mixing) is

$$\frac{\sqrt{2}\varepsilon}{m_\phi - m_\omega},$$

or about 0.036. The vector meson masses are given in Table 5.3.

Why the annihilation terms are so much smaller for the V nonet compared to the P nonet is unclear. The kinds of diagrams are different because of the opposite charge conjugation of the states. The simplest examples are shown in Figure 5.3.

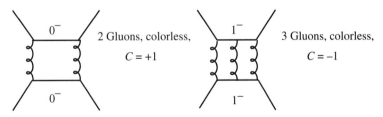

FIGURE 5.3 Simple annihilation diagrams leading to mixing in 0^- and 1^- systems.

The mixing patterns we have discussed above imply a significantly larger strange quark mass than the u and d masses. In addition, we expect some mixing from terms that violate the $SU(2)$ isospin symmetry, for example, electromagnetism and the difference in u and d masses. These mix states that differ in isospin but are similar otherwise. This is the analog of the $SU(3)$ mixing, which mixes the $I = 0$ states in the singlet and octet. Isospin breaking therefore mixes π^0, η, η' and ρ^0, ω, ϕ. The real states do not have precisely equal mixtures of $|u, \bar{u}\rangle$ and $|d, \bar{d}\rangle$.

Perhaps surprisingly, there is also mixing due to the weak interactions, which we will look at in detail in Chapter 9. It involves mixing between K^0 and $\bar{K}^0$. We expect weak effects to be very small, but since K^0 and $\bar{K}^0$ are degenerate and decay only through the weak interaction, even very tiny mixing terms can lead to very large effects.

Finally, we mention some of the limitations of the above approach to mixing. We have assumed that the states in a first-order approximation all have similar masses, which is a poor approximation for the 0^- mesons, where the masses differ greatly. In the mixing, we have ignored all states other than those in the given nonet. However, the vector mesons have large widths, so they have significant transitions to other states. In the next section we discuss a more general approach, which will better handle some of these issues. Otherwise, we actually have an inconsistency that was glossed over. For the vector mesons, the value $\varepsilon \simeq 6$ MeV is comparable to the mass difference of the $|u, \bar{u}\rangle$ and $|d, \bar{d}\rangle$ systems (that is, to the isospin breaking). Including such an isospin mass difference in H_m leads to much more mixing than is seen between

$$I = 1 \quad \rho^0 = \frac{|u, \bar{u}\rangle - |d, \bar{d}\rangle}{\sqrt{2}} \quad \text{and} \quad I = 0 \quad \omega = \frac{|u, \bar{u}\rangle + |d, \bar{d}\rangle}{\sqrt{2}}.$$

The reader should think about how this problem can be resolved. We discuss it in Section 6.4.3.

5.3 ■ BREIT–WIGNER PROPAGATORS AND MESON LIFE HISTORIES

We turn next to a systematic discussion of the life history of the various meson states. We know that none of them are fully stable, so they all decay by one interaction or another. In this chapter we look at strong decays only; in subsequent chapters we will look at decays due to the other interactions. The latter do not respect the approximate $SU(3)$ flavor symmetry of the strong decays. Because of the inability to deal quantitatively with the nonperturbative strong interactions, we typically cannot calculate the value of the decay width for strong decays. The discussion will thus focus on relations between rates that follow from $SU(3)$ symmetry or the constraints coming from conservation rules such as angular momentum or parity. The decays of the mesons conserve the various quantum numbers appropriate to the symmetries of the interaction responsible for the decay. For the strong decays these will include the internal symmetries arising from $SU(3)$ (isospin and strangeness conservation), charge conjugation symmetry, and parity conservation. We will find that we can introduce one last symmetry called

G-parity (or generalized parity), which is related to isospin symmetry. We will look at this symmetry later in the chapter.

The simplest systems that decay predominantly via the strong interactions are the vector mesons, which decay into pseudoscalar mesons. We list the primary (strong) decays and particle widths in Table 5.4 (we can get antiparticle decays by changing particles to antiparticles everywhere in the table). Note that all the decays conserve the additive strong quantum numbers I_3 and S. The perfect $SU(3)$ symmetry expectation, that all members of the vector meson octet have the same width, is badly broken because of the large $SU(3)$ violation in the mass values for the vector mesons and the pseudoscalars. The pseudoscalar mesons produced in the vector meson decay are constrained by energy conservation to decay subsequently via the weak or electromagnetic interactions. The vector mesons also have small, but interesting, branching ratios due to electromagnetic decays, and in principle (but much too small to measure) weak decays. We will look at the electromagnetic decays in the next chapter.

TABLE 5.4 Vector Meson Decay Properties.

State	Γ	Final states for strong decays
K^{*+}	51 MeV	$\frac{2}{3}(\pi^+ K^0) + \frac{1}{3}(\pi^0 K^+)$
K^{*0}	51 MeV	$\frac{1}{3}(\pi^0 K^0) + \frac{2}{3}(\pi^- K^+)$
ρ^+	150 MeV	$\pi^+ \pi^0$
ρ^0	150 MeV	$\pi^+ \pi^-$
ω^0	8.4 MeV	$\pi^+ \pi^- \pi^0$
ϕ^0	4.3 MeV	$.49(K^+ K^-) + .34(K^0 \bar{K}^0) + .15(\pi^+ \pi^- \pi^0)$

The various widths in Table 5.4 correspond to very short lifetimes and the 1^- mesons are therefore not observed as distinct objects, but rather as resonances in multiparticle distributions. Examples are shown in Figure 5.4. These data were taken using $\pi^+ p$ reactions, yielding various multibody final states. The mass distributions in Figure 5.4 can be understood in terms of contributions of decaying resonances and an incoherent particle mix with a mass distribution given primarily by the available phase space.

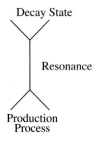

Before continuing to look specifically at the decays in Table 5.4, we must understand a little better how an unstable particle appears in a given invariant mass distribution. It is an intermediate state between a production process and a final decay state. Thus, taking the simplest situation, it must arise using the process illustrated at the left.

This looks no different from a propagator in a Feynman diagram, and in fact it isn't. For example, the virtual photon in $e^+ e^- \rightarrow$ hadrons also behaves as a resonance. What we would like to calculate is the correct propagator to use

5.3 Breit–Wigner Propagators and Meson Life Histories

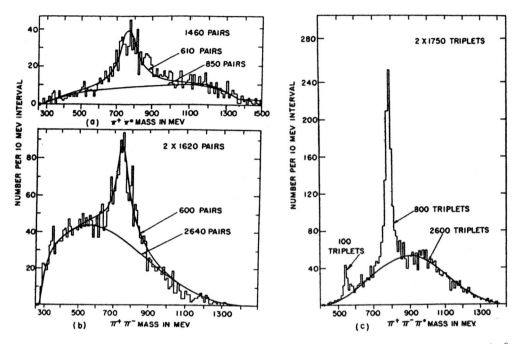

FIGURE 5.4 Production of vector meson resonances. (a) Mass distribution of $\pi^+\pi^0$ pairs produced in reaction $\pi^+ p \to \pi^+ \pi^0 p$. Peak corresponds to ρ^+. (b) Mass distribution of $\pi^+\pi^-$ pairs produced in reaction $\pi^+ p \to \pi^+ \pi^+ \pi^- p$. Peak corresponds to ρ^0. (c) Mass distribution of $\pi^+\pi^0\pi^-$ triplets produced in reaction $\pi^+ p \to \pi^+ \pi^- \pi^0 p$. Large peak corresponds to ω^0, tiny peak to η. The various smooth curves correspond to phase space. [From C. Alff et al., *Phys. Rev. Letters* 9, 322 (1962).]

given that there are dynamics leading to the decay. For simplicity we take a spin 0 resonance (or, imagine doing the calculation for a given helicity intermediate state, and then summing over helicities). The resonance is assumed to live long enough that it leaves the production region. Subsequently, during propagation, it interacts only via processes it can participate in as an isolated particle. This gives, however, corrections to the propagator, which we can picture as shown in Figure 5.5. If we write the amplitude for one interaction as a function $\pi(q^2)$,

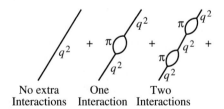

FIGURE 5.5 Particle propagation including interactions shown as a bubble. The amplitude to interact, denoted as the function π, is a function of q^2.

then in terms of this function we get a series of contributing diagrams. Using the perturbation series, the full amplitude is:

$$\frac{i}{q^2 - m^2 + i\varepsilon} + \frac{i}{q^2 - m^2 + i\varepsilon}\left(-i\pi(q^2)\right)\frac{i}{q^2 - m^2 + i\varepsilon} + \cdots.$$

We can sum this series, which gives for the corrected propagator

$$\frac{i}{q^2 - m^2 - \pi(q^2)}.$$

We will see that the function $\pi(q^2)$ typically has an imaginary part, so we can drop the $i\varepsilon$ in those cases where it does.

The corrections due to interactions while propagating are called loop corrections. They have both a real part, which shifts (or renormalizes) the apparent mass of the resonance, and an imaginary part, which is related to the decay width. We can calculate the imaginary part in a general way, using unitarity. Note that π is just M_{ii}—the amplitude for the state to go into itself due to the interactions, rather than into a new final state.

We write the S matrix for the resonance to go into a specific final state as

$$S_{fi} = \delta_{fi} - i(2\pi)^4 \delta^4(p_f - p_i) M_{fi}.$$

S is a unitary matrix, which means that

$$SS^\dagger = I \quad \text{or} \quad \sum_n S_{nf}^* S_{ni} = \delta_{fi}.$$

(Here $\sum_n$ is the integral over the multiparticle invariant phase space for state n summed over all choices for n).

Substituting for S_{fi} in terms of M_{fi} and multiplying terms gives the general unitarity relation:

$$i(M_{if}^* - M_{fi}) = -\sum_n M_{nf}^* M_{ni} (2\pi)^4 \delta^4(p_n - p_i),$$

or taking $i = f$, the imaginary part satisfies

$$\frac{-\operatorname{Im}\pi(q^2)}{m_R} = \frac{1}{2m_R} \sum_n |M_{ni}|^2 (2\pi)^4 \delta^4(p_n - p_i), \quad m_R^2 = q^2. \qquad (5.6)$$

The expression on the right is just the usual formula for the width for $q^2 = m_R^2$ in the sum. It is a width calculation with the particle virtual, that is, with a variable mass of m_R instead of m. In particular, if q^2 is small enough (or negative) so that no decay channels exist, the width vanishes. As an example, the Z^0 is produced in e^+e^- annihilation at M_Z with a width of about 2.5 GeV. The Z^0 can also mediate low-energy neutrino scattering. Here $q^2 < 0$ and the width of the Z^0 is zero.

For a narrow resonance, $\Gamma \ll m$, we can absorb the real part of π, Re π, into a new definition of the resonance mass squared, m_R^2 (which is approximately constant in the vicinity of the resonance), and approximate $\Gamma(q^2)$ as $\Gamma(m_R^2) = \Gamma_R$

5.3 Breit–Wigner Propagators and Meson Life Histories

for $q^2 \approx m_R^2$. Then the propagator is

$$\frac{i}{q^2 - m_R^2 + i m_R \Gamma_R}. \tag{5.7}$$

This is called a relativistic Breit–Wigner and its square describes the shapes in Figure 5.4, provided that the resolution of the measuring apparatus doesn't distort the shape, and $\Gamma_R \ll m_R$. In fact, Γ_R is extracted by fitting the shape. Note that for higher precision, or wide states, we must keep the q^2 dependence of $\mathrm{Im}\,\pi(q^2)$, which requires knowing the behavior of $|M_{ni}|^2$ with q^2, as well as the change in phase space for the decays that enter into Γ_R.

We can also look at the Fourier transform of the Breit–Wigner propagator in terms of the exchange of particles, as was done in Section 2.11. We assume a narrow resonance ($\Gamma_R \ll m_R$) and a fixed-width approximation. Defining $E = \sqrt{|\vec{q}|^2 + m_R^2}$, the propagator

$$\frac{i}{q^2 - m_R^2 + i m_R \Gamma_R}$$

has poles at locations

$$q_0 = E - \frac{i m_R}{E} \frac{\Gamma_R}{2} \quad \text{and} \quad q_0 = -\left(E - \frac{i m_R}{E} \frac{\Gamma_R}{2}\right),$$

to first order in Γ_R. Thus, integrating over q_0 we get the expression:

$$\int \frac{i}{q^2 - m_R^2 + i m_R \Gamma_R} e^{-i q \cdot x} \frac{d^4 q}{(2\pi)^4}$$
$$= \int \frac{d^3 p}{2E (2\pi)^3} \left[e^{-i p \cdot x} e^{-\frac{m_R}{E} \frac{\Gamma_R}{2} t} \theta(t) + e^{i p \cdot x} e^{\frac{m_R}{E} \frac{\Gamma_R}{2} t} \theta(-t) \right].$$

Here we have kept Γ_R only in the exponential. The Fourier transform of the Breit–Wigner corresponds to the exchange of particles, which, however, decay exponentially from the production point. The decay rate in space-time, as opposed to the amplitude, has the expected form $e^{-\Gamma_R t}$ in the rest frame, and for a moving particle includes the relativistic time dilation.

5.3.1 ■ Narrow Resonances and Independent Events

An amplitude dominated by a narrow resonance can be thought of as two separated independent events, for example, production of the resonance in an interaction and then later, outside the initial interaction volume, decay of the resonance. This is necessary since a very small Γ_R corresponds to a long lifetime, in which case we can in fact measure the intermediate state, which can fly a long distance. For example, a π meson produced in a reaction eventually decays, but it lives

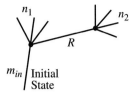

FIGURE 5.6 Decay of initial state to $n_1 + R$ with subsequent decay of R to n_2.

long enough that we can make π beams. The Breit–Wigner has exactly this property, as we now demonstrate for a spinless state. We look at the process shown in Figure 5.6.

We consider a decay of an initial single particle state; however, the calculation works the same way for a two particle scattering initial state. We assume below that n_1 and n_2 stand for the final states at the vertices shown in Figure 5.6. The reader can calculate analogous results when there are a number of possible final states. We want to show that $\Gamma(in \to n_1 + R) = \Gamma(in \to n_1 + n_2)$ through the resonance, that is, we can equally well calculate the decay as happening to $n_1 + R$ or into $n_1 + n_2$. Stated another way, we can think of the decay $R \to n_2$ as happening much later, which then can't affect the decay width of the initial state.

We write the decay formula for the diagram in Figure 5.6:

$$\Gamma_{in} = \frac{1}{2m_{in}} \sum_n \frac{|M(R \to n_2)|^2 |M(in \to n_1 + R)|^2}{|q^2 - m_R^2 + im_R\Gamma_R|^2} (2\pi)^4 \delta^4(p_f - p_{in}).$$

This formula consists of amplitudes at the two vertices and one propagator, where

$$\sum_n = \int \prod_{n_1} \frac{d^3 p_n}{2E_n(2\pi)^3} \prod_{n_2} \frac{d^3 p_n}{2E_n(2\pi)^3}.$$

The Breit–Wigner propagator squared is

$$\frac{1}{(q^2 - m_R^2)^2 + m_R^2 \Gamma_R^2}.$$

For Γ_R very narrow ($\Gamma_R \ll m_R$), we get contributions only for q^2 near m_R^2. We can therefore approximate the propagator as a δ function, which is the key factor in relating the rate formulas. By integrating, we can check that the correspondence is

$$\frac{1}{(q^2 - m_R^2)^2 + m_R^2 \Gamma_R^2} \simeq \frac{2\pi \delta(q^2 - m_R^2)}{2m_R \Gamma_R}.$$

We also make one more change, which is

$$[\ ] \delta^4(p_f - p_{in}) = \int [\ ] \delta^4(p_{n_1} + p_R - p_{in}) \delta^4(p_{n_2} - p_R) d^4 p_R,$$

where [ ] is any function of the momenta. Here $p_R^2 = q^2$.

With these changes we have

$$\Gamma_{in} = \frac{1}{2m_{in}} \int \sum_{n_1} \sum_{n_2} |M(R \to n_2)|^2 |M(in \to n_1 + R_2)|^2$$

$$\times (2\pi)^4 \delta^4(p_{n_1} + p_R - p_{in}) \frac{2\pi \delta(q^2 - m_R^2)}{2m_R \Gamma_R} \frac{(2\pi)^4 \delta^4(p_{n_2} - p_R)}{(2\pi)^4} d^4 p_R.$$

5.3 Breit–Wigner Propagators and Meson Life Histories

Next, we use

$$\int \delta(q^2 - m_R^2) d^4 p_R = \frac{d^3 p_R}{2 E_R}$$

after integrating over the energy component of p_R. Thus we can write

$$\Gamma_{\text{in}} = \frac{1}{2m_{\text{in}}} \int_{p_R} \sum_{n_1} \underbrace{\frac{d^3 p_R}{2 E_R (2\pi)^3} |M(in \to n_1 + R)|^2 (2\pi)^4 \delta^4(p_{n_1} + p_R - p_{\text{in}})}_{\text{Independent of } n_2 \text{ variables}}$$

$$\times \frac{1}{\Gamma_R} \underbrace{\left(\frac{1}{2 m_R} \sum_{n_2} |M(R \to n_2)|^2 (2\pi)^4 \delta^4(p_{n_2} - p_R) \right)}_{\Gamma_R}$$

Note that the $\sum_{n_2}$ term is invariant, so we can calculate it in the m_R rest frame giving the normal expression for Γ_R. Thus:

$$\Gamma_{\text{in}} = \frac{1}{2m_{\text{in}}} \sum_{n_1, R} |M(in \to n_1 + R)|^2 (2\pi)^4 \delta^4(p_{n_1} + p_R - p_{\text{in}}).$$

This is just the standard calculation of the decay of the initial particle to $n_1 + R$, with R treated the same way as the other stable particles.

The reader should look at the case where R has, for example, spin 1. This case is interesting, since in calculating Γ_{in} with R a real final state we use $\sum_{\lambda_R} |M(n_1, R, \lambda_R)|^2$ (that is, the helicity λ_R is a measurable); while treating R as an intermediate state, we use

$$\left| \sum_{\lambda_R} M(R, \lambda_R \to n_2) M(n_1, R, \lambda_R) \right|^2$$

(that is, λ_R is summed over in the amplitude). For the two approaches to give the same width for the initial state requires a linear dependence of M on the particle spins, the orthogonality of the spin states, and that the decay rate of R is independent of λ_R (rotational invariance). Thus, the equality depends on some of our very basic assumptions for how quantum mechanics works.

5.3.2 ■ Propagators and Mass Eigenstates

For $\text{Re}\,\pi$ and $\text{Im}\,\pi \ll m^2$, we can approximate

$$q^2 - m^2 - \pi(q^2) \simeq q^2 - \left(m_R - \frac{i \Gamma_R}{2} \right)^2, \tag{5.8}$$

where

$$\frac{\pi(q^2)}{2m_R} = \Delta m - \frac{i\Gamma_R}{2} \quad \text{for } q^2 \simeq m_R^2, \quad \text{and} \quad m_R = m + \Delta m.$$

The operator H_m, which we diagonalized in Section 5.2 to arrive at the meson masses and eigenstates, is $m + \Delta m$, considered as a matrix in the $SU(3)$ space of states for a given spin and parity. This should work if we ignore Γ_R and assume that Δm is small for all the states. This is clearly very approximate; for example, Γ_R is not small for the vector mesons and the pseudoscalar mass variation is large.

From the propagator point of view, mixing is caused by diagrams shown in Figure 5.7. The gluon annihilation diagrams of Figure 5.3 would, for example, provide a contribution to π_{ij}. Assuming a negligible width, as for the pseudoscalar mesons, these diagrams generate a matrix (the mass-squared matrix) $m_{ij}^2 + \pi_{ij}(q^2)$, in the space of states that can turn into each other via the interactions. If we diagonalize the matrix for a given q^2, we get eigenvalues $m_R^2(q^2)$. An eigenvalue will represent a physical resonance when $q^2 = $ one of the $m_R^2(q^2)$. These states propagate without mixing and are the mass eigenstates we find in an experiment.

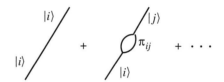

FIGURE 5.7 Mixing in particle propagation.

The mass-squared matrix is often used to calculate mass relations for the $SU(3)$ pseudoscalar mesons, rather than the more approximate mass matrix H_m. Given the measured mass eigenvalues and the makeup of the eigenstates in terms of quark and antiquark types, we can try to construct π_{ij}. Since the states in the propagator are virtual and all have the same q^2, we have to make an assumption for the q^2 dependence of $\pi_{ij}(q^2)$, for example, that it is approximately constant over the mass range of the mesons. The general conclusions using such a mass-squared matrix are similar to those reached earlier in Section 5.2 using H_m.

The real part of $\pi(q^2)$ has, however, additional far-reaching effects. The renormalization from the real part arises for all interactions, because our previous discussion for resonances of the strong interactions would apply equally well to higher-order diagrams for any particle that has an interaction vertex. This topic is very interesting and is related to some very important ideas we look at later— renormalization, including which theories make sense in perturbation theory, and running coupling constants, that is, the interaction strength for gauge theories. The resulting renormalization is often infinite if just calculated in a direct way. The methods for dealing with these infinities, combined with our earlier

5.4 ■ VECTOR MESON DECAYS

We return now to the strong decays of the ground-state vector mesons. The widths and final states were listed in Table 5.4. Notice first that states in the same $SU(2)$ multiplet (K^{*+} and K^{*0} or ρ^+ and ρ^0) have the same width. This follows from the $SU(2)$ symmetry being nearly unbroken for the case where the decay is due to the strong interactions. We expect also that all analogous decays are related at least approximately by the full $SU(3)$ symmetry. In general, symmetry breaking can occur in the decay dynamics and certainly occurs when the phase space is different due to the various mass differences. We return to this question after we look at the decays from the point of view of J^P conservation.

The vector mesons decay to two or three pseudoscalar mesons; we look at the decays to two first, as shown in Figure 5.8. The two mesons in the final state must be in a relative state of one unit of angular momentum (a p-wave decay), since a $1^- \to 0^- + 0^-$ requires $\ell = 1$ to conserve angular momentum and parity. The neutral nonstrange vector mesons are C eigenstates of negative charge conjugation, which must lead to a negative C final state. For the case where the final mesons are $\pi^+\pi^-$, a charge conjugation transformation exchanges the particles so that C equals $(-1)^\ell$, which is negative for the p-wave state. This matches the C of the initial state. Since $C|\pi^0\rangle = |\pi^0\rangle$, the final state $|\pi^0\pi^0\rangle$ is not produced (as seen in Table 5.4) since it would violate C invariance (in fact, $\ell = 1$ is not allowed for two identical final state bosons).

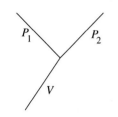

FIGURE 5.8 Vector decay to two pseudoscalars.

The matrix element for the decay is:

$$M_{fi} = g_{VP_1P_2} e_V^\mu (p_1^\mu - p_2^\mu).$$

This is the only nonvanishing invariant linear in the spin, since $e_V^\mu(p_1^\mu + p_2^\mu) = e_V^\mu p_V^\mu = 0$. Thus the dynamics are reflected entirely in the one complex number $g_{VP_1P_2}$. If we calculate the width from M_{fi}, the result is

$$\Gamma_{V \to P_1 P_2} = \frac{|g_{VP_1P_2}|^2}{6\pi} \frac{|\vec{p}_1|^3}{m_V^2}, \tag{5.9}$$

where $\vec{p}_1$ is the final-state momentum in the V rest frame. Note that for e_μ^V a given state of $|\ell = 1, m\rangle$ in the rest frame of V, we get the expected angular distribution:

$$\frac{d\Gamma_{V \to P_1 P_2}}{d\Omega} \propto |Y_{1,m}(\theta, \phi)|^2.$$

The decay to three pions, for example from the ω^0, yields a more complicated final state. To just write the matrix element in momentum space is easiest, although we first make a few comments from the angular momentum point of view. We can make up a $\pi^+\pi^-\pi^0$ final state by taking a $\pi^+\pi^-$ state with an angular momentum ℓ in the $\pi^+\pi^-$ relative center of mass and then adding a π^0 with angular momentum L between the π^0 and the overall $\pi^+\pi^-$ system. The spatial contribution to the parity is $(-1)^{\ell+L}$ and J is obtained through vector addition of ℓ and L. The negative intrinsic parity for each pion will contribute an additional factor of -1 to the overall parity. For the ω^0 decay we want a final state of $C = -1$ to match the charge conjugation of the initial state. This means that $3\pi^0$ is not allowed as a final state; $\pi^+\pi^-\pi^0$ does occur. Since $C|\pi^0\rangle = +|\pi^0\rangle$ and $C|\pi^+\pi^-\rangle = (-1)^\ell|\pi^+\pi^-\rangle$ for a state of given ℓ, we see that ℓ must be odd. For overall parity conservation, $(-1)^{\ell+L} = +1$; thus L is also odd. The simplest choice is $\ell = L = 1$, in order to get $J^{PC} = 1^{--}$. For ℓ, L both odd, the state changes sign under exchange of any pair of pions. This property must also hold for M_{fi} written in terms of momentum space variables.

In momentum space, M_{fi} will be a function of the independent variables needed to describe the final phase space distribution. For example, for $V \to P_1 + P_2$ discussed earlier, the final momenta of P_1 and P_2 are uniquely determined in the center of mass and the only degree of freedom is the angular distribution, which is determined by angular momentum conservation. Looking at a three-body decay in the initial-state center of mass, a main constraint from momentum conservation is that the final particles must lie in a plane. The decay plane configuration is shown in Figure 5.9.

To describe a three-body decay requires

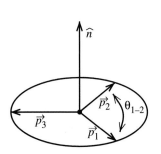

FIGURE 5.9 Momentum-space configuration in three-particle decay.

1. Specifying the orientation of the decay plane. There are several ways to do this, for example by specifying (θ, ϕ) of the plane normal $\hat{n}$, which we can define via the direction of $\vec{p}_1 \times \vec{p}_2$ in a given coordinate system.

2. For fixed energies and relative angles of the three final particles in the decay plane, the configuration can be rigidly rotated around the axis $\hat{n}$, providing another valid decay configuration. Specifying the rotation of all the vectors around the axis $\hat{n}$ requires another azimuthal angle. This third angle is needed to describe the decay configuration for a given external frame. The distribution of the three angles is determined by the particle spins; for example, if nothing has spin or we sum over all spins, then $|M_{fi}|^2$ does not depend on the three angles, since the orientation of the plane in space must be equally likely for all choices.

3. The configuration within the plane is specified by two variables—for example, E_1 and E_2. E_3 and $\cos\theta_{ij}$ are then determined by 4-momentum conservation.

A three-body decay is therefore described by five variables, of which two determine the configuration in the decay plane and reflect the dynamics, and three depend on helicities. We can see another way that five variables determine the

5.4 Vector Meson Decays

final state, since the phase space integral involves

$$\underbrace{\frac{d^3p_1}{2E_1}\frac{d^3p_2}{2E_2}\frac{d^3p_3}{2E_3}}_{9} \delta^4(p_1+p_2+p_3-p_{\text{in}})$$

$$9 \quad - \quad 4 \quad = 5 \text{ variables.}$$

Often the variables chosen are invariant masses that are related to the center of mass energies; for example:

$$m_{23}^2 = (p_2+p_3)^2 = (p_{\text{in}}-p_1)^2 = m_{\text{in}}^2 + m_1^2 - 2m_{\text{in}}E_1.$$

Thus E_1 and m_{23}^2 are linearly related; the same is true for E_2 and m_{13}^2, and E_3 and m_{12}^2. A plot of the density of events versus E_1, E_2, shown schematically in Figure 5.10, gives information on the matrix element squared and is called a Dalitz plot.

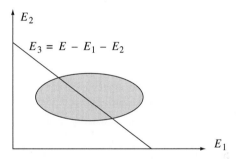

FIGURE 5.10 Plot of event density versus E_1 and E_2. Line is a curve of constant E_3. Shaded area is the kinematically allowed region for E_1 and E_2. This is called a Dalitz plot.

We will calculate the density to see what the relationship is to the square of the matrix element. Integrating over d^3p_3 leaves us with the expression:

$$d\Gamma = \frac{|M_{fi}|^2}{16m_{\text{in}}(2\pi)^5} \frac{d^3p_1}{E_1} \frac{d^3p_2}{E_2} \frac{\delta(E_1+E_2+E_3-E_{\text{in}})}{E_3},$$

where

$$\frac{d^3p_1}{E_1}\frac{d^3p_2}{E_2} = (p_1 dE_1 p_2 dE_2)d\Omega_1 d\Omega_2.$$

Ω_1, Ω_2 are defined with respect to an external axis. We change variables to θ_1, ϕ_1, relative to an external coordinate system (the same as before), and for the other variable we use the angles of 2 relative to 1 with the direction of $\vec{p}_1$ chosen as a z-axis. This defines

$$\phi_{1-2} = \phi_1 - \phi_2 \quad \text{and} \quad \cos\theta_{1-2} = \frac{\vec{p}_1 \cdot \vec{p}_2}{|\vec{p}_1||\vec{p}_2|}.$$

The relation between the solid angle elements is

$$d\Omega_1 d\Omega_2 = d\Omega_1 d\Omega_{1-2}.$$

In the absence of polarization of the initial state, the matrix element squared cannot depend on an arbitrary direction in space, so we can integrate over $d\Omega_1$ and $d\phi_{1-2}$, which have nothing to do with the relative configuration of the particles. Defining (where an average over initial spins has been performed):

$$\int \sum_{\text{spins}} |\overline{M_{fi}}|^2 d\Omega_1 d\phi_{12} = 4\pi \times 2\pi \times F(E_1, E_2),$$

$$d\Gamma = \frac{1}{8M_{\text{in}}(2\pi)^3} F(E_1, E_2) p_1 dE_1 p_2 dE_2 \frac{d\cos\theta_{12}}{E_3} \delta(E_1 + E_2 + E_3 - E).$$

Finally, we have:

$$\frac{d\Gamma}{dE_1 dE_2} = \frac{1}{8m_{\text{in}}(2\pi)^3} F(E_1, E_2) \frac{p_1 p_2}{E_3(\frac{dE_3}{d\cos\theta_{12}})}$$

after integrating the δ function against $d(\cos\theta_{12})$. Writing $E_3^2 = p_3^2 + m_3^2 = [(\vec{p}_1 + \vec{p}_2)^2 + m_3^2]$ in the m_{in} rest frame gives

$$E_3^2 = p_1^2 + p_2^2 + m_3^2 + 2\vec{p}_1 \cdot \vec{p}_2 = 2|\vec{p}_1||\vec{p}_2|\cos\theta_{12} + p_1^2 + p_2^2 + m_3^2.$$

Thus

$$\frac{E_3 dE_3}{d\cos\theta_{12}} = p_1 p_2 \quad \text{and} \quad \frac{d\Gamma}{dE_1 dE_2} = \frac{1}{8m_{\text{in}}(2\pi)^3} F(E_1, E_2).$$

The Dalitz plot density is directly proportional to the spin averaged matrix element squared and provides a nice link between data and the dynamics.

The ω^0 Dalitz plot (using somewhat different coordinates, here $T_i =$ kinetic energy $= E_i - m_i$) is shown in Figure 5.11. Dashed contour lines indicate regions of equal area, which would contain equal numbers of events if $|M_{fi}|^2 =$ constant. The plot is relatively smooth (although depleted near the outer boundary), since there are no resonances with mass below $m_\omega - 3m_\pi$, which would enhance the amplitude for a given π, π combination. It is interesting to compare this to the decay of the J/ψ to $\pi^+\pi^-\pi^0$, whose Dalitz plot is shown in Figure 5.12. The J/ψ is the analogous 1^- meson made of $c\bar{c}$. Note the strong dominance by $J/\psi \to \rho\pi \to 3\pi$, which can be seen directly from the bands in the Dalitz plot. A similar result occurs for $J/\psi \to K\bar{K}\pi$, also shown in Figure 5.12. The three bands in $J/\psi \to 3\pi$ correspond to $\rho^+\pi^-$, $\rho^0\pi^0$, and $\rho^-\pi^+$, while $J/\psi \to K^+K^-\pi^0$ has only two bands for $K^{*+}K^-$, $K^{*-}K^+$. The third band,

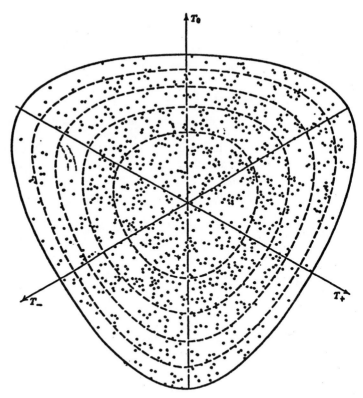

FIGURE 5.11 Dalitz plot for the decay $\omega \to \pi^+\pi^-\pi^0$. Dashed bands contain equal areas. Axes shown are the kinetic energies of each pion in the ω^0 rest frame; all points satisfy the constraint that the sum of the three kinetic energies is a constant. [From C. Alff et al., *Phys. Rev. Letters* 9, 325 (1962).]

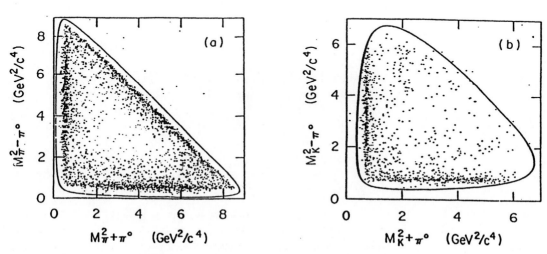

FIGURE 5.12 Dalitz plots for (a) $J/\psi \to \pi^+\pi^-\pi^0$ and (b) $J/\psi \to K^+K^-\pi^0$. [From R. Baltrusaitis et al., *Phys. Rev.* D32, 2883 (1985).]

which would correspond to $J/\psi \to \phi^0 \pi^0$, is not allowed by isospin conservation since $I(J/\psi) = I(\phi) = 0$, and $I(\pi^0) = 1$.

5.4.1 ■ Form of Meson Decay Amplitudes

We return now to the $\omega^0 \to 3\pi$ matrix element. It must be linear in the ω^0 spin, change sign under exchange of any of the π pairs, and be a scalar under Lorentz transformations. The unique momentum space amplitude of this form can be written in terms of one arbitrary function $g(E_1, E_2)$:

$$M_{fi} = g(E_1, E_2) \varepsilon_{\mu\nu\sigma\rho} e^{\omega^0}_\mu p^{\pi^+}_\nu p^{\pi^-}_\sigma p^{\pi^0}_\rho.$$

Note that $g(E_1, E_2)$ is a fairly smooth function based on the smooth Dalitz plot. It is a symmetric function under interchange of any pair of pions, that is, a symmetric function of E_{π^+}, E_{π^-}, and E_{π^0}, and we can choose any pair as E_1, E_2. M_{fi} vanishes if two pion momenta are collinear; this corresponds to the Dalitz plot boundary and accounts for the depletion at the edges. The fully anti-symmetric symbol, $\varepsilon_{\mu\nu\sigma\rho}$, generalizes the cross product to the case of a 4-dimensional space.

We can calculate M_{fi} in the ω^0 rest frame. To get a simple expression, we make the substitution $p^{\pi^0} = p^{\omega^0} - (p^{\pi^+} + p^{\pi^-})$, which implies that we can also write $M_{fi} = g(E_1, E_2) \varepsilon_{\mu\nu\sigma\rho} e^{\omega^0}_\mu p^{\pi^+}_\nu p^{\pi^-}_\sigma p^{\omega^0}_\rho$. In the ω^0 rest frame, this is

$$M_{fi} = m_{\omega^0} g(E_{\pi^+}, E_{\pi^-})[\hat{e}_{\omega^0} \cdot (\vec{p}_{\pi^+} \times \vec{p}_{\pi^-})],$$

which is a spatial pseudoscalar. We will call the 4-dimensional expression a momentum-space Lorentz pseudoscalar. Note that the expression $\hat{e}_{\omega^0} \cdot (\vec{p}_{\pi^+} \times \vec{p}_{\pi^-}) = (\hat{e}_{\omega^0} \cdot \hat{n})|\vec{p}_{\pi^+}||\vec{p}_{\pi^-}|\sin\theta_{12}$. We see that the polarization affects only the angular distribution of the normal, $\hat{n}$, to the decay plane. It is distributed as $|Y_{1,m}(\theta, \phi)|^2$ for an initial ω^0 polarization state $|1, m\rangle$. For an unpolarized ω^0,

$$|\overline{M_{fi}}|^2 = \frac{m^2_{\omega^0}}{3}|g(E_{\pi^+}, E_{\pi^-})|^2|\vec{p}_{\pi^+}|^2|\vec{p}_{\pi^-}|^2 \sin^2\theta_{12},$$

which now depends only on the variables within the decay plane.

We can generalize the idea of a Lorentz pseudoscalar, so that it is not necessary to look at individual two-body angular momenta to derive the properties of the matrix element for several particles in the final state, but instead can directly find general forms of a matrix element in momentum space. We consider a decay that conserves parity and involves mesons only. This means, for the S matrix, $\langle Pf|S|Pi\rangle = \langle f|S|i\rangle$, where $|Pi\rangle$ and $|Pf\rangle$ are the parity reflected states of $|i\rangle$ and $|f\rangle$, respectively. For the corresponding matrix elements this gives

$$M_{fi}(-\vec{V}_m)(-1)^n = M_{fi}(\vec{V}_m),$$

where $\vec{V}_m$ is the set of all space vectors appearing in M_{fi} (the time components of these are unchanged under the parity transformation) and n = number of unnatural spin parity particles in the initial plus final state. For n odd, M_{fi}, with regard to the various 4-momenta, is a momentum space Lorentz pseudoscalar. It must be a Lorentz scalar under proper Lorentz transformations, but must involve a product of an odd number of spatial momenta. Dot products of 4-vectors are Lorentz scalars, but involve an even number of spatial momenta. Therefore we cannot make a suitable M_{fi} out of such dot products for n odd. We can, however, accomplish what we want by having a matrix element that is a function of the Lorentz scalars multiplied by

$$\varepsilon_{\mu\nu\sigma\rho} V_\mu^{(1)} V_\nu^{(2)} V_\sigma^{(3)} V_\rho^{(4)}.$$

The $V_\mu^{(i)}$ can be spin or 4-momentum vectors. All four must be different or the sum will vanish, since $\varepsilon_{\mu\nu\sigma\rho}$ is antisymmetric. This forbids a number of reactions, since no valid matrix element can be constructed. For example, the following processes are forbidden if parity is conserved (where J^P for the initial and the final particles are indicated), since n is odd and the process does not have four independent 4-vectors:

$$0^- \to 0^+ + 0^+, \; 0^- \to 0^- + 0^-, \; 0^+ \to 0^- + 0^- + 0^-, \; 1^- \to 0^- + 0^+.$$

For n even, M_{fi} is a momentum space Lorentz scalar, which can be constructed from constants or dot products of 4-momenta; such processes therefore are allowed by J^P conservation.

5.5 ■ PHYSICAL PICTURE OF DECAY PROCESS

We turn next to the ϕ, which is predominantly an $|s, \bar{s}\rangle$ state. It is the only vector meson with significant decays into both two and three pseudoscalars. Using the relative rates and a comparison to the ω^0 allows some insight into possible decay mechanisms. The largest branching ratio is to final states that contain the two initial strange quarks. These decays must happen mainly through the creation of new $q\bar{q}$ pairs, which combine with those already there to form new mesons, rather than through annihilation into gluons that would, in the subsequent creation of particles, not favor strange particles. If the latter were responsible for ω^0 and $\phi^0 \to 3\pi$ we would expect that $\Gamma_{\phi^0 \to 3\pi} > \Gamma_{\omega \to 3\pi}$, whereas it is about 15 times smaller. The possible diagrams are of the type shown below and are labeled (a) and (b).

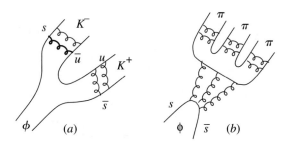

The data indicate that $(a) \gg (b)$. This is called the Okubo–Zweig–Iizuka rule, although the complexity of the strong interactions means that a quantitative calculation is not available. This rule, which is presumably why the vector multiplet is nearly ideally mixed, works progressively better for systems made of heavier quarks. For example, the 1^- heavy quark states J/ψ and Υ are very narrow, even though decays of type (b) are available. For these mesons the analog of (a) is not allowed by energy conservation. For example, the analog of the $K^+ K^-$ final state, called $D^+ D^-$ for the charm case, is heavier than J/ψ. The width $\Gamma(J/\psi \to \text{hadrons})$ is only about 75 keV.

We could ask, what would we expect for $\phi \to 3\pi$? From the mass matrix in Eq. 5.5, ϕ is expected to have a probability $\sim 10^{-3}$ to contain

$$\frac{|u, \bar{u}\rangle + |d\bar{d}\rangle}{\sqrt{2}}.$$

The

$$\frac{|u, \bar{u}\rangle + |d\bar{d}\rangle}{\sqrt{2}}$$

state can go into 3π via a diagram analogous to (a) and doesn't require annihilation as in (b). In addition, the ω would decay via the same diagram. We might expect the ϕ to 3π decay to go through this term, implying:

$$\frac{\Gamma_{\phi \to 3\pi}}{\Gamma_{\omega \to 3\pi}} \sim \text{few} \times 10^{-3}.$$

In reality, the ratio, which is $\frac{1}{15}$, is larger since the ϕ is heavy enough that a resonant channel, $\phi \to \rho\pi$, exists and dominates the decay. For the analogous ω^0 amplitude, a calculation involving the ρ, which is virtual instead of real, allows a reconciliation of the two rates.

The dominant decay pattern for the ϕ allows us a rather physical picture for how the decay occurs. The color field between the s and $\bar{s}$ breaks by producing a new pair from the vacuum, allowing formation of two new color singlets, each of lower mass.

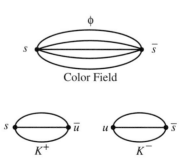

5.5 Physical Picture of Decay Process

Since the singlets are "neutral," they can fly off separately, that is, the confining potential does not keep the states together. We will call such a "picture" a decay diagram. We can use these to calculate relative rates, by imposing the $SU(2)$ symmetry (or with less accuracy, the $SU(3)$ symmetry) on processes that differ only by the quark flavors involved in the diagram. These diagrams give us a physical picture by which we can get the $SU(2)$ or $SU(3)$ Clebsch–Gordan coefficients relating such decays. In addition, since some of the neutral particles are mixtures of octet and singlet components, the diagrams are an explicit picture of how the decays are related for the two components, which is not a priori given just by the group symmetry.

For example, for the ϕ decay diagram shown above, by $SU(2)$ symmetry the process where $d\bar{d}$ is created from the vacuum instead of $u\bar{u}$, must have the same amplitude. Thus in the symmetry limit we expect that

$$|M_{fi}(\phi \to K^0 \bar{K}^0)| = |M_{fi}(\phi \to K^+ K^-)|,$$

where in the two matrix elements the analogous kaons have the same momentum. Because the mass of two kaons is so close to that of the ϕ, the final rates actually differ because of the $SU(2)$ violation. This means we have to figure out how to apply the symmetry, since M_{fi} contains both the constant $g_{VP_1P_2}$ and momentum factors. It is not completely clear without a full theory, but we typically assume that dimensionless quantities like $g_{\phi K^+ K^-}$ and $g_{\phi K^0 \bar{K}^0}$ obey the symmetry best. This choice is reasonable for quantities that have a finite limit as

$$\frac{m_q}{\lambda_{QCD}} \to 0,$$

and vary slowly from this value for m_q/λ_{QCD} small. The symmetry assumption amounts to ignoring the variation of the correction for different values of m_q. With this assumption we get the prediction, using Eq. 5.9:

$$\frac{\Gamma_{\phi \to K^+ K^-}}{|\vec{p}_{K^+}|^3} = \frac{\Gamma_{\phi \to K^0 \bar{K}^0}}{|\vec{p}_{K^0}|^3}.$$

Using the correct kaon masses, this gives the prediction

$$\frac{\Gamma_{\phi \to K^+ K^-}}{\Gamma_{\phi \to K^0 \bar{K}^0}} = 1.53,$$

which compares reasonably well to the experimental value of 1.46 ± 0.03. Our decay model has not only given us the expectation in the symmetry limit, but also the kinematic factors that break the symmetry.

We can use the decay diagrams to look at the relative rates for the various vector meson decays. In general, for diagrams where the initial and final states are made of mixtures of quarks, for example,

$$\omega^0 = \frac{|u, \bar{u}\rangle + |d, \bar{d}\rangle}{\sqrt{2}},$$

the amplitude to go from the initial to the final hadrons will involve a product of

- the amplitude to find an initial quark-antiquark pair in $|i\rangle$,
- the amplitude to create a given $q\bar{q}$ pair from the vacuum,
- the amplitude to find the resulting set of quarks and antiquarks in the final state $|f\rangle$,

summed over all the options for the quarks that can link $|i\rangle$ and $|f\rangle$. The resulting relations are for full matrix elements in the symmetry limit and will be applied to the coupling constants in the broken symmetry case. The calculation will be clearer as we do some examples, and is particularly simple for several pions in the same angular momentum final state where we can use the exchange symmetry of the amplitude.

We start with K^{*+} and draw the ϕ decay diagrams as well for comparison (K^{*0} is equally straightforward). The resulting amplitudes are shown in the accompanying figure.

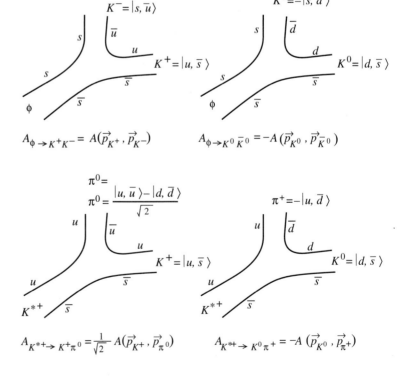

5.5 Physical Picture of Decay Process

In arriving at an overall amplitude, note the factor of $1/\sqrt{2}$ for $K^{*+} \to K^+\pi^0$, due to the amplitude to find $u\bar{u}$ in the π^0 state, and the minus sign in $K^{*+} \to K^0\pi^+$, since π^+ is $-|u, \bar{d}\rangle$, not $|u, \bar{d}\rangle$. There is a similar minus sign in the $\phi \to K^0\bar{K}^0$ amplitude because the $\bar{K}^0$ state is $-|s, \bar{d}\rangle$. The amplitudes are all written in terms of one function (called A) times the constants coming from the projections onto the states listed above. The single function, from Section 5.4,

$$A(\vec{p}_1, \vec{p}_2) = g e_V^\mu \cdot (p_1^\mu - p_2^\mu),$$

contains the constant g, which is determined by the dynamics of how quarks are created from the vacuum for the vector mesons and overlap to form pseudoscalar final state mesons. Our symmetry assumption is that one value of g describes all the amplitudes. We see immediately that we expect:

$$\frac{\Gamma_{K^{*+} \to K^+\pi^0}}{\Gamma_{K^{*+} \to K^0\pi^+}} = \frac{1}{2},$$

if we ignore the isospin-violating small mass differences—a good assumption in this case. This ratio is the same as the result we would get by using Clebsch–Gordan coefficients and the vector addition of the $I = 1$ pion multiplet with the $I = \frac{1}{2}$ kaon multiplet to get the $I = \frac{1}{2} K^*$. By energy conservation, only $u\bar{u}$ and $d\bar{d}$ created from the vacuum contribute to K^* decay. For more massive states $s\bar{s}$ could also contribute, with equal amplitude in the limit of $SU(3)$ symmetry.

We look at the ρ^+ decay next—it has two amplitudes, called A_1 and A_2, shown below, which contribute coherently.

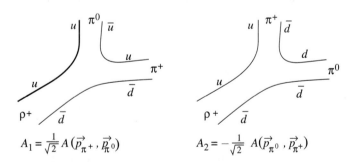

$$A_1 = \tfrac{1}{\sqrt{2}} A(\vec{p}_{\pi^+}, \vec{p}_{\pi^0}) \qquad A_2 = -\tfrac{1}{\sqrt{2}} A(\vec{p}_{\pi^0}, \vec{p}_{\pi^+})$$

A_1 and A_2 interfere since they give the same final state. A is antisymmetric under exchange (that is, $A(\vec{p}_{\pi^0}, \vec{p}_{\pi^+}) = -A(\vec{p}_{\pi^+}, \vec{p}_{\pi^0})$); thus, adding the two gives:

$$A_{\rho^+ \to \pi^+\pi^0} = \sqrt{2} A(\vec{p}_{\pi^+}, \vec{p}_{\pi^0}).$$

Looking at the ρ^0 decay next; we have four diagrams giving two distinct final states. The amplitudes A_1 and A_2 for the $\pi^+\pi^-$ final state are:

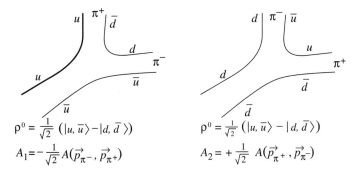

These interfere to give $A_{\rho^0 \to \pi^+\pi^-} = \sqrt{2} A(\vec{p}_{\pi^+}, \vec{p}_{\pi^-})$.

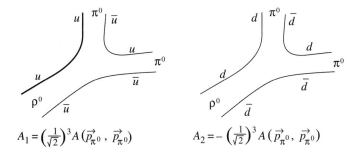

For the $\pi^0\pi^0$ final state, $A_1 + A_2 = 0$. Thus $A_{\rho^0 \to \pi^0\pi^0} = 0$. Actually each term A_1 or A_2 will vanish if we correctly include Bose symmetrization, since we should add an amplitude where the two π^0 are exchanged prior to squaring and integrating over the solid angle, corresponding to indistinguishable configurations. This correctly Bose-symmetrized amplitude vanishes for a p-wave. That is, we should write the Bose-symmetrized amplitude:

$$A_1 = \left(\frac{1}{\sqrt{2}}\right)^3 \left[A(\vec{p}_{\pi^0}^{(1)}, \vec{p}_{\pi^0}^{(2)}) + A(\vec{p}_{\pi^0}^{(2)}, \vec{p}_{\pi^0}^{(1)})\right], \quad \text{which} = 0.$$

Using the nonvanishing amplitudes we find, as expected, $\Gamma_{\rho^+} = \Gamma_{\rho^0}$, a general consequence of the good $SU(2)$ symmetry.

To calculate ω^0 to $\pi\pi$, we have to take for the initial state:

$$\frac{|u, \bar{u}\rangle + |d, \bar{d}\rangle}{\sqrt{2}} \quad \text{instead of} \quad \frac{|u, \bar{u}\rangle - |d, \bar{d}\rangle}{\sqrt{2}},$$

which changes the sign of the $d\bar{d}$ contribution relative to the $u\bar{u}$ term in the calculation for the ρ^0. This amounts to changing the sign of A_2 and leaving A_1 alone. This gives a width $\Gamma_{\omega^0 \to \pi^+\pi^-}$ that vanishes because the A_1 and A_2 contributions cancel, while $\Gamma_{\omega^0 \to \pi^0\pi^0} = 0$ after Bose symmetrization.

5.5 Physical Picture of Decay Process

We can now make a table of expectations, where we interpret symmetry breaking as meaning that we use the physical momenta and masses in the function A and in the phase space. We delete the small contributions, other than $V \to P_1 + P_2$, and add together the decays of this type. Table 5.5 shows good agreement between the decay widths and expectations. Apparently, the dimensionless constant

$$\frac{|g|^2}{6\pi} \simeq 1.$$

Note that it is not small; these are strong processes!

TABLE 5.5 $SU(3)$ Decay Width Expectations for $V \to P_1 + P_2$.

V	Prediction for $\Gamma_{V \to P_1+P_2}$	Measured $\dfrac{m_V^2 \Gamma_{V \to P_1+P_2}}{|\vec{p}|^3}$	Expected
ϕ	$\dfrac{2|g|^2|\vec{p}_K|^3}{6\pi m_\phi^2}$	2.29	$2\dfrac{|g|^2}{6\pi}$
K^*	$\dfrac{3}{2}\dfrac{|g|^2|\vec{p}_K|^3}{6\pi m_{K^*}^2}$	1.60	$1.5\dfrac{|g|^2}{6\pi}$
ρ	$\dfrac{2|g|^2|\vec{p}_\pi|^3}{6\pi m_\rho^2}$	1.93	$2\dfrac{|g|^2}{6\pi}$

The symmetry expectations work to $\sim 15\%$ for the rate and we can understand the pattern of widths, which vary by a large factor, in terms of changes in the various kinematically determined momenta. This level of accuracy for decays is typical of $SU(3)$ symmetry comparisons—even though the masses are very different for the various pseudoscalars.

We return now to the decays to 3π using the decay diagrams. As an example, we look at ρ^0 and $\omega^0 \to \pi^+\pi^-\pi^0$. We can write eight contributing decay diagrams, of which six lead to $\pi^+\pi^-\pi^0$, and two to $3\pi^0$. The rate to $3\pi^0$, however, vanishes because of the antisymmetric final state. We use the same notation A for an overall amplitude, although it now represents an amplitude of the form suitable for a $1^- \to 0^- + 0^- + 0^-$ decay. We draw the six diagrams for $\pi^+\pi^-\pi^0$ and group them as shown on page 176.

The individual amplitudes have been written in terms of one of them, but have not yet been summed over $u\bar{u}$ and $d\bar{d}$ initial contributions. We have also left out an overall (-1) from $\pi^+ = -|u, \bar{d}\rangle$. For

$$\omega^0 = \frac{|u\bar{u}\rangle + |d\bar{d}\rangle}{\sqrt{2}},$$

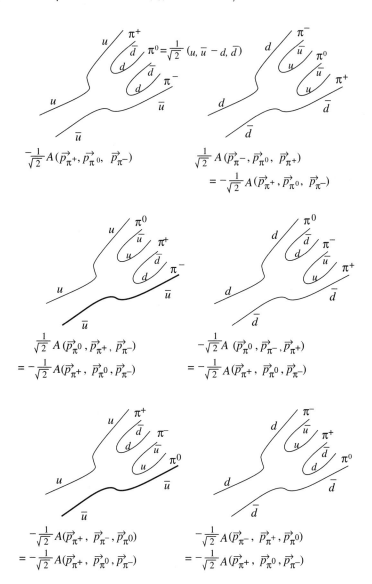

we add all 6, multiply by -1, and divide by $\sqrt{2}$, giving a nonzero result equal to $3A(\vec{p}_{\pi^+}, \vec{p}_{\pi^0}, \vec{p}_{\pi^-})$. For

$$\rho^0 = \frac{|u\bar{u}\rangle - |d\bar{d}\rangle}{\sqrt{2}},$$

each $u\bar{u}$ term cancels a $d\bar{d}$ term and we therefore find that $\rho^0 \not\to 3\pi$, which agrees with what is seen experimentally. The decay $\rho^+ \to \pi^+\pi^+\pi^-$ also does

5.5 Physical Picture of Decay Process

not occur. In this case the antisymmetry of the final state forbids a decay that includes two identical bosons.

5.5.1 ■ G Parity

We see an interesting pattern:

$$\rho \to 2\pi, \quad \omega \not\to 2\pi$$
$$\rho \not\to 3\pi, \quad \omega \to 3\pi,$$

independent of the magnitude of the basic amplitudes. Often, when rates vanish, it is due to symmetries. This is true here as well; the symmetry is called G parity and it applies to the annihilation diagrams through gluons as well as the decay diagrams. It works for interactions involving mesons with u and d quarks and antiquarks only or where the mesons are self charge-conjugate states like $\phi = |s, \bar{s}\rangle$. For these states the only nonzero internal, strong interaction, quantum numbers involve isospin. G parity is a symmetry in the limit of perfect isospin; thus it is violated at a small level, for example, by electromagnetic interactions. G is made up out of C and isospin operators. The Hamiltonian H, in the symmetry limit, commutes with C and T_1, T_2, or T_3. If we can find a G such that $[G, T_i] = 0, [G, H] = 0$, it will be conserved over time, if we start in a G eigenstate. Of course, this is true of a C eigenstate as well. The advantage of a G eigenstate is that more particles are G eigenstates than C eigenstates, for example, π^+, π^0, π^- are all G eigenstates, whereas only π^0 is a C eigenstate.

Look at C first for the pion multiplet. It leaves us within the same multiplet, so it can be represented as a 3×3 matrix operating on the π isotriplet states. The correct matrix is:

$$C = \begin{pmatrix} 0 & 0 & -1 \\ 0 & 1 & 0 \\ -1 & 0 & 0 \end{pmatrix}.$$

We can see this from the following: For the 0^- onium state,

$$|\pi^0\rangle = \frac{|u, \bar{u}\rangle - |d, \bar{d}\rangle}{\sqrt{2}},$$

we know that $C|\pi^0\rangle = |\pi^0\rangle$; therefore, we have $C|u, \bar{d}\rangle = |d, \bar{u}\rangle$ and $C|d, \bar{u}\rangle = |u, \bar{d}\rangle$ for these analogous bound states. Since, however, $|\pi^+\rangle = -|u, \bar{d}\rangle$, $|\pi^-\rangle = |d, \bar{u}\rangle$, we get the extra minus signs in the C matrix. For the C matrix,

$$C = C^\dagger, \quad C^2 = \begin{pmatrix} 1 & 0 & 0 \\ 0 & 1 & 0 \\ 0 & 0 & 1 \end{pmatrix}.$$

Within the $I = 1$ multiplet we can represent the T_i by 3×3 matrices we call I_i. These are:

$$\vec{I} = \left[\frac{1}{\sqrt{2}} \begin{pmatrix} 0 & 1 & 0 \\ 1 & 0 & 1 \\ 0 & 1 & 0 \end{pmatrix}, \frac{1}{\sqrt{2}} \begin{pmatrix} 0 & -i & 0 \\ i & 0 & -i \\ 0 & i & 0 \end{pmatrix}, \begin{pmatrix} 1 & 0 & 0 \\ 0 & 0 & 0 \\ 0 & 0 & -1 \end{pmatrix} \right]$$

The matrix C satisfies $C I_i C^\dagger = -I_i^*$ as expected for a change from particles to antiparticles.

Consider next an isospin rotation with respect to I_2:

$$R(\theta) = e^{i\theta I_2} = \begin{pmatrix} \frac{1+\cos\theta}{2} & \frac{-\sin\theta}{\sqrt{2}} & \frac{1-\cos\theta}{2} \\ \frac{\sin\theta}{\sqrt{2}} & \cos\theta & \frac{-\sin\theta}{\sqrt{2}} \\ \frac{1-\cos\theta}{2} & \frac{\sin\theta}{\sqrt{2}} & \frac{1+\cos\theta}{2} \end{pmatrix}.$$

For $\theta = \pi$:

$$R(\pi) = \begin{pmatrix} 0 & 0 & 1 \\ 0 & -1 & 0 \\ 1 & 0 & 0 \end{pmatrix} = -C.$$

Thus, defining

$$G = C e^{i\pi I_2} = \begin{pmatrix} -1 & 0 & 0 \\ 0 & -1 & 0 \\ 0 & 0 & -1 \end{pmatrix}.$$

This clearly commutes with I_i and we see that $G |\pi\rangle = -|\pi\rangle$, that is, each π state is a G eigenstate. G commutes with H in the $SU(2)$ symmetry limit since both C and I_2 do.

What about the other mesons? For the ρ everything works the same way as for the π except that $C|\rho^0\rangle = -|\rho^0\rangle$. This gives a matrix representation for C for the ρ multiplet, which is minus the matrix used for the π multiplet. It still satisfies $C I_i C^\dagger = -I_i^*$. The isospin rotation, however, is the same for π and ρ. Thus, $G|\rho\rangle = |\rho\rangle$. What about ω^0 or ϕ^0? For these, $I = 0$, so $G|\omega^0\rangle = C|\omega^0\rangle = -|\omega^0\rangle$ and similarly $G|\phi^0\rangle = C|\phi^0\rangle = -|\phi^0\rangle$. For the $|\eta\rangle$ and $|\eta'\rangle$, the G parity (which for these is C) is positive. Thus we can now use G parity conservation to see that:

$$\begin{array}{cccc} \rho \to & 2\pi & \rho \not\to & 3\pi \\ G = +1 & G = (-1)^2 & G = +1 & G = (-1)^3 \\ \omega^0 \text{ or } \phi^0 & \not\to 2\pi & \omega^0 \text{ or } \phi^0 & \to 3\pi \\ G = -1 & G = (-1)^2 & G = -1 & G = (-1)^3. \end{array}$$

Note that earlier, we looked at the Dalitz plot for $J/\psi \to \rho\pi$. J/ψ, like ϕ, has $G = -1$. Thus $J/\psi \to \rho\pi$ strongly, but $J/\psi \to \pi^+\pi^-$ would have to come about through isospin breaking, for example, electromagnetically. Thus G is generally useful for understanding such decays into pions.

5.5.2 ■ Large Invariant Mass Processes and Hadronic Jets

The quark decay diagrams also provide a physical picture of how hadronic jets are formed. We take as an example the process $e^+e^- \to$ hadrons at a reasonably large energy away from thresholds for one resonance. At short times the process is viewed as occurring by the diagram in Figure 5.13(a). The $q_i, \bar{q}_i$ are produced with a large invariant mass. As they fly apart the color interaction grows, since it becomes increasingly stronger with separation.

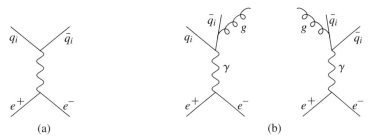

FIGURE 5.13 Short distance e^+e^- diagrams that lead to hadrons. (a) Quark-antiquark production in e^+e^- annihilation. (b) Lowest order processes with gluons radiated.

On a longer time scale the color field between the quarks breaks by pair production from the vacuum, yielding quark-antiquark pairs of smaller invariant mass. This is repeated many times until we are left with particles of small mass, like pions, kaons, and nucleons, which no longer interact and can fly off from the interaction region. The final-state hadrons tend to lie along the directions of the initial quark and antiquark, as long as this hadron formation process generates little transverse momentum compared to the initial momentum.

The formation of hadrons occurs on a longer time scale than the initial reaction and does not change the rate—rather analogous to the later decay of a resonance not changing the production rate of the resonance. The e^+e^- rate is usually quantified by a cross section ratio called R, using the analogous cross section to $\mu^+\mu^-$ pairs at the same invariant mass for normalization:

$$R = \frac{\sigma_{\text{Hadrons}}}{\sigma_{\mu^+\mu^-}} = \sum_{\text{QuarkTypes}} e_{q_i}^2,$$

if the energy is large enough so that all quark masses and the muon mass can be ignored and the only diagrams are those in Figure 5.13(a). Here the quark electric charges e_{q_i} are measured in units of e and the prediction for R is given below for several choices of the quarks that can be produced, corresponding to several energy ranges. Note that in the sum we get a factor of 3 due to color! These ideas give a good description of the rates to hadrons. Adding in extra short-distance diagrams where gluons can be radiated, as shown in Figure 5.13(b), we

Quark type	R
u, d, s	$3\left(\left(\frac{2}{3}\right)^2 + \left(\frac{1}{3}\right)^2 + \left(\frac{1}{3}\right)^2\right) = 2.$
u, d, s, c	$2 + \frac{4}{3} = 3\frac{1}{3}.$
u, d, s, c, b	$3\frac{2}{3}.$

get a quantitatively excellent description. The extra gluons can give rise to 3-jet configurations in the final state. The data for $e^+e^- \to$ hadrons will be discussed in Chapter 12, along with several other processes whose rates can be understood using simple Feynman diagrams like Figure 5.13.

5.6 ■ BARYON STATES OF THREE QUARKS

We conclude this chapter by working out the ground-state baryon quark configurations. These are all color-singlet three-quark systems. We restrict the discussion to states made of the three light quarks only, so they again exhibit the approximate $SU(3)$ flavor symmetry. Additional baryonic states exist where light quarks are replaced by one or more of the heavier quarks. For the baryons, the number of states is restricted by the exclusion principle, unlike the situation for the meson states. Since the color state is antisymmetric, the quark configuration must be symmetric for the exchange of the other degrees of freedom for the identical quarks in a given baryon. This symmetry for the space and spin degrees of freedom was a paradox, until clarified by the color hypothesis.

The amplitude for finding a quark should be spherically symmetric for the ground state. Including the spin, there will be two choices for the angular momentum for each quark, which we indicate by $\uparrow$ and $\downarrow$ for a given direction in space. For all three quarks in the ground state, the spatial state will be fully symmetric and the constraint coming from the exclusion principle affects mainly the spin correlations for given flavors in the state. We find the types of states by focusing on the spin correlations.

We start by adding spins of two of the quarks, which will provide a two-particle state with spin 0 or 1. The spin 1 state is symmetric and the spin 0 state antisymmetric. For any state with two or more identical quarks, the identical quark pair must be in the spin 1 state based on the exclusion principle. Adding a third quark, we have the options shown in the diagram below, where the exchange symmetries are indicated.

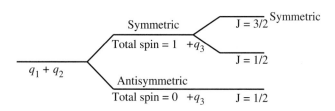

5.6 Baryon States of Three Quarks

The $J = \frac{3}{2}$ state is the only fully symmetric choice and is therefore the only possible state for three identical quarks. The states are listed below:

Flavors	Choices for two-particle spin	Choices for total spin
uuu, ddd, sss	1	$\frac{3}{2}$
$uud, uus, ddu, dds, ssu, ssd$	1	$\frac{3}{2}$ or $\frac{1}{2}$
uds	0,1	$\frac{3}{2}, \frac{1}{2}, \frac{1}{2}$

Adding together the number of states of given total angular momentum, we find 10 flavor combinations with $J = \frac{3}{2}$ (the 10 of $SU(3)$, called a decuplet) and eight flavor combinations (an octet of baryons) with $J = \frac{1}{2}$. All states have positive parity (the antibaryons would have negative parity and the same spin choices).

The decuplet states with $J = \frac{3}{2}$, $J_z = \frac{3}{2}$ are as follows:

Strangeness					Mass (MeV)
$S = 0$	$d \uparrow d \uparrow d \uparrow$ Δ^-	$\{u \uparrow d \uparrow d \uparrow\}$ Δ^0	$\{u \uparrow u \uparrow d \uparrow\}$ Δ^+	$u \uparrow u \uparrow u \uparrow$ Δ^{++}	1232
$S = -1$	$\{d \uparrow d \uparrow s \uparrow\}$ Σ^{*-}	$\{u \uparrow d \uparrow s \uparrow\}$ Σ^{*0}	$\{u \uparrow u \uparrow s \uparrow\}$ Σ^{*+}		1385
$S = -2$	$\{d \uparrow s \uparrow s \uparrow\}$ Ξ^{*-}	$\{u \uparrow s \uparrow s \uparrow\}$ Ξ^{*0}			1533
$S = -3$	$s \uparrow s \uparrow s \uparrow$ Ω^-				1672

Indicated are the particle names, the approximate masses, and strangeness values. The symmetrization notation indicates a symmetrized state:

$$\{u \uparrow u \uparrow d \uparrow\} = \tfrac{1}{\sqrt{3}}[\psi_u(\vec{x}_1) \uparrow \psi_u(\vec{x}_2) \uparrow \psi_d(\vec{x}_3) \uparrow$$
$$+ \psi_u(\vec{x}_1) \uparrow \psi_d(\vec{x}_2) \uparrow \psi_u(\vec{x}_3) \uparrow$$
$$+ \psi_d(\vec{x}_1) \uparrow \psi_u(\vec{x}_2) \uparrow \psi_u(\vec{x}_3) \uparrow],$$

for example, assuming a product wave function. This state, like all the states in the decuplet, is explicitly symmetric under exchange of any pair.

We can also arrive at the states in the multiplet by starting with $\Delta^{++} = u \uparrow u \uparrow u \uparrow$ and systematically applying lowering operators in the $SU(3)$ space. For example, Δ^{++} is part of an isospin multiplet with $I = \frac{3}{2}$. We thus expect

$$I_-\Delta^{++} = I_- \left|I = \tfrac{3}{2}, I_3 = \tfrac{3}{2}\right\rangle$$
$$= \sqrt{(I+I_3)(I-I_3+1)}\left|I = \tfrac{3}{2}, I_3 = \tfrac{3}{2} - 1\right\rangle = \sqrt{3}\Delta^+,$$

using the $SU(2)$ formula for the lowering operator. We can write I_- in terms of the individual quark lowering operators $I_- = I_-^{(1)} + I_-^{(2)} + I_-^{(3)}$. Using this expression we indeed obtain $I_-\Delta^{++} = \sqrt{3}\Delta^+$, where the Δ^{++} and Δ^+ states are $u\uparrow u\uparrow u\uparrow$ and $\{u\uparrow u\uparrow d\uparrow\}$, respectively.

We turn next to the octet of states with $J = \tfrac{1}{2}$. To be specific, we look at the state with $J_z = \tfrac{1}{2}$ and take as an example the state made of uud (proton state). Adding two u quark states gives the following possible states with total spin $S = 1$:

$$|S = 1, S_z = 1\rangle = u\uparrow u\uparrow$$
$$|S = 1, S_z = 0\rangle = \frac{u\uparrow u\downarrow + u\downarrow u\uparrow}{\sqrt{2}} = \{u\uparrow u\downarrow\}$$

Adding to the above a d quark, which can come in a state $|S = \tfrac{1}{2}, S_z = \tfrac{1}{2}\rangle$ or $|S = \tfrac{1}{2}, S_z = -\tfrac{1}{2}\rangle$, and using the correct Clebsch–Gordan coefficients to get the state $|J = \tfrac{1}{2}, J_z = \tfrac{1}{2}\rangle$, gives:

$$\left|\tfrac{1}{2}, \tfrac{1}{2}\right\rangle = \sqrt{\tfrac{2}{3}}|1,1\rangle\left|\tfrac{1}{2}, -\tfrac{1}{2}\right\rangle - \sqrt{\tfrac{1}{3}}|1,0\rangle\left|\tfrac{1}{2}, \tfrac{1}{2}\right\rangle$$
$$= \sqrt{\tfrac{2}{3}}u\uparrow u\uparrow d\downarrow - \frac{1}{\sqrt{3}}\left(\frac{u\uparrow u\downarrow + u\downarrow u\uparrow}{\sqrt{2}}\right)d\uparrow$$
$$= \frac{1}{\sqrt{6}}[2u\uparrow u\uparrow d\downarrow - u\uparrow u\downarrow d\uparrow - u\downarrow u\uparrow d\uparrow].$$

This is the proton spin state in terms of its quark constituents.

We can proceed similarly to generate the six $J = \tfrac{1}{2}$ states that contain two identical quarks. To generate the two states with u, d, s quarks, we add the $u + d$ spins first, giving a state with $S = 0$ or 1, to which the s quark will be added. For $S = 1$ the relevant spin states are:

$$|S = 1, S_z = 1\rangle = \frac{u\uparrow d\uparrow + d\uparrow u\uparrow}{\sqrt{2}} = \{u\uparrow d\uparrow\}$$
$$|S = 1, S_z = 0\rangle = \frac{\{u\uparrow d\downarrow\} + \{u\downarrow d\uparrow\}}{\sqrt{2}}$$

Adding the strange quark spin state with the appropriate Clebsch–Gordan coefficients, as in the case of the proton state, gives the state with $J = \tfrac{1}{2}, J_z = \tfrac{1}{2}$.

5.6 Baryon States of Three Quarks

Since the s quark is a singlet under isospin rotations, the isospin of the state is given by the ud part of the state. The above symmetric state with regard to $u \leftrightarrow d$ exchange is a state with $I = 1, I_3 = 0$. The other states of this $I = 1$ multiplet contain uu and dd (instead of ud) with the same spin correlations.

For the state where the $u + d$ spins couple to $S = 0$, we have:

$$|S = 0, S_z = 0\rangle = \frac{\{u \uparrow d \downarrow\} - \{u \downarrow d \uparrow\}}{\sqrt{2}}$$

$$= \frac{1}{\sqrt{2}}\left[\frac{u \uparrow d \downarrow - d \uparrow u \downarrow}{\sqrt{2}} + \frac{d \downarrow u \uparrow - u \downarrow d \uparrow}{\sqrt{2}}\right].$$

This state is an isospin singlet. We can now add the s quark with spin up to get a state with overall spin $\frac{1}{2}$ and spin up: $\frac{1}{\sqrt{2}}(\{u \uparrow d \downarrow\}s \uparrow -\{u \downarrow d \uparrow\}s \uparrow)$. The two states made of u, d, s have different internal spin correlations and also different isospins. We can now display the baryon octet of states:

$S = 0$ •n •p

$S = -1$ •Σ^- •Σ^0 •Σ^+
                         •Λ^0

$S = -2$ •Ξ^- •Ξ^0

The spin states and masses are given in Table 5.6. Note that the states in a given isospin multiplet are very nearly degenerate, while the $SU(3)$ symmetry is not as good. This holds for the two u, d, s states (Λ^0 and Σ^0), which are related by the $SU(3)$ symmetry but not by the isospin symmetry.

To write the states as full product wave-functions with explicit $SU(3)$ transformation properties, the particles need to be assigned position states in a symmetric fashion, leaving the spin states unchanged. This gives, for example, the proton final state to within an overall sign:

$$|p \uparrow\rangle = \frac{1}{\sqrt{6}} \times \frac{1}{\sqrt{3}}$$

$[(2\psi_u(\vec{x}_1) \uparrow \psi_u(\vec{x}_2) \uparrow \psi_d(\vec{x}_3) \downarrow + 2\psi_u(\vec{x}_1) \uparrow \psi_d(\vec{x}_2) \downarrow \psi_u(\vec{x}_3) \uparrow$
$+ 2\psi_d(\vec{x}_1) \downarrow \psi_u(\vec{x}_2) \uparrow \psi_u(\vec{x}_3) \uparrow - \psi_u(\vec{x}_1) \uparrow \psi_u(\vec{x}_2) \downarrow \psi_d(\vec{x}_3) \uparrow$
$- \psi_u(\vec{x}_1) \uparrow \psi_d(\vec{x}_2) \uparrow \psi_u(\vec{x}_3) \downarrow - \psi_d(\vec{x}_1) \uparrow \psi_u(\vec{x}_2) \uparrow \psi_u(\vec{x}_3) \downarrow$
$- \psi_u(\vec{x}_1) \downarrow \psi_u(\vec{x}_2) \uparrow \psi_d(\vec{x}_3) \uparrow - \psi_u(\vec{x}_1) \downarrow \psi_d(\vec{x}_2) \uparrow \psi_u(\vec{x}_3) \uparrow$
$- \psi_d(\vec{x}_1) \uparrow \psi_u(\vec{x}_2) \downarrow \psi_u(\vec{x}_3) \uparrow)].$

TABLE 5.6 Octet Baryon Spin States and Masses.

State	Spin state	Mass (MeV)
p	$-\frac{1}{\sqrt{6}}(2u\uparrow u\uparrow d\downarrow - u\uparrow u\downarrow d\uparrow - u\downarrow u\uparrow d\uparrow)$	938.27
n	$\frac{1}{\sqrt{6}}(2d\uparrow d\uparrow u\downarrow - d\uparrow d\downarrow u\uparrow - d\downarrow d\uparrow u\uparrow)$	939.57
Σ^+	$\frac{1}{\sqrt{6}}(2u\uparrow u\uparrow s\downarrow - u\uparrow u\downarrow s\uparrow - u\downarrow u\uparrow s\uparrow)$	1189.4
Σ^0	$\frac{1}{\sqrt{6}}(2\{u\uparrow d\uparrow\}s\downarrow - \{u\uparrow d\downarrow\}s\uparrow - \{d\uparrow u\downarrow\}s\uparrow)$	1192.6
Σ^-	$\frac{1}{\sqrt{6}}(2d\uparrow d\uparrow s\downarrow - d\uparrow d\downarrow s\uparrow - d\downarrow d\uparrow s\uparrow)$	1197.4
Λ^0	$\frac{1}{\sqrt{2}}(\{u\uparrow d\downarrow\}s\uparrow - \{u\downarrow d\uparrow\}s\uparrow)$	1115.7
Ξ^0	$-\frac{1}{\sqrt{6}}(2s\uparrow s\uparrow u\downarrow - s\uparrow s\downarrow u\uparrow - s\downarrow s\uparrow u\uparrow)$	1314.9
Ξ^-	$-\frac{1}{\sqrt{6}}(2s\uparrow s\uparrow d\downarrow - s\uparrow s\downarrow d\uparrow - s\downarrow s\uparrow d\uparrow)$	1321.3

In Table 5.6, the signs for the states within an isospin multiplet have been chosen so that the isospin raising or lowering operators satisfy the standard $SU(2)$ relations. Analogously, the four other shift operators in the $SU(3)$ space are denoted

U_+ and U_-, which transform between $s \leftrightarrow d$.
V_+ and V_-, which transform between $u \leftrightarrow s$.

The relative signs for the states with different strangeness has been chosen so that V_+ and V_-, which link Ξ^0 and Σ^+, and Σ^- and n, satisfy the same phase relations as I_+ and I_- do for isospin multiplets. Since the weak interactions involve transitions $d \to u$ and $s \to u$, they act as I_+ and V_+ raising operators in the $SU(3)$ space. The phase conventions treat these similarly.

The baryons in the decuplet typically decay strongly to the baryons in the octet with the same strangeness and a pion. Decays involving kaons are forbidden by energy conservation. Relative branching fractions for states in given isospin multiplets can be determined accurately in terms of $SU(2)$ Clebsch–Gordan coefficients. The ratio of widths can be calculated more approximately for states that differ in strangeness in terms of the analogous $SU(3)$ coefficients. Widths vary from about 10 MeV to 120 MeV for these strong decays, a range rather similar to the range for the vector mesons. States within an isospin multiplet (for example, the four types of Δ) have nearly the same width.

An interesting exception to the above pattern is the decay of the Ω^-. As the only baryon with strangeness $= -3$, it is forbidden by energy conservation to decay strongly, and therefore decays weakly via an $s \to u$ transition. The resulting

5.6 Baryon States of Three Quarks

lifetime of 0.82×10^{-10} sec is about 12 orders of magnitude longer than for the strong decays!

For the baryon octet the proton is the only stable particle. Since the electromagnetic interactions violate isospin conservation, the decay $\Sigma^0 \to \Lambda^0 + \gamma$ occurs. The mean lifetime for the Σ^0 is about 7×10^{-20} sec, about three orders of magnitude longer than for typical strong decays. The other six baryons in the octet all decay weakly, mostly via the transitions $s \to u$ or $d \to u$ with emission of a virtual W. The virtual W can then, constrained by energy conservation, go into $e^- \bar{\nu}_e$, $\mu^- \bar{\nu}_\mu$, or a $u\bar{d}$ quark final state system. For the latter, the multiple quarks in the final state interact strongly, producing a baryon with one less unit of strangeness and a pion. The decays involving leptons (called semileptonic decays, since a hadron is also present in the final state) allow a nice verification of the underlying transition pattern. For example, the transition $d \to u$ allows the decays:

$$n \to p e^- \bar{\nu}_e, \quad \text{and} \quad \Sigma^- \to \Lambda e^- \bar{\nu}_e,$$

while the $s \to u$ transition allows:

$$\Sigma^- \to n e^- \bar{\nu}_e, \quad \Xi^0 \to \Sigma^+ e^- \bar{\nu}_e, \quad \Xi^- \to \Lambda e^- \bar{\nu}_e,$$

as illustrated in Figure 5.14. Transitions that change strangeness by two units, or decays such as $\Sigma^+ \to n e^+ \nu_e$, are not seen. The decay $\Sigma^+ \to \Lambda^0 e^+ \nu_e$ does occur and involves the transition $u \to d$. This is the only system for the baryons where $u \to d$ is the energetically allowed decay rather than $d \to u$. The transitions to Λ^0 for Σ^+ and Σ^- are illustrated in Figure 5.15.

The neutron lifetime is about 886 sec. It is very long because of the very limited phase space for the final state. The other weakly decaying baryons have lifetimes typically $\sim 10^{-10}$ sec.

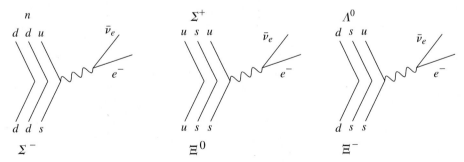

FIGURE 5.14 Examples of diagrams for semileptonic decay of strange baryons. Note that certain processes, for example $\Xi^0 \to \Sigma^- e^+ \nu_e$, although allowed by charge conservation, do not occur given the quark content of the baryons.

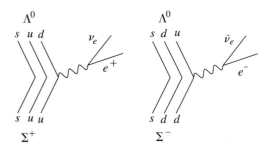

FIGURE 5.15 Weak transitions $u \to d$ and $d \to u$ leading to semileptonic Σ^+ and Σ^- decays to a Λ^0. The amplitudes for these two are equal.

CHAPTER 5 HOMEWORK

5.1. Consider the mass matrix discussed in the text:

$$H_m = \begin{pmatrix} m+\varepsilon & \varepsilon & \varepsilon' \\ \varepsilon & m+\varepsilon & \varepsilon' \\ \varepsilon' & \varepsilon' & m'+\varepsilon'' \end{pmatrix}$$

for the basis $|u, \bar{u}\rangle, |d, \bar{d}\rangle, |s, \bar{s}\rangle$.

(a) Show that if we use the ideally-mixed base states

$$\frac{|u,\bar{u}\rangle - |d,\bar{d}\rangle}{\sqrt{2}}, \frac{|u,\bar{u}\rangle + |d,\bar{d}\rangle}{\sqrt{2}}, |s,\bar{s}\rangle,$$

that

$$H_m = \begin{pmatrix} m & 0 & 0 \\ 0 & m+2\varepsilon & \sqrt{2}\varepsilon' \\ 0 & \sqrt{2}\varepsilon' & m'+\varepsilon'' \end{pmatrix}.$$

(b) Show that the masses of the three eigenstates are

$$m, \quad \frac{m+m'+\varepsilon''+2\varepsilon+\Delta}{2}, \quad \frac{m+m'+\varepsilon''+2\varepsilon-\Delta}{2},$$

where $\Delta = \sqrt{[(m+2\varepsilon)-(m'+\varepsilon'')]^2 + 8\varepsilon'^2}$.

(c) For $\varepsilon, \varepsilon', \varepsilon'' \ll m' - m$, show that these masses are approximately

$$m, \quad m+2\varepsilon - \frac{2\varepsilon'^2}{[(m'+\varepsilon'')-(m+2\varepsilon)]}, \quad m'+\varepsilon'' + \frac{2\varepsilon'^2}{[(m'+\varepsilon'')-(m+2\varepsilon)]},$$

with eigenstates

$$\frac{|u,\bar{u}\rangle - |d,\bar{d}\rangle}{\sqrt{2}}, \quad \frac{|u,\bar{u}\rangle + |d,\bar{d}\rangle}{\sqrt{2}} - \alpha|s,\bar{s}\rangle, \quad |s,\bar{s}\rangle + \alpha\left(\frac{|u,\bar{u}\rangle + |d,\bar{d}\rangle}{\sqrt{2}}\right),$$

where

$$\alpha = \frac{\sqrt{2}\varepsilon'}{[(m'+\varepsilon'')-(m+2\varepsilon)]}.$$

(d) Assuming this describes the masses of the vector mesons, ρ^0, ω, ϕ, find m, m', ε, α. Assume that $\varepsilon \simeq \varepsilon' \simeq \varepsilon''$.

5.2. Consider using a mass-squared matrix to look at the pseudoscalar mesons. In a basis given by

$$\frac{|u,\bar{u}\rangle - |d,\bar{d}\rangle}{\sqrt{2}}, \quad \frac{|u,\bar{u}\rangle + |d,\bar{d}\rangle}{\sqrt{2}} \quad \text{and} \quad |s,\bar{s}\rangle$$

we can write this matrix as

$$\begin{pmatrix} m_\pi^2 & 0 & 0 \\ 0 & m_\pi^2 + 2\Delta & \sqrt{2}\Delta' \\ 0 & \sqrt{2}\Delta' & m_{s\bar{s}}^2 + \Delta'' \end{pmatrix}.$$

(a) Suppose we know that the eigenstates for the nondiagonal 2×2 submatrix are:

$$|\eta\rangle = X_\eta \left(\frac{|u,\bar{u}\rangle + |d,\bar{d}\rangle}{\sqrt{2}}\right) + Y_\eta |s,\bar{s}\rangle$$

$$|\eta'\rangle = X_{\eta'} \frac{|u,\bar{u}\rangle + |d,\bar{d}\rangle}{\sqrt{2}} + Y_{\eta'} |s,\bar{s}\rangle.$$

Show that we can write the submatrix as

$$\begin{pmatrix} m_\pi^2 + 2\Delta & \sqrt{2}\Delta' \\ \sqrt{2}\Delta' & m_{s\bar{s}}^2 + \Delta'' \end{pmatrix}$$

$$= \begin{pmatrix} X_\eta^2 m_\eta^2 + X_{\eta'}^2 m_{\eta'}^2 & X_\eta Y_\eta m_\eta^2 + X_{\eta'} Y_{\eta'} m_{\eta'}^2 \\ X_\eta Y_\eta m_\eta^2 + X_{\eta'} Y_{\eta'} m_{\eta'}^2 & Y_\eta^2 m_\eta^2 + Y_{\eta'}^2 m_{\eta'}^2 \end{pmatrix}.$$

Given the values $m_\eta = 547$ MeV, $m_{\eta'} = 958$ MeV, $X_\eta = Y_{\eta'} = .8$, $X_{\eta'} = -Y_\eta = .6$, find

$$\Delta, \, \Delta', \, m_{s\bar{s}}^2 + \Delta''.$$

How large is the $SU(3)$ violation in Δ' compared to Δ?

(b) What is the physical significance of $m_{s\bar{s}}$? If we estimate Δ'' as

$$\Delta'' = \Delta' \left(\frac{\Delta'}{\Delta}\right),$$

calculate $m_{s\bar{s}}$.

5.3. Consider a particle that decays via the decay sequence: initial state $\to n_1 + R \to n_1 + n_2$, where $R \to n_2$. Assume that the resonance R has spin zero. Suppose we integrate over all variables except the invariant mass of R, which we call q^2, in a calculation of the decay width. Show that if it is sufficiently accurate to replace q^2 by a constant value m_R^2 in all parts of the calculation except in the resonance propagator, that

$$\frac{1}{\Gamma_{in}} \frac{d\Gamma_{in}}{dq^2} = \frac{m_R \Gamma_R/\pi}{(q^2 - m_R^2)^2 + m_R^2 \Gamma_R^2}.$$

Γ_{in} is the width of the initial particle. By plotting the data in terms of the invariant mass squared of the resonance, summing over all other variables describing the final state, we can directly measure the Breit–Wigner distribution.

5.4. Consider the elastic scattering of two scalar particles through a narrow spin zero resonance. Assume the resonance decays only into the initial particles.

(a) Calculate the differential and total cross sections.

(b) For a center of mass energy equal to the resonance mass, show that the result above corresponds to the maximum allowed cross section for the angular momentum zero partial wave (derived in nonrelativistic quantum mechanics using phase shifts). Note, this is again a constraint from unitarity.

5.5. The ρ meson has a large width so that the variation of the width with the invariant mass q^2 is a significant effect. Assuming that we can take $g_{\rho\pi\pi}$ constant over a significant range in q^2, find the expression for $\Gamma_\rho(q^2)$ over this range. The width enters the propagator through the expression

$$\frac{1}{q^2 - m_\rho^2 + im_\rho \Gamma_\rho(q^2)},$$

as discussed in the text.

5.6. The vector meson mass matrix discussed in the text is the simplest choice for such a matrix, conceptually arising from gluon exchange, which is an $SU(3)$ flavor singlet interaction. Since the vectors decay, an alternative mass matrix can be generated by exchange of pairs of virtual pseudoscalar mesons. The terms in the matrix would involve couplings at each of the two vertices (g_{VP_1,P_2} at each) and propagation between vertices of the virtual p-wave state made of pseudoscalars P_1 and P_2.

(a) For vector base states $|u, \bar{u}\rangle, |d, \bar{d}\rangle, |s, \bar{s}\rangle$, the amplitudes to $\pi^+\pi^-$, K^+K^-, $K^0\bar{K}^0$ are given by $A(\vec{p}_{\pi^+}, \vec{p}_{\pi^-})$, $A(\vec{p}_{K^+}, \vec{p}_{K^-})$, and $A(\vec{p}_{K^0}, \vec{p}_{\bar{K}^0})$, respectively, times a constant. Show that the constants are given by the values in the table below.

State	Coupling factor multiplying amplitude		
	$\pi^+\pi^-$	K^+K^-	$K^0\bar{K}^0$
$|u, \bar{u}\rangle$	-1	1	0
$|d, \bar{d}\rangle$	1	0	-1
$|s, \bar{s}\rangle$	0	-1	1

Why do intermediate states involving π^0, η, η' (for example, $\pi^0\pi^0, \pi^0\eta$) not contribute?

(b) Using these vector base states, construct the mass matrix, H_m, in the neutral vector meson sector. Assume that the mixing in the matrix is given, taking $|u, \bar{u}\rangle$ to $|u, \bar{u}\rangle$ as an example, by the sum over terms $|u, \bar{u}\rangle \rightarrow \pi^+\pi^- \rightarrow |u, \bar{u}\rangle$ and

$|u, \bar{u}\rangle \to K^+K^- \to |u, \bar{u}\rangle$. Since we do not know how to calculate the propagation factor for the virtual mesons, we will parameterize these via phenomenological constants. Including the number $|g|^2$ and momentum factors from the vertices in the parameters below, we define the contributions to H_m, excluding the coupling factors from the table, as follows:

- $-\varepsilon$ for $\pi^+\pi^-$ propagating between $|u, \bar{u}\rangle$ and $|d, \bar{d}\rangle$,
- $-\varepsilon'$ for $K\bar{K}$ propagating between $|u, \bar{u}\rangle$ and $|d, \bar{d}\rangle$,
- $-\varepsilon''$ for $K\bar{K}$ propagating between $|u, \bar{u}\rangle$, $|d, \bar{d}\rangle$ and $|s, \bar{s}\rangle$,
- $-\varepsilon'''$ for $K\bar{K}$ propagating between $|s, \bar{s}\rangle$ and $|s, \bar{s}\rangle$.

We have chosen the sign to agree with the data for which ρ^0 is the lightest meson. Using m and m' as the $|u, \bar{u}\rangle$, $|d, \bar{d}\rangle$ and $|s, \bar{s}\rangle$ masses, respectively, in the absence of mixing, and the relative coupling factors in the table in (a), show that H_m is:

$$H_m = \begin{pmatrix} m - \varepsilon - \varepsilon' & \varepsilon & \varepsilon'' \\ \varepsilon & m - \varepsilon - \varepsilon' & \varepsilon'' \\ \varepsilon'' & \varepsilon'' & m' - 2\varepsilon''' \end{pmatrix}.$$

(c) For the matrix above, show that

$$\rho^0 = \frac{|u, \bar{u}\rangle - |d, \bar{d}\rangle}{\sqrt{2}}$$

is always an eigenstate with eigenvalue $m_{\rho^0} = m - 2\varepsilon - \varepsilon'$. By looking at the analogous diagrams for ρ^+, that is, $\rho^+ \to \pi^+\pi^0 \to \rho^+$ and $\rho^+ \to K^+\bar{K}^0 \to \rho^+$ via virtual pseudoscalar pairs, show that the charged ρ^+ mass is renormalized so it is $m_{\rho^+} = m - 2\varepsilon - \varepsilon'$. Unlike the model involving virtual gluons (an $SU(3)$ singlet) the two pseudoscalars represent octet transitions and the ρ masses are changed. The isospin symmetry, however, is maintained so that the three ρ mesons remain degenerate.

(d) The $SU(3)$ symmetry limit is obtained by $\varepsilon = \varepsilon' = \varepsilon'' = \varepsilon'''$, $m = m'$. In this limit show that both octet states have a mass $m - 3\varepsilon$ while the singlet has mass m. Thus the singlet mass is not renormalized for the interaction chosen, which is not an $SU(3)$ singlet.

(e) Change to a basis given by the ideally mixed vector meson states. Show that this gives the mixing submatrix for the isospin zero states:

$$\begin{pmatrix} m - \varepsilon' & \sqrt{2}\varepsilon'' \\ \sqrt{2}\varepsilon'' & m' - 2\varepsilon''' \end{pmatrix}.$$

Could you have anticipated that ε would not appear? Show that for $m' - m \gg \varepsilon', \varepsilon'', \varepsilon'''$, this gives for the masses $m_\rho = m - 2\varepsilon - \varepsilon'$, $m_\omega = m - \varepsilon'$, $m_\phi = m' - 2\varepsilon'''$ and the amplitude to find

$$\frac{|u, \bar{u}\rangle + |d, \bar{d}\rangle}{\sqrt{2}}$$

in the ϕ equal to

$$\frac{\sqrt{2}\varepsilon''}{m_\phi - m_\omega}.$$

5.7. Suppose the $SU(3)$ flavor symmetry were much more exact, such that
1. octet of pseudoscalar meson states each have a mass = 140 MeV,
2. singlet pseudoscalar meson has a mass = 900 MeV,
3. the neutral nonstrange vector meson octet eigenstates are ρ^0 and ω_8 (octet state), each with mass = 770 MeV.

(a) What are the possible decays of ρ^0 and ω_8 into two pseudoscalar mesons?
(b) What are the relative branching ratios for these decays?
(c) Considering only the decays into two pseudoscalars, how large is Γ_{ω_8} compared to Γ_ρ?

5.8. For a two-body system we can write the total angular momentum $\vec{L} = \vec{x}_1 \times \vec{p}_1 + \vec{x}_2 \times \vec{p}_2$ as $\vec{L} = \vec{R}_{cm} \times \vec{p}_{cm} + \vec{r}_{rel} \times \vec{p}_{rel}$, where in the nonrelativistic case,

$$\vec{R}_{cm} = \frac{m_1 \vec{x}_1 + m_2 \vec{x}_2}{m_1 + m_2}$$

is the center of mass coordinate and $\vec{r}_{rel} = \vec{x}_1 - \vec{x}_2$ is the relative coordinate describing the internal motion. Here $\vec{p}_{cm} = \vec{p}_1 + \vec{p}_2$ and

$$\vec{p}_{rel} = \frac{m_2 \vec{p}_1 - m_1 \vec{p}_2}{m_1 + m_2}$$

are the momenta corresponding to the new coordinates. In studying the decay to three pions we assumed that we could describe the angular momentum in the center of mass in terms of the relative angular momentum of a pair and the angular momentum of the third with respect to the center of mass of the initial pair. Check that this works.

Defining:

$$\vec{R}_{cm} = \frac{m_1 \vec{x}_1 + m_2 \vec{x}_2 + m_3 \vec{x}_3}{m_1 + m_2 + m_3}$$

$$\vec{r}_{1-2} = \vec{x}_1 - \vec{x}_2, \quad \vec{r}_{3 \text{ versus } 1-2} = \left[\frac{m_1 \vec{x}_1 + m_2 \vec{x}_2}{m_1 + m_2}\right] - \vec{x}_3;$$

find the corresponding momenta for these coordinates and show that

$$\vec{L} = \vec{x}_1 \times \vec{p}_1 + \vec{x}_2 \times \vec{p}_2 + \vec{x}_3 \times \vec{p}_3 = \vec{L}_{cm} + \vec{L}_{1-2} + \vec{L}_{3 \text{ versus } 1-2},$$

where

$$\vec{L}_{cm} = \vec{R}_{cm} \times \vec{p}_{cm}, \ \vec{L}_{1-2} = \vec{r}_{1-2} \times \vec{p}_{1-2}, \ \vec{L}_{3 \text{ versus } 1-2}$$
$$= \vec{r}_{3 \text{ versus } 1-2} \times \vec{p}_{3 \text{ versus } 1-2}.$$

Note that, in the relativistic situation, all masses m_i above are replaced by E_i, but the relations for the angular momentum decomposition are unchanged.

5.9. Consider a hypothetical uncharged particle of $J^P = 0^+$. Write the particle as h^0. Assuming the following are allowed by energy conservation and charge conjugation invariance, which of the following decays are allowed by conservation of angular momentum and parity? For the allowed decays write down the most general possible matrix element in terms of an overall constant.

(a) $h^0 \to \pi^+\pi^-$
(b) $\pi^0 \to h^0 + h^0$
(c) $\pi^0 \to h^0 + h^0 + h^0$
(d) $\rho^+ \to \pi^+ + h^0$

5.10. Meson excited states exist where the quark-antiquark system has nonzero angular momentum L. For $L = 1, S = 1$ we can have states with $J = 0, 1, 2$. Consider the positively charged $a_1^+(1260)$ particle, which is made of $u\bar{d}$ with $L = 1, S = 1, J = 1$. Its mass is about 1260 MeV.

(a) What is its parity?
(b) Can it decay strongly into $\pi\pi$?
(c) Can it decay strongly into $\rho\pi$?
(d) For the decay that can occur, how many angular momentum channels exist and therefore how many independent amplitudes describe the decay?

5.11. For $L = 1, S = 1, J = 2$ we have a meson, called $f_2(1270)$, which is made of

$$\frac{|u, \bar{u}\rangle + |d, \bar{d}\rangle}{\sqrt{2}}.$$

(a) What are its quantum numbers?
(b) Can it decay strongly to $\pi\pi$?
(c) Can it decay strongly to $\rho\pi$?
(d) Can it decay to $\gamma\pi^0$?
(e) Can it decay to $\gamma\gamma$?
(f) For the strong decay that occurs, find the relative branching ratios for the different charge states in the final state.

5.12. Suppose the meson f_2 discussed in the problem above had quark content:

$$X_{f_2} \frac{|u, \bar{u}\rangle + |d, \bar{d}\rangle}{\sqrt{2}} + Y_{f_2} |s, \bar{s}\rangle.$$

Using quark decay diagrams with $s\bar{s}$, $u\bar{u}$, and $d\bar{d}$ having equal amplitudes for creation show that the relative branching ratios are:

$$\frac{\Gamma(f_2 \to \pi\pi)}{\Gamma(f_2 \to K\bar{K})} = \frac{3X_{f_2}^2}{2\left(Y_{f_2} + \frac{X_{f_2}}{\sqrt{2}}\right)^2}.$$

Here $\pi\pi$ is a sum over $\pi^+\pi^-$ and $\pi^0\pi^0$ and $K\bar{K}$ is a sum over K^+K^- and $K^0\bar{K}^0$. To arrive at the ratio, we have ignored the effect of the different K and π masses in the amplitudes and phase space, which is not a good approximation. Note that the real $f_2(1270)$ has an 85% branching ratio to $\pi\pi$ and 5% to $K\bar{K}$.

5.13. In QCD, as a consequence of confinement, we expect the existence of mesons made of two gluons in a color singlet state. The simplest such state would have $J^P = 0^+$ and even charge conjugation. For this state:

(a) What would be the strangeness?
(b) What would be the G parity?

Assuming that it is sufficiently massive:
(c) Could it decay strongly into $\pi\pi$?
(d) Could it decay strongly into $\pi\eta$?
(e) Could it decay strongly into $\pi\rho$?
(f) Could it decay electromagnetically into $\pi\eta$?
(g) Could it decay electromagnetically into $\pi\rho$?

5.14. The strong interactions are charge conjugation invariant, so that for a strong decay:
$$\langle Cf|\, S\, |Ci\rangle = \langle f|\, S\, |i\rangle.$$

Once the transformation properties of the fields are known, this provides a constraint on the behavior of the matrix element in momentum space.

(a) What is the constraint for $\rho^0 \to \pi^+\pi^-$? For $\omega^0 \to \pi^+\pi^-\pi^0$?
(b) Do the matrix elements we have used in the text satisfy these constraints?

5.15. Suppose there was only one light quark flavor (all others have mass $\gg \lambda_{QCD}$). What would be the quark content, charge, and J^P of the ground state baryon in this case? Assume the light quark is the u quark (which comes in the usual three colors).

5.16. The four states in the Δ multiplet decay into a nucleon and a pion. Write down the possible decays for each Δ state. Using Clebsch–Gordan coefficients, find the branching ratios in the cases where more than one charge combination is possible for the nucleon and pion. Ignore the small mass differences within the isospin multiplets.

CHAPTER 6

The Constituent Quark Model

6.1 ■ CONSTITUENT QUARKS

The strong decays and strong interaction particle multiplets discussed in Chapter 5 illustrate the approximate $SU(3)$ flavor symmetry displayed by the light hadrons. The underlying quarks, however, have different masses as well as different values for their electric charge. We focus in this chapter on the consequences of these differences, as revealed, for example, in a variety of electromagnetic interactions. The electromagnetic processes serve as a good probe of the charge and spin structure of the hadron constituents. We will look at several different processes that are sensitive to these features. Among the electromagnetic processes we study will be those that contribute to the decays of the lightest hadrons. Many of the calculations depend on the superposition principle, illustrating the linearity of quantum mechanics within the world of hadrons. Since there are no free quarks, the matrix elements measured always involve hadrons. The underlying quarks will usually influence the matrix elements in surprisingly direct and simple ways. This behavior, called the "Constituent Quark Model," ultimately needs to be explained in terms of the QCD interaction that creates the states involved in the matrix elements.

6.2 ■ BARYON MAGNETIC MOMENTS

The first properties we look at are the magnetic moments of the various light baryons. These moments serve to probe the spin structure of the baryons, as described in Section 5.6. The magnetic moments are measured by the energy difference for the baryon at rest with spin aligned or anti-aligned with a static magnetic field. This is a low-energy, long-distance phenomenon. All constituents of the baryon experience the same magnetic field and contribute coherently to the interaction energy, which is small compared to the baryon mass.

Since no extra degrees of freedom are excited, we can write the interaction completely in terms of the baryon and electromagnetic fields. For the baryon at rest, its momentum $\vec{p} = 0$ and the static magnetic field can be described in terms of a (classical) vector potential $\vec{A}$. Taking the proton as our example, the most general gauge invariant interaction we can write that conserves parity and has no derivatives, since these will give vanishing contributions for $\vec{p} = 0$, is:

$$\mathcal{H} = e\bar{\psi}\gamma_\mu\psi A_\mu + \frac{ek}{2m_p}\left[\frac{1}{2}F_{\mu\nu}\bar{\psi}\sigma_{\mu\nu}\psi\right].$$

The normalization of the first term is specified by the value of the proton charge. The second has been divided by the proton mass, m_p, to have the correct dimensions. Each term in $\mathcal{H}$ is separately gauge invariant. The only unknown is the constant k, which is determined by the structure of the proton.

Using the expression for the γ matrices, the second term in the interaction can be written as

$$\frac{-ek}{2m_p}\bar{\psi}(\vec{\Sigma}\cdot\vec{B})\psi,$$

where $\vec{\Sigma} = \begin{pmatrix}\vec{\sigma} & 0 \\ 0 & \vec{\sigma}\end{pmatrix}$ in the Dirac–Pauli representation. This term is called the anomalous moment interaction, since it does not come directly out of the Dirac equation.

We next want to separate out the spin-dependent part of the first term in $\mathcal{H}$. This can be accomplished by the Gordon decomposition. Writing:

$$\bar{\psi}\gamma_\mu\psi = \tfrac{1}{2}(\bar{\psi}\gamma_\mu\psi + \bar{\psi}\gamma_\mu\psi),$$

we replace ψ in the first term in the parentheses by using the expression

$$\frac{1}{m_p}i\gamma_\mu\frac{\partial\psi}{\partial x_\mu} = \psi,$$

and $\bar{\psi}$ in the second term using the adjoint of the above. The magnetic field is being treated as a perturbation, so the proton field satisfies the free field equation. These substitutions give expressions where two γ matrices appear. To simplify these expressions, we use the identity

$$\gamma_\mu\gamma_\nu = \tfrac{1}{2}(\gamma_\mu\gamma_\nu + \gamma_\nu\gamma_\mu) + \tfrac{1}{2}(\gamma_\mu\gamma_\nu - \gamma_\nu\gamma_\mu) = g_{\mu\nu} - i\sigma_{\mu\nu}.$$

This allows replacement of $\gamma_\mu\gamma_\nu$, with the sum of a term with no γ matrices and one with an explicit spin operator. Collecting the terms resulting from the above substitutions gives the Gordon decomposition:

$$\bar{\psi}\gamma_\mu\psi = \frac{i}{2m_p}\left[\bar{\psi}\frac{\partial\psi}{\partial x_\mu} - \frac{\partial\bar{\psi}}{\partial x_\mu}\psi\right] + \frac{1}{2m_p}\frac{\partial}{\partial x_\nu}(\bar{\psi}\sigma_{\mu\nu}\psi).$$

The term involving derivatives and no spin operator looks similar to the current in nonrelativistic quantum mechanics. For a particle at rest in a static $\vec{B}$ field this term doesn't contribute. In constructing the Hamiltonian from the density, we can integrate the term involving $\sigma_{\mu\nu}$ by parts and discard the piece that is a perfect

6.2 Baryon Magnetic Moments

differential. This results in a term where the derivative acts on the vector potential, which corresponds to a Hamiltonian density:

$$\frac{e}{2m_p}\left[\frac{1}{2}F_{\mu\nu}\bar{\psi}\sigma_{\mu\nu}\psi\right].$$

Finally, adding the two spin terms together gives, for the static magnetic interaction,

$$\mathcal{H} = -\mu_p \bar{\psi}\vec{\Sigma}\cdot\vec{B}\psi, \quad \text{where } \mu_p = \frac{e}{2m_p}(1+k).$$

The density yields an interaction Hamiltonian by integrating over space. Including the normalization factor to arrive at an energy for one particle in a volume V and integrating the density over V, gives the interaction Hamiltonian in terms of the two-component spinors for the particle at rest:

$$H = -\mu_p \chi^\dagger \vec{\sigma}\cdot\vec{B}\chi.$$

This is the standard Hamiltonian used for the nonrelativistic interaction of a spin in a magnetic field.

We would next like to calculate this interaction in terms of the quark constituents. To do this we will assume the following:

1. The states in Table 5.6 are an adequate description of the baryons in terms of constituent quarks in the low energy static limit. The spin correlations, in particular, are correct.
2. For each quark within the baryon, the energy in a static magnetic field is proportional to $-e_q \chi_q^\dagger \vec{\sigma}_q \cdot \vec{B}\chi_q$, where e_q is the charge for quark type q. To make this dimensionally correct requires division by a mass. We therefore take for each quark, by analogy with the electron moment interaction:

$$H_q = \frac{-e_q}{2m_q}\chi_q^\dagger \vec{\sigma}_q \cdot \vec{B}\chi_q = -\mu_q \chi_q^\dagger \vec{\sigma}_q \cdot \vec{B}\chi_q.$$

We expect that $m_q \simeq \lambda_{QCD}$, since this is the primary mass scale for the light quarks. We call m_q the constituent quark mass. For a given baryon, $H = \sum_{i=1}^{3} H_{qi}$, where the sum is over the three quark types in the baryon.

We will, for now, approximate $m_u = m_d$, since isospin is a very good symmetry. Thus the moments should be describable in terms of two mass parameters, the constituent quark masses m_u and m_s, or one mass parameter m_u and a ratio $r = m_u/m_s$, whose deviation from 1 measures the violation of the $SU(3)$ symmetry. Taking the magnetic field direction to point along the z-axis, we can calculate the baryon moments in terms of the quark moments by equating the expectation values of H expressed in terms of the baryon moment or the various contributing quark terms. As an example, for the proton, the spin state:

$$\tfrac{-1}{\sqrt{6}}(2u\uparrow u\uparrow d\downarrow - u\uparrow u\downarrow d\uparrow - u\downarrow u\uparrow d\uparrow)$$

leads to the prediction:

$$\mu_p = \frac{1}{6}[4(2\mu_u - \mu_d) + 2(\mu_d)] = \frac{1}{3}(4\mu_u - \mu_d)$$

$$= \frac{1}{3}\left(4\left(\frac{2}{3}\right) + \frac{1}{3}\right)\frac{e}{2m_u} = \frac{e}{2m_u}.$$

Performing a similar calculation for the other spin $\tfrac{1}{2}$ baryons gives the result in Table 6.1. For the predictions in the table, the measured values of μ_p and μ_{Λ^0} are used to determine the following parameters:

$$m_u = \frac{m_p}{2.79} = 336 \text{ MeV}, \quad r = .66,$$

implying $m_s = 509$ MeV. The agreement with the measured values is reasonably good; in particular, the model provides an understanding of the sign and magnitude of the moment for each neutral particle, which we might naively expect to have no moment at all. The agreement is, however, not perfect and typical of a phenomenologically motivated model; we do not know how to improve the

TABLE 6.1 Baryon Magnetic Moments.

Baryon	Predicted moment in units of $e/2m_u$	Predicted in units of $e/2m_p$	Measured in units of $e/2m_p$
p	1	2.79	2.79
n	$-\tfrac{2}{3}$	-1.86	-1.91
Λ^0	$-\tfrac{r}{3}$	-0.613	$-0.613 \pm .004$
Σ^+	$\left[1 - \tfrac{(1-r)}{9}\right]$	2.68	2.46 ± 0.01
Σ^0	$\tfrac{1}{3}\left[1 - \tfrac{(1-r)}{3}\right]$	0.82	Not measured
Σ^-	$-\tfrac{1}{3}\left[1 + \tfrac{(1-r)}{3}\right]$	-1.04	-1.16 ± 0.03
Ξ^0	$-\tfrac{2}{3}\left[1 - \tfrac{2}{3}(1-r)\right]$	-1.44	-1.25 ± 0.01
Ξ^-	$-\tfrac{1}{3}\left[1 - \tfrac{4}{3}(1-r)\right]$	-0.51	$-.651 \pm 0.003$

6.2.1 ■ Σ^0 Decay to $\Lambda^0 + \gamma$, Magnetic Dipole Transition

The Σ^0 decays electromagnetically via the transition $\Sigma^0 \to \Lambda^0 + \gamma$. It has a lifetime of $7.4 \pm 0.7 \times 10^{-20}$ sec, or $\Gamma = 8.9 \pm 0.8 \times 10^{-3}$ MeV. The photon carries off an energy and momentum $= 74$ MeV, so that the Λ^0 barely recoils. We can think of the transition as a spin rearrangement within the baryon, which arrives at a lower energy state with the same quark content. The photon wavelength is large compared to the hadron size, so that this is a first-order magnetic dipole transition for which we can use the same quark magnetic interactions as in Section 6.2. Using the expression for a first-order transition derived in Section 2.9, $M_{fi} = \mathcal{H}_{fi}(0)$. We take the Λ^0 to be approximately at rest in the Σ^0 rest frame, resulting in:

$$\mathcal{H}_{fi}(0) = -\sqrt{(2m_{\Sigma^0})(2m_{\Lambda^0})} \left\langle \Lambda^0 \left| \sum_{i=1}^{3} \frac{e_{q_i} \vec{\sigma}_{q_i}}{2m_{q_i}} \cdot \vec{B}(0) \right| \Sigma^0 \right\rangle.$$

The factor of $\sqrt{(2m_{\Sigma^0})(2m_{\Lambda^0})}$ from the covariant spinors appears explicitly, so that the spin states for Σ^0 and Λ^0 are two-component spinors normalized to 1. The sum over i in the magnetic dipole operator runs over the u, d, s constituents in the baryon. Using the curl of the vector potential plane wave, $\vec{B}(0) = i\vec{k} \times \hat{e}^*$, where $\vec{k}$ is the outgoing photon momentum and $\hat{e}$ its polarization. Using the two-body phase space formula from Section 3.2.3, taking $m_{\Lambda^0} \simeq m_{\Sigma^0}$, and defining:

$$|M|^2 = \frac{1}{2} \sum_{\text{all spins}} \left| \left\langle \Lambda^0 \left| \sum_{i=1}^{3} \frac{e_{q_i} \vec{\sigma}_{q_i}}{2m_{q_i}} \cdot \vec{B}(0) \right| \Sigma^0 \right\rangle \right|^2, \tag{6.1}$$

gives the result

$$\Gamma = \frac{k}{8\pi^2} \int |M|^2 d\Omega. \tag{6.2}$$

As an alternative way to arrive at the width, we can calculate $|M|^2$ by leaving out the $\frac{1}{2}$ and fixing the Σ^0 spin direction in Eq. 6.1. We shall do the calculation for a fixed spin-up initial state. The final Λ^0 spin can be up or down, and we have to sum the rate for both. Comparing the spin states in Table 5.6 for a spin-up Σ^0 and Λ^0, the corresponding terms in these baryons either have the same spin orientations for the u, d, s or have two spins flipped. The sum of single quark spin operators in Eq. 6.1 can either leave the spins alone (σ_z term) or flip one spin at a time (σ_x and σ_y terms). Thus if we start with a spin-up Σ^0, only the σ_z terms have nonvanishing matrix elements with a spin-up Λ^0. The calculation of the transition to a spin-down final state requires using the Λ^0 state with spin-down, which is

$$\tfrac{1}{\sqrt{2}}(\{u\uparrow d\downarrow\}s\downarrow - \{u\downarrow d\uparrow\}s\downarrow).$$

It has nonvanishing matrix elements only for the spin-flip terms in Eq. 6.1.

Evaluating first the matrix element for Σ^0 and Λ^0 with spin up, we need keep only the σ_z piece of the expression:

$$\left\langle \Lambda^0 \uparrow \left| \frac{2}{3}\frac{e}{2m_u}\vec{\sigma}_u - \frac{1}{3}\frac{e}{2m_u}\vec{\sigma}_d - \frac{1}{3}\frac{e}{2m_s}\vec{\sigma}_s \right| \Sigma^0 \uparrow \right\rangle = \frac{-1}{\sqrt{3}}\frac{e}{2m_u}\hat{e}_z. \quad (6.3)$$

The term multiplying $\hat{e}_z$ is called the transition moment $\mu_{\Sigma\Lambda}$. In units of $e/2m_p$, it is predicted to be -1.61, which we find is in excellent agreement with the experimental data.

A similar calculation for the spin-down Λ^0 final state gives for the expression analogous to that in Eq. 6.3,

$$\frac{-1}{\sqrt{3}}\frac{e}{2m_u}(\hat{e}_x + i\hat{e}_y).$$

Thus for $|M|^2$ we get:

$$|M|^2 = \frac{1}{3}\left(\frac{e}{2m_u}\right)^2 \sum_\lambda |\hat{e}_z \cdot (\vec{k}\times \hat{e}^*(\lambda))|^2 + |\hat{e}_x \cdot (\vec{k}\times \hat{e}^*(\lambda))|^2$$
$$+ |\hat{e}_y \cdot (\vec{k}\times \hat{e}^*(\lambda))|^2$$
$$= \frac{1}{3}\left(\frac{e}{2m_u}\right)^2 \sum_\lambda |(\vec{k}\times \hat{e}^*(\lambda))|^2 = \frac{2}{3}\left(\frac{e}{2m_u}\right)^2 k^2.$$

This gives finally:

$$\Gamma = \frac{1}{3}\left(\frac{e^2}{4\pi}\right)\frac{k^3}{m_u^2}.$$

Using

$$\frac{e^2}{4\pi} = \frac{1}{137}, \quad k = 74 \text{ MeV}, \quad m_u = 336 \text{ MeV},$$

we get the prediction $\Gamma = 8.7 \times 10^{-3}$ MeV, which is in excellent agreement with the data. Note that even for a polarized initial state, the decay is isotropic if we sum over final spins.

6.2.2 ■ ϕ Decay to $\eta' + \gamma$

The Σ^0 electromagnetic decay discussed above is the predominant decay for this baryon and is the only one of this type for the octet of baryons. We next look at a meson decay that is rather analogous to the baryon decay, but now represents

6.2 Baryon Magnetic Moments

only a small part of the decay width for the initial particle. We look at the decay $\phi \to \eta' + \gamma$. This can again be thought of as a spin rearrangement within the state, leading to a state of lower energy. In this case the photon carries away 61.6 MeV and the transition is again of the magnetic dipole type. We will ignore the small recoil momentum of the η' in the following.

Using the same procedure as in Section 6.2.1 gives, for the width for this decay:

$$\Gamma = \frac{k}{8\pi^2} \int |M|^2 \, d\Omega,$$

where now

$$|M|^2 = \frac{1}{3} \sum_{\text{all spins}} \left| \left\langle \eta' \left| \sum_i \frac{e_i \vec{\sigma}_i}{2m_i} \cdot \vec{B}(0) \right| \phi \right\rangle \right|^2.$$

Here i runs over both quarks and antiquarks, since the mesons have both. The constituents in common for the ϕ and η' are $s\bar{s}$, so the spin operator that contributes is:

$$-\frac{1}{3} \frac{e}{2m_s} [\vec{\sigma}_s - \vec{\sigma}_{\bar{s}}].$$

Here we have explicitly put in the opposite charges for the quark and antiquark.

We do the calculation starting with a ϕ state with $J_z = 1$. The rate is, of course, independent of the initial spin direction. The ϕ spin state is given in terms of the constituents by $|\uparrow, \uparrow\rangle$. The first term is the spin of the s, the second is the spin of the $\bar{s}$. The final state has spin zero, which is

$$\frac{|\uparrow, \downarrow\rangle - |\downarrow, \uparrow\rangle}{\sqrt{2}}.$$

The operator σ_z produces no spin flips and does not contribute. Thus only σ_x and σ_y contribute. Applying the spin operator, we have

$$-\frac{1}{3} \frac{e}{2m_s} [\vec{\sigma}_s - \vec{\sigma}_{\bar{s}}] |\uparrow, \uparrow\rangle = \frac{2}{3} \frac{e}{2m_s} \left[\frac{\hat{e}_x + i\hat{e}_y}{\sqrt{2}} \right] \left[\frac{|\uparrow, \downarrow\rangle - |\downarrow, \uparrow\rangle}{\sqrt{2}} \right]. \quad (6.4)$$

The spin-zero η' state has a mix of $q, \bar{q}$ constituents:

$$\eta' = X_{\eta'} \frac{|u, \bar{u}\rangle + |d, \bar{d}\rangle}{\sqrt{2}} + Y_{\eta'} |s, \bar{s}\rangle.$$

Taking the matrix element with the ϕ picks off the $s\bar{s}$ piece, so that we have

$$\left\langle \eta' \left| \sum_i \frac{e_i \vec{\sigma}_i}{2m_i} \right| \phi \right\rangle = \frac{2}{3} \frac{e}{2m_s} \left(\frac{\hat{e}_x + i\hat{e}_y}{\sqrt{2}} \right) Y_{\eta'}.$$

We can foresee the factor

$$\frac{-[\hat{e}_x + i\hat{e}_y]}{\sqrt{2}}$$

appearing in the matrix element of the dipole operator; it is the polarization vector of the initial ϕ for spin along the z-axis. For momentum zero it is the only available vector along which the matrix element of the dipole moment operator can point. Note that, although $\vec{\sigma}$ is an axial vector operator, the transition to a pseudoscalar particle results in a matrix element that is a vector. Calling the polarization vector $\hat{e}_\phi$, we get for a given photon helicity λ, the contribution to $|M|^2$:

$$\frac{4}{9}\left(\frac{e}{2m_s}\right)^2 |\hat{e}_\phi \cdot (\vec{k} \times \hat{e}^*(\lambda))|^2 |Y_{\eta'}^2.$$

For a polarized ϕ the angular distribution is not isotropic, so the easiest way to calculate the rate is to average over initial polarizations. Averaging $|\hat{e}_\phi \cdot (\vec{k} \times \hat{e}^*(\lambda))|^2$ over ϕ polarizations gives $\frac{1}{3}|\vec{k} \times \hat{e}^*(\lambda)|^2$. Summing this over the two photon polarizations gives $\frac{2}{3}|\vec{k}|^2$. Thus:

$$|M|^2 = \frac{8}{27}\left(\frac{e}{2m_s}\right)^2 |\vec{k}|^2 |Y_{\eta'}|^2 \quad \text{and} \quad \Gamma = \frac{4}{27}\left(\frac{e^2}{4\pi}\right) \frac{|\vec{k}|^3}{m_s^2} |Y_{\eta'}|^2. \quad (6.5)$$

Using $|\vec{k}| = 61.6$ MeV, $m_s = 509$ MeV, $Y_{\eta'} = .8$ gives a prediction that $\Gamma = 6.2 \times 10^{-4}$ MeV. The measured value is $3.0 \pm 0.6 \times 10^{-4}$ MeV corresponding to a branching ratio $\sim 10^{-4}$. This is a small branching ratio since the ϕ decays predominantly via the strong interaction.

It is nice that we can roughly understand the meson electromagnetic decay rate in terms of the quark constituents. We have, however, made two tacit assumptions in the calculation. The first is that the mass scale in the denominator for the quark magnetic moments is the same in the mesons as in the baryons. We have taken the constituent quark masses for baryons and mesons to be the same. The second is that the vector and pseudoscalar spatial wave functions are identical. This is not just a consequence of the $SU(3)$ flavor symmetry, since these two particles are in different multiplets. We will return to this decay in a more systematic fashion in Section 6.5.

6.3 ■ MESON AND BARYON MASSES

In Section 6.2 we looked at baryon and meson interactions with an external field that is static or slowly varying over the dimensions of the hadron. We turn now to the interactions internal to a hadron. These interactions are responsible for the

6.3 Meson and Baryon Masses

mass spectrum. Our approach will again be phenomenological. By analogy with electromagnetic interactions in an atom, we can expect terms that do not depend on the quark spin orientations (analogous to electric terms) and terms that depend on the orientations (analogous to magnetic terms). The spin-dependent terms are not small, so they are not just a perturbation, as in the atom. We can get a good description of the masses of the lightest hadrons in terms of a small number of phenomenological parameters that correspond to the strength of the physical effects. The model will describe the effect of the gluon exchange terms introduced in Section 5.2, but will not describe the annihilation terms discussed there. It will provide a common approach for describing the masses of the lightest mesons and baryons.

We look at the lightest hadrons made of u and d quarks first, to see what we might expect. The masses for these should be described in terms of λ_{QCD} and the correction due to the small u and d quark masses, which for now we take to be identical. For the nonspin-dependent terms, we might expect the mass to be $\sim$ sum of the constituent quark masses. This would be $2m_u$ for the mesons and $3m_u$ for the baryons. The spin-dependent terms should be proportional to $\vec{\sigma}_i \cdot \vec{\sigma}_j$ for pair i and j. There is only one pair for the mesons, so the splitting between pseudoscalars and vectors comes about because of the different expectation values for this spin-operator. Taking $\vec{J} = \vec{S}_1 + \vec{S}_2$ for the mesons, we have by squaring that

$$\vec{S}_1 \cdot \vec{S}_2 = \frac{J^2 - S_1^2 - S_2^2}{2}.$$

This gives, taking expectation values,

$$\left\langle \vec{S}_1 \cdot \vec{S}_2 \right\rangle = -\tfrac{3}{4} \quad \text{for } J = 0, \tag{6.6}$$

$$\left\langle \vec{S}_1 \cdot \vec{S}_2 \right\rangle = \tfrac{1}{4} \quad \text{for } J = 1. \tag{6.7}$$

Assuming a positive coefficient for this interaction, as would be true for an s-wave perturbation derived from *QED* by replacing the product of electric charges with the net color charge for a q-$\bar{q}$ color singlet ($-\tfrac{4}{3}\alpha_s$ from Section 4.5), the pseudoscalars are pushed down in energy while the vectors are raised.

For the baryons the analogous spin interaction arises from three pairs interacting. Thus the spin dependent terms depend on

$$\sum_{3 \text{ pairs}} \vec{\sigma}_i \cdot \vec{\sigma}_j = 4 \sum_{3 \text{ pairs}} \vec{S}_i \cdot \vec{S}_j.$$

Using now

$$\vec{J} = \vec{S}_1 + \vec{S}_2 + \vec{S}_3, \quad \sum_{\text{pairs}} \vec{S}_i \cdot \vec{S}_j = \frac{1}{2}[J^2 - (S_1^2 + S_2^2 + S_3^2)].$$

Taking expectation values:

$$\sum_{\text{pairs}} \langle \vec{S}_i \cdot \vec{S}_j \rangle = -\frac{3}{4} \quad \text{for } J = \frac{1}{2}, \tag{6.8}$$

$$\sum_{\text{pairs}} \langle \vec{S}_i \cdot \vec{S}_j \rangle = \frac{3}{4} \quad \text{for } J = \frac{3}{2}. \tag{6.9}$$

Again, assuming a positive coefficient for this interaction, as would be the case for an s-wave perturbation arising from a term with strength proportional to $-\frac{2}{3}\alpha_s$ for each q-q in a color singlet baryon, the spin $\frac{1}{2}$ baryons (proton and neutron) are lowered in energy while the spin $\frac{3}{2}$ baryons (Δ multiplet) are raised. This simple picture is shown in Figure 6.1.

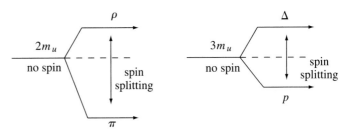

FIGURE 6.1 Simple mass pattern expected for lightest hadrons. The spin interaction splits the initial mass, which is given in terms of the constituent quark masses.

Using this pattern we can solve for the constituent quark masses implied by the physical particle masses.

For the mesons: $m_u \simeq \frac{1}{2}\left(\frac{3m_\rho + m_\pi}{4}\right) = 307$ MeV.

For the baryons: $m_u \simeq \frac{1}{3}\left(\frac{m_p + m_\Delta}{2}\right) = 362$ MeV.

These are rather close to the constituent quark mass, 336 MeV, derived from the magnetic moments.

We would like to discuss one last issue before turning to our phenomenological model. We are close to a situation where the u mass in the underlying QCD theory is $\simeq 0$. This is different from the constituent quark mass that describes a bound quark and includes the interaction energy. The constituent quark mass $\simeq \lambda_{QCD}$. We expect that the hadron masses can each be written as a given function of λ_{QCD} plus a small correction coming from the nonzero u mass in the underlying theory. In the case of the pion, the physical mass is small and we believe that the term proportional to λ_{QCD} (that is, the term for a massless quark theory) vanishes. This

6.3 Meson and Baryon Masses

general result follows from understanding the consequences of an extra symmetry in the massless limit of *QCD*. This symmetry is called a chiral symmetry. We will discuss it in Chapter 7. Our phenomenological model will result in a massless pion for a constituent quark mass reduced by only a small amount from the 336 MeV we found using the magnetic moments.

6.3.1 ■ Meson Masses

For the pseudoscalar mesons, we use a mass formula that is just the sum of the constituent quark masses minus a spin-dependent term, which is inversely proportional to the product of the constituent quark masses. The inverse dependence on the constituent quark masses is expected from a spin-spin interaction, whose strength for each quark is proportional to the magnetic moment measured in an external magnetic field. Thus in terms of the two constituents of mass m_1 and m_2, the pseudoscalar mass is

$$m_P = m_1 + m_2 - \frac{2m_0^3}{m_1 m_2}. \quad (6.10)$$

The constant in the spin-dependent term, $2m_0^3$, is chosen to indicate how close the pion is to being massless. For a nonrelativistic system, this constant is proportional to the square of the wave function at the origin; for example, it is $2\pi\alpha|\psi(0)|^2$ for the electromagnetic spin-spin interaction in the positronium ground state. Taking $m_u = 336$ MeV, and solving for m_0 using the pion mass, gives $m_0 = 311$ MeV. The value of m_π would be zero if m_u were equal to m_0. In this picture, m_0 would be the constituent quark mass in a theory where the underlying quark masses vanish. The change of m_0 to m_u coming from the finite underlying quark masses in the real world, is thus 25 MeV in this model.

For the vector mesons, we will use the following mass formula:

$$m_V = m_1 + m_2 + \frac{2}{3}\frac{m_0^3}{m_1 m_2} - \Delta. \quad (6.11)$$

The spin-dependent term has the opposite sign and is $\frac{1}{3}$ the size of the term for the pseudoscalars, as given by Eqs. 6.6 and 6.7. Since this term is not really a perturbation, we expect the meson spatial state to be a little different from the pseudoscalar state in order to arrive at an energy minimum. The constant Δ represents the effect of such a change. Using the physical masses, we get $\Delta \simeq 75$ MeV. In terms of these parameters we get the masses listed in Table 6.2. Note that the entry $m_{s\bar{s}}$ is the mass of a pseudoscalar state that has no annihilation terms. It goes into the determination of the η and η' masses, along with the annihilation terms, as discussed in Section 5.2. The calculation of the vector meson masses assumes that such annihilation terms are very small for the vectors, as seen experimentally.

The model gives an excellent description of the data. We have ignored the splitting within an isospin multiplet for both the data and the model. We come back to this splitting in Section 6.3.3.

Chapter 6 The Constituent Quark Model

TABLE 6.2 Meson Masses.

Particle	Mass from formula	Expected value (MeV)	Measured value (MeV)
π	$2m_u - \dfrac{2m_0^3}{m_u^2}$	139	139
K	$m_u + m_s - \dfrac{2m_0^3}{m_u m_s}$	493	494
$m_{s\bar{s}}$	$2m_s - \dfrac{2m_0^3}{m_s^2}$	786	
ρ	$2m_u + \dfrac{2}{3}\dfrac{m_0^3}{m_u^2} - \Delta$	775	771
K^*	$m_u + m_s + \dfrac{2}{3}\dfrac{m_0^3}{m_u m_s} - \Delta$	887	892
ϕ	$2m_s + \dfrac{2}{3}\dfrac{m_0^3}{m_s^2} - \Delta$	1020	1019

Parameters used: $m_u = 336$ MeV, $m_s = 509$ MeV, $m_0 = 311$ MeV, $\delta = 75$ MeV.

6.3.2 ■ Baryon Masses

The observed splitting in the baryons, for example $m_\Delta - m_p = 294$ MeV, is much smaller than in the mesons, where $m_\rho - m_\pi = 631$ MeV. On the basis of color factors alone, we would have expected a reduction by a factor of $\frac{3}{4}$. We will make a model for the baryons with the same ingredients as for the mesons, but the model parameters will be matched to the baryon data. First, we must have a term equal to the sum of the constituent quark masses. The spin-dependent term includes a contribution for each pair that is proportional to the expectation value of $\vec{\sigma}_i \cdot \vec{\sigma}_j$ and inversely proportional to the product of the constituent quark masses $m_i m_j$. For the proton and the Δ baryon, this gives the expressions (if we include no other terms)

$$m_p = 3m_u - \frac{\delta^3}{m_u^2}, \quad m_\Delta = 3m_u + \frac{\delta^3}{m_u^2}.$$

Using the value of $m_\Delta - m_p$, we can solve for δ, which is 255 MeV. It is analogous to the m_0 for the meson case. Solving also for m_u from the data, these expressions do not agree with the choice of $m_u = 336$ MeV. We have to add another mass term, which represents an average shift in the binding energy per quark in the baryon versus the meson. To get agreement with the data, we must add another 79 MeV, or 26 MeV per quark. This is comparable to the correction term we had to add to the vector mesons when compared to the pseudoscalars. In terms of the

6.3 Meson and Baryon Masses

spin operator for each pair, the baryon mass formula is now

$$m_b = m_1 + m_2 + m_3 + \delta' + \frac{4}{3}\delta^3 \sum_{\text{pairs}} \frac{\langle \vec{S}_i \cdot \vec{S}_j \rangle}{m_i m_j}, \qquad (6.12)$$

where, from the previous discussion, $\delta' = 79$ MeV, $\delta = 255$ MeV. The parameters δ and δ' have been chosen so that the mass formula agrees exactly with the p and Δ masses.

To evaluate Eq. 6.12 for each baryon, we need to know the expectation values for spin operators of each quark pair within the baryon. Since the states were derived in Chapter 5 in terms of first coupling a quark pair and then adding a third quark, we can use the first pair coupling to derive the expectation values. To illustrate this, consider the Λ^0, Σ^0, and Σ^{0*} states. The spin term we have to average for these baryons, which are each made of u, d, s, is

$$\frac{\vec{S}_u \cdot \vec{S}_d}{m_u^2} + \frac{\vec{S}_u \cdot \vec{S}_s + \vec{S}_d \cdot \vec{S}_s}{m_u m_s}.$$

Consider $\vec{J} = \vec{S}_u + \vec{S}_d + \vec{S}_s$. Then:

$$\vec{S}_u \cdot \vec{S}_s + \vec{S}_d \cdot \vec{S}_s = \tfrac{1}{2}\left[J^2 - \left(S_u^2 + S_d^2 + S_s^2 \right) \right] - \vec{S}_u \cdot \vec{S}_d.$$

The average of $J^2 - (S_u^2 + S_d^2 + S_s^2)$ is just $J(J+1) - 3[\tfrac{1}{2}(\tfrac{1}{2}+1)]$ in terms of the spin J of the overall state. If we knew $\langle \vec{S}_u \cdot \vec{S}_d \rangle$ we could complete the calculation. The u and d are coupled to give total spin 1 in the Σ^0 and Σ^{*0} and total spin 0 in the Λ^0. Thus, writing $\vec{J}_{u,d} = \vec{S}_u + \vec{S}_d$, we have $\langle \vec{S}_u \cdot \vec{S}_d \rangle = \tfrac{1}{2}\langle J_{u,d}^2 - \vec{S}_u^2 - \vec{S}_d^2 \rangle$. This gives $\langle \vec{S}_u \cdot \vec{S}_d \rangle = \tfrac{1}{4}$ for the Σ^0 and Σ^{*0}, and $-\tfrac{3}{4}$ for the Λ^0.

Proceeding in this way, we can figure out the predicted mass for each baryon. The results are given in Table 6.3. The mass formula is accurate to about 10 MeV. We can see the $SU(3)$ symmetry, that is, equal masses in a multiplet, that would result for $r = (m_u/m_s) = 1$. From the table, we can see that the difference in mass between the Λ^0 and Σ^0 arises from the spin-spin interaction differences. For Λ^0, the u, d are coupled to give $J_{u,d} = 0$; therefore these two quarks do not give a spin-dependent interaction term with the s quark. The spin-spin attractive term comes entirely from the u and d interaction. In the Σ^0 we have contributions from each pair, including the weaker attractive term when an s quark is involved. Hence the average spin attraction is larger for the Λ^0 and it is the lighter of the two baryons.

6.3.3 ■ Meson Isospin Violating Mass Splittings

We turn next to the isospin violating mass splittings. The electromagnetic interactions and the u, d mass difference give four types of contributions to the meson masses. These are as follows:

TABLE 6.3 Baryon Masses.

Baryon	Formula	Prediction (MeV)	Measured value (MeV)
p	$3m_u + \delta' - \dfrac{\delta^3}{m_u^2}$	938	938
Λ	$2m_u + m_s + \delta' - \dfrac{\delta^3}{m_u^2}$	1111	1116
Σ	$2m_u + m_s + \delta' - \dfrac{\delta^3}{m_u^2}\left(\dfrac{4r}{3} - \dfrac{1}{3}\right)$	1178	1193
Ξ	$m_u + 2m_s + \delta' - \dfrac{\delta^3}{m_u^2}\left(\dfrac{4r}{3} - \dfrac{r^2}{3}\right)$	1323	1318
Δ	$3m_u + \delta' + \dfrac{\delta^3}{m_u^2}$	1232	1232
Σ^*	$2m_u + m_s + \delta' + \dfrac{\delta^3}{m_u^2}\left(\dfrac{1}{3} + \dfrac{2r}{3}\right)$	1372	1384
Ξ^*	$m_u + 2m_s + \delta' + \dfrac{\delta^3}{m_u^2}\left(\dfrac{2r}{3} + \dfrac{r^2}{3}\right)$	1517	1532
Ω	$3m_s + \delta' + \dfrac{\delta^3}{m_u^2}r^2$	1668	1672

Parameters used: $m_u = 336$, $m_s = 509$, $r = (m_u/m_s) = .66$, $\delta' = 79$ MeV, $\delta = 255$ MeV.

1. A spin-independent term, which would correspond to the Coulomb electric interaction energy if the system were nonrelativistic. Using units of e for the quark charges, this term is $e_i e_j m_E$ for a quark-pair, where m_E is a phenomenological electric parameter determined by the details of the state. The value of m_E should be $\sim \alpha \lambda_{QCD}$, which is ~ 3 MeV.

2. A magnetic spin-spin interaction term. We can write this as

$$-\left[e_i e_j \left(\frac{m_u^2}{m_i m_j}\right) \frac{4}{3}\langle \vec{S}_i \cdot \vec{S}_j \rangle\right] m_B,$$

where m_B is a phenomenological magnetic parameter. Since the system is relativistic, we expect $m_B \sim m_E$. The sign has been chosen so that opposite charges in an s-wave state with $J = 0$ have an attractive interaction. The various factors have been included to provide a simple normalization for m_B, such that the magnetic term for the pion is $e_i e_j m_B$.

3. For a d quark, we need to use the correct mass, which is slightly different from m_u. We define the mass difference by $m_d = m_u + \Delta m$. For every d quark, we get a constituent quark contribution that should be increased by Δm.

4. In the spin-spin strong interaction term, we should use m_d in the denominator, instead of m_u, for each d quark. This means replacing

$$\frac{1}{m_u} \quad \text{by} \quad \frac{1}{m_u + \Delta m} \simeq \frac{1}{m_u} - \frac{\Delta m}{m_u^2}$$

in the spin-spin terms everywhere a d quark contributes.

We can calculate the various isospin splittings in terms of the above four pieces. The success of this simple model in describing the measured mass differences is the clearest evidence that Δm is nonzero. For the pseudoscalars the predicted splittings are

$$m_{\pi^+} - m_{\pi^0} = \frac{1}{2}(m_E + m_B)$$

$$m_{K^0} - m_{K^+} = \Delta m \left(1 + 2\left(\frac{m_0}{m_u}\right)^3 r\right) - \frac{1}{3}(m_E + r m_B).$$

Note that the π^+, π^0 difference doesn't depend at all on the u, d mass difference, while the sign of the K^0, K^+ difference is incorrect without such a term. With $m_{\pi^+} - m_{\pi^0} = 4.6$ MeV, we get $m_E + m_B = 9.2$ MeV. We could solve for Δm from $m_{K^0} - m_{K^+} = 4.0$ MeV, if we knew how to apportion the separate contributions of m_E and m_B for the pion. The result for Δm is, however, very insensitive to the energy split. Taking, for example, $m_E = m_B = 4.6$ MeV, gives $\Delta m = 3.2$ MeV from the kaon mass difference. We can better pin down these parameters by using the baryon mass splittings also. We look at these next.

6.3.4 ■ Baryon Isospin Violating Mass Splittings

We expect the baryons to have the same four types of isospin violating mass contributions as the mesons. In choosing phenomenological parameters, we use the same values of Δm and m_E for the baryons as for the mesons. This choice is the simplest and is motivated by the similar values for the spin independent strong interaction terms for the baryons and mesons. We assume that the electromagnetic spin dependent term for the baryons is suppressed relative to the mesons by the same factor as for the spin dependent strong interaction terms in Sections 6.3.1 and 6.3.2. Defining the baryon magnetic parameter as m'_B, this assumption implies that

$$m'_B = \left(\frac{\delta}{m_0}\right)^3 m_B = .55 m_B.$$

Finally, to calculate the baryon magnetic terms we need the expectation values for the pairwise spin-spin operators, as we saw in Section 6.3.2. For the $J = \frac{1}{2}$ multiplet, $\langle \vec{S}_i \cdot \vec{S}_j \rangle = \frac{1}{4}$ for the first two quarks that are paired to have total angular momentum 1, and $\langle \vec{S}_i \cdot \vec{S}_j \rangle = -\frac{1}{2}$ for the third quark with either of the first two quarks. For the $J = \frac{3}{2}$ multiplet: $\langle \vec{S}_i \cdot \vec{S}_j \rangle = \frac{1}{4}$ for any pair. The equal value for any pair is expected, since the $J = \frac{3}{2}$ spin state is fully symmetric.

Using the above relations and Eqs. 6.10, 6.11, and 6.12 gives the splitting formulas in Table 6.4. A good description of the data is obtained with the choices

$$\Delta m = 3.0 \text{ MeV}, \quad m_E = 3.0 \text{ MeV}, \quad m_B = 6.0 \text{ MeV}, \quad m'_B = 3.3 \text{ MeV}.$$

TABLE 6.4 Isospin Violating Mass Splittings.

Particular pair	Formula	Prediction (MeV)	Measured value (MeV)
$m_{\pi^+} - m_{\pi^0}$	$\frac{1}{2}(m_E + m_B)$	4.5	4.6
$m_{K^0} - m_{K^+}$	$\Delta m \left(1 + 2\left(\frac{m_0}{m_u}\right)^3 r\right) - \frac{1}{3}(m_E + r m_B)$	3.9	4.0
$m_{K^{*0}} - m_{K^{*+}}$	$\Delta m \left(1 - \frac{2}{3}\left(\frac{m_0}{m_u}\right)^3 r\right) - \frac{1}{3}\left(m_E - \frac{r}{3} m_B\right)$	1.4	4.4 ± 0.4
$m_n - m_p$	$\Delta m \left(1 - \frac{2}{3}\left(\frac{\delta}{m_u}\right)^3\right) - \frac{1}{3}\left(m_E - \frac{m'_B}{3}\right)$	1.5	1.3
$m_{\Sigma^-} - m_{\Sigma^0}$	$\Delta m \left(1 + \left(\frac{2r-1}{3}\right)\left(\frac{\delta}{m_u}\right)^3\right) + \frac{2}{3}\left(m_E + \frac{(2r-1)}{6} m'_B\right)$	5.2	4.8
$m_{\Sigma^0} - m_{\Sigma^+}$	$\Delta m \left(1 + \left(\frac{2r-1}{3}\right)\left(\frac{\delta}{m_u}\right)^3\right) - \frac{1}{3}\left(m_E - \frac{2}{3}(1+r) m'_B\right)$	3.4	3.3 ± 0.1
$m_{\Xi^-} - m_{\Xi^0}$	$\Delta m \left(1 + \frac{4}{3} r \left(\frac{\delta}{m_u}\right)^3\right) + \frac{2}{3}\left(m_E + \frac{2r}{3} m'_B\right)$	7.1	6.5 ± 0.2
$m_{\Sigma^{*-}} - m_{\Sigma^{*0}}$	$\Delta m \left(1 - \frac{(1+r)}{3}\left(\frac{\delta}{m_u}\right)^3\right) + \frac{2}{3}\left(m_E - \frac{(1+r)}{6} m'_B\right)$	3.7	3.5 ± 1.1
$m_{\Sigma^{*0}} - m_{\Sigma^{*+}}$	$\Delta m \left(1 - \frac{(1+r)}{3}\left(\frac{\delta}{m_u}\right)^3\right) - \frac{1}{3}\left(m_E - \frac{m'_B}{3}(2-r)\right)$	1.8	0.9 ± 1.1
$m_{\Xi^{*-}} - m_{\Xi^{*0}}$	$\Delta m \left(1 - \frac{2}{3}\left(\frac{\delta}{m_u}\right)^3 r\right) + \frac{2}{3}\left(m_E - \frac{r m'_B}{3}\right)$	3.9	3.2 ± 0.6

Parameters used: $\Delta m = 3.0$ MeV, $m_E = 3.0$ MeV, $m_B = 6.0$ MeV, $m'_B = .55 \, m_B$, $r = .66$, $m_0 = 311$ MeV, $\delta = 255$ MeV.

6.3 Meson and Baryon Masses

Predictions using these parameters, as well as the experimentally measured mass splittings, are listed in the table. The splittings for the particles with very large widths (ρ meson and Δ baryon) have not been accurately determined and are not included. The formulas are accurate, typically, to a few tenths of an MeV and indicate the need for choosing m'_B significantly smaller than m_B, as was done. Note that the general pattern is explained rather well by the formulas: for example, why $m_n - m_p$ is one of the smallest splittings, while $m_{\Xi^-} - m_{\Xi^0}$ is the largest, why sometimes the neutral particle is heavier than its charged partner, and why sometimes lighter. The only puzzling mass difference is $m_{K^{*0}} - m_{K^{*+}}$, for which the measured value is in very poor agreement with the prediction (including any reasonable uncertainty in the parameters as determined by the other mass differences).

6.3.5 ■ Decays of the η Meson

The isospin-violating mass terms above provide a mechanism for mixing analogous states with different isospins. In the case of the pseudoscalars, we get mixing that generates a small $I = 1$ piece in the predominantly $I = 0$ η and η'. To evaluate the mixing we first imagine a situation where $\Delta m, m_E, m_B$ are all zero. The resulting *QCD* Hamiltonian generates eigenstates:

$$|\pi\rangle = \frac{|u, \bar{u}\rangle - |d, \bar{d}\rangle}{\sqrt{2}}$$

$$|\eta\rangle = X_\eta \frac{|u, \bar{u}\rangle + |d, \bar{d}\rangle}{\sqrt{2}} + Y_\eta |s, \bar{s}\rangle$$

$$|\eta'\rangle = X_{\eta'} \frac{|u, \bar{u}\rangle + |d, \bar{d}\rangle}{\sqrt{2}} + Y_{\eta'} |s, \bar{s}\rangle .$$

Using these states, we can take into account $\Delta m, m_E, m_B$ as perturbations. Indicating the physical states as $|\pi\rangle_P, |\eta\rangle_P, |\eta'\rangle_P$, we get

$$|\pi\rangle_P = |\pi\rangle + \lambda_{\pi\eta} |\eta\rangle + \lambda_{\pi\eta'} |\eta'\rangle$$

$$|\eta\rangle_P = |\eta\rangle - \lambda_{\pi\eta} |\pi\rangle$$

$$|\eta'\rangle_P = |\eta'\rangle - \lambda_{\pi\eta'} |\pi\rangle .$$

We ignore the small mixing term for η, η', since we are only interested in the $I = 1$ piece generated in these mesons. Using first-order perturbation theory to estimate the mixing coefficients:

$$\lambda_{\pi\eta} = \frac{\langle\pi| H |\eta\rangle}{m_\pi - m_\eta}, \quad \lambda_{\pi\eta'} = \frac{\langle\pi| H |\eta'\rangle}{m_\pi - m_{\eta'}},$$

where H is the isospin-violating Hamiltonian discussed in Section 6.3.3. This Hamiltonian does not link states with different quark content, since it has no annihilation terms. We will see below that electromagnetic annihilation terms are very small and can be ignored. The strong interaction annihilation terms are large, but have already been included in defining the initial *QCD* eigenstates.

The only matrix element of H that we need for calculating the mixing parameters is

$$\left[\frac{\langle u,\bar{u}| - \langle d,\bar{d}|}{\sqrt{2}}\right] H \left[\frac{|u,\bar{u}\rangle + |d,\bar{d}\rangle}{\sqrt{2}}\right] = \frac{\langle u,\bar{u}| H |u,\bar{u}\rangle}{2} - \frac{\langle d,\bar{d}| H |d,\bar{d}\rangle}{2}.$$

Using the mass terms defined in Section 6.3.3, this is:

$$-\left[\Delta m \left(1 + 2\left(\frac{m_0}{m_u}\right)^3\right) + \frac{1}{6}(m_E + m_B)\right] = -9.3 \text{ MeV}.$$

Finally taking $X_\eta = .8$, $X_{\eta'} = .6$,

$$\lambda_{\pi\eta} = 2.2 \times 10^{-2} X_\eta = 1.8 \times 10^{-2}, \quad \lambda_{\pi\eta'} = 1.1 \times 10^{-2} X_{\eta'} = 6.8 \times 10^{-3}.$$

We look below at the η decays in more detail; we will look at a few η' decays later in the chapter.

The η width is $\Gamma = 1.18$ keV, a small value, because it cannot decay strongly via isospin-conserving decays. The decay to 2π is not allowed by conservation of J^P, the decay to 3π violates G parity, and the mass of 4π is too large to allow any significant decays (only $4\pi^0$ is barely energetically allowed). We list in Table 6.5 the primary decays and their branching ratios.

TABLE 6.5 Primary η Branching Ratios.

Channel	Branching ratios
2γ	39.4%
$3\pi^0$	32.5%
$\pi^+\pi^-\pi^0$	22.6%
$\pi^+\pi^-\gamma$	4.7%

The hadronic decays to 3π can come from the $I = 1$ admixture in the η state, which has negative G parity like the final state. Decay from this admixture will have a suppression factor of $|\lambda_{\pi\eta}|^2 = 3 \times 10^{-4}$, based on our calculation of the mixing. Dividing this into the measured width into 3π gives a value of 2 MeV, which is of strong interaction strength.

The assumption that the 3π decays occur because of a

$$\frac{|u,\bar{u}\rangle - |d,\bar{d}\rangle}{\sqrt{2}}$$

piece in the η allows an estimate of the relative rates to $3\pi^0$ and $\pi^+\pi^-\pi^0$. Assuming a predominantly fully symmetric s-wave final state, we can calculate the relative rates. For example, using the decay diagram method of Section 5.5, we get a prediction that the relative rate for $3\pi^0$ versus $\pi^+\pi^-\pi^0$ is $\frac{3}{2}$. This is very close to the measured ratio.

The decay $\eta \to 2\gamma$ is an example of the interesting 2γ decays for the pseudoscalar mesons π^0, η, and η'. We will look at all three later in this chapter. We make two points here. First, the overall size of $\eta \to 2\gamma$ compared to $\eta \to 3\pi$ is not surprising, since α and $\lambda_{\pi\eta}$ are both $\sim 10^{-2}$. Secondly, since these are comparable, the contribution to the mixing amplitudes from $\eta \to 2\gamma \to \pi^0$ is expected to be small, that is, of order α smaller than $\lambda_{\pi\eta}$.

The last large η decay in Table 6.5 is to $\pi^+\pi^-\gamma$. The simplest π^+-π^- system that can be produced has $J^P = 1^-$. This system would be dominated by the ρ if the π^+-π^- mass were large enough to produce a real ρ. The diagram is shown in Figure 6.2.

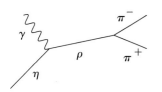

FIGURE 6.2 η decay via virtual ρ.

The simplest assumption for the decay is that it is still dominated by the ρ, which is now virtual. This hypothesis provides a reasonably good description of the rate and shape of the π^+-π^- mass spectrum. It requires that we specify the basic η-γ-ρ coupling, which can be measured in the related process $\rho \to \eta + \gamma$. We will discuss such radiative decays in general later in the chapter. Before turning to the radiative decays, we will look at the simpler coupling of the ρ meson to the photon and how the production of the π^+-π^- system with small invariant mass and $J^P = 1^-$ is dominated by the ρ. This hypothesis is called vector meson dominance. Several consequences of this hypothesis will be discussed in the next few sections.

6.4 ■ PHOTON COUPLING TO THE VECTOR MESONS

In Section 4.7, we looked at the decay of a vector meson made of a heavy quark-antiquark pair into e^+e^-. Defining the current matrix element $\langle 0| J_\mu |V\rangle = e_{q_i} g_V e_\mu^V$, we found that

$$\Gamma_{V \to e^+e^-} = |e_{q_i}|^2 \frac{4\pi}{3} \alpha^2 \frac{g_V^2}{m_V^3},$$

where e_{q_i} is the quark charge in units of e. This current matrix element defines the coupling of a virtual photon to the vector meson at the vector meson mass. For a nonrelativistic system, $g_V^2 = 12 m_V |\psi(0)|^2$. Note that g_V has dimensions of mass squared.

We want to extend these results to vector mesons composed of a light quark and antiquark. These are made of mixtures of various $q_i \bar{q}_i$ constituents. We take the decay amplitude for a vector meson to a virtual photon to be proportional to the amplitude to find a given $q_i \bar{q}_i$ pair in the vector meson times the quark charge e_{q_i}, summed over constituents. For ρ^0 and ω^0, mixtures of $|u, \bar{u}\rangle$ and $|d, \bar{d}\rangle$ are the constituents of these mesons (we will ignore small mixing terms with $|s, \bar{s}\rangle$

or isospin breaking mixing). We assume that the remaining terms in the current matrix element, other than the charge couplings, are the same for $|u, \bar{u}\rangle$ and $|d, \bar{d}\rangle$. Replacing e_{qi} by an averaged charge, we can write the current matrix element as:

$$\langle e_{q_i}\rangle g_{\gamma V} m_V^2 e_\mu^V. \quad (6.13)$$

Here $\langle e_{q_i}\rangle$ is the quark charge averaged over the amplitude to find $q_i \bar{q}_i$ in the vector V. With this parameterization of the current matrix element, $g_{\gamma V}$ is a dimensionless coupling constant. Factoring out e from the quark charge, we find for the vector meson width:

$$\Gamma_{V\to e^+e^-} = |\langle e_{q_i}\rangle|^2 \frac{4\pi}{3} \alpha^2 g_{\gamma V}^2 m_V. \quad (6.14)$$

We list in Table 6.6 the measured widths and values extracted for $g_{\gamma V}$ from the widths for the three light vector mesons. Note that the dominant factor governing the relative size of the widths are the values for $|\langle e_{q_i}\rangle|^2$, which are predicted nicely from the quark-antiquark states. The factor $g_{\gamma V}$ shows a small $SU(3)$ violation and, in fact, the simple formula $\Gamma_{V\to e^+e^-} \simeq |\langle e_{q_i}\rangle|^2 \Gamma$, with Γ a constant, agrees rather well with the data.

TABLE 6.6 Light Vector Meson Decays to e^+e^-.

Meson	State	$|\langle e_{q_i}\rangle|^2$	Measured width (keV)	Calculated $g_{\gamma V}$
ρ^0	$\frac{1}{\sqrt{2}}(|u, \bar{u}\rangle - |d, \bar{d}\rangle)$	$\frac{1}{2}$	6.85 ± 0.11	0.28
ω^0	$\frac{1}{\sqrt{2}}(|u, \bar{u}\rangle + |d, \bar{d}\rangle)$	$\frac{1}{18}$	0.60 ± 0.02	0.25
ϕ^0	$|s, \bar{s}\rangle$	$\frac{1}{9}$	1.26 ± 0.04	0.22

6.4.1 ■ Vector Meson Dominance

In Section 3.6.2 we discussed the Compton scattering cross section. This cross section, the simplest for an initial photon and charged particle, decreases with increasing energy and eventually becomes very small. In the lepton sector, the photon cross section that is largest at very high energies is given by the diagrams shown in Figure 6.3.

The target is shown as a charge distribution with net charge Z. This process is called pair production. For energies $E_\gamma \gg m_e$, the exchanged internal photon can have a very small invariant mass yielding a large cross section. For this reason, pair production is the dominant process at high energies in the lepton sector, despite the fact that it is of higher order than the Compton diagram.

Looking at photons incident on a proton target, we can consider what will change if we replace the e^+e^- in Figure 6.3 by a quark-antiquark pair. The

6.4 Photon Coupling to the Vector Mesons

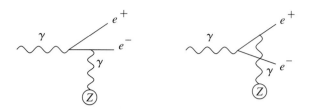

FIGURE 6.3 Pair production diagrams.

first change is that the $q, \bar{q}$ system will interact strongly via gluon exchange. This exchange of gluons can result in binding of the $q, \bar{q}$ into a meson system, which then scatters on the target. The second change is that the interaction with the target does not have to occur through photon exchange; rather, the meson system can interact strongly with the target. This picture for $\gamma + p \to$ hadrons is called Vector Meson Dominance, since the $q, \bar{q}$ meson system must have $J^P = 1^-$. The lightest mesons will be least virtual, so the greatest contributions to the cross section are expected to come from ρ, ω, ϕ. This is shown in Figure 6.4.

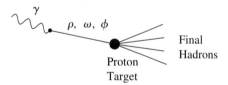

FIGURE 6.4 Simple Vector Meson Dominance picture of the cross section to produce hadrons for a real photon incident on a proton.

We look at the ρ meson contribution to the cross section, since the ρ is the largest contributor. The terms in the amplitude are coupling of the γ to the ρ, given by Eq. 6.13, the virtual ρ propagator for $q^2 = 0$, and the ρ-p scattering amplitude. Combining the first two pieces gives a factor of

$$\langle e_{q_i} \rangle g_{\gamma V} m_V^2 \left[\frac{1}{-m_V^2} \right] \delta_{\lambda_V, \lambda_\gamma} = -\langle e_{q_i} \rangle g_{\gamma V} \delta_{\lambda_V, \lambda_\gamma}.$$

Note that the virtual ρ has the same transverse helicity as that of the incident photon. Measuring the quark charges in units of e, and using the ρ coupling from Table 6.6, we can calculate the probability for

$$(\gamma \to \rho) = |\langle e_{q_i} \rangle|^2 g_{\gamma V}^2 e^2 = 4\pi\alpha |\langle e_{q_i} \rangle|^2 g_{\gamma V}^2 = \frac{1}{278}.$$

This then multiplies the ρ-p cross section, providing the ρ contribution to the cross section for the incident photon.

Adding in the other vector meson contributions, based on the analogous calculation for $V = \omega$ and ϕ, we get the prediction that at high energies:

$$\sigma_{\text{tot}}(\gamma\text{-}p) \simeq \frac{1}{200}\sigma_{\text{tot}}(\rho\text{-}p).$$

Taking the ρ-p cross section equal to the measured $\sigma_{\text{tot}}(\pi\text{-}p)$ then gives a very good description of the measured high-energy photoproduction cross section of hadrons.

6.4.2 ■ ρ Dominance in the π^+, π^- Channel with $J^P = 1^-$

The ρ meson is directly measured as an intermediate state in the production of a π^+-π^- system with $J^P = 1^-$. The simplest such process occurs through the coupling of a virtual photon to ρ, as we have discussed. Focusing on the π^+-π^- final state system, we describe the coupling to the photon with four-momentum q_μ via a current matrix element:

$$\langle \pi^+\pi^- | J_\mu | 0 \rangle.$$

The only four-vector that maintains gauge invariance (that is, $q_\mu \langle \pi^+\pi^- | J_\mu | 0 \rangle = 0$) is $P_\mu^{\pi^+} - P_\mu^{\pi^-}$. Thus $\langle \pi^+\pi^- | J_\mu | 0 \rangle$ must be proportional to this. It can, however, be multiplied by a Lorentz invariant function, maintaining the gauge invariance. The only invariant is the mass-squared of π^+-π^-, which is just q^2. Factoring out the charge e gives a general way to write the matrix element:

$$\langle \pi^+\pi^- | J_\mu | 0 \rangle = eF_\pi(q^2)(P_\mu^{\pi^+} - P_\mu^{\pi^-}). \tag{6.15}$$

$F_\pi(q^2)$ is called the form factor of the pion.

For q^2 in the vicinity of the ρ meson mass squared, we expect that the ρ dominates the π^+-π^- production. This gives us an explicit result for $F_\pi(q^2)$. Putting together the ρ production amplitude from the photon, $\langle e_{q_i} \rangle g_{\gamma\rho} m_\rho^2 e_\mu^{\rho*}$, the ρ propagator, and the ρ decay amplitude, $g_{\rho\pi\pi} e_\mu^\rho (P_\mu^{\pi^+} - P_\mu^{\pi^-})$, gives, after summing over the virtual ρ spins:

$$F_\pi(q^2) = \langle e_{q_i} \rangle g_{\gamma\rho} \left[\frac{-m_\rho^2}{q^2 - m_\rho^2 + im_\rho \Gamma_\rho(q^2)} \right] g_{\rho\pi\pi}. \tag{6.16}$$

Here $\langle e_{q_i} \rangle$ is measured in units of e. Following the results of Sections 5.3 and 5.4, and taking $g_{\rho\pi\pi}$ independent of q^2 gives:

$$\Gamma_\rho(q^2) = \left[\frac{\frac{|\vec{P}_\pi(q)|^3}{m(q)}}{\frac{|\vec{P}_\pi(m_\rho)|^3}{m_\rho}} \right] \Gamma_0,$$

where

$$\Gamma_0 = 150 \text{ MeV}, \quad |\vec{P}_\pi(q)|^2 = \frac{q^2 - 4m_\pi^2}{4}, \quad m(q) = \sqrt{q^2}.$$

6.4 Photon Coupling to the Vector Mesons

The ρ is sufficiently wide that we must allow for the q^2 dependence of the width to get good agreement with data. This specifies the expectations for π^+-π^- production via a ρ meson. Using measured rates (Tables 6.6 and 5.5, and note $g^2_{\rho\pi\pi} = 2|g|^2$ in Table 5.5), we can calculate the product of the dimensionless constants:

$$\langle e_{q_i}\rangle g_{\gamma\rho} g_{\rho\pi\pi} = 1.19.$$

Figure 6.5 shows the data for $e^+e^- \to \pi^+\pi^-$ expressed as $|F_\pi(q^2)|^2$. It shows the large ρ meson contribution, a sharp dip due to $\omega \to \pi^+\pi^-$ (ρ, ω interference) and at $q^2 > 1$ GeV^2, the contribution from heavier resonances.

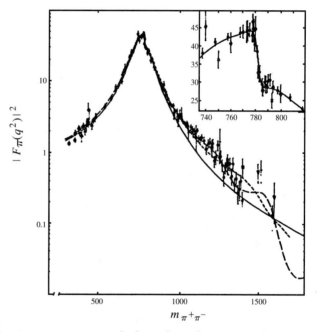

FIGURE 6.5 Data showing $|F_\pi(q^2)|^2$ for $q^2 = m^2_{\pi^+\pi^-}$. Data are taken using the reaction $e^+e^- \to \pi^+\pi^-$. Solid line shows ρ^0 contribution, including ρ, ω interference. Inset shows ρ, ω interference region. Dashed curves show models with higher mass resonances contributing. [From L. Barkov et al. *Nucl. Phys.* B256, 365(1985).]

The form factor idea can be extended to the case of $q^2 < 0$, which corresponds to scattering of a pion by a virtual photon. In this case we have:

$$\langle\pi^+|J_\mu|\pi^+\rangle = eF_\pi(q^2)\left(P^{\pi^+}_{\mu\,\text{out}} + P^{\pi^+}_{\mu\,\text{in}}\right). \tag{6.17}$$

The charge of the π^+ requires $F_\pi(0) = 1$. The $q^2 = 0$ value corresponds to the case where the π is a static source in the very nonrelativistic limit. A simple point-like spin zero particle would have $F_\pi(q^2) = 1$ for all q^2, not just at $q^2 = 0$. The pion is, however, not point-like, and has a structure.

A simple assumption for $F_\pi(q^2)$, at least in the region $|q^2| \lesssim 1$ GeV^2, is that the ρ meson dominates the amplitude not only for $q^2 > 0$ but also for $q^2 < 0$. The Feynman diagram corresponding to this idea, which again is called Vector Meson Dominance, is shown in Figure 6.6. Taking $q^2 = 0$, Eq. 6.16 then gives the prediction that $F_\pi(0) = \langle e_{q_i} \rangle g_{\gamma\rho} g_{\rho\pi\pi} = 1$. This relation between couplings is roughly satisfied, but not exact, if we use the values for the couplings seen at $q^2 = m_\rho^2$.

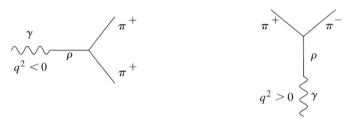

FIGURE 6.6 Vector Meson Dominance picture for virtual photons with $q^2 < 0$ and $q^2 > 0$. Time runs upward.

The ρ exchange picture gives an interesting prediction. Moving away from $q^2 = 0$, but constraining $F_\pi(0) = 1$, we get from the model the prediction

$$F_\pi(q^2) = \frac{1}{1 - \frac{q^2}{m_\rho^2}}.$$

We can use this for q^2 slightly negative, which describes the Coulomb scattering from a nearly static pion source. The form factor we are looking at for this nonrelativistic limit is the Fourier transform of the spatial charge density in the pion center of mass. This gives $F_\pi(q^2) = \int e^{i\vec{q}\cdot\vec{x}} \tilde{J}_0(\vec{x}) d^3x$, where $q = (0, \vec{q})$ and $\tilde{J}_0(\vec{x})$ is the spatial charge density normalized to one particle after integrating over the volume. For a spherically symmetric charge distribution, we can use a series expansion for small $\vec{q}^{\,2}$ to show that

$$F_\pi(q^2) \simeq 1 - \tfrac{1}{6} |\vec{q}^{\,2}| \langle r^2 \rangle_\pi,$$

where $\langle r^2 \rangle_\pi$ is the pion mean-squared radius calculated by averaging over $\tilde{J}_0(\vec{x})$. Comparing this to the Vector Meson Dominance formula for $F_\pi(q^2)$, where we use a series expansion again, we predict

$$\langle r^2 \rangle_\pi = \frac{6}{m_\rho^2} \quad \text{or} \quad \sqrt{\langle r^2 \rangle_\pi} = \frac{\sqrt{6}}{m_\rho}.$$

6.4 Photon Coupling to the Vector Mesons

This predicts the value of the mean-square charge radius of $.63 \times 10^{-13}$ cm. This is in good agreement with what is measured directly in π^+-e^- scattering, which is $.66 \pm .02 \times 10^{-13}$ cm. It also agrees with the extrapolated behavior near $q^2 = 0$ of the data in Figure 6.5.

6.4.3 ■ ρ, ω Mixing

The discussion of the γ couplings to the vector mesons has so far assumed that these mesons are ideally mixed, although the data in Figure 6.5 show the need for a contribution from $\omega^0 \to \pi^+\pi^-$. We expect in fact to have isospin violating mixing effects, analogous to those discussed for the π, η system, which will imply the existence of $\omega^0 \to \pi^+\pi^-$. We will look at the mixing among the vector mesons next. The branching ratio for the π^+-π^- decay is measured to be 1.7%.

We start with the general mixing denominator of the propagator of Section 5.3.2, which gives for the ρ-ω-ϕ system in the basis $|u, \bar{u}\rangle, |d, \bar{d}\rangle, |s, \bar{s}\rangle$, the matrix

$$q^2 - m_{ij}^2 - \pi_{ij}(q^2). \tag{6.18}$$

Here m_{ij}^2 is a diagonal matrix calculated ignoring mixing due to strong interaction annihilation effects, which occur, for example, through three gluons or virtual pseudoscalar meson pairs. These, as well as the decay widths, contribute to π_{ij}. The eigenstates are obtained by diagonalizing Eq. 6.18. We will ignore the real part of π_{ij} and the I-spin violation, when defining initial eigenstates. The initially ignored terms can then be added as perturbations. The small size of the real part of π_{ij} (as opposed to the widths) can be inferred from the near equality of m_ρ and m_ω, so it is appropriate to treat these as well as the I-spin violating terms as perturbations.

With the above assumptions, $|u, \bar{u}\rangle$ and $|d, \bar{d}\rangle$ start out degenerate, with mass m, and the correct eigenstates will be the mix that diagonalizes the imaginary part of π_{ij}. We can, however, figure out the correct mix immediately since there is a symmetry, G parity, that tells us which strong states will decay such that they have no final states in common. Defining

$$\left|\rho^0\right\rangle = \frac{|u,\bar{u}\rangle - |d,\bar{d}\rangle}{\sqrt{2}}, \quad \left|\omega^0\right\rangle = \frac{|u,\bar{u}\rangle + |d,\bar{d}\rangle}{\sqrt{2}},$$

results in

$$\pi_{\rho\omega} = \left\langle \rho^0 \middle| \pi \middle| \omega^0 \right\rangle = 0,$$

if we keep only I-spin invariant terms. In fact, we can include all the I-spin invariant mixing terms in π_{ij}, leaving $\pi_{\rho\omega} = 0$, because these terms satisfy G-parity symmetry (they will, however, mix ω, ϕ slightly). The eigenvalues in the propagation denominator will then be $m^2 - \pi_{\rho\rho} = m_\rho^2 - im_\rho \Gamma_\rho$ and $m^2 - \pi_{\omega\omega} = m_\omega^2 - im_\omega \Gamma_\omega$.

Ignoring the mixing with ϕ, we define the real eigenstates as

$$|\rho\rangle_V = \left|\rho^0\right\rangle + \lambda_{\rho\omega}\left|\omega^0\right\rangle,$$

$$|\omega\rangle_V = \left|\omega^0\right\rangle - \lambda_{\rho\omega}\left|\rho^0\right\rangle.$$

Including the I-spin violating terms, $\pi_{\rho\omega} \neq 0$. We can then calculate, using perturbation theory,

$$\lambda_{\rho\omega} = \frac{\pi_{\rho\omega}}{\pi_{\rho\rho} - \pi_{\omega\omega}}.$$

Using Eq. 5.8, with $m_\rho \simeq m_\omega$, and the definition

$$\Delta m_{\rho\omega} = \frac{\pi_{\rho\omega}}{2m_\rho},$$

we can rewrite $\lambda_{\rho\omega}$ in terms of the physical ρ and ω parameters:

$$\lambda_{\rho\omega} = \frac{\Delta m_{\rho\omega}}{\left(m_\rho - \frac{i\Gamma_\rho}{2}\right) - \left(m_\omega - \frac{i\Gamma_\omega}{2}\right)}. \qquad (6.19)$$

From this equation, we see that $\lambda_{\rho\omega}$ is small mainly because Γ_ρ is large. The large ρ width results in the physical ρ and ω being very close to ideally mixed after I-spin violation is introduced. The particle widths are irrelevant for a π, η mixing calculation, but they are very important for the ρ, ω case.

We next calculate $\Delta m_{\rho\omega}$. From the mass operator, we get a contribution analogous to that for π, η mixing, namely, a contribution to $\Delta m_{\rho\omega}$ due to the m_u and m_d mass difference:

$$\frac{\langle u, \bar{u}| H |u, \bar{u}\rangle}{2} - \frac{\langle d, \bar{d}| H |d, \bar{d}\rangle}{2},$$

where this is now evaluated for vector meson states. Following the results and notation of Section 6.3.3, this term is estimated to be

$$-\left[\Delta m\left(1 - \frac{2}{3}\left(\frac{m_0}{m_u}\right)^3\right) + \frac{1}{6}\left(m_E - \frac{m_B}{3}\right)\right] = -1.6 \text{ MeV}.$$

The electromagnetic interactions of the vector mesons also contribute a term, through the single-photon annihilation diagram $\rho \to \gamma \to \omega$. The contribution of this process is most easily calculated using Feynman diagrams and the matrix elements in Section 6.4. Using Eq. 3.5, with the same helicity for the ρ and ω, and $q^2 \simeq m_\rho^2 \simeq m_\omega^2$ for the virtual photon:

$$\pi_{\rho\omega}^\gamma = \frac{-\langle \rho| J_\mu |0\rangle \langle 0| J_\mu |\omega\rangle}{q^2} = \langle e_i\rangle_\rho \langle e_i\rangle_\omega g_{\gamma\rho} g_{\gamma\omega} m_\rho^2.$$

6.4 Photon Coupling to the Vector Mesons

This gives

$$\Delta m^\gamma_{\rho\omega} = \langle e_i \rangle_\rho \langle e_i \rangle_\omega g_{\gamma\rho} g_{\gamma\omega} \frac{m_\rho}{2} = \frac{\pi}{3} \alpha g_{\gamma\rho} g_{\gamma\omega} m_\rho.$$

Using the values of the couplings from Table 6.6, $\Delta m^\gamma_{\rho\omega} = 0.4$ MeV. Adding together the two terms contributing to $\Delta m_{\rho\omega}$, we predict that $\Delta m_{\rho\omega} \simeq -1.2$ MeV.

Using the physical vector meson eigenstates implies that ω will decay to π^+-π^- through mixing (that is, through the $I = 1$ component of the ω state). The I-spin violation, however, provides an additional possibility, which is that

$$|\omega^0\rangle = \frac{|u, \bar{u}\rangle + |d, \bar{d}\rangle}{\sqrt{2}}$$

can decay directly into π^+-π^-. The two diagrams that contribute, as discussed in Section 5.5 and shown below, will no longer cancel for the ω^0 if the amplitude to create a $\bar{d}d$ from the vacuum is not equal to the amplitude for $\bar{u}u$. In fact, with unequal masses for u and d we expect a small difference. The same phenomenon is evident for $\bar{s}s$ in very high energy jets, where there is a significant suppression factor for creating strange quarks. With the $\bar{d}d$ amplitude a little smaller than $\bar{u}u$ (by a factor of $1 - 2r$), we expect a direct amplitude for ω^0:

$$M_{fi}(\omega^0 \to \pi^+\pi^-) = r M_{fi}(\rho^0 \to \pi^+\pi^-).$$

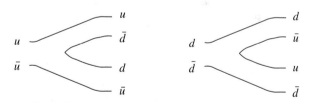

Given a direct amplitude, we must, however, also include this process in the mixing term $\pi_{\rho\omega}$, since it provides an additional way for ω^0 to turn into a ρ^0, via $\omega^0 \to \pi^+\pi^- \to \rho^0$. Assuming that $\pi_{\rho\rho}$ is predominantly due to $\rho^0 \to \pi^+\pi^- \to \rho^0$, we see that the contribution to $\pi_{\rho\omega}$ from direct ω^0 decay to $\pi^+\pi^-$ is $r\pi_{\rho\rho}$. Collecting the contributions to $|\omega\rangle_V \to \pi^+\pi^-$ from the direct decay and the mixing term stemming from the direct decay, gives a contribution to $M_{fi}(|\omega\rangle_V \to \pi^+\pi^-)$:

$$\left[r - \frac{r\pi_{\rho\rho}}{\pi_{\rho\rho} - \pi_{\omega\omega}}\right] M_{fi}(\rho^0 \to \pi^+\pi^-) = \left(\frac{-r\pi_{\omega\omega}}{\pi_{\rho\rho} - \pi_{\omega\omega}}\right) M_{fi}(\rho^0 \to \pi^+\pi^-).$$

Note that the direct decay and the mixing term it induces, tend to cancel.

Adding together all the various terms we have identified:

$$M_{fi}(|\omega\rangle_V \to \pi^+\pi^-) = \left[\frac{-2\Delta m_{\rho\omega}m_\rho - r\pi_{\omega\omega}}{\pi_{\rho\rho} - \pi_{\omega\omega}}\right] M_{fi}(\rho^0 \to \pi^+\pi^-).$$

The terms contributing are due to the ρ^0 component in the ω^0 state, coming from quark mass differences and electromagnetic transitions, and the contributions from the direct decay. For a reasonable estimate of r, for instance,

$$r \sim \frac{m_d - m_u}{\lambda_{QCD}} \simeq 0.01,$$

the term proportional to r gives a small contribution to M_{fi} and the term involving $\Delta m_{\rho\omega}$ dominates—that is, we can ignore the direct decay term. Using the expression for the decay amplitude with $\Delta m_{\rho\omega}$ negative explains the contribution from the ω Breit–Wigner in the ρ, ω interference region in Figure 6.5. The $\omega \to \pi^+\pi^-$ interfering with the dominant ρ resonance gives a positive enhancement to $|F_\pi|^2$ for $q^2 < m_\omega^2$, negative interference for $q^2 > m_\omega^2$, and an incoherent small contribution for $q^2 \simeq m_\omega^2$. Note that the phase of the ω amplitude is determined by the denominator term of Eq. 6.19, where

$$\frac{\pi_{\rho\rho} - \pi_{\omega\omega}}{2m_\rho} \simeq \left(m_\rho - \frac{i\Gamma_\rho}{2}\right) - \left(m_\omega - \frac{i\Gamma_\omega}{2}\right) \simeq -\frac{i\Gamma_\rho}{2}.$$

This general picture agrees with the data. Using the interference we can calculate from the data that $\Delta m_{\rho\omega} = -2.1$ MeV. Thus our calculation of the I-spin violating mass mixing term is too small by ~ 1 MeV.

In Chapter 9, we will analyze analogous mixing for the $K^0, \bar{K}^0$ mesons (which are analogous to the $|u, \bar{u}\rangle$ and $|d, \bar{d}\rangle$ states discussed in this section). In that case the mixing is due to weak transitions. Again, the initial eigenstates can immediately be delineated because of a symmetry. In the $K^0, \bar{K}^0$ case, CP plays the role that G parity did in the ρ, ω discussion. Surprisingly, it turns out that CP is also broken. The breaking changes the mix a little from the "ideal" mixing solution. The CP violation, which is visible through a forbidden π^+-π^- decay, can be classified as coming from the propagation matrix (which turns out to be very dominant) or as a direct decay, as we did for the ρ, ω system. The mechanism and consequences of the CP violation are of great interest, since they imply a difference between particle and antiparticle behavior.

6.5 ■ RADIATIVE TRANSITIONS BETWEEN PSEUDOSCALAR AND VECTOR MESONS

Section 6.4 focused on a number of topics that arise when looking at the coupling of a photon to a light vector meson. Another simple first-order electromagnetic process involves the pseudoscalar and vector mesons, which can undergo

6.5 Radiative Transitions between Pseudoscalar and Vector Mesons

transitions in which a final state photon is emitted. An example is the transition $\phi \to \eta' + \gamma$ discussed in Section 6.2.2, where the recoiling η' is nonrelativistic in the center of mass. The mass difference between the other mesons that have analogous decays are much larger and the decays are therefore relativistic. We will therefore treat these in a relativistically covariant way. The radiative decays discussed in this section serve to probe the quark content of the various mesons.

We define the variables that can appear in the amplitude as follows: k_γ is the photon four-momentum, e_μ^γ its spin; P_V is the vector meson four-momentum, e_μ^V its spin; P_P is the pseudoscalar four-momentum. The matrix element must be a Lorentz pseudoscalar, linear in the various spins. In addition we require gauge invariance; that is, substituting k_μ^γ for e_μ^γ should give a vanishing result. The above constraints uniquely determine the matrix element for a given meson pair in terms of one constant.

For $P \to V + \gamma$: $M_{fi} = f_{VP\gamma} \varepsilon_{\mu\nu\sigma\rho} e_\mu^{\gamma*} k_\nu^\gamma e_\sigma^{V*} P_\rho^V$.

For $V \to P + \gamma$: $M_{fi} = f_{VP\gamma} \varepsilon_{\mu\nu\sigma\rho} e_\mu^{\gamma*} k_\nu^\gamma e_\sigma^V P_\rho^V$.

Note that we could replace the momentum of the vector meson by that of the pseudoscalar in M_{fi}, since the difference is k_γ, which already appears once in the fully antisymmetric expression for M_{fi}. To calculate the rate we go to the rest frame of the decaying particle.

For $P \to V + \gamma$: $M_{fi} = m_P \vec{e}_V^{\,*} \cdot (\hat{e}_\gamma^* \times \vec{k}_\gamma) f_{VP\gamma}$.

For $V \to P + \gamma$: $M_{fi} = m_V \hat{e}_V \cdot (\hat{e}_\gamma^* \times \vec{k}_\gamma) f_{VP\gamma}$.

With these matrix elements, we can calculate the decay rates summed over final state helicities, which are

$$\Gamma(P \to V + \gamma) = \frac{f_{VP\gamma}^2 |\vec{k}_\gamma|^3}{4\pi}, \tag{6.20}$$

$$\Gamma(V \to P + \gamma) = \frac{f_{VP\gamma}^2 |\vec{k}_\gamma|^3}{12\pi}. \tag{6.21}$$

Note that $f_{VP\gamma}$ has dimensions of (mass)$^{-1}$.

The decays for various meson pairs are expected to differ because of the differing quark and antiquark charges and masses of their constituents. We want to make a plausible model of how the charges and masses enter and affect the rate. To do this, we recall the case discussed in Section 6.2.2 where m_V and m_P are nearly equal. For V and P made of a quark of type i and antiquark j, the magnetic dipole transition discussed in Section 6.2.2 gives

$$f_{VP\gamma} = \frac{e_{q_i}}{m_{q_i}} - \frac{e_{\bar{q}_j}}{m_{q_j}}.$$

For example, for the $\phi \to \eta' + \gamma$ decay this gives

$$-\frac{2}{3}\frac{e}{m_s} \simeq -\frac{4}{3}\frac{e}{m_\phi}.$$

To make an explicit model we replace $2m_{q_i}$ in the magnetic dipole formula by m_{V_i}, where m_{V_i} is the mass of the vector meson made of $q_i \bar{q}_i$. This gives a dimensionally correct formula. Each term e_{q_i}/m_{V_i} then has to be multiplied by a constant, which depends on i, to account for the violation of the $SU(3)$ symmetry. It turns out that the data are approximately described in terms of one constant, independent of quark type, whose value is close to 1. Thus our formula for mesons with quark content $q_i \bar{q}_j \to q_i \bar{q}_j + \gamma$ is

$$f_{VP\gamma} = 2\left[\frac{e_{q_i}}{m_{V_i}} - \frac{e_{\bar{q}_j}}{m_{V_j}}\right] f_o.$$

The data indicate that $|f_o|^2 \simeq \frac{3}{4}$.

For states made of linear combinations of different quark-antiquark pairs we have to weight the various $f_{VP\gamma}$ by the amplitudes to find the given $q_i \bar{q}_j$ in each meson. We write this average as

$$f_{VP\gamma} = 2\left\langle \frac{e_{q_i}}{m_{V_i}} - \frac{e_{\bar{q}_j}}{m_{V_j}} \right\rangle f_o.$$

Factoring out a charge factor e, we can then write the rates as

$$\Gamma(P \to V + \gamma) = 4\alpha k_\gamma^3 |f_o|^2 \left|\left\langle \frac{e_{q_i}}{m_{V_i}} - \frac{e_{\bar{q}_j}}{m_{V_j}} \right\rangle\right|^2, \qquad (6.22)$$

$$\Gamma(V \to P + \gamma) = \frac{4\alpha}{3} k_\gamma^3 |f_o|^2 \left|\left\langle \frac{e_{q_i}}{m_{V_i}} - \frac{e_{\bar{q}_j}}{m_{V_j}} \right\rangle\right|^2. \qquad (6.23)$$

We use these formulas to tabulate the various predicted and measured rates in Table 6.7. We take ideal mixing for the ρ and ω,

$$|\eta\rangle = X_\eta \frac{|u,\bar{u}\rangle + |d,\bar{d}\rangle}{\sqrt{2}} + Y_\eta |s,\bar{s}\rangle$$

and the analogous expression for $|\eta'\rangle$. $|f_o|^2$ is assumed to be $\frac{3}{4}$. The ϕ decay to π^0 is expected to be dominated by the small mixture of

$$\frac{|u,\bar{u}\rangle + |d,\bar{d}\rangle}{\sqrt{2}}$$

in the ϕ, so we write this mixture as $\lambda_{\omega\phi}$. Finally, for the predicted rates in the table we choose $X_\eta = Y_{\eta'} = .8$, $X_{\eta'} = -Y_\eta = .6$.

6.5 Radiative Transitions between Pseudoscalar and Vector Mesons

TABLE 6.7 Pseudoscalar–Vector Radiative Widths.

Process	$\left\|\left\langle \dfrac{e_{q_i}}{m_{V_i}} - \dfrac{e_{\bar{q}_j}}{m_{V_j}} \right\rangle\right\|^2$	Predicted rate (keV)	Measured rate (keV)
$\omega^0 \to \pi^0 + \gamma$	$\dfrac{1}{m_\omega^2}$	650	734 ± 34
$\rho^0 \to \pi^0 \gamma,\ \rho^\pm \to \pi^\pm \gamma$	$\dfrac{1}{9 m_\rho^2}$	71	70 ± 8
$\omega^0 \to \eta + \gamma$	$\dfrac{1}{9 m_\omega^2} X_\eta^2$	6.7	5.5 ± 0.8
$\rho^0 \to \eta + \gamma$	$\dfrac{1}{m_\rho^2} X_\eta^2$	53	57 ± 11
$\phi \to \eta + \gamma$	$\dfrac{4}{9 m_\phi^2} Y_\eta^2$	54	55 ± 1
$\phi \to \eta' + \gamma$	$\dfrac{4}{9 m_\phi^2} Y_{\eta'}^2$	0.47	0.30 ± 0.06
$\phi \to \pi^0 + \gamma$	$\dfrac{1}{m_\omega^2}\|\lambda_{\omega\phi}\|^2$		5.3 ± 0.4
$K^{*+} \to K^+ + \gamma$	$\dfrac{1}{9}\left\|\dfrac{2}{m_\omega} - \dfrac{1}{m_\phi}\right\|^2$	59	50 ± 5
$K^{*0} \to K^0 + \gamma$	$\dfrac{1}{9}\left\|\dfrac{1}{m_\omega} + \dfrac{1}{m_\phi}\right\|^2$	124	116 ± 10
$\eta' \to \rho + \gamma$	$\dfrac{1}{m_\rho^2} X_{\eta'}^2$	64	60 ± 5
$\eta' \to \omega + \gamma$	$\dfrac{1}{9 m_\omega^2} X_{\eta'}^2$	5.9	6.1 ± 0.8

The model predicts the general pattern very well, for example, why $\omega^0 \to \pi^0 + \gamma$ is much larger than $\rho^0 \to \pi^0 + \gamma$; why $\eta' \to \rho + \gamma$ is much larger than $\eta' \to \omega + \gamma$; and why $K^{*+} \to K^+ + \gamma$ is smaller than $K^{*0} \to K^0 + \gamma$, even though the latter involves only uncharged particles. Taking $X_\eta, Y_\eta, X_{\eta'}, Y_{\eta'}$ as coupled parameters to be extracted from the data allows their determination with uncertainty arising from the errors in the data and limitations in the model, which are not easy to quantify. This kind of analysis leads to the determination that $X_\eta \simeq .8$, $X'_\eta \simeq .6$. Note, finally, that the measured rate for $\phi \to \pi^0 + \gamma$ corresponds to $\lambda_{\omega\phi} = .06$. This value for the parameter is of the same order or magnitude, but somewhat larger than the value obtained from the simplest mass model in Section 5.2.

6.6 ■ PSEUDOSCALAR MESON DECAYS TO TWO PHOTONS

The final topic we will consider is the decay of a neutral pseudoscalar meson to two photons. This is the dominant decay for a π^0 and provides the largest branching ratio for η^0. We start again with the most general gauge invariant decay amplitude for $P \to \gamma\gamma$, where P is the pseudoscalar. Labeling the photons (1) and (2), we get:

$$M_{fi} = f_{P\gamma\gamma} \varepsilon_{\mu\nu\sigma\rho} e_\mu^{(1)*} k_\nu^{(1)} e_\sigma^{(2)*} k_\rho^{(2)}$$
$$+ \text{ term from exchanging (1) and (2)}. \quad (6.24)$$

The constant $f_{P\gamma\gamma}$ is the only dynamically-determined dimensional constant that governs the rate for a given pseudoscalar meson decay.

Writing the matrix element in the pseudoscalar center of mass frame gives

$$M_{fi} = f_{P\gamma\gamma} m_P \left[\hat{e}_{(2)}^* \cdot (\hat{e}_{(1)}^* \times \vec{k}_{(1)}) \right] + \text{exchange term}. \quad (6.25)$$

The rate calculation is analogous to that for $P \to V + \gamma$; however, we have to take into account the fact that we have identical bosons in the final state. The final result, after summing over helicities, is:

$$\Gamma(P \to \gamma + \gamma) = \frac{f_{P\gamma\gamma}^2 |\vec{k}_\gamma|^3}{2\pi}. \quad (6.26)$$

This is a factor of two larger than the analogous expression for $P \to V + \gamma$ because of the identical particles in the final state.

Again, we want to make a model that accounts for the charged objects within P. Since two photons are emitted, each $q_i \bar{q}_i$ component of P will contribute an amplitude proportional to $e_{q_i}^2$. If there is no $SU(3)$ violation, we must then have an overall amplitude proportional to the sum over the amplitudes to find $q_i \bar{q}_i$ times $e_{q_i}^2$. We write this sum as $\langle e_i^2 \rangle$. With $SU(3)$ violation, the terms involving $|s, \bar{s}\rangle$ can have a somewhat different weight in the amplitude than the others. For simplicity we will ignore this and let the reader explore the likely changes due to the $SU(3)$ violation.

With this simplifying assumption we can write

$$\Gamma(P \to \gamma + \gamma) = |\langle e_i^2 \rangle|^2 \frac{f^2 |\vec{k}_\gamma|^3}{2\pi},$$

where f^2 is now independent of meson type P. The data and expressions for $|\langle e_i^2 \rangle|^2$ are given in Table 6.8.

Defining r_η as the ratio

$$\frac{|\langle e_i^2 \rangle|_\eta^2}{|\langle e_i^2 \rangle|_{\pi^0}^2},$$

6.6 Pseudoscalar Meson Decays to Two Photons

TABLE 6.8 Pseudoscalar Radiative Widths.

Meson	$\|\langle e_i^2 \rangle\|^2$	Measured width
π^0	$\left\|\dfrac{1}{3\sqrt{2}}\right\|^2$	7.7 ± 0.6 eV
η	$\left\|\dfrac{1}{9}\left(\dfrac{5X_\eta}{\sqrt{2}} + Y_\eta\right)\right\|^2$	0.47 ± 0.04 keV
η'	$\left\|\dfrac{1}{9}\left(\dfrac{5X_{\eta'}}{\sqrt{2}} + Y_{\eta'}\right)\right\|^2$	4.3 ± 0.4 keV

with the analogous definition for η',

$$r_\eta = \frac{25}{9}\left|X_\eta + \frac{\sqrt{2}}{5}Y_\eta\right|^2. \tag{6.27}$$

Using the data we can calculate the measured values for r_η and $r_{\eta'}$ as

$$r_\eta = \frac{25}{9}\left|X_\eta + \frac{\sqrt{2}}{5}Y_\eta\right|^2 = .92 \pm .11,$$

$$r_{\eta'} = \frac{25}{9}\left|X_{\eta'} + \frac{\sqrt{2}}{5}Y_{\eta'}\right|^2 = 1.56 \pm .20.$$

Choosing $X_\eta = Y_{\eta'} = 0.78$ and $X_{\eta'} = -Y_\eta = 0.626$ gives $r_\eta = 1$ and $r_{\eta'} = 2$, which roughly agrees with the data. Note that this is the first calculation so far where X_η and Y_η interfere. The data indicate that Y_η and X_η have opposite signs, as we have chosen. The $SU(3)$ violation gives an uncertainty that is difficult to estimate; we therefore have generally used values for $X_\eta, Y_\eta, X_{\eta'}, Y_{\eta'}$ with only one significant figure.

6.6.1 ■ Vector Meson Dominance and Radiative Decays

In Section 6.4.2 we saw that reactions of a photon with the π^+-π^- system are mediated by the ρ^0, which couples to the photon. This description seems to work well for a virtual ρ (at least down to $q^2 \approx 0$) and not just near the ρ mass, where the strong interaction coupling of ρ to π^+-π^- should dominate the rate.

An analogous strong coupling should exist in the ρ-ω-π system. Assuming a virtual ρ, we can in fact describe ω decay via $\omega \to \rho + \pi$ with subsequent decay of the virtual ρ to π-π. The existence of this coupling, along with the vector meson photon couplings, provides a mechanism for radiative decays. Extending the vector meson dominance hypothesis to these reactions provides a relationship between a number of decays in terms of the single ρ-ω-π coupling. The reactions are shown in Figure 6.7. From the diagrams we can calculate the various decay constants in terms of one coupling, $g_{\rho\omega\pi}$, and the various vector meson-photon couplings.

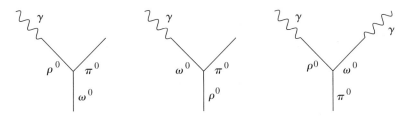

FIGURE 6.7 Vector meson dominance for several radiative decays.

Using the diagrams and the notation of Section 6.4, we get the following relations from the couplings:

$$f_{\omega\pi\gamma} = g_{\rho\omega\pi} g_{\gamma\rho} \langle e_i \rangle_\rho, \quad f_{\rho\pi\gamma} = g_{\rho\omega\pi} g_{\gamma\omega} \langle e_i \rangle_\omega,$$

$$f_{\pi\gamma\gamma} = g_{\rho\omega\pi} g_{\gamma\rho} \langle e_i \rangle_\rho g_{\gamma\omega} \langle e_i \rangle_\omega.$$

These formulas give an excellent prediction of the ratio of widths for $\omega^0 \to \pi^0 + \gamma$ and $\rho^0 \to \pi^0 + \gamma$. We can also use them to relate the widths for $\pi^0 \to \gamma + \gamma$ decay and the radiative decay $\omega^0 \to \pi^0 + \gamma$. The relation for the widths is

$$\frac{\Gamma(\pi^0 \to \gamma + \gamma)}{\Gamma(\omega^0 \to \pi^0 + \gamma)} = 6 \frac{|\vec{k}_\gamma^{\pi^0}|^3}{|\vec{k}_\gamma^{\omega^0}|^3} |g_{\gamma\omega} \langle e_i \rangle_\omega|^2.$$

Here $\vec{k}_\gamma^{\pi^0}$ and $\vec{k}_\gamma^{\omega^0}$ are the final state photon momenta in the respective decay processes. Using

$$|\langle e_i \rangle_\omega|^2 = \frac{e^2}{18} = \frac{4\pi\alpha}{18}, \quad g_{\gamma\omega} = .25,$$

and the measured value for $\Gamma(\omega^0 \to \pi^0 + \gamma)$, we get the prediction:

$$\Gamma(\pi^0 \to \gamma + \gamma) = 7.7 \text{ eV},$$

which is in excellent agreement with the measured decay width.

6.7 ■ CONCLUSION

We have considered a number of static properties and electromagnetic interactions for the lightest mesons and baryons. Despite the lack of a simple equation with which to make quantitative predictions, the existence of quarks as the backbone of the various hadrons and their low energy interactions is evident in all the quantities investigated. This quark model hypothesis allows the correlation of a large amount of data in terms of a small number of parameters. These parameters are often the constituent quark masses, which are defined using experimental data but lack a rigorous theoretical definition, and the quark charges that are well defined. In the process of correlating some of the data, we also found an interesting dynamical relationship called Vector Meson Dominance. It relates the photon coupling to hadrons to the photon coupling to vector mesons and the vector-hadron interaction.

CHAPTER 6 HOMEWORK

6.1. Work through the details of the calculations of the widths for the decays $\Sigma^0 \to \Lambda^0 + \gamma$ and $\phi \to \eta' + \gamma$.

6.2. Unfortunately, we do not have a simple method for calculating the meson and baryon masses in the real relativistic theory. A nonrelativistic approximation is, however, instructive. The simplest confining potential is the harmonic oscillator, which gives a Hamiltonian for systems made of particles with one given mass:

For the meson: $\quad H = \dfrac{1}{2m}\left(p_1^2 + p_2^2\right) + \dfrac{k}{2}|\vec{x}_1 - \vec{x}_2|^2.$

For the baryon: $\quad H = \dfrac{1}{2m}\left(p_1^2 + p_2^2 + p_3^2\right) + \dfrac{1}{2}\left(\dfrac{k}{2}\right)\sum_{\text{pairs}}|\vec{x}_i - \vec{x}_j|^2.$

The extra factor of $\frac{1}{2}$ in the baryon potential energy term is a result of the smaller color factor per pair in the baryon, compared to a meson.

(a) Show that the ratio of ground-state binding energies for the baryon compared to the meson is $\sqrt{3}$. This is close to the factor of 1.5 that we would get by assuming an energy per quark (constituent quark energy) that is the same for both systems.

(b) Show that the ratio of the average spacing between particle pairs, defined by $\langle|\vec{x}_i - \vec{x}_j|^2\rangle^{1/2}$, is $(4/3)^{1/4} = 1.07$ for the baryon relative to the meson. Thus, both states have nearly the same size when expressed in terms of the pairwise spacing.

You will need to separate out the center of mass and relative motions. This calculation should be familiar for the meson. For the baryon it can be done by defining new coordinates

$$\vec{x}_1 = \vec{R} - \sqrt{\dfrac{2}{3}}\vec{\xi}, \quad \vec{x}_2 = \vec{R} + \dfrac{1}{\sqrt{6}}\vec{\xi} - \dfrac{1}{\sqrt{2}}\vec{\eta}, \quad \vec{x}_3 = \vec{R} + \dfrac{1}{\sqrt{6}}\vec{\xi} + \dfrac{1}{\sqrt{2}}\vec{\eta}.$$

$\vec{R}$ is the center of mass coordinate. Besides the center of mass motion, the Hamiltonian consists of 3-dimensional harmonic oscillators for the $\vec{\xi}$ and $\vec{\eta}$ coordinates.

6.3. For a nonrelativistic system such as positronium, the s-wave states experience an electromagnetic spin-spin interaction given by

$$\frac{8\pi\alpha}{3}\frac{\langle \vec{S}_i \cdot \vec{S}_j \rangle}{m_e^2}\delta^3(\vec{x}_i - \vec{x}_j).$$

We can expect a *QCD* analog for singlet states in the nonrelativistic limit:

$$\text{For a meson:} \quad \frac{8\pi}{3}\left(\frac{4}{3}\alpha_s\right)\frac{\langle \vec{S}_i \cdot \vec{S}_j \rangle}{m_i m_j}\delta^3(\vec{x}_i - \vec{x}_j).$$

$$\text{For a baryon:} \quad \frac{8\pi}{3}\left(\frac{2}{3}\alpha_s\right)\sum_{\text{pairs}}\frac{\langle \vec{S}_i \cdot \vec{S}_j \rangle}{m_i m_j}\delta^3(\vec{x}_i - \vec{x}_j).$$

In the notation of our mass model, the nonrelativistic approximation implies

$$\left(\frac{\delta}{m_0}\right)^3 = \frac{\langle \delta^3(\vec{x}_i - \vec{x}_j) \rangle_{\text{baryon}}}{\langle \delta^3(\vec{x}_i - \vec{x}_j) \rangle_{\text{meson}}}.$$

For the harmonic oscillator model of the previous problem, show that this ratio is $(\frac{\sqrt{3}}{2})^{3/2} = .8$. In the nonrelativistic limit, (δ/m_0) is approximately the inverse of the ratio of the typical spacing between pairs in the state.

6.4. Consider the model in the text for the isospin violating mass differences.
(a) Show that $m_{\pi^+} - m_{\pi^0} = \frac{1}{2}(m_E + m_B)$.
(b) Show that

$$\frac{\langle u, \bar{u}| H |u, \bar{u}\rangle - \langle d, \bar{d}| H |d, \bar{d}\rangle}{2} = -\left[\Delta m\left(1 + 2\left(\frac{m_0}{m_u}\right)^3\right) + \frac{1}{6}(m_E + m_B)\right]$$

for the ground-state pseudoscalars.

6.5. Some types of electromagnetic mass terms seen for the hadrons also contribute in other systems.

(a) In the text we discussed annihilation into a single photon as a term contributing to ρ, ω mixing. This diagram will also change the ρ, ω, and ϕ masses by a small amount. The analogous mechanism contributes to the mass of the nonrelativistic 1^- positronium ground state via $e^+e^- \to \gamma \to e^+e^-$. Show that the change in energy of this state is

$$\Delta E = 2\pi\alpha \frac{|\psi(0)|^2}{m_e^2} = \frac{\alpha^4 m_e}{4},$$

using the Coulomb ground state wave function. This will contribute to splitting the 0^- and 1^- ground states, as discussed in Section 3.5.

(b) The second term contributing to the positronium ground state splitting is the spin-spin electromagnetic interaction. For s-wave states this is given by a term for each state, of the form

$$\Delta E = \frac{8\pi\alpha}{3}\frac{\langle \vec{S}_1 \cdot \vec{S}_2 \rangle}{m_e^2}\langle \delta^3(\vec{x}_1 - \vec{x}_2) \rangle.$$

Show that this contributes a term to the 0^- and 1^- ground state splitting (with 1^- heavier):

$$\Delta E = \tfrac{1}{3}\alpha^4 m_e.$$

(c) Adding together the two terms above, show that $E_{1^-} - E_{0^-} = \tfrac{7}{12}\alpha^4 m_e = 8.45 \times 10^{-4}$ eV. This number agrees with the data to about 1%, which is the size of the next-order correction.

6.6. For a nonrelativistic meson system we can expect the color analog of the electromagnetic spin-spin interaction. For the s-wave ground state, this gives an interaction Hamiltonian:

$$\frac{8\pi}{3}\left(\frac{4}{3}\alpha_s\right)\frac{\langle \vec{S}_i \cdot \vec{S}_j \rangle}{m_i m_j}\delta^3(\vec{x}_i - \vec{x}_j).$$

(a) Show that this gives a splitting between the 0^- and 1^- state, if the Hamiltonian is treated as a perturbation, of:

$$\Delta E = \frac{8\pi}{3}\left(\frac{4}{3}\alpha_s\right)\frac{\langle \delta^3(\vec{x}_i - \vec{x}_j)\rangle}{m_i m_j}$$

(b) The most nonrelativistic system we have is the $b\bar{b}$ system. Assuming that the ground state is described fairly well by a Coulomb bound state in the potential

$$-\frac{4}{3}\frac{\alpha_s}{r},$$

show that

$$\Delta E = \tfrac{1}{3}\left(\tfrac{4}{3}\alpha_s\right)^4 m_b.$$

(c) In the Coulombic approximation, $\alpha_s = .3$. Calculate ΔE for this approximation.

6.7. Consider the $\eta^0 \to 3\pi$ decays.

(a) Show that the matrix elements describing this decay are of the form

$$M_{fi} = A_1\left(m^2_{\pi^0_1 \pi^0_2}, m^2_{\pi^0_2 \pi^0_3}, m^2_{\pi^0_1 \pi^0_3}\right) \quad \text{for } 3\pi^0$$

$$M_{fi} = A_2\left(m^2_{\pi^0 \pi^+}, m^2_{\pi^0 \pi^-}, m^2_{\pi^+ \pi^-}\right) \quad \text{for } \pi^+\pi^-\pi^0.$$

(b) Based on Bose symmetry for the $3\pi^0$ case and charge conjugation symmetry for $\pi^+\pi^-\pi^0$, show that A_1 is a symmetric function of the three invariant masses and A_2 is a symmetric function of $m^2_{\pi^0\pi^+}$ and $m^2_{\pi^0\pi^-}$.

(c) Assuming that A_1 and A_2 are nearly constant, a first-order expansion in the masses-squared consistent with (b) gives:

$$A_1 = C_1$$
$$A_2 = C_2 + C_3 m^2_{\pi^+\pi^-},$$

where C_1, C_2, C_3 are constants.

(d) For $C_3 \simeq 0$, we have a fully symmetric final state in both cases. Using decay diagrams, but ignoring the Bose symmetrization, show that

$$\frac{C_1}{C_2} = -\frac{1}{2}.$$

Including Bose symmetrization, this implies a relative rate of $3!(\frac{1}{2})^2 = \frac{3}{2}$. Assume the decay is due to η, π^0 mixing.

(e) An example of dynamics that can influence the rates, is the contribution of an intermediate hadronic resonance. An example consistent with the expected states in the quark model is an intermediate isoscalar state with $J^P = 0^+$. This gives contributions of

$$\eta^0 \to \pi^0 S, \ S \to \pi^0 \pi^0 \quad \text{or} \quad \pi^+ \pi^-.$$

Here S is the scalar with mass m_s. Write the amplitudes for the 3π decays in terms of the S propagator and coupling constants at the two vertices. The relative coupling for $S \to \pi^0 \pi^0$ and $\pi^+ \pi^-$ can be obtained by using Clebsch–Gordan coefficients or decay diagrams.

(f) For $m_s \gg m_\eta$, show that we get the fully symmetric situation of part (d). More generally, show that the result is consistent with the symmetric behavior of part (b).

(g) An alternative mechanism to η, π^0 mixing, that would allow η decay to 3π, is isospin violation in the decay amplitude. Assuming a fully symmetric final state, we can use decay diagrams to calculate the decay for an isoscalar η. Assuming a small suppression for creating $d\bar{d}$ compared to $u\bar{u}$, show that this gives the same ratio of amplitudes as in (d) above. Thus either mechanism gives the same prediction, which agrees with the data.

6.8. In terms of the pion form factor $F_\pi(q^2)$, calculate the differential and total cross section for $e^+ e^- \to \pi^+ \pi^-$ via exchange of a single photon, for an unpolarized initial state.

6.9. Starting with the matrix elements in the text, complete the calculation and show that:

$$\Gamma(P \to V + \gamma) = \frac{f_{VP\gamma}^2 |\vec{k}_\gamma|^3}{4\pi}$$

$$\Gamma(V \to P + \gamma) = \frac{f_{VP\gamma}^2 |\vec{k}_\gamma|^3}{12\pi}$$

$$\Gamma(P \to \gamma + \gamma) = \frac{f_{P\gamma\gamma}^2 |\vec{k}_\gamma|^3}{2\pi}.$$

CHAPTER 7

The Full Color Gauge Theory

7.1 ■ LOCAL GAUGE SYMMETRY

We introduced color symmetry of the Lagrangian in Chapter 4, in various stages. The simplest idea was a symmetry under the interchange of one particle type (color) with another. This corresponds to an operation we can imagine doing, even in a classical physical system. This idea was broadened to a quantum operation through the introduction of a symmetry under a change of states given by a unitary transformation. We envisioned the transformation as being the same everywhere, that is, a change of the basis used in the Lagrangian. This is specified by a global non-Abelian gauge transformation, as opposed to just a global phase change (Abelian transformation), which we discussed for the simpler theories of Chapter 2. The color interaction term for quarks contained no derivatives and we introduced gluons in a way that preserved the global symmetry under color rotations of the fields in the interaction term.

The non-Abelian gauge symmetry can be broadened to a local symmetry by considering the behavior of the Lagrangian under rotations of the color fields via a continuous spatially-dependent unitary transformation, as was first considered for $SU(2)$ by Yang and Mills in 1954. This corresponds to a choice of basis for the quarks that changes as we move from place to place in space-time. Under such a transformation, the quark fields are rotated and the gluon fields transformed in a coordinated fashion to be specified. The color gauge theory and the transformations are chosen so that the Lagrangian is invariant under these combined local operations.

We look at the Lagrangian for one quark flavor that comes in three colors. We write the three colored quark fields in terms of the one-column vector $\psi(x)$. The local gauge transformation then corresponds to:

$$\psi(x) \to e^{i\theta_a(x) T_a} \psi(x).$$

A global transformation corresponds to the functions $\theta_a(x)$ equal to constants θ_a. T_a are the eight generators for $SU(3)$ color rotations and the index a is summed over. The interaction term in Chapter 4 corresponds to a Lagrangian:

$$\begin{aligned} \mathcal{L} &= \bar{\psi}\left(i\gamma_\mu \frac{\partial}{\partial x_\mu} - m\right)\psi - g\bar{\psi}\gamma_\mu T_a \psi A^a_\mu \\ &= \bar{\psi}\gamma_\mu \left(\frac{i\partial}{\partial x_\mu} - g T_a A^a_\mu\right)\psi - m\bar{\psi}\psi. \end{aligned} \quad (7.1)$$

Chapter 7 The Full Color Gauge Theory

The prescription for including the interaction in Eq. 7.1 looks very much like the minimal substitution used in electrodynamics to introduce electromagnetism, with the electron charge $-e$ replaced by g.

We know how ψ transforms under the gauge transformation, but what about A_μ^a? For a global transformation, we expect that A_μ^a must rotate in a fashion consistent with being in an octet representation of $SU(3)$. However, for the local transformation, it also needs to change so as to compensate for the changes introduced in the Lagrangian by the derivative term operating on ψ in Eq. 7.1—the only change to A_μ in the simpler electromagnetic theory.

To work out the appropriate transformations, we look at an infinitesimal local gauge transformation and its effect on the Lagrangian in Eq. 7.1. We can imagine building up a finite transformation from repeated infinitesimal transformations. Under the infinitesimal transformation, we have for the quark fields:

$$\psi \to (1 + i\theta_a(x)T_a)\psi,$$

$$\frac{i\partial\psi}{\partial x_\mu} \to (1 + i\theta_a(x)T_a)\frac{i\partial\psi}{\partial x_\mu} - \frac{\partial\theta_a}{\partial x_\mu}T_a\psi.$$

The derivative term

$$-\bar\psi \frac{\partial\theta_a}{\partial x_\mu} T_a \psi$$

appearing in the transformed Lagrangian will be cancelled by choosing the transformation for the gluons:

$$A_\mu^a \to A_\mu^a - \frac{1}{g}\frac{\partial\theta_a(x)}{\partial x_\mu}.$$

But this is not sufficient, because it would imply no change for a global transformation, whereas we expect that A_μ^a should rotate like an $SU(3)$ octet for this case. In fact, we left out a term in the transformed Lagrangian by considering only the derivative part, because we ignored the fact that $-g\bar\psi\gamma_\mu T_a\psi$ does not go into itself under non-Abelian transformations. In particular, for the infinitesimal transformation:

$$\bar\psi\gamma_\mu T_a\psi \to \bar\psi\gamma_\mu T_a\psi + i\theta_b(x)\bar\psi\gamma_\mu(T_aT_b - T_bT_a)\psi.$$

Since the T_a operators do not commute, the second term does not vanish. Using the formula for the commutator, Eq. 4.4:

$$\bar\psi\gamma_\mu T_a\psi \to \bar\psi\gamma_\mu T_a\psi - f_{abc}\theta_b(x)\bar\psi\gamma_\mu T_c\psi. \tag{7.2}$$

We can cancel the contribution of this second term in the transformed Lagrangian if we also make an additional change to A_μ^a, since the second term in Eq. 7.2 involves the eight currents with which we started.

7.1 Local Gauge Symmetry

Defining now the infinitesimal gauge transformation for the A^a_μ via

$$A^a_\mu \to A^a_\mu - \frac{1}{g}\frac{\partial \theta_a}{\partial x_\mu} - f_{abc}\theta_b A^c_\mu, \qquad (7.3)$$

the Lagrangian in Eq. 7.1 is locally gauge invariant. Note that the antisymmetry of the f_{abc} under exchange of indices is used to arrive at the cancellation between terms coming from Eqs. 7.2 and 7.3.

For a global gauge transformation, both the current in Eq. 7.2 and the gluon fields in Eq. 7.3 transform identically as octets. The eight gluon field operators are Hermitian, so there are no separate gluons and anti-gluons. The gluons are, however, not charge conjugation eigenstates. The fields can also be combined in linear combinations to make complex field operators. This will be important for understanding the charge structure of the weak interactions, which have an $SU(2)$ non-Abelian group symmetry rather similar to the $SU(3)$ color symmetry.

A given set of spin-one gauge particles interacts with a variety of fermions. Each quark flavor requires a term of the form given by Eq. 7.1 in the QCD Lagrangian. Each flavor also has a term for the electromagnetic interactions. In the case of electromagnetism, we can scale both the phase factor in the gauge transformation for a given fermion, as well as its coupling, by the electric charge to arrive at a gauge invariant Lagrangian for any charge value. Thus, the value of the electric charge is not constrained by the gauge symmetry. This does not work for the gluons because the gluon transformation in Eq. 7.3 has two terms, for which g and θ_a do not appear in the same ratio. This is a consequence of not all the generators commuting for the color interaction. As a consequence, the gauge symmetry requires that all quark triplets have the same coupling. Thus all quark flavors have identical strong interactions.

To complete the Lagrangian we still need to add the pure gluon term to Eq. 7.1, which is analogous to the term $-\frac{1}{4}F_{\mu\nu}F_{\mu\nu}$ from electromagnetism. The relevant field replacing $F_{\mu\nu}$ should come in eight varieties so as to provide equations for the propagation of the eight gluons. Writing these fields as $G^a_{\mu\nu}$, the additional term in the Lagrangian is

$$-\tfrac{1}{4}G^a_{\mu\nu}G^a_{\mu\nu}. \qquad (7.4)$$

We can guess, based on the electromagnetic case, that one term that will be included in $G^a_{\mu\nu}$ is

$$\frac{\partial A^a_\nu}{\partial x_\mu} - \frac{\partial A^a_\mu}{\partial x_\nu}. \qquad (7.5)$$

The requirement of gauge symmetry now is that Eq. 7.4 is invariant under the infinitesimal transformation in Eq. 7.3. The antisymmetric combination of derivatives in Eq. 7.5 eliminates the derivative term from Eq. 7.3 but not the second term stemming from the $SU(3)$ rotation. By choosing

$$G^a_{\mu\nu} = \frac{\partial}{\partial x_\mu} A^a_\nu - \frac{\partial}{\partial x_\nu} A^a_\mu - g f_{abc} A^b_\mu A^c_\nu, \qquad (7.6)$$

the expression in Eq. 7.4 is gauge invariant and can be added to the Lagrangian.

The terms in Eq. 7.4 that are independent of the coupling constant g, after substituting for $G^a_{\mu\nu}$ from Eq. 7.6, give free field equations for eight massless gluons. The remaining terms represent gluon interactions that are uniquely determined by the local gauge invariance in terms of a single coupling constant. The interaction terms that are first order in g involve three gluon fields at a point in space-time. A term quadratic in g also occurs. This term involves four gluon fields at a point in space-time. Thus, these are the types of vertices for pure gluonic interactions. This is a major change from electromagnetism, where the photon has no direct self-interactions. The coupling g is uniquely determined to be identical to the value for the coupling of gluons to quarks.

7.2 ■ PARADOX OF NO SCALES

We defined an $SU(3)$ invariant Lagrangian that yields field equations for the colored quarks and gluons. It does not contain the parameter λ_{QCD} anywhere and therefore provides no hint regarding how the parameter comes about. We noted that the u and d quark masses are very small, and we can imagine setting their masses to zero in the Lagrangian and ignoring the heavier quarks. In this case, the Lagrangian has *no* parameters with a mass scale, and we would expect that all states should be massless, reflecting the lack of a scale. This is, of course, incorrect and reflects again the fact that we do not know the source of λ_{QCD}. From the Lagrangian we can see the types of vertices in the theory directly. But it is not clear what the higher-order diagrams do for the full quantum theory. These hold the key to λ_{QCD} and we turn to them next. To start with a simpler (but also very interesting) situation, we will look first at electrodynamics for electrons and photons.

7.2.1 ■ The Running Coupling Constant in Electrodynamics

For the meson resonances, we saw that the meson propagator is modified through higher-order interactions. For a spin 0 particle, the interactions result in the change that

$$\frac{i}{q^2 - m^2 + i\varepsilon} \quad \text{goes into} \quad \frac{i}{q^2 - m^2 - \pi(q^2)}.$$

For a photon, the simple higher-order interaction shown in Figure 3.5 is the creation of an electron-positron pair that then reconverts to a photon. This will modify the photon propagator in all processes where a photon is exchanged. For $q^2 > (2m_e)^2$ the pair can be real and we expect the analogue of $\pi(q^2)$ for a photon to have an imaginary part; for $q^2 < (2m_e)^2$ the imaginary part vanishes by the unitarity relation, and $\pi(q^2)$ is real. Since $m^2 = 0$ for the photon, the denominator in the propagator is equal to $q^2 - \pi(q^2)$. If $\pi(q^2) \neq 0$ at $q^2 = 0$, then the photon has a mass generated by higher-order corrections to the propagator. This would violate the gauge invariance of the theory; in fact, $\pi(q^2) = 0$ at $q^2 = 0$ as a consequence of gauge invariance. Thus, factoring out a q^2, we see that the denominator in the propagator can be written $q^2(1 + I(q^2))$. If we include the electric charge coupling factors, a photon propagator always appears in the combination

$$\frac{-ig_{\mu\nu}e^2}{q^2(1+I(q^2))}.$$

We can consider the effect of the higher-order diagrams to correspond to a charge squared, or value for α, which is

$$\alpha(q^2) = \frac{\alpha_0}{[1+I(q^2)]}. \tag{7.7}$$

In particular, if $I(0) \neq 0$ (which we will see is correct), $\alpha(0)$ is not just α_0. Thus the charge, which appeared in the Lagrangian and field equations, is not the charge measured experimentally in low energy processes through a potential (which is the Fourier transform of the renormalized matrix element due to photon exchange). Instead, we experimentally determine a renormalized value for the charge.

The function $\alpha(q^2)$ is called a running coupling constant. It is a particular way in which the higher-order corrections show up in a gauge theory for the massless gauge particles. The other higher-order corrections, for example, to the electron propagator and the electron vertex, can be shown not to change this relation. The charge renormalization is universal, and in fact is the same for all charged particles that share in common the photon propagator as the carrier of the electromagnetic interaction. Thus, the equality of the magnitude of the proton and electron charge remains even after the inclusion of higher-order corrections. Since the renormalized charge is the experimentally measured charge, we will write it as e, calling the charge in the initial Lagrangian e_0.

By measuring scattering at a q^2 not equal to zero, we can determine the running of $\alpha(q^2)$. To express this in terms of the measured charge e, we write in the propagator:

$$1 + I(q^2) = 1 + I(0) + I(q^2) - I(0) = 1 + I(0)\left[1 + \frac{I(q^2) - I(0)}{1 + I(0)}\right].$$

The factor in front, $1 + I(0)$, can then be absorbed to change e_0 to e as for the case of $q^2 = 0$. Note, however, that the loop correction $I(q^2)$ also contains a factor e_0^2 from the two vertices involved. Thus, we can also absorb the factor $1 + I(0)$ in the denominator of the second term by changing e_0^2 to e^2 in $I(q^2) - I(0)$. The measurable quantities are then the renormalized charge and coupling constant α

at $q^2 = 0$, and its change with q^2 through the expression

$$\alpha(q^2) = \frac{\alpha}{[1 + I(q^2, \alpha) - I(0, \alpha)]}, \quad (7.8)$$

where we've indicated explicitly that the renormalized value of α is to be used in the function $I(q^2)$. Note that now, the value e_0 does not appear anywhere and is in fact not directly measurable. Similarly, $I(0)$ by itself is also unmeasurable. The writing of the rates for processes in terms of a small number of measurable constants, coupling constants, and masses, and the formulas for their running, is called the renormalization of the theory. If this can be applied to all orders in the perturbation expansion, the theory is said to be renormalizable. If not, the theory is nonrenormalizable and does not have a meaningful perturbation expansion. It can be proved that quantum electrodynamics is a renormalizable theory.

7.2.2 ■ Expression for $\alpha(q^2)$ in Electrodynamics

We can evaluate the contribution of the electron-positron loop to the photon propagator using the Feynman rules. The result for the propagator is

$$\frac{-ig_{\mu\nu}}{q^2} + \frac{-i}{q^2}(-i\pi_{\mu\nu})\frac{-i}{q^2} + \cdots \quad (7.9)$$

where

$$-i\pi_{\mu\nu} = -\int \frac{d^4p}{(2\pi)^4} \operatorname{Tr}\left[(ie_0\gamma_\mu)\frac{i(\slashed{p} + m_e)}{p^2 - m_e^2 + i\varepsilon}(ie_0\gamma_\nu)\frac{i(\slashed{p} - \slashed{q} + m_e)}{(q-p)^2 - m_e^2 + i\varepsilon}\right]. \quad (7.10)$$

The charge appearing in Eq. 7.10 is the unrenormalized charge, now indicated as e_0, based on the discussion in Section 7.2.1 above. The factor of -1 in front of the integral is associated with the positron propagator, discussed in Section 3.6, and the trace results from writing the expression in terms of matrices resulting from $u, \bar{u}$ and $v, \bar{v}$, summed over helicities of the virtual fermions.

The integral for $\pi_{\mu\nu}$ can be evaluated using general techniques developed for integrals resulting from higher-order Feynman diagrams. The result is that we can write $i\pi_{\mu\nu} = -ig_{\mu\nu}q^2I(q^2)$, as anticipated earlier, with

$$I(q^2) = \frac{\alpha_0}{3\pi}\int_{m_e^2}^{\infty}\frac{dp^2}{p^2} - \frac{2\alpha_0}{\pi}\int_0^1 dz\, z(1-z)\log\left(1 - \frac{q^2 z(1-z)}{m_e^2 - i\varepsilon}\right). \quad (7.11)$$

The $i\varepsilon$ in the second term gives an imaginary part for $q^2 > (2m_e)^2$. The second term, still written as an integral, vanishes at $q^2 = 0$, so the first term is $I(0)$. $I(0)$ is not equal to zero, so the charge is indeed renormalized. Unfortunately $I(0)$ is infinite, since it diverges logarithmically as we integrate over dp^2 to infinity. This divergence is an indication of the incompleteness of our knowledge of all the physics at very short distance scales corresponding to very large momenta. To cut off the integral, extra physics at very short distances would either have to modify the idea of a "point-like particle" or provide additional vertices that cancel the divergence. Note that the other charged fermions will also contribute loops, which add similar terms to $I(q^2)$, but with larger fermion masses in the integrals. For small q^2 the electron term dominates the value of the second term, which provides the measurable effect after renormalization. We turn to this term next.

Absorbing the $1+I(0)$ into the value of α at $q^2 = 0$, we have for the propagator to the single-loop level, keeping only the electron term:

$$\frac{-ig_{\mu\nu}}{q^2}\left(1 + \frac{2\alpha}{\pi}\int_0^1 dz\, z(1-z) \log\left(1 - \frac{q^2 z(1-z)}{m_e^2 - i\varepsilon}\right)\right). \tag{7.12}$$

For $|q^2| \ll m_e^2$ the logarithm is approximately:

$$\log\left(1 - \frac{q^2 z(1-z)}{m_e^2 - i\varepsilon}\right) \simeq -q^2 \frac{z(1-z)}{m_e^2}. \tag{7.13}$$

The running with q^2 is inhibited by the existence of the fixed-scale electron mass, so that for $|q^2| \ll m_e^2$ this term is small. Thus for classical physics at very small q^2, the electron charge can be adequately described just by a fixed constant. For $|q^2| \gg m_e^2$ we will see that the logarithm is large (although when multiplied by α it is still small, except at enormously large q^2).

Using the approximation in Eq. 7.13, we can calculate the integral in Eq. 7.12, giving for small $|q^2|$,

$$\frac{-ig_{\mu\nu}}{q^2}\left(1 - \frac{\alpha q^2}{15\pi m_e^2}\right). \tag{7.14}$$

In the nonrelativistic limit, the interaction is determined by a potential that is the Fourier transform of a matrix element whose q^2 dependence comes entirely from the propagator. The second term in Eq. 7.14 gives a q^2 independent constant, whose Fourier transform is a δ-function at the origin. The corresponding term in the potential is therefore very short-range, which typically has very little effect. It gives a very small but measurable shift of the s-wave energy levels in the hydrogen atom, which serves to experimentally verify the existence and size of this term.

If we evaluate the integral in Eq. 7.12 for a scattering problem where $|q^2| \gg m_e^2$, we have, approximately,

$$\int_0^1 dz\, z(1-z) \log\left(1 - \frac{q^2 z(1-z)}{m_e^2 - i\varepsilon}\right) \simeq \frac{1}{6}\log\frac{|q^2|}{m_e^2}.$$

We can then substitute this into Eq. 7.8 to arrive at an $\alpha(q^2)$ that runs logarithmically. For the extension of such a formula to QCD, it is convenient to take the point of view that we will not always refer $\alpha(q^2)$ to its value at $q^2 = 0$. Rather we can refer it to any scale μ^2. We can accomplish this by subtracting $I(\mu^2)$ instead of $I(0)$ in Eq. 7.8, in which case the α at $q^2 = 0$ that appears in the equation is replaced by $\alpha(\mu^2)$. In renormalizing the coupling constant we divided α_0 by $1 + I(\mu^2)$ instead of $1 + I(0)$.

The resulting logarithmically running coupling constant at large $|q^2|$ is

$$\alpha\left(|q^2|\right) = \frac{\alpha(\mu^2)}{1 - \frac{\alpha(\mu^2)}{3\pi} \log\left(\frac{|q^2|}{\mu^2}\right)}. \tag{7.15}$$

The electron mass, which is irrelevant at very high q^2, no longer appears; but for the formula to be useful we must know $\alpha(\mu^2)$ at some scale μ^2. Note that $\alpha(|q^2|)$ grows with increasing $|q^2|$; that is, the interaction is stronger at short distances and weaker at long distances. This behavior is the opposite to the behavior of the strong color force, as seen in the potential models in Chapter 4. Finally, for a more correct result we must add in the effect of loops coming from the other charged particles.

7.2.3 ■ Running Coupling Constant for *QCD*

In Chapter 4, gluon exchange was introduced as the mediator of the strong interactions. As in the photon case, we expect higher-order corrections due to loops. Since gluons interact with gluons, the loops can involve both quarks and gluons. The lowest-order diagrams are shown in Figure 7.1.

FIGURE 7.1 Loops due to $q, \bar{q}$ and multiple gluon coupling.

The loops again require renormalization of the coupling to yield a sensible result. Since the gluons have no mass, there is no natural scale at which to renormalize the multi-gluon contribution. The gluon terms always yield logarithmically-running contributions with mass scale changes, after renormalization at a given scale. The quark pairs contribute in a manner analogous to their contribution in electrodynamics. For $|q^2| \ll$ quark mass squared the contribution is very small; for $|q^2| \gg$ quark mass squared (which defines a "light" flavor for the given q^2) the quark flavor contributes, to good approximation, a term that runs logarithmically. Keeping only contributions from the gluons and n_f "light" flavors, the running coupling can be shown to be (after renormalization, and keeping

7.2 Paradox of No Scales

only the largest logarithmic term):

$$\alpha_s(|q^2|) = \frac{\alpha_s(\mu^2)}{1 + \frac{\alpha_s(\mu^2)}{12\pi}(33 - 2n_f)\log\left(\frac{|q^2|}{\mu^2}\right)}. \tag{7.16}$$

Here $\alpha_s(|q^2|)$ is defined in terms of its value $\alpha_s(\mu^2)$ at some scale μ^2. The strong coupling decreases as $|q^2|$ increases; this is called asymptotic freedom. The common convention is to pick as a reference scale $\mu = M_Z$, the mass of the Z weak vector boson. Experimentally, $\alpha_s(M_Z^2) = 0.118 \pm 0.002$. This is determined by using processes for which a perturbation expansion in $\alpha_s(\mu^2)$ is a good approximation, and running α_s from the scale μ of the given measurement to $\mu = M_Z$. We will look at some of the relevant measurements in subsequent chapters. Note that for $\mu = M_Z$, the number of "light" flavors is 5.

An alternative way to express $\alpha_s(|q^2|)$ is to introduce a mass scale λ that replaces the dependence on μ. Defining:

$$\lambda^2 = \mu^2 \exp\left[\frac{-12\pi}{(33 - 2n_f)\alpha_s(\mu^2)}\right], \tag{7.17}$$

we can write in a more compact form

$$\alpha_s(|q^2|) = \frac{12\pi}{(33 - 2n_f)\log\left(\frac{|q^2|}{\lambda^2}\right)}. \tag{7.18}$$

From this equation it is clear that $\alpha_s \to 1$, as $|q^2|$ becomes of order λ^2. Thus λ reveals the scale at which the theory becomes nonperturbative. This form for α_s is the motivation for the phenomenological Richardson color potential discussed in Section 4.6.

The scale λ enters, because the coupling is not a constant but runs instead. Still, this does not tell us where the scale comes from, which is a question outside the QCD theory itself. If we had a more comprehensive theory, which fixed the value of α_s at some (perhaps very high energy) scale, then the value of λ would be just a result of running from that scale to the scale where α_s becomes ~ 1.

In the discussion above we ignored quark mass effects. The coupling constant is a smooth function, so that as a quark goes from being "heavy" (say, $m_q \gg \mu$) to being "light" (say, $m_q \ll \mu$) the running of α_s changes (that is, the derivative changes), but the value is continuous through the transition region. A reasonable prescription is to match the formulas for α_s so that $\alpha_s(\mu, n_f - 1) = \alpha_s(\mu, n_f)$ at $\mu = m_q$. This results in a λ that is not a constant, but rather changes with the mass region of interest. Thus λ should have a superscript $\lambda^{(n_f)}$ that indicates the number of active flavors. The λ_{QCD} used earlier to describe the scale relevant for u, d, s quark interactions is most analogous to $\lambda^{(3)}$. The change in α_s between scales μ_1 and μ_2 is thus sensitive to all strongly-interacting fields that can be excited over the given mass region, while high-mass physics, with masses $\gg \mu_1$ and μ_2, largely cannot be seen.

The expression given by Eq. 7.18 is a good approximation over a range of $|q^2|$ provided the logarithm in the denominator is large. In practice, to compare to data, we obtain a somewhat better approximation by keeping additional terms in the expression for α_s after renormalization. For example, keeping terms involving the logarithm of the logarithm also (but still ignoring heavy quark masses) gives an expression of the form:

$$\alpha_s(|q^2|) = \frac{12\pi}{(33 - 2n_f)\log\left(\frac{|q^2|}{\lambda^2}\right)}\left[1 - \frac{\beta \log\left(\log\left(\frac{|q^2|}{\lambda^2}\right)\right)}{\log\left(\frac{|q^2|}{\lambda^2}\right)}\right], \quad (7.19)$$

where

$$\beta = \frac{3(306 - 38n_f)}{(33 - 2n_f)^2}.$$

The term in brackets turns out to be ~ 0.7 to 0.8 for scales from a few GeV to M_Z. Using data to solve for λ, the value for λ versus n_f is approximately:

$$\lambda^{(3)} \simeq 400 \text{ MeV},$$
$$\lambda^{(4)} \simeq 300 \text{ MeV}, \quad (7.20)$$
$$\lambda^{(5)} \simeq 200 \text{ MeV}.$$

Note that α_s runs from a value of about 0.32 at 2 GeV to about 0.12 at 100 GeV. By running from M_Z down in mass, the value in the perturbative regime predicts correctly the nonperturbative mass region. The strong interactions allow us to study both perturbative and nonperturbative physics, depending on the energy scale. The lower-energy nonperturbative physics, discussed in Chapters 5 and 6, was studied first. A number of higher-energy processes, where perturbative methods are applicable will be discussed later, primarily in Chapter 12.

7.3 ■ APPROXIMATE CHIRAL SYMMETRY OF THE STRONG INTERACTIONS

The bound states of the strong interactions, that is, the mesons and baryons, have a pattern of quantum numbers rather independent of the quark flavors from which the states are constructed. For example, the lightest mesons are 0^- and 1^- states for both the nonrelativistic heavy quark bound states and the ultra-relativistic light quark bound states. We expect that this pattern of states would be unchanged if the light quark masses were zero instead of just very small. The consistency of the pattern of states with an underlying *QCD* theory in this limit has an important consequence, which is that the 0^- mesons with nonzero flavor quantum numbers have to be massless. The pions made of the lightest quarks are, in fact, close to being massless. This results from an extra symmetry, called chiral symmetry (after *chiros*, the Greek word for hand), in the massless quark limit. This symmetry in-

7.3 Approximate Chiral Symmetry of the Strong Interactions

volves unitary transformations among the different flavors, rather than the colors, but also involves the quark helicities.

To understand the idea of chiral symmetry, we start with the *QCD* Lagrangian for several massless quark flavors and examine the flavor-dependent part:

$$\mathcal{L} = \bar{\psi}_i \gamma_\mu \left(i \frac{\partial}{\partial x_\mu} - g T_a A_\mu^a \right) \psi_i.$$

Each ψ_i has three colored components and i runs over three flavor choices if we consider u, d, s as the three light quarks (or over u and d if we want to stick to the very light quarks only). This $\mathcal{L}$ is invariant under rotations among the three flavors, which would be true even if we had an $SU(3)$ symmetric mass term $m\bar{\psi}_i\psi_i$. The lack of a mass term, however, leads to a larger global symmetry. If we look separately at the left-handed and right-handed components of the fermion fields, which we can project via

$$\psi_i^L = \left(\frac{1 - \gamma_5}{2} \right) \psi_i, \quad \psi_i^R = \left(\frac{1 + \gamma_5}{2} \right) \psi_i,$$

$$\mathcal{L} = \bar{\psi}_i^L \gamma_\mu \left(i \frac{\partial}{\partial x_\mu} - g T_a A_\mu^a \right) \psi_i^L + \bar{\psi}_i^R \gamma_\mu \left(i \frac{\partial}{\partial x_\mu} - g T_a A_\mu^a \right) \psi_i^R. \quad (7.21)$$

$\mathcal{L}$ has separated into two terms, each of which has the same flavor symmetry structure. This separation would not be possible with a mass term. Thus, completely separate $SU(3)$ flavor transformations on ψ_i^L and ψ_i^R leave $\mathcal{L}$ invariant, implying a larger symmetry than we had with one transformation acting simultaneously on all components of ψ_i. The extended flavor symmetry is an $SU(3)_L \times SU(3)_R$ symmetry. This symmetry will lead to conserved flavor octet right-handed and left-handed currents. The $SU(3)_L \times SU(3)_R$ symmetry is called a chiral symmetry. The conserved currents are:

$$J_{\mu,\alpha}^L = \bar{\psi}_L \gamma_\mu t_\alpha \psi_L, \quad J_{\mu,\alpha}^R = \bar{\psi}_R \gamma_\mu t_\alpha \psi_R.$$

Here t_α (as opposed to T_a, which acts on color indices) are the eight flavor $SU(3)$ generators, and the flavor indices on ψ_L and ψ_R are not explicitly indicated. Alternatively, by adding and subtracting, the conserved currents can be written as

$$J_{\mu,\alpha}^V = \bar{\psi} \gamma_\mu t_\alpha \psi, \quad J_{\mu,\alpha}^A = \bar{\psi} \gamma_\mu \gamma_5 t_\alpha \psi,$$

which are just the vector and axial-vector currents that can be constructed from the Dirac fields. The symmetry then implies:

$$\frac{\partial}{\partial x_\mu} J_{\mu,\alpha}^V = 0 \quad \text{and} \quad \frac{\partial}{\partial x_\mu} J_{\mu,\alpha}^A = 0. \quad (7.22)$$

7.3.1 ■ Goldberger–Treiman Relation

The foregoing discussion focused on the quarks that appear in the Lagrangian. If these were the physical states, the massless left- and right-handed quarks would

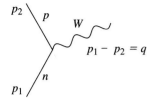

FIGURE 7.2 Momentum flow in neutron decay to a proton and a virtual W.

provide representations of the symmetry group and the conserved vector and axial currents. Instead, the physical states are the hadrons. We can look at the current conservation question for hadrons such as n and p using experimentally-measured currents, since the weak decay of $n \to p e^- \bar{\nu}_e$ occurs through matrix elements of both vector and axial vector currents. These currents are the same as those conserved by the QCD chiral symmetry, which is a flavor symmetry. We will look at the weak decays in the next chapter; these decays involve flavor transitions explicitly. Here, we consider only the currents involved. The decay vertex for neutron decay is shown in Figure 7.2.

The vector current, which involves an isospin raising operator in the $SU(3)$ flavor space, can be written generally in terms of two form factors:

$$\langle p | J^V_\mu | n \rangle = \bar{u}_p [F^V_1(q^2)\gamma_\mu + i\sigma_{\mu\nu} q_\nu F^V_2(q^2)] u_n.$$

Dotting this with q_μ, the resulting expression vanishes provided that $m_n = m_p$, which is expected in the limit where the u and d quarks are massless and therefore degenerate in mass.

Turning to the axial vector current, which again involves an isospin raising operator in the flavor space, we can try by analogy with the vector current:

$$\langle p | J^A_\mu | n \rangle = \bar{u}_p [F^A_1(q^2)\gamma_\mu \gamma_5 + i\sigma_{\mu\nu} q_\nu \gamma_5 F^A_2(q^2)] u_n.$$

Dotting this with q_μ gives

$$q_\mu \langle p | J^A_\mu | n \rangle = -(m_n + m_p) F^A_1(q^2) \bar{u}_p \gamma_5 u_n. \tag{7.23}$$

This would vanish if $F^A_1(q^2) = 0$; however, the neutron decay data indicate that this is not the case. For q^2 near 0, $F^A_1(q^2)$, defined to be $-g_A$, equals -1.26. So how can the divergence of the current vanish? The solution requires noticing that the bound states include not only the baryons but also mesons. So additional diagrams can contribute, for example, when the baryon emits a meson that then couples to the current. The diagrams are shown in Figure 7.3. The axial current conservation will be a cooperative phenomenon among the physical states.

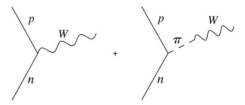

FIGURE 7.3 Direct axial vector coupling for nucleons and, shown separately, the pseudoscalar meson contribution to the weak interaction vertex.

7.3 Approximate Chiral Symmetry of the Strong Interactions

The diagram due to pion exchange can be calculated in terms of:

1. Amplitude for $n \to p + \pi$, which is a strong interaction process.
2. π propagator.
3. Amplitude for the π to turn into a W.

These are given by:

1. $g_{\pi pn} \bar{u}_p i \gamma_5 u_n$. This pseudoscalar amplitude is given in terms of one constant.
2. The pion propagator is $i/(q^2 - m_\pi^2)$.
3. The weak interaction operator is a current dotted into the W vector field, analogous to the electromagnetic interaction operator. For π, which is a pseudoscalar, the relevant current is the axial vector current. The axial current matrix element is $\langle 0| J_\mu^A |\pi\rangle = i f_\pi q_\mu$, which is also responsible for the weak decay of the charged pion, as discussed in Chapter 8. This current matrix element is a vector, and the only vector to which it can be proportional is the pion momentum, which is q_μ in the present case. The proportionality constant is defined to be f_π for $q^2 = m_\pi^2$, and in general is a function of q^2 for a virtual pion.

Putting all the factors together (including the $-i$ from the extra vertex), the pion exchange term gives the following contribution to the current matrix element:

$$-f_\pi g_{\pi pn} \frac{q_\mu}{q^2 - m_\pi^2} \bar{u}_p \gamma_5 u_n. \qquad (7.24)$$

Dotting this with q_μ gives

$$-f_\pi g_{\pi pn} \frac{q^2}{q^2 - m_\pi^2} \bar{u}_p \gamma_5 u_n. \qquad (7.25)$$

This vanishes at $q^2 = 0$ and is no help with current conservation at low q^2 unless $m_\pi^2 = 0$. Since the QCD Lagrangian implies axial vector current conservation in the massless quark limit, the pion must be massless in this limit if it contributes to the nucleon axial vector matrix element at low q^2. This general conclusion is remarkable. In the real world, where the pion is not quite massless, the pion contribution does vanish at $q^2 = 0$.

In addition to requiring a massless pion, the coefficients of the terms in Eqs. 7.23 and 7.25 must add to zero to give a vanishing divergence of the current. Requiring this for $q^2 \simeq 0$ gives the relation, called the Goldberger–Treiman relation:

$$g_A(m_n + m_p) = f_\pi g_{\pi pn}.$$

The various constants in this relation have been measured and the equality, which should be exact for massless quarks, is good to about 5%. The constant $g_{\pi pn}$ is approximately 20 and is measured in various hadronic processes, such as low energy nucleon scattering. The exchange of a pion, as the lightest hadron, dominates many very low energy hadronic scattering processes at very small momentum transfer. This allows the measurement of $g_{\pi pn}$, called the residue at the pion pole.

The vanishing of the divergence of the axial vector current for the nucleon system in the chiral symmetry limit involves both nucleons and pions. We can look at the same issue for the pion itself, where the axial vector current is responsible for the weak decay of a real pion. The conclusion that the pion mass vanishes in the massless quark limit, is sufficient to make the divergence of $\langle 0| J_\mu^A |\pi\rangle$, which is proportional to $q^2 = m_\pi^2$, also vanish. The approximately conserved axial current, which becomes an exact result in the massless pion limit, is called PCAC (partially conserved axial current).

7.4 ■ SPONTANEOUSLY BROKEN SYMMETRY

In addition to current conservation, we expect a continuous symmetry to be reflected in the spectrum of states through multiplets of degenerate particles. We saw this for the approximate $SU(3)$ flavor symmetry in Chapter 5. In the massless quark limit, the symmetry $SU(3)_L \times SU(3)_R$ is larger, however, and we would expect larger multiplets. But this is not the case; the light quark bound state spectrum consists of $SU(3)$ flavor multiplets with the same J^P as the spectrum of states for the heavy quarks. The real physical states are eigenstates of parity, which is an operator that does not commute with the chiral transformations. We look next at the parity eigenstates that could exist in a chiral theory, and see that the known states do not form complete chiral representations.

We can write a multiplet of a product group as a set of states $|i, j\rangle$, which transform via separate unitary transformations for the first label and separate transformations for the second label. For example, for a transformation on the first label, the rotated state is

$$|i, j\rangle' = \sum_k U_{ik} |k, j\rangle.$$

The overall multiplet forms several separate multiplets for each group in the product. For example, for each j, the set of states enumerated by i forms a multiplet under the first transformation group.

Returning to the chiral symmetry rotations, we consider $SU(2)_L \times SU(2)_R$ of the u, d system in order to use the simplest and most familiar group as well as the quarks that best obey the symmetry. We label the multiplet by the number of states in the individual $SU(2)_L$ and $SU(2)_R$ representations. Thus a multiplet is labeled as:

$$(n_L, n_R) \quad \text{with} \quad n_L = 2I_L + 1, \ n_R = 2I_R + 1$$

7.4 Spontaneously Broken Symmetry

in terms of the value of the individual Casimir operators, which are each analogous to J for angular momentum. We write the generators of the chiral rotations for the states as $\vec{I}_L$ and $\vec{I}_R$. The chiral generators are not wholly independent, in that parity is an operation that relates the generators via

$$P\vec{I}_L P^\dagger = \vec{I}_R \quad \text{and} \quad P\vec{I}_R P^\dagger = \vec{I}_L.$$

Parity thus does not commute with these generators; however, it does commute with $\vec{I}_L + \vec{I}_R$, which are the generators of the normal isospin. Parity is an operator that also commutes with H, provided the left- and right-handed groups have the same interaction, as they do in QCD, given in Eq. 7.21. Although this seems natural, it is not true for the weak interactions, giving rise to parity violation.

The left- and right-handed massless u and d quark states give rise to the representations (2, 1) and (1, 2), respectively. These are clearly not parity eigenstates. The hadrons are, however, multi-particle bound states of fixed parity and could belong to more complicated representations. If we find a parity eigenstate, what representation could it belong to? One possibility is that the state could be a linear combination of states from different, degenerate, chiral multiplets. We look here at the possibility that the state is contained in one multiplet, which requires the smallest total number of degenerate states. Using the relations for the generators, the left- and right-handed Casimir operators satisfy $PI_L^2 = I_R^2 P$. If we have a state that is a linear combination of the states in (n_L, n_R) and is a parity eigenstate, we can apply the operator above to give us $n_L = n_R$. Note, however, that the states in (n_L, n_R) cannot all have the same parity (except for the singlet (1, 1)) since P does not commute with $\vec{I}_L$ and $\vec{I}_R$. The simplest multiplet that could have parity eigenstates by making appropriate linear combinations of the chiral states is (2, 2). In the symmetry limit the states in (2, 2) should be degenerate in mass.

We can now examine the real states seen in nature. The lightest are the π^+, π^0, π^- mesons, which are parity eigenstates. The (2, 2) requires four states, so we are missing a state needed for an $SU(2)_L \times SU(2)_R$ multiplet. The missing state would be a scalar meson, degenerate in mass with the pions in the massless quark limit. We thus see that the physical states do not seem to fall into a full chiral representation. On the other hand, the pion states do form a representation of the $SU(2)$ formed from the generators $\vec{I}_L + \vec{I}_R$, which commute with parity, and provide the normal isospin discussed in Chapter 5.

7.4.1 ■ The Role of the Vacuum

In Section 4.3 we showed rather generally that states related by unitary transformations that commute with the Hamiltonian are degenerate. This led to multiplets of degenerate states. This situation does not seem to work for the chiral symmetry. The resolution of this discrepancy lies in the role of the vacuum. We tacitly assumed that the vacuum is a maximally symmetric state—that is, applying any symmetry operator to $|0\rangle$ gives us back $|0\rangle$. A particle state is then an excitation relative to the vacuum. In going from the quarks and gluons to the hadrons, the spectrum of states (which are nonperturbative) looks very different from the initial state spectrum. The new vacuum, which represents a minimum energy state, could also look different; in particular it could be less symmetric. In such a situation, the symmetries of the Lagrangian do not have to be realized by the spectrum of states, which are defined relative to the vacuum. Such a situation is called spontaneous symmetry breaking. The Higgs mechanism, which we will discuss as a model for symmetry breaking of the weak gauge symmetry in Chapter 10, will provide another example. In that case, the simplest model is perturbative and the relation of symmetry breaking to a minimum energy vacuum is easier to understand. For the hadrons, the experimental information indicates that the vacuum does not satisfy the chiral symmetry, but rather only the simpler $SU(3)$ flavor symmetry displayed by the mesons and baryons.

Consider the above discussion a little more mathematically. We consider a state that is created relative to the vacuum by a creation operator $a^\dagger$. Thus, $|A\rangle = a^\dagger |0\rangle$. The operator $a^\dagger$ is part of a field operator that transforms under a global internal symmetry group. We now imagine applying a symmetry transformation U; $U|A\rangle = Ua^\dagger |0\rangle = Ua^\dagger U^\dagger U |0\rangle$. We define $Ua^\dagger U^\dagger = b^\dagger$, which is the operator arising from transforming the field. The state $|B\rangle = b^\dagger |0\rangle$ has the same mass as $|A\rangle$ if U commutes with the Hamiltonian, which is required for a symmetry transformation, and also $U|0\rangle = |0\rangle$. If $|0\rangle$ is not symmetric under the symmetry transformation, it spoils our expectation that the states form multiplets of the group of transformations.

Despite the asymmetric vacuum, the divergence of the currents defined from the Lagrangian still vanish. The massless spin zero particles like the pion (in the massless quark limit) that contribute cooperatively to the current, are called Goldstone bosons.

7.5 ■ QUARK MASSES IN THE LAGRANGIAN

The discussion of chiral symmetry has been in the limit of massless quarks. Including the mass term in the Lagrangian, the divergence of the $SU(3)$ flavor currents does not vanish. We can, however, calculate the divergence explicitly using the QCD Lagrangian for quarks and gluons. Taking the currents that contribute to π^- and K^- decay:

$$\frac{\partial}{\partial x_\mu} J^A_{\mu I_+} = (m_u + m_d)\bar{\psi}_u i\gamma_5 \psi_d, \quad (7.26)$$

where $J^A_{\mu I_+} = \bar{\psi}_u \gamma_\mu \gamma_5 \psi_d$ corresponds to an isospin raising current, and

$$\frac{\partial}{\partial x_\mu} J^A_{\mu V_+} = (m_s + m_u)\bar{\psi}_u i\gamma_5 \psi_s, \quad (7.27)$$

7.6 Other Issues

where $J^A_{\mu V_+} = \bar{\psi}_u \gamma_\mu \gamma_5 \psi_s$ corresponds to a current that changes $s \to u$ quarks (the V_+ operator). Note, the masses that appear here are the masses that appear in the Lagrangian and should not be confused with the constituent quark masses, although we use the same notation for both. The relations above do not involve the gluon fields explicitly and would hold for noninteracting quark fields. Taking matrix elements between the vacuum and the physical mesons $|\pi^-\rangle$ or $|K^-\rangle$ for Eqs. 7.26 or 7.27, and using the axial current matrix elements in terms of f_π and an analogous constant f_K, we get equations:

$$f_\pi m_\pi^2 = (m_u + m_d) \langle 0 | \bar{\psi}_u i \gamma_5 \psi_d | \pi^- \rangle,$$
$$f_K m_K^2 = (m_s + m_u) \langle 0 | \bar{\psi}_u i \gamma_5 \psi_s | K^- \rangle. \quad (7.28)$$

If the mesons involved were heavy quark nonrelativistic systems, these equations could be evaluated in terms of the quark masses and the wave functions at the origin. For the light quarks, however, we expect instead that the various dimensional quantities that do not vanish as the quark masses approach zero are determined mainly by λ_{QCD}. Assuming that this applies to the quantities

$$\frac{\langle 0 | \bar{\psi}_u i \gamma_5 \psi_d | \pi^- \rangle}{f_\pi}, \quad \frac{\langle 0 | \bar{\psi}_u i \gamma_5 \psi_s | K^- \rangle}{f_K},$$

and that they are equal for an approximate $SU(3)$ symmetry, we get a prediction:

$$\frac{m_K^2}{m_\pi^2} = \frac{m_s + m_u}{m_u + m_d} = 12.5. \quad (7.29)$$

This implies that $m_s \gg m_u, m_d$, and in particular that

$$m_s \simeq 25 \left(\frac{m_u + m_d}{2} \right).$$

If we assume that m_s is at least a few times smaller than λ_{QCD}, then this relation would imply that m_u and m_d are in the neighborhood of 5 to 10 MeV. One possibility is that the constituent quark mass is a constant plus the mass that appears in the Lagrangian. Although how this comes about is not clear, it is consistent with the measured constituent quark masses. In such a picture,

$$m_s \simeq 170 \text{ MeV}, \quad m_u \simeq 5 \text{ MeV}, \quad m_d \simeq 8 \text{ MeV}, \quad (7.30)$$

and the constant added to each mass above to get the constituent quark mass is about 330 MeV.

7.6 ■ OTHER ISSUES

In this chapter we looked at some of the issues surrounding renormalization, with the non-Abelian gauge theory of the strong interaction as the prime example.

These topics are quite complicated and we have just introduced the subject. We did not look explicitly at the amplitudes for diagrams with gluon loops where the requirements of gauge invariance, unitarity, and Lorentz invariance require some care, although we stated the final result for the running coupling constant. After renormalization, the predictions of the theory include a small number of constants that have to be extracted from the data. The requirements for a renormalizable theory are quite restrictive and most theories are nonrenormalizable, requiring the absorption of additional new infinite constants in each order of perturbation theory. The various individual renormalizable interactions will come together in Chapter 10 to make up the Standard Model for electroweak symmetry breaking.

In general, renormalizability limits the spin of the fundamental particles to low values, with higher spins requiring more symmetry to avoid divergences. This comes about because the propagator for particles of high spin have more powers of momentum (compared to spin 0), which create divergences in Feynman diagram calculations. As an example, compare the propagator for a massive spin 1 particle, which has a term in the numerator that is proportional to two powers of momentum, with that for spin 0, which has no such term, or the photon, where gauge invariance results in a propagator without such a term. In addition to constraints on the spin for point-like particles, the vertices that are allowed are also constrained. Vertices with too many particles or too many derivatives of fields, which bring in powers of momentum in Feynman diagrams, yield nonrenormalizable theories. Such problems make a simple theory of gravity, based on a spin 2 massless particle, nonrenormalizable. Modern approaches tend to derive gravity from theories with a higher degree of symmetry, although approaches such as string theory represent a significant departure from the idea of local point-like fields and interactions. Such an approach may be needed to solve the divergence problems that we have seen arise in perturbation theory. All of the Standard Model interactions, except gravity, are based on renormalizable gauge theories. To explain electroweak symmetry breaking, we will have to introduce scalar fields that couple in a gauge-invariant manner to the gauge bosons.

In this chapter we have not discussed an important technique for dealing with certain strong interaction calculations, which involves solving *QCD* for quarks and gluons on a lattice. Choosing a large enough lattice and taking the limit of small lattice spacing compared to 1 fermi, allows the extraction of quantities expected in the continuum limit. This allows a nonperturbative solution for quantities involving a single hadron, such as the masses and decay constants calculated in Sections 4.6 and 4.7 using the color potential. The lattice gauge approach provides a technique for performing such calculations even when potential models are inadequate, for example, for mesons containing light quarks. In Chapters 8 and 9, results stated for meson decay constants relevant to weak decays, when not directly measured, are based on lattice gauge calculations.

CHAPTER 7 HOMEWORK

7.1. The result of an infinitesimal global transformation for the color octet currents and gluons is given in Eqs. 7.2 and 7.3, respectively. What are the generators for the octet color multiplet?

7.2. We would like to look at charge conjugation for a non-Abelian gauge theory. The simplest case is a color $SU(2)$, which we consider because of its familiarity (since it is like angular momentum), but there are analogous results for QCD. For $SU(2)$ there are 3 gauge bosons, which we call gluons as in the $SU(3)$ theory.

The $SU(2)$ theory has interactions for the fermions involving the operator $T_i W_\mu^i$ (W_μ^1, W_μ^2, W_μ^3 are the three gluon field operators, $T_i = \sigma_i/2$). The field tensor for the gluons is

$$G_{\mu\nu}^i = \frac{\partial}{\partial x_\mu} W_\nu^i - \frac{\partial}{\partial x_\nu} W_\mu^i - g\varepsilon_{ijk} W_\mu^j W_\nu^k.$$

(a) Under charge conjugation, $T_i \to -T_i^*$ and $C W_\mu^i C^\dagger = W_\mu^{i'}$. Then $W_\mu^{i'} = \pm W_\mu^i$, with the sign depending on the value of i. For C invariance we want

$$T_i W_\mu^i = (-T_i^*) W_\mu^{i'}.$$

What sign do we choose for $i = 1, 2$, or 3, for the above identity to work?

(b) With the sign choices in (a), show that $-\frac{1}{4} G_{\mu\nu}^i G_{\mu\nu}^i$ is invariant; thus the various terms coming from the gauge bosons are C invariant.

(c) Defining fields that are the analog of the circular polarization states

$$W_\mu^{(+)} = -\frac{\left(W_\mu^1 + i W_\mu^2\right)}{\sqrt{2}}, \quad W_\mu^{(0)} = W_\mu^3, \quad W_\mu^{(-)} = \frac{W_\mu^1 - i W_\mu^2}{\sqrt{2}},$$

show that

$$W_\mu^{(0)} \to -W_\mu^{(0)}, \quad W_\mu^{(+)} \to W_\mu^{(-)}, \quad W_\mu^{(-)} \to W_\mu^{(+)}$$

under charge conjugation. (Note that these are the same relations as for the three ρ mesons).

(d) If the $SU(2)$ color is confined, we can only have singlets under $SU(2)$. From your knowledge of the addition of angular momentum for states with $J = 1$, can we have a singlet created from two W_μ^i operators? From three W_μ^i operators? Such singlet states made only of gluons are called glueballs.

(e) The two and three gluon singlets in $SU(2)$ can be calculated from your knowledge of spatial vectors. For spatial vectors $\vec{V}_1$, $\vec{V}_2$, $\vec{V}_3$, the singlets (that is, scalars) are:

$$\vec{V}_1 \cdot \vec{V}_2, \quad \vec{V}_1 \cdot (\vec{V}_2 \times \vec{V}_3).$$

What is the behavior under charge conjugation of the analogous two and three gluon singlet operators? Note that this is an example where $SU(3)$ differs from $SU(2)$. For $SU(3)$, three gluon operators of both even and odd C can be constructed. For specific states we also need to specify the behavior of the spin and spatial degrees of freedom under charge conjugation, to arrive at this quantum number for the state.

7.3. Consider the $SU(2)_L \times SU(2)_R$ symmetry and (2,2) multiplet. Writing the states in terms of the I_3^L and I_3^R values, we can specify a state as $|I_3^L, I_3^R\rangle$ where I_3^L and I_3^R can equal $\pm \frac{1}{2}$. The parity operator satisfies $P\vec{I}_L = \vec{I}_R P$, so, for example, $PI_3^L = I_3^R P$ and the raising and lowering operators satisfy a similar relation. Assuming that applying the parity operator leaves us in the same multiplet, show that:

(a) $|\frac{1}{2}, \frac{1}{2}\rangle$ and $|-\frac{1}{2}, -\frac{1}{2}\rangle$ are parity eigenstates.

(b) More generally:

$$\left|\tfrac{1}{2}, \tfrac{1}{2}\right\rangle, \quad \frac{\left|\tfrac{1}{2}, -\tfrac{1}{2}\right\rangle + \left|-\tfrac{1}{2}, \tfrac{1}{2}\right\rangle}{\sqrt{2}}, \quad \left|-\tfrac{1}{2}, -\tfrac{1}{2}\right\rangle$$

all have the same parity, while

$$\frac{\left|\tfrac{1}{2}, -\tfrac{1}{2}\right\rangle - \left|-\tfrac{1}{2}, \tfrac{1}{2}\right\rangle}{\sqrt{2}}$$

has opposite parity.

7.4. The σ-model of Gell–Mann and Levy is the simplest field theoretic model including mesons, nucleons, and chiral symmetry. It allows for spontaneous symmetry breaking, but is not a model based on quarks, and its spectrum is simpler than that of the bound quark states. The Lagrangian looks as follows:

$$\mathcal{L} = i\bar{\psi}\gamma_\mu \frac{\partial}{\partial x_\mu}\psi - g\bar{\psi}_L \Sigma \psi_R - g\bar{\psi}_R \Sigma^\dagger \psi_L + \mathcal{L}(\Sigma).$$

We will specify $\mathcal{L}(\Sigma)$ later. ψ is a flavor doublet, for example, the proton and neutron isospin flavor doublet.

$\Sigma = \sigma I + i\vec{\tau}\cdot\vec{\pi}$, where I is the 2×2 identity matrix, $\vec{\tau}$ are the three Pauli matrices, and $\vec{\pi}$ are a triplet of fields that are identified with the pions. The σ and $\vec{\pi}$ fields are Hermitian and are the four spinless constituents in the theory.

ψ_L and ψ_R are the left- and right-handed components of ψ. There is no fermion mass term in a chirally symmetric theory. Under $SU(2)_L \times SU(2)_R$ transformations the fields transform as follows:

$$\psi_L \to U_L \psi_L, \quad \psi_R \to U_R \psi_R,$$

where U_L and U_R can be written as

$$U_L = U(\vec{\alpha}_L) = e^{i\vec{\alpha}_L \cdot \vec{\tau}} \quad \text{and} \quad U_R = U(\vec{\alpha}_R) = e^{i\vec{\alpha}_R \cdot \vec{\tau}}.$$

To make the interaction term chirally invariant requires

$$\Sigma \to U_L \Sigma U_R^\dagger \quad \text{and} \quad \Sigma^\dagger \to U_R \Sigma^\dagger U_L^\dagger.$$

(a) Under a parity transformation,

$$\psi_L \to \psi_R, \quad \psi_R \to \psi_L.$$

Show that taking $\Sigma \to \Sigma^\dagger$, $\Sigma^\dagger \to \Sigma$ under parity gives a parity invariant $\mathcal{L}$ for the terms written above. Show that this means that σ creates scalar particles and π_i pseudoscalar particles. Applying parity after a chiral transformation, show that the parity transformation rule is consistent with the behavior of U_L and U_R under parity (that is, $\vec{\alpha}_L \to \vec{\alpha}_R, \vec{\alpha}_R \to \vec{\alpha}_L$ under parity).

(b) Write out the interaction term in $\mathcal{L}$ involving the fermions and show that it is

$$-g[\sigma \bar{\psi}\psi + \bar{\psi} i \gamma_5 (\vec{\tau} \cdot \vec{\pi}) \psi].$$

This is a scalar interaction for σ and a pseudoscalar interaction for the triplet of pions.

(c) An isospin transformation is given by $\vec{\alpha}_L = \vec{\alpha}_R = \vec{\alpha}/2$. Show that, for an infinitesimal isospin transformation,

$$\sigma \to \sigma$$
$$\vec{\pi} \to \vec{\pi} + \vec{\alpha} \times \vec{\pi},$$

which are the correct transformations for isospin 0 and isospin 1 fields, respectively.

(d) Show that

$$\Sigma^\dagger \Sigma \quad \text{and} \quad \frac{\partial}{\partial x_\mu} \Sigma^\dagger \frac{\partial}{\partial x_\mu} \Sigma$$

are the identity matrix times:

$$\sigma^2 + \vec{\pi} \cdot \vec{\pi} \quad \text{and} \quad \frac{\partial \sigma}{\partial x_\mu} \frac{\partial \sigma}{\partial x_\mu} + \frac{\partial \vec{\pi}}{\partial x_\mu} \cdot \frac{\partial \vec{\pi}}{\partial x_\mu},$$

respectively. Therefore, under the global $SU(2)_L \times SU(2)_R$ transformations, these two expressions are invariant and can be used to construct an invariant $\mathcal{L}(\Sigma)$. Taking $V(\sigma^2 + \vec{\pi} \cdot \vec{\pi})$ to be a function of the invariant $\sigma^2 + \vec{\pi} \cdot \vec{\pi}$,

$$\mathcal{L}(\Sigma) = \frac{1}{2} \left(\frac{\partial \sigma}{\partial x_\mu} \frac{\partial \sigma}{\partial x_\mu} + \frac{\partial \vec{\pi}}{\partial x_\mu} \cdot \frac{\partial \vec{\pi}}{\partial x_\mu} \right) - V(\sigma^2 + \vec{\pi} \cdot \vec{\pi}).$$

Is this parity invariant?

(e) If $V(\sigma^2 + \vec{\pi} \cdot \vec{\pi})$ is a minimum for $\sigma = 0, \vec{\pi} = 0$, we have a normal symmetric vacuum state. If $V(\sigma^2 + \vec{\pi} \cdot \vec{\pi})$ is a minimum for $\sigma \neq 0, \vec{\pi} = 0$, the chiral symmetry is spontaneously broken (but not the isospin symmetry). An example of the former is

$$V = \frac{m^2}{2}(\sigma^2 + \vec{\pi} \cdot \vec{\pi}).$$

Ignoring the baryons, show that this leads to free field equations for σ and $\vec{\pi}$ corresponding to four particles of mass m.

(f) The simplest form leading to spontaneous symmetry breaking is

$$V = \frac{\lambda}{4}\left[\left(\sigma^2 + \vec{\pi} \cdot \vec{\pi}\right) - f_\pi^2\right]^2.$$

V is minimized for a field configuration where the fields are not all zero. This generates a vacuum that has a nonzero expectation value for some of the fields. Assuming that the field σ has an expectation value $\langle \sigma \rangle = f_\pi$ and that we expand around this point via $\sigma' = \sigma - f_\pi$, show that the $\vec{\pi}$ fields have no mass term in $\mathcal{L}$. Looking at the interaction with the baryons, show that they have a mass = $g \langle \sigma \rangle = g f_\pi$, which has been generated through spontaneous symmetry breaking.

(g) Ignoring the baryons for simplicity, we write the three conserved vector and axial vector currents, respectively:

$$(J_\mu^V)_i = \varepsilon_{ijk} \pi^j \frac{\partial}{\partial x_\mu} \pi^k$$

$$(J_\mu^A)_i = \pi^i \frac{\partial \sigma}{\partial x_\mu} - \sigma \frac{\partial \pi^i}{\partial x_\mu}.$$

Show that

$$\frac{\partial}{\partial x_\mu}(J_\mu^V)_i = \frac{\partial}{\partial x_\mu}(J_\mu^A)_i = 0.$$

These currents, like the π field, transform like isospin 1 operators.

CHAPTER 8

Weak Interactions of Fermions

8.1 ■ WEAK GAUGE GROUP

Weak interactions were discovered through the second-order transitions between fermions, where the fermion charge is changed by one unit at each vertex. The charge is carried between the vertices by a heavy spin 1 particle that comes with charge $+1$ or -1 in units of e. An example of such a process is $n \to p + e^- + \bar{\nu}_e$.

The spin 1 particles, called W bosons, are the quanta of an $SU(2)$ gauge theory. Since QCD is an $SU(3)$ gauge theory, we already know some of the features of such a non-Abelian gauge theory from looking at QCD. However, the weak theory has a number of crucial differences:

1. Different W couplings to left- and right-handed fermions. This violates parity and requires massless fermions.
2. Breaking of the gauge symmetry, generating masses for the gauge bosons and fermions.
3. Mixing of a separate Abelian gauge group with the $SU(2)$ group due to the symmetry breaking, yielding electromagnetism as the only unbroken gauge symmetry at low energies.
4. A complicated mixing pattern among the fermions due to a complicated mass matrix from the symmetry breaking. This gives rise to CP violation and various mixing phenomena.
5. The requirement of extra interactions of an (at present) unknown nature to explain the symmetry breaking. These interactions are characterized by a nonzero vacuum expectation value for one or more scalar fields.

We will develop these features as we explore the many measurements made for systems interacting weakly. We start with the $SU(2)$ symmetry for the W^+ and W^- interactions with the fermions.

The interaction Hamiltonian density for the $SU(2)$ symmetric fermion interaction is

$$-\mathcal{L}_{\text{Int}} = \mathcal{H} = g\bar{\psi}_L \gamma_\mu T_k \psi_L W^k_\mu. \qquad (8.1)$$

Here g is the coupling constant, T_k are the $SU(2)$ generators $\sigma_k/2$, W^k_μ are three vector boson fields for $k = 1, 2,$ or 3, and ψ_L is a doublet of fermion fields. For several doublets we get several similar terms in $\mathcal{H}$.

We consider the $SU(2)$ properties first and then the Lorentz structure, at which time we will explain the subscript L on the fermion doublet. Leaving out the Lorentz index:

$$T_k W_k = \frac{1}{2} \begin{pmatrix} W_3 & W_1 - iW_2 \\ W_1 + iW_2 & -W_3 \end{pmatrix}.$$

We can also rewrite this in terms of raising and lowering operators for the fermion field. Using

$$T^{(\pm)} = T_1 \pm iT_2 \quad \text{gives} \quad T^{(+)} = \begin{pmatrix} 0 & 1 \\ 0 & 0 \end{pmatrix}, \quad T^{(-)} = \begin{pmatrix} 0 & 0 \\ 1 & 0 \end{pmatrix},$$

therefore

$$\mathcal{H} = g \bar{\psi}_L \gamma_\mu \left[T_3 W_3 + \frac{1}{\sqrt{2}} T^{(+)} W^{(+)} + \frac{1}{\sqrt{2}} T^{(-)} W^{(-)} \right] \psi_L. \tag{8.2}$$

Here the conventional definition is

$$W^{(+)} = \frac{W_1 - iW_2}{\sqrt{2}}, \quad W^{(-)} = \frac{W_1 + iW_2}{\sqrt{2}}.$$

If we write the W fields in terms of creation and annihilation operators, those for the fields $W^{(+)}$, $W^{(-)}$, and W_3 have an analogous normalization.

The symmetry breaking, which results in a large W mass and a weak interaction, prevents the formation of weakly bound composite fermion states. The fermion representations we see are doublets or noninteracting singlets. Based on experiment, three different doublets involve leptons, which we will look at in the next sections. These are

$$\begin{pmatrix} \nu_e \\ e^- \end{pmatrix}, \begin{pmatrix} \nu_\mu \\ \mu^- \end{pmatrix}, \begin{pmatrix} \nu_\tau \\ \tau^- \end{pmatrix},$$

giving three analogous independent terms in $\mathcal{H}$. There are also three doublets of quarks for each color of quark. Since an $SU(n)$ non-Abelian gauge theory has only one coupling constant, the coupling for all doublets must be identical. This is called universality, and is analogous to the color coupling being the same for all quark flavors. Given the electric charges for the doublet members, the field $W^{(+)}$ corresponds to the emission of a W^- particle or absorption of a W^+ particle, while $W^{(-)}$ corresponds to the emission of a W^+ particle or absorption of a W^- particle. These terms in $\mathcal{H}$ are called charged current interactions. The W_3 field operator gives rise to absorption or emission of a neutral particle, W^0. We will see that W^+ and W^- particles remain after $SU(2)$ symmetry breaking is included, but a W^0 degenerate with the charged particles does not. Z^0 is the mass eigenstate we find, with couplings and mass related to but not identical to the W^0.

Turning next to the Lorentz structure of the interaction, we find the first major difference between the weak $SU(2)$ theory and QCD. We indicated the doublet

8.1 Weak Gauge Group

that interacts as ψ_L. This field corresponds in the relativistic limit to the interaction of left-handed particles and right-handed antiparticles only. For example, if we direct a very high energy beam of left-handed electrons on a target, they interact via W exchange while a similar beam of right-handed electrons passes through the target without suffering any weak transitions due to W exchange. This striking prediction is not typically apparent because of the weakness of the weak interactions. Assuming the target to be unpolarized, this is also a clear violation of invariance under parity.

To write a left-handed field in the relativistic limit we multiply each of the two $SU(2)$ components of ψ by $(1 - \gamma_5)/2$. Thus:

$$\psi_L = \left(\frac{1 - \gamma_5}{2}\right)\psi \quad \text{and} \quad \bar{\psi}_L = \bar{\psi}\left(\frac{1 + \gamma_5}{2}\right),$$

where ψ is a doublet of Dirac spinor fields. The interaction Hamiltonian density is then

$$g\bar{\psi}\left(\frac{1 + \gamma_5}{2}\right)\gamma_\mu T_k \left(\frac{1 - \gamma_5}{2}\right)\psi W_\mu^k = g\bar{\psi}\gamma_\mu \frac{(1 - \gamma_5)}{2} T_k \psi W_\mu^k.$$

The interaction is due to a vector current, like electromagnetism, but because of the participation of only left-handed fermion fields, it can be thought of as the sum of a vector (the γ_μ term) and an axial vector (the $-\gamma_\mu \gamma_5$ term) interaction involving the full Dirac spinors ψ. The perturbation expansion, for example, for leptonic interactions, starts with the free field solutions, which are the Dirac spinors.

In writing weak matrix elements of the charged current, we usually just write the correct fields and do not explicitly write raising or lowering operators, since the states connected are easy to remember. Thus for a vertex involving, for example, $e^- \to W^- + \nu_e$, the vertex factor resulting from the Hamiltonian density is obtained from Eq. 8.2:

$$\frac{-ig}{2\sqrt{2}} \bar{u}_{\nu_e} \gamma_\mu (1 - \gamma_5) u_e e_\mu^{W*}. \tag{8.3}$$

The W propagator is that of a massive spin 1 particle,

$$\frac{-i\left(g_{\mu\nu} - \frac{q_\mu q_\nu}{M_W^2}\right)}{q^2 - M_W^2 + i\varepsilon}, \tag{8.4}$$

which for $q^2 \ll M_W^2$, reduces to $ig_{\mu\nu}/M_W^2$. This is the situation for the decays of particles much lighter than M_W, for example, decays of μ, τ, π, K. We show in Figure 8.1 a second-order charged current process for low q^2. The corresponding matrix element, assuming the fermions are all leptons (for example, $\psi_n = u_e$ and $\psi_m = u_{\nu_e}$), is

$$M_{fi} = \frac{g^2}{8M_W^2} \left[\bar{\psi}_m \gamma_\mu (1 - \gamma_5)\psi_n\right]\left[\bar{\psi}_l \gamma_\mu (1 - \gamma_5)\psi_k\right]. \tag{8.5}$$

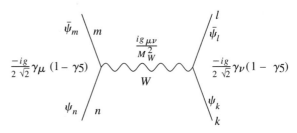

FIGURE 8.1 Second-order weak process for $q^2 \ll M_W^2$. It involves four fermions with a change of one unit of charge at each vertex. Shown are the various factors that go into a calculation for this diagram.

Written as an operator, this is called the effective weak Hamiltonian. This is the basic low energy charged current weak interaction. For the interaction written in space-time, it corresponds to a point-like four-fermion coupling. For quarks the matrix element is modified by the strong interactions, which bind the quarks into hadrons. Consequently matrix elements describing measureable processes are always between hadronic states, rather than the quark states.

The combination

$$\frac{g^2}{8M_W^2} = \frac{G_F}{\sqrt{2}} \qquad (8.6)$$

defines the Fermi four-fermion coupling G_F. The $1/\sqrt{2}$ is historical, arising from the fact that the initial guess for the interaction (by Fermi in 1934) had only vector currents interacting at a point in space-time. The measured value for G_F is

$$G_F = 1.16639 \times 10^{-5} \text{ GeV}^{-2}. \qquad (8.7)$$

Using G_F and the value $M_W = 80.4$ GeV implies that

$$\frac{g^2}{4\pi} = \frac{1}{30}.$$

The intrinsic interaction coupling is therefore larger than the coupling for electromagnetism and the interaction is weak only because of the large value of M_W. For $q^2 \gg M_W^2$, the interaction is not small compared to electromagnetism. Note that because the gauge particles are massive, the higher-order corrections at low q^2 do not create a running coupling constant as for a massless theory, but can be calculated using higher-order diagrams.

Before turning to calculations using the four-fermion coupling, we want to consider the consistency of using the left-handed fields. If we look at a given fermion, taking the electron as an example, the free field Lagrangian is:

$$\mathcal{L}_{\text{Free}} = i\bar{\psi}_e \gamma_\mu \frac{\partial}{\partial x_\mu} \psi_e - m_e \bar{\psi}_e \psi_e.$$

8.1 Weak Gauge Group

Writing

$$\psi_e = \left(\frac{1-\gamma_5}{2}\right)\psi_e + \left(\frac{1+\gamma_5}{2}\right)\psi_e,$$

we have $\psi_e = \psi_L^e + \psi_R^e$. Note that the fields here all refer to one fermion and are not $SU(2)$ doublets. Substituting for ψ_e into $\mathcal{L}_{\text{Free}}$ gives

$$\mathcal{L}_{\text{Free}} = i\bar{\psi}_L^e \gamma_\mu \frac{\partial}{\partial x_\mu} \psi_L^e + i\bar{\psi}_R^e \gamma_\mu \frac{\partial}{\partial x_\mu} \psi_R^e - m_e \left[\bar{\psi}_L^e \psi_R^e + \bar{\psi}_R^e \psi_L^e\right]. \quad (8.8)$$

For $m_e = 0$, $\mathcal{L}_{\text{Free}}$ separates into two separate terms and these can have different interactions. For $m_e \neq 0$, ψ_L^e and ψ_R^e are not individual solutions of the resulting Dirac equation and their dynamics are linked. Thus a theory involving only ψ_L^e is possible, but only for massless fermions.

For the massless case we have a free field Lagrangian for several fermions:

$$\mathcal{L}_{\text{Free}} = \sum_j \left[\bar{\psi}_L^j \gamma_\mu \frac{i\partial}{\partial x_\mu} \psi_L^j + \bar{\psi}_R^j \gamma_\mu \frac{i\partial}{\partial x_\mu} \psi_R^j\right].$$

This Lagrangian is invariant under separate unitary transformations among the ψ_L^j and ψ_R^j, analogous to the chiral symmetry for the quarks discussed in Chapter 7. The left-handed free fields can be chosen to be doublets under the weak $SU(2)$ symmetry, which can be extended to be a local gauge symmetry. For example, we want $j = 1, 2$ to correspond to e and ν_e, respectively. The right-handed fields have no weak interactions and thus must be singlets under the $SU(2)$ symmetry. They undergo no unitary rotations when we make a weak $SU(2)$ gauge transformation. The eigenvalues for the weak diagonal operator T_3 are therefore

$$\psi_L^e : T_3 = -\frac{1}{2}, \quad \psi_L^{\nu_e} : T_3 = \frac{1}{2}, \quad \psi_R^e : T_3 = 0, \quad \psi_R^{\nu_e} : T_3 = 0.$$

Thinking of T_3 as a generalized charge, the left- and right-handed fermions have different weak charges!

A mass term violates the weak $SU(2)$ symmetry since the term $m_e(\bar{\psi}_L^e \psi_R^e + \bar{\psi}_R^e \psi_L^e)$ is not an $SU(2)$ scalar. ψ_L^e is a piece of a doublet, while ψ_R^e is a singlet. Thus the weak $SU(2)$ gauge symmetry requires that all fermions that interact weakly are massless. The mass must somehow be dynamically generated, through an $SU(2)$ scalar term in the Lagrangian (to maintain the symmetry) which somehow "turns" into the $SU(2)$ violating mass term. Similarly the W mass must somehow arise dynamically, since this also violates the gauge symmetry. How this can be done will be the subject of later discussion. For the time being we accept the theory with a broken symmetry and look at its low-energy consequences.

8.2 ■ MUON DECAY

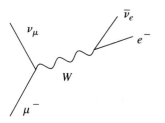

FIGURE 8.2 Lowest-order diagram for muon decay.

We are now ready to proceed with a calculation using the charged current matrix element of Eq. 8.5. The prototypical example is muon decay,

$$\mu^- \to e^- + \bar{\nu}_e + \nu_\mu,$$

with the Feynman diagram given in Figure 8.2. The muon mass is 105.66 MeV, so the final state electron is typically relativistic. The neutrinos are not detectable in individual decays; what we can measure is the e^- direction and energy for a muon decaying at rest with a given spin direction. With more effort, the e^- helicity can also be measured. The muons come typically from π decay, which we discuss later, and are naturally polarized. They can be stopped in a material that allows measuring the decay at rest. To avoid depolarizing the muons, the stopping material has to be chosen carefully; or the experiment can be performed in a large longitudinal magnetic field, which suppresses spin flip transitions. To measure the electron helicity, the electron can be scattered in a magnetized foil, for which the cross section depends on the helicity. Experiments have borne out the correctness of the left-handed four-fermion matrix element.

We label the four-momenta as follows: p_μ for μ, p_e for e, p_1 for $\bar{\nu}_e$, and p_2 for ν_μ. Then

$$M_{fi} = \frac{G_F}{\sqrt{2}} \left[\bar{u}_2 \gamma_\lambda (1 - \gamma_5) u_\mu \right] \left[\bar{u}_e \gamma_\lambda (1 - \gamma_5) v_1 \right],$$

$$|M_{fi}|^2 = \frac{G_F^2}{2} \text{Tr} \left[u_2 \bar{u}_2 \gamma_\lambda (1 - \gamma_5) u_\mu \bar{u}_\mu \gamma_\sigma (1 - \gamma_5) \right]$$
$$\times \text{Tr} \left[u_e \bar{u}_e \gamma_\lambda (1 - \gamma_5) v_1 \bar{v}_1 \gamma_\sigma (1 - \gamma_5) \right].$$

Summing over the final ν spins, for which only one helicity actually contributes, we can use:

$$\sum_{\text{spins}} u_2 \bar{u}_2 = \not{p}_2 \quad \text{and} \quad \sum_{\text{spins}} v_1 \bar{v}_1 = \not{p}_1.$$

Note that we ignore the presence of any very small neutrino masses. We will consistently do this throughout our calculations until we get to the issue of neutrino oscillations in Chapter 9. The neutrino masses are sufficiently small as to have no measurable effect on any decay rate.

For the charged leptons we will calculate for fixed helicities. For this case:

$$u_e \bar{u}_e = (\not{p}_e + m_e) \left[\frac{1 + \gamma_5 \not{s}_e}{2} \right]$$

$$u_\mu \bar{u}_\mu = (\not{p}_\mu + m_\mu) \left[\frac{1 + \gamma_5 \not{s}_\mu}{2} \right]. \tag{8.9}$$

8.2 Muon Decay

Note that we never need to use explicit spinors, just the measurables p and s, where s is the spin 4-vector for the given state.

We will perform the calculation in the muon rest frame, for which $s_\mu = (0, \hat{s}_\mu)$. We take $\hat{s}_\mu = \hat{e}_z$ to define the z-axis. For the electron, a helicity state has

$$s_e = \left(\frac{\vec{p}_e \cdot \hat{s}_e}{m_e}, \frac{E_e \hat{s}_e}{m_e} \right), \tag{8.10}$$

resulting from boosting the rest-frame spin $\hat{s}_e$ along $\vec{p}_e$. The two electron helicity states correspond to the choice of the initial $\hat{s}_e$ along or opposite $\vec{p}_e$. With the spin choices above:

$$|M_{fi}|^2 = \frac{G_F^2}{2} \mathrm{Tr}\left[\slashed{p}_2 \gamma_\lambda (1 - \gamma_5)(\slashed{p}_\mu + m_\mu)\left(\frac{1 + \gamma_5 \slashed{s}_\mu}{2}\right) \gamma_\sigma (1 - \gamma_5) \right]$$
$$\times \mathrm{Tr}\left[(\slashed{p}_e + m_e)\left(\frac{1 + \gamma_5 \slashed{s}_e}{2}\right) \gamma_\lambda (1 - \gamma_5) \slashed{p}_1 \gamma_\sigma (1 - \gamma_5) \right].$$

For each trace, terms with an odd number of γ matrices (excluding γ_5) will vanish. Keeping nonvanishing terms this gives for the first trace:

$$\mathrm{Tr}\left[\slashed{p}_2 \gamma_\lambda (1 - \gamma_5) \frac{(\slashed{p}_\mu + m_\mu \gamma_5 \slashed{s}_\mu)}{2} \gamma_\sigma (1 - \gamma_5) \right]$$
$$= \mathrm{Tr}\left[\slashed{p}_2 \gamma_\lambda (\slashed{p}_\mu - m_\mu \slashed{s}_\mu) \gamma_\sigma (1 - \gamma_5) \right]$$

after commuting the first $(1 - \gamma_5)$ all the way to the right and using $(1 - \gamma_5)^2 = 2(1 - \gamma_5)$. After the same simplifications for the second trace,

$$|M_{fi}|^2 = \frac{G_F^2}{2} \mathrm{Tr}\left[\slashed{p}_2 \gamma_\lambda (\slashed{p}_\mu - m_\mu \slashed{s}_\mu) \gamma_\sigma (1 - \gamma_5) \right]$$
$$\times \mathrm{Tr}\left[(\slashed{p}_e - m_e \slashed{s}_e) \gamma_\lambda \slashed{p}_1 \gamma_\sigma (1 - \gamma_5) \right].$$

Finally,

$$\mathrm{Tr}\left[\gamma_\alpha \gamma_\lambda \gamma_\beta \gamma_\sigma (1 - \gamma_5) \right] \mathrm{Tr}\left[\gamma_\theta \gamma_\lambda \gamma_\phi \gamma_\sigma (1 - \gamma_5) \right] = 64 g_{\alpha\theta} g_{\beta\phi};$$
$$|M_{fi}|^2 = 32 \, G_F^2 \left[(p_e - m_e s_e) \cdot p_2 \right] \left[(p_\mu - m_\mu s_\mu) \cdot p_1 \right].$$

The theory tells us which 4-vector to dot into which other one, among the several possible choices, to make the invariant matrix element squared.

We specify the electron direction relative to the μ helicity direction (z-axis) by a unit vector $\hat{n}$ in the muon rest-frame. Since $m_e/m_\mu = 1/210$, the electron is relativistic over almost the whole phase space for the decay. Making the relativistic approximation for the electron, $p_e = (E_e, E_e \hat{n})$. With this approximation, using Eq. 8.10:

$$(p_e - m_e s_e) = E_e(1, \hat{n}) - m_e \left(\frac{E_e}{m_e}, \frac{E_e}{m_e} \hat{n} \right) (\hat{s}_e \cdot \hat{n}) = (1 - \hat{s}_e \cdot \hat{n}) p_e.$$

Therefore, the matrix element vanishes if $\hat{s}_e$ and $\hat{n}$ are in the same direction, restricting the electron to be left-handed, as expected from the earlier discussion. The same formalism with $1 + \gamma_5$ instead of $1 - \gamma_5$ leads to a right-handed electron. Choosing the left-handed electron, we have

$$|M_{fi}|^2 = 64 G_F^2 [(p_e \cdot p_2)(p_\mu - m_\mu s_\mu) \cdot p_1].$$

We next need to integrate this over the phase space and divide by $2m_\mu$. This gives

$$d\Gamma = \frac{32 G_F^2}{(2\pi)^5 m_\mu} \left[(p_e \cdot p_2)(p_\mu - m_\mu s_\mu) \cdot p_1\right] \frac{d^3 p_e}{2 E_e} \frac{d^3 p_1}{2 E_1} \frac{d^3 p_2}{2 E_2} \delta^4(p_f - p_i).$$

Leaving this differential in the electron variables, which are the ones measured,

$$d\Gamma = \frac{4 G_F^2}{(2\pi)^5 m_\mu} \frac{d^3 p_e}{E_e} (p_e)_\alpha (p_\mu - m_\mu s_\mu)_\beta I_{\alpha\beta},$$

where the tensor

$$I_{\alpha\beta} = \int p_{2\alpha} p_{1\beta} \frac{d^3 p_1}{E_1} \frac{d^3 p_2}{E_2} \delta^4(p_1 + p_2 - k)$$

and $k = p_\mu - p_e$. We can derive $I_{\alpha\beta}$ by using its invariance properties. $I_{\alpha\beta}$ is a Lorentz invariant tensor, symmetric in α, β, with quadratic dependence on only one 4-vector k. Thus $I_{\alpha\beta}$ has to be of the form

$$I_{\alpha\beta} = A k^2 g_{\alpha\beta} + B k_\alpha k_\beta,$$

with A and B Lorentz invariant scalars. Since $k = p_\mu - p_e$ is time-like, a simple frame in which to do the integral is the frame where $k = (k_0, 0)$. In this frame $\vec{p}_1 = -\vec{p}_2$ and $E_1 = E_2 = k_0/2$. The integral in this frame is

$$\int p_{2\alpha} p_{1\beta} \frac{p_1 \, d\Omega_1}{E_2} dE_1 \, \delta(2 E_1 - k_0) = \int p_{2\alpha} p_{1\beta} \frac{d\Omega_1}{2},$$

with $p_{2\alpha}, p_{1\beta}$ given in terms of k_0.

We can derive two simple relations for A and B by calculating

$$g_{\alpha\beta} I_{\alpha\beta} = (4A + B) k^2 = \int (p_1 \cdot p_2) \frac{d\Omega_1}{2} = \int \frac{k_0^2}{4} d\Omega_1 = \pi k^2$$

$$k_\alpha k_\beta I_{\alpha\beta} = (A + B) k^4 = \int \frac{k_0^4}{8} d\Omega_1 = \frac{\pi}{2} k^4.$$

Solving for A and B, we have $A = \pi/6$, $B = \pi/3$, and therefore $I_{\alpha\beta} = \frac{1}{6}\pi(k^2 g_{\alpha\beta} + 2 k_\alpha k_\beta)$, where $k = p_\mu - p_e$.

8.2 Muon Decay

We next substitute $I_{\alpha\beta}$ into the rate expression. Ignoring m_e everywhere, we have the following relations for the resulting 4-vector dot products:

$$k \cdot p_e = m_\mu E_e, \quad k \cdot p_\mu = m_\mu^2 - E_e m_\mu,$$

$$k^2 = m_\mu^2 - 2m_\mu E_e, \quad p_e \cdot s_\mu = -E_e(\hat{n} \cdot \hat{s}_\mu), \quad p_\mu \cdot s_\mu = 0.$$

Using these we get, taking $\hat{n} \cdot \hat{s}_\mu = \cos\theta_e$,

$$d\Gamma = \frac{2G_F^2}{(2\pi)^5}(dE_e \, d\Omega_e)\left(\frac{\pi}{3}\right) E_e^2 \left[3m_\mu^2 - 4m_\mu E_e + (m_\mu^2 - 4m_\mu E_e)\cos\theta_e\right].$$

Defining

$$x = \frac{2E_e}{m_\mu} = \frac{E_e}{E_e^{\max}},$$

x varies from 0 to 1. In terms of x,

$$d\Gamma = \frac{G_F^2 m_\mu^5}{192\pi^3}\left[2x^2(3-2x)\right]\left[1 + \left(\frac{1-2x}{3-2x}\right)\cos\theta_e\right] dx \frac{d\Omega_e}{4\pi}.$$

The second factor in brackets gives the correlation of the electron direction with the muon spin. Note that it vanishes at $x = 1$ and $\cos\theta_e = 1$. This is a configuration where the electron is collinear with the muon spin and the neutrinos are opposite. The rate vanishes because of angular momentum conservation, given the fixed helicities of the various particles. Integrating over $d\Omega_e$, the term dependent on $\cos\theta_e$ averages to zero. The energy spectrum is then given by the first factor in brackets. The spectrum peaks at $x = 1$, that is, the electron tends to carry a lot of energy. Integrating over x gives the muon decay formula

$$\Gamma = \frac{G_F^2 m_\mu^5}{192\pi^3}. \tag{8.11}$$

Corrections to this formula come from electromagnetic effects and the neglect of the electron mass. A much smaller correction is due to taking the W propagator as a constant. Figure 8.3 shows the data and theoretical expectation. The latter has been corrected for photon radiation, which rounds off the spectrum at $x = 1$.

It turns out that the formula for the width can be calculated in closed form, even for a massive electron. Defining the function $f(x) = 1 - 8x + 8x^3 - x^4 + 12(x^2 \log(1/x))$, the muon width, including lowest-order radiative corrections (term in brackets) is

$$\Gamma = \frac{G_F^2 m_\mu^5}{192\pi^3}\left[1 - \frac{\alpha}{2\pi}\left(\pi^2 - \frac{25}{4}\right)\right] f\left(\frac{m_e^2}{m_\mu^2}\right). \tag{8.12}$$

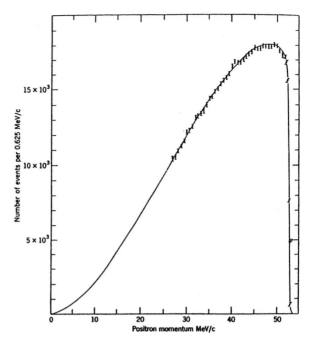

FIGURE 8.3 Experimental spectrum of positrons in $\mu^+ \to e^+ + \nu_e + \bar{\nu}_\mu$. The solid line is the theoretically predicted spectrum, including the correction for electromagnetic effects. [From M. Bardon et al., *Phys. Rev. Lett.* 14, 449(1965).]

The corrections are smaller than 1%. The measured muon lifetime is $\tau_\mu = 2.19703 \pm 0.00004 \times 10^{-6}$ sec. Using the lifetime and Eq. 8.12 gives

$$G_F = 1.16639 \pm 0.00001 \times 10^{-5} \text{ GeV}^{-2}.$$

Note that the muon lifetime is long enough so that relativistic muons travel very long distances in one lifetime. Muons were in fact discovered as remnants of cosmic ray showers after traveling through the whole atmosphere.

8.3 ■ DECAYS OF THE TAU LEPTON

The τ lepton is a member of the third doublet of leptons. Through the indirect measurement of the rate for Z^0 decay to $\nu\bar{\nu}$ of all types, it has been possible to rule out the existence of any additional families with ν lighter than about 45 GeV. Thus the τ is expected to be the last and the heaviest lepton forming a doublet with a light neutrino. The τ mass is $m_\tau = 1777$ MeV. It was discovered in the annihilation reaction $e^+e^- \to \tau^+\tau^-$ and its mass was accurately measured by the behavior of the cross section with energy near the threshold for this reaction. It has a lifetime $\tau_\tau = 2.906 \pm 0.011 \times 10^{-13}$ sec, which is a factor about 10^7

8.3 Decays of the Tau Lepton

shorter than that of the muon. The value of $c\tau_\tau = 87 \times 10^{-6}$ m is small, but still measurable for high momentum τ's using precision detectors placed close to the τ production point. The accuracy of the lifetime measurement directly reflects our ability to measure short flight paths accurately.

The τ is the only lepton heavy enough to decay to hadrons. Thus, its decays are of the type

$$\tau^- \to \nu_\tau e^- \bar{\nu}_e, \; \nu_\tau \mu^- \bar{\nu}_\mu, \; \tau^- \to \nu_\tau h^-$$

where h^- is a hadronic system of the given charge. Given the value of the τ mass, h^- can be composed of quark types u, d, and s and their antiparticles only.

The decay to leptons proceeds as for muon decay; the expected widths are given by the analog of Eq. 8.12. To lowest order, ignoring m_e,

$$\Gamma_{\tau^- \to e^- \bar{\nu}_e \nu_\tau} = \frac{G_F^2 m_\tau^5}{192\pi^3},$$

$$\Gamma_{\tau^- \to \mu^- \bar{\nu}_\mu \nu_\tau} = \Gamma_{\tau^- \to e^- \bar{\nu}_e \nu_\tau} f\left(\frac{m_\mu^2}{m_\tau^2}\right) = 0.973 \Gamma_{\tau^- \to e^- \bar{\nu}_e \nu_\tau}.$$

Using the value of G_F and m_τ, we can predict the branching ratio, B_e, into $e^- \bar{\nu}_e \nu_\tau$ in terms of the measured lifetime. B_e and the lifetime are the two quantities that are measured directly. Including the small radiative corrections, the relation is

$$\frac{\Gamma_{\tau^- \to e^- \bar{\nu}_e \nu_\tau}}{\Gamma_{\text{tot}}} = B_e = \frac{2.906 \pm 0.011 \times 10^{-13} \text{ sec}}{16.32 \pm 0.01 \times 10^{-13} \text{ sec}} = 17.81 \pm 0.06\%.$$

Note that the partial width is calculated with an uncertainty coming from the uncertainties in m_τ and G_F; the uncertainty in the lifetime comes from the experimental error in the direct measurement of this quantity. The measured value for B_e is $17.84 \pm 0.06\%$, in excellent agreement with the theory. B_μ is measured to be $17.37 \pm 0.06\%$, also in excellent agreement with expectations.

If quarks were free final states and the doublets were

$$\begin{pmatrix} u \\ d \end{pmatrix}, \; \begin{pmatrix} c \\ s \end{pmatrix}, \; \begin{pmatrix} t \\ b \end{pmatrix},$$

we would expect to find one type of decay involving quarks $\tau^- \to \nu_\tau \bar{u} d$. The other two doublets have particle pairs that are too heavy to be produced. The decay to quarks is obtained by the replacement in the leptonic decay $e^- \to d$ and $\bar{\nu}_e \to \bar{u}$. The corresponding particles have the same weak $SU(2)$ properties. We will call the $SU(2)$ group weak isospin to emphasize the analogy with isospin. For the first quark doublet, weak isospin corresponds to the flavor isospin discussed in Chapter 5.

Including the factor of three from color, the simple quark picture implies that $B_{\bar{u}d} = 3B_e$ if we ignore quark masses. This predicts, using the measured value of B_e, that $B_{\bar{u}d} = 0.534$, whereas the branching ratio to hadrons is $B_h = 1 - B_e -$

$B_\mu = 0.649$. The lowest-order decay diagram, however, must be supplemented by diagrams where gluons are emitted. Although analogous to the radiative corrections in the lepton case, the effect on the rate is much larger because $\alpha_s \gg \alpha$. The gluonic corrections have been calculated to third order, giving a predicted branching ratio (which we now identify with the full hadronic branching ratio) of

$$B_h = 3B_e \left[1 + \frac{\alpha_s(m_\tau^2)}{\pi} + 5.20 \left(\frac{\alpha_s(m_\tau^2)}{\pi} \right)^2 + 26.4 \left(\frac{\alpha_s(m_\tau^2)}{\pi} \right)^3 \right].$$

Using this formula to calculate α_s gives $\alpha_s(m_\tau^2) = 0.35$ with an uncertainty of $\sim 10\%$ from next-order corrections and the use of perturbation theory. This value for $\alpha_s(m_\tau^2)$ is in excellent agreement with the other reliable methods used to determine $\alpha_s(q^2)$.

The formula for the hadronic branching ratio is a perturbative short-distance calculation. Over longer time scales, the system evolves into hadronic final states. Viewing the expansion of the time-evolution operator in terms of quantum events, we have the effective weak Hamiltonian that creates the short-distance system, which then evolves nonperturbatively according to the strong interaction dynamics. The effective weak Hamiltonian is constructed from a vector and axial vector hadronic current dotted into the leptonic current for τ and its neutrino. In the flavor isospin symmetry limit, the hadronic vector current $\bar{\psi}_d \gamma_\mu \psi_u$ is conserved. As in the case of $e^+ e^-$ production of hadrons, this implies that only states with $J^P = 1^-$ are produced from the vacuum by this current. Writing the vector current as J_μ^V, $\langle h^- | J_\mu^V | 0 \rangle \neq 0$ only for an h^- system that has $J^P = 1^-$ in its rest frame.

The axial vector current is not conserved (except in the limit of massless quarks, as discussed in Chapter 7) so that the system h^- produced via $\langle h^- | J_\mu^A | 0 \rangle$ can be a state with $J^P = 0^-$ or 1^+. Higher spins for h^- are not allowed, since we cannot construct a vector out of the spin and momentum of h^- for higher spins. One other additional constraint follows from G parity, which applies here because the weak isospin and flavor isospin are the same for the lightest quark doublet. J_μ^V transforms under G like the ρ meson, while J_μ^A transforms like the pion. That is,

$$G J_\mu^V G^{-1} = J_\mu^V, \quad G J_\mu^A G^{-1} = -J_\mu^A.$$

The G parity property for the currents means that, to good approximation, $\langle n\pi | J_\mu^V | 0 \rangle$ only produces states where the number of pions n is even, while $\langle n\pi | J_\mu^A | 0 \rangle$ only produces states with n odd. Based on the measured branching ratios, the vector and axial vector currents contribute nearly equally to the overall width, after summing over states, as is the case for the purely leptonic τ decays.

Table 8.1 gives the largest measured branching ratios into various hadronic final states.

Our picture of the transitions of the W into $\bar{u}d$ is inadequate in one respect, since it cannot account for the small rate at which decays into strange particles are observed, although the bulk of the decays are correctly explained. In the next

8.3 Decays of the Tau Lepton

TABLE 8.1 Hadronic Tau Decays, $\tau^- \to \nu_\tau h^-$.

Final state h^-	Branching ratio (%)
π^-	11.06 ± 0.11
$\pi^-\pi^0$	25.41 ± 0.14
$\pi^-\pi^0\pi^0$	9.17 ± 0.14
$\pi^-\pi^+\pi^-$	9.22 ± 0.10
$\pi^-\pi^+\pi^-\pi^0$	4.24 ± 0.10
$\pi^-\pi^0\pi^0\pi^0$	1.08 ± 0.10
K^-	0.69 ± 0.02
$K^-\pi^0$	0.45 ± 0.03

section we will address the question of the strange particle weak couplings, which first arose historically in the weak decays of the strange particles. For example, the decays $K^- \to \pi^-\pi^0$, $K^- \to \mu^-\bar{\nu}_\mu$, $K^- \to \pi^0 e^- \bar{\nu}_e$ all require the transition of the strange quark to a u quark or the annihilation of the quarks within the K^-.

Comparing a few of the τ branching ratios with other measurements allows a stringent test of the overall framework. For example, the matrix elements $\langle \pi^- | J_\mu^A | 0 \rangle$ and $\langle K^- | J_\mu^A | 0 \rangle$ are directly related to the matrix elements $\langle 0 | J_\mu^A | \pi^- \rangle$ and $\langle 0 | J_\mu^A | K^- \rangle$, which account for π^- or $K^- \to \mu^- \bar{\nu}_\mu$ or $e^- \bar{\nu}_e$. The matrix element $\langle \rho^- | J_\mu^V | 0 \rangle$ is related (by an isospin rotation) directly to the matrix element of the electromagnetic current $\langle \rho^0 | J_\mu^{EM} | 0 \rangle$ (the decay $\rho^- \to \mu^- \bar{\nu}_\mu$, which involves the weak current, is too small to be measured). The relations between rates implied by the above have been borne out very nicely by the data. As an example, we look at the ρ^- current matrix element responsible for the process $\tau^- \to \rho^- \nu_\tau$.

Writing the most general current matrix element $\langle \rho^- | J_\mu^V | 0 \rangle = g_{W\rho} m_\rho^2 e_\mu^{\rho*}$, the amplitude for the process is:

$$M_{fi} = \frac{G_F}{\sqrt{2}} g_{W\rho} m_\rho^2 \bar{u}_{\nu_\tau} \not{e}_\rho (1 - \gamma_5) u_\tau,$$

where we choose a real basis for e_μ^ρ. The coupling $g_{W\rho}$ is dimensionless and defined analogously to the coupling to a photon discussed in Section 6.4. Squaring, summing over final spins, and averaging over τ spins, leads to the expression

$$\frac{1}{2} \sum_{\text{spins}} |M_{fi}|^2 = \frac{G_F^2 g_{W\rho}^2 m_\rho^4}{2} \sum_{\rho \text{ spins}} \text{Tr}\left[\not{p}_{\nu_\tau} \not{e}_\rho \not{p}_\tau \not{e}_\rho (1 - \gamma_5)\right].$$

The γ_5 term gives no contribution and the trace can be simply evaluated to give

$$\frac{1}{2} \sum_{\text{spins}} |M_{fi}|^2 = 2 G_F^2 g_{W\rho}^2 m_\rho^4 \sum_{\rho \text{ spins}} \left[p_{\nu_\tau} \cdot p_\tau + 2(p_{\nu_\tau} \cdot e_\rho)(p_\tau \cdot e_\rho)\right].$$

Finally, summing over ρ spins using

$$\sum_{\rho \text{ spins}} e_\mu^\rho e_\nu^\rho = -g_{\mu\nu} + \frac{p_\mu^\rho p_\nu^\rho}{m_\rho^2}$$

gives

$$\frac{1}{2}\sum_{\text{spins}} |M_{fi}|^2 = 2G_F^2 g_{W\rho}^2 m_\rho^4 \left[p_{\nu_\tau} \cdot p_\tau + 2(p_\rho \cdot p_{\nu_\tau})(p_\rho \cdot p_\tau)\right].$$

In the τ rest frame, the various 4-vector products are

$$p_{\nu_\tau} \cdot p_\tau = p_\rho \cdot p_{\nu_\tau} = \frac{m_\tau^2 - m_\rho^2}{2}, \qquad p_\rho \cdot p_\tau = \frac{m_\tau^2 + m_\rho^2}{2}.$$

Integrating over the final two-body phase space and dividing by $2m_\tau$ gives

$$\Gamma_{\tau^- \to \rho^- \nu_\tau} = \frac{G_F^2 g_{W\rho}^2}{16\pi}\left(\frac{m_\rho}{m_\tau}\right)^2 (m_\tau^2 - m_\rho^2)^2 \left[m_\tau + \frac{2m_\rho^2}{m_\tau}\right]. \tag{8.13}$$

Dividing by the expression for $\Gamma_{\tau^- \to e^- \bar{\nu}_e \nu_\tau}$ gives the relation

$$\frac{\Gamma_{\tau^- \to \rho^- \nu_\tau}}{\Gamma_{\tau^- \to e^- \bar{\nu}_e \nu_\tau}} = 12\pi^2 g_{W\rho}^2 \left(\frac{m_\rho}{m_\tau}\right)^2 \left(1 - \frac{m_\rho^2}{m_\tau^2}\right)^2 \left[1 + \frac{2m_\rho^2}{m_\tau^2}\right] = 20 g_{W\rho}^2.$$

We next want to relate $g_{W\rho}^2$ to the coupling that appeared in $e^+ e^- \to \rho^0$. The electromagnetic current operator in the light quark sector is (in units of e):

$$J_\mu^{EM} = \frac{2}{3}\bar{\psi}_u \gamma_\mu \psi_u - \frac{1}{3}\bar{\psi}_d \gamma_\mu \psi_d$$

$$= \frac{1}{\sqrt{2}}\left[\frac{\bar{\psi}_u \gamma_\mu \psi_u - \bar{\psi}_d \gamma_\mu \psi_d}{\sqrt{2}}\right] + \frac{1}{3\sqrt{2}}\left[\frac{\bar{\psi}_u \gamma_\mu \psi_u + \bar{\psi}_d \gamma_\mu \psi_d}{\sqrt{2}}\right].$$

The first term in brackets transforms as an $I = 1$ operator (like the ρ^0 meson) and the second like an $I = 0$ operator (like the ω^0 meson). The square of the ratio of coupling coefficients gives the ratio of rates to ρ^0 and ω^0 in the ideal mixing case. Using the isospin symmetry we have:

$$\langle \rho^- | \bar{\psi}_u \gamma_\mu \psi_d | 0 \rangle = \langle \rho^0 | \left[\frac{\bar{\psi}_u \gamma_\mu \psi_u - \bar{\psi}_d \gamma_\mu \psi_d}{\sqrt{2}}\right] | 0 \rangle = \sqrt{2} \langle \rho^0 | J_\mu^{EM} | 0 \rangle.$$

In Section 6.4 we defined $\langle \rho^0 | J_\mu^{EM} | 0 \rangle = \langle e_i \rangle g_{\gamma\rho} m_\rho^2 e_\mu^{\rho*} = \frac{1}{\sqrt{2}} g_{\gamma\rho} m_\rho^2 e_\mu^{\rho*}$, given the value of the ρ charge average. Therefore $g_{W\rho} = g_{\gamma\rho} = 0.28$ based on the ρ^0 entry in Table 6.6.

8.4 Charged Weak Currents for Quarks

In the ρ^- production calculation, we left out one factor related to the decay to strange quarks, namely, a small suppression of the light quark rate so that the sum of both has unit strength. This provides a 5% reduction of the expected ρ^- rate, as discussed in the next section. Including this small suppression, we expect that

$$B_{\rho^-} = 19 g_{W\rho}^2 B_e.$$

Using the values of B_e and $g_{W\rho} = 0.28 \pm 0.01$, we therefore expect that

$$B_{\rho^-} = 0.265 \pm 0.019.$$

Our calculation ignored the width of ρ^-, which is not a negligible effect since m_τ is not very large compared to m_ρ. Including this effect, by integrating over the ρ^- Breit–Wigner shape, lowers the prediction of the branching ratio to about 0.24 and also increases the uncertainty of the result, which depends on understanding the mass spectrum over a broad range. The measured value is about 0.25 (the bulk of the branching ratio into $\pi^-\pi^0$), which is in good agreement with the expected value.

8.4 ■ CHARGED WEAK CURRENTS FOR QUARKS

We turn now to the fermionic Lagrangian for the three doublets of quarks. In particular, we want to understand how the doublets enter the weak interactions, and to understand decays involving strange particles. To simplify the notation, we indicate the upper doublet member as ψ^U and the lower members as ψ^D. Thus the indices U and D take on three values for the three families. In this notation the weak currents are: $\frac{1}{2}(\bar{\psi}_L^U \gamma_\mu \psi_L^U) - \frac{1}{2}(\bar{\psi}_L^D \gamma_\mu \psi_L^D)$ coupled to W_3, $\frac{1}{\sqrt{2}} \bar{\psi}_L^U \gamma_\mu \psi_L^D$ coupled to $W^{(+)}$ and $\frac{1}{\sqrt{2}} \bar{\psi}_L^D \gamma_\mu \psi_L^U$ coupled to $W^{(-)}$. We want to look at the other parts of the Lagrangian involving these fields, which are the weak eigenstate fields.

The first terms in $\mathcal{L}$ are the fermion kinetic energy terms:

$$i\bar{\psi}_L^U \gamma_\mu \frac{\partial}{\partial x_\mu} \psi_L^U + i\bar{\psi}_R^U \gamma_\mu \frac{\partial}{\partial x_\mu} \psi_R^U + i\bar{\psi}_L^D \gamma_\mu \frac{\partial}{\partial x_\mu} \psi_L^D + i\bar{\psi}_R^D \gamma_\mu \frac{\partial}{\partial x_\mu} \psi_R^D.$$

To arrive at the free field $\mathcal{L}$, we next add the mass terms. For one fermion, we have written this earlier as $-m(\bar{\psi}_L \psi_R + \bar{\psi}_R \psi_L)$ and we might expect that we only need to add together six such terms. This is, however, incorrect, that is, the mass terms are not the simplest choice we might make. The reasons are not understood. Note, however, that in the massless theory the left- and right-handed fields are unrelated objects, so that how they are married into 4-component fields does not have to be simple. Since the masses are dynamically determined, the unknown dynamics apparently couples together the quarks in a complicated manner, reflected in a complicated set of mass terms.

The most general mass term we can have couples together all quarks with the same colors and electric charges. Taking the fields now as column vectors containing the three flavor families, the mass term is

$$-\left[(\bar{\psi}_L^U M_U \psi_R^U + \bar{\psi}_R^U M_U^\dagger \psi_L^U) + (\bar{\psi}_L^D M_D \psi_R^D + \bar{\psi}_R^D M_D^\dagger \psi_L^D)\right]. \quad (8.14)$$

This term has been chosen to be Hermitian. Not indicated are the color degrees of freedom—each mass matrix actually comes in three identical versions, for the three colors. From the point of view of QCD, the mass matrices could have been larger, coupling also the U and D fields. However, if we include electromagnetism, only fields of the same charge can couple, as we have chosen.

The eigenstates of the mass term are quark fields that are the basis, when we add the gluon interactions, for forming the hadrons discussed earlier. To look at weak transitions involving these hadronic states, we need to rewrite the weak interaction Hamiltonian in terms of these mass eigenstate fields.

In general, a given matrix can be diagonalized in terms of two unitary matrices and M_U and M_D can therefore be diagonalized by four such matrices. We write these as U_L^U, U_R^U, U_L^D, and U_R^D, chosen such that $M'_U = U_L^U M_U U_R^{U\dagger}$ and $M'_D = U_L^D M_D U_R^{D\dagger}$ are diagonal.

$$M'_U = \begin{pmatrix} m_u & 0 & 0 \\ 0 & m_c & 0 \\ 0 & 0 & m_t \end{pmatrix} \quad \text{and} \quad M'_D = \begin{pmatrix} m_d & 0 & 0 \\ 0 & m_s & 0 \\ 0 & 0 & m_b \end{pmatrix}.$$

The fields $\psi_L^{'U} = U_L^U \psi_L^U$, $\psi_R^{'U} = U_R^U \psi_R^U$, $\psi_L^{'D} = U_L^D \psi_L^D$, and $\psi_R^{'D} = U_R^D \psi_R^D$ then give a mass term, after substitution into Eq. 8.14, that is merely the sum over six individual quark mass terms.

We can now rewrite the weak currents in terms of the ψ' fields, by substituting $\psi_L^U = U_L^{U\dagger} \psi_L^{'U}$, $\psi_L^D = U_L^{D\dagger} \psi_L^{'D}$. The current that couples to W_3 has the same form in terms of ψ', as it had in terms of ψ, because the transformations between ψ and ψ' are unitary and this current does not link U and D quarks. So the neutral current does not change the flavor content in an interaction of mass eigenstates. It is still possible to have higher-order processes whose net effect is electrically neutral (for example, through two charge changing currents) leading to flavor changing transitions, but these will be small, occurring at higher order.

The charged current interaction is:

$$\frac{g}{\sqrt{2}} \left[\left(\bar{\psi}_L^{'U} U_L^U \gamma_\mu U_L^{D\dagger} \psi_L^{'D}\right) W_\mu^{(+)} + \left(\bar{\psi}_L^{'D} U_L^D \gamma_\mu U_L^{U\dagger} \psi_L^{'U}\right) W_\mu^{(-)} \right].$$

We can combine the matrices $U_L^U U_L^{D\dagger} = V_{UD}$, where

$$V_{UD} = \begin{pmatrix} V_{ud} & V_{us} & V_{ub} \\ V_{cd} & V_{cs} & V_{cb} \\ V_{td} & V_{ts} & V_{tb} \end{pmatrix} \quad (8.15)$$

is a unitary matrix that indicates how much of the coupling is shared between the various U and D quarks.

8.4 Charged Weak Currents for Quarks

V_{UD} is called the Cabibbo–Kobayashi–Maskawa (CKM) matrix. In terms of V_{UD} we can write, dropping the primes from the ψ fields to simplify the notation (in the future we will always use mass eigenstate fields), the charged current coupling in terms of

$$\frac{1}{\sqrt{2}} \bar{\psi}_L^U \gamma_\mu V_{UD} \psi_L^D$$

and its Hermitian conjugate. The τ decays listed in Table 8.1 involve the terms in V_{UD} that couple $u, d\, (V_{ud})$ and $u, s\, (V_{us})$. The τ decay pattern can then be understood if $V_{ud} \sim 1$ and $V_{us} \ll 1$. The rate to nonstrange final states is proportional to $|V_{ud}|^2$, the rate to strange final states is proportional to $|V_{us}|^2$.

We consider matrices for the lepton masses that are analogous to the quark mass matrices. As long as the neutrino masses are all very small ($\ll m_e$), then our rate calculations to neutrinos in the final state do not depend on whether a unique neutrino or a mix of neutrinos is produced. Measurable consequences of a CKM type of matrix in the lepton sector requires nonzero neutrino mass, so that U_L^U cannot be chosen equal to U_L^D for the leptons, and implies that the individual lepton numbers are not conserved. However, explicit rate calculations of lepton number violating processes (for example, $\mu^- \to e^- + \gamma$) among the charged leptons are very small for very small neutrino masses. Since the neutrinos do not have strong or electromagnetic interactions, the only interactions in which we can observe them participating are weak interactions. Thus, for neutrinos we produce or detect lepton flavor eigenstates, but it is the mass eigenstates that propagate in between. This gives rise to the phenomenon of neutrino oscillations. We will discuss the lepton sector mixing matrix in Chapter 9, where such oscillations are quantified. The nature of the interaction responsible for neutrino mass generation is at least as mysterious as for the other particles.

8.4.1 ■ Cabibbo–Kobayashi–Maskawa Matrix

The terms in V_{UD} have been determined experimentally, since there is as yet no theory that gives us this matrix or the various fermion masses. What is found is that the largest terms in V_{UD} are the diagonal terms, with the off-diagonal terms linking the first and second generation much greater than off-diagonal terms linking the first or second with the third generation. Thus for some situations, for example K decay or the τ decays discussed earlier, it is sufficient to look at a two-generation CKM matrix, assuming the third generation is decoupled.

For n generations, a complex matrix has $2n^2$ parameters. The unitary constraint, however, amounts to n^2 relations, leaving n^2 free parameters for a unitary matrix. Thus, for example, a two-generation unitary matrix has four parameters. We can write the matrix in general as

$$\begin{pmatrix} e^{i\phi_1} \cos\theta & e^{i(\phi_1+\phi_2)} \sin\theta \\ -e^{i\phi_3} \sin\theta & e^{i(\phi_2+\phi_3)} \cos\theta \end{pmatrix}.$$

The four parameters are $\phi_1, \phi_2, \phi_3, \theta$. The reader can check that this matrix is unitary. However, not all the parameters have physical significance. We can rewrite the 2 × 2 matrix as:

$$\begin{pmatrix} e^{i\phi_1} & 0 \\ 0 & e^{i\phi_3} \end{pmatrix} \begin{pmatrix} \cos\theta & \sin\theta \\ -\sin\theta & \cos\theta \end{pmatrix} \begin{pmatrix} 1 & 0 \\ 0 & e^{i\phi_2} \end{pmatrix}.$$

The phases ϕ_1, ϕ_2, ϕ_3 just multiply individual quark fields when we write the weak interaction Lagrangian. However, the other terms in the Lagrangian are invariant under the replacement of a quark field by a quark field times a phase factor. Thus we can redefine the fields, absorbing the phase factors, and not change the physics of the states involved in the interaction. Thus the phases are not physically relevant and we can use the simplified matrix:

$$\begin{pmatrix} \cos\theta & \sin\theta \\ -\sin\theta & \cos\theta \end{pmatrix}. \tag{8.16}$$

The angle in this matrix for the two-generation approximation is called the Cabibbo angle, θ_c. Based on data that we discuss in Section 8.6,

$$\sin\theta_c = 0.221 \pm 0.002, \quad \text{implying that} \quad \cos\theta_c = 0.975 \pm 0.001.$$

Weak decay rates involving a u, s transition are suppressed by a factor $\sin^2\theta_c = 0.049$. Of course this factor is determined experimentally by just such decays, which are described in a very consistent way in terms of the one parameter. Note that a stringent test of the overall framework is that the independently measured values of $\sin^2\theta_c$ and $\cos^2\theta_c$ add up to 1. The consequences of the small value of the off-diagonal terms of V_{UD} is called Cabibbo suppression.

For the case of n generations, we can eliminate $2n-1$ parameters of the unitary matrix by redefining individual quark fields. Thus an n-generation CKM matrix has $n^2 - (2n-1) = (n-1)^2$ meaningful free parameters. For comparison, a real orthogonal n-generation rotation matrix has

$$\frac{n(n-1)}{2}$$

angles, while the n-generation CKM matrix has

$$\frac{n(n-1)}{2}$$

angles and

$$\frac{(n-1)(n-2)}{2}$$

phases. For $n = 3$ this gives 3 angles and 1 phase. The convention for writing the matrix is to define three angles $\theta_{12}, \theta_{13}, \theta_{23}$, and a phase δ, in terms of which we

8.4 Charged Weak Currents for Quarks

can write

$$V_{UD} = \begin{pmatrix} 1 & 0 & 0 \\ 0 & c_{23} & s_{23} \\ 0 & -s_{23} & c_{23} \end{pmatrix} \begin{pmatrix} c_{13} & 0 & s_{13}e^{-i\delta} \\ 0 & 1 & 0 \\ -s_{13}e^{i\delta} & 0 & c_{13} \end{pmatrix} \begin{pmatrix} c_{12} & s_{12} & 0 \\ -s_{12} & c_{12} & 0 \\ 0 & 0 & 1 \end{pmatrix}.$$

Multiplying the three simple matrices:

$$V_{UD} = \begin{pmatrix} c_{12}c_{13} & s_{12}c_{13} & s_{13}e^{-i\delta} \\ -s_{12}c_{23} - c_{12}s_{23}s_{13}e^{i\delta} & c_{12}c_{23} - s_{12}s_{23}s_{13}e^{i\delta} & s_{23}c_{13} \\ s_{12}s_{23} - c_{12}c_{23}s_{13}e^{i\delta} & -c_{12}s_{23} - s_{12}c_{23}s_{13}e^{i\delta} & c_{23}c_{13} \end{pmatrix}. \quad (8.17)$$

Here $s_{ij} = \sin(\theta_{ij})$ and $c_{ij} = \cos(\theta_{ij})$, with $j > i$. If an angle θ_{ij} vanishes, then the coupling for $D_j \to U_i$ for generations i and j will vanish.

We can write the matrix in a rather accurate approximation by using the fact that the off-diagonal elements become progressively smaller as we increase the generation number. Defining:

$$s_{12} = \sin\theta_c = \lambda$$

$$s_{23} = A\lambda^2 \ll s_{12}$$

$$s_{13}e^{-i\delta} = A\lambda^3(\rho - i\eta) \ll s_{23} \text{ in magnitude,}$$

$$V_{UD} = \begin{pmatrix} 1 - \frac{\lambda^2}{2} & \lambda & A\lambda^3(\rho - i\eta) \\ -\lambda & 1 - \frac{\lambda^2}{2} & A\lambda^2 \\ A\lambda^3(1 - \rho - i\eta) & -A\lambda^2 & 1 \end{pmatrix}. \quad (8.18)$$

The measured values for A and $\rho^2 + \eta^2$ are each ~ 1, so the powers of λ tell us roughly how large each term in the matrix is. The coupling for the first two generations is given approximately by the Cabibbo matrix written previously. The phase factor that occurs in the matrix appears only in the coupling between the first and third generations in the approximation above.

The terms in the matrix are measured in various weak decay processes. For example, the term $V_{cb} = A\lambda^2$. The decays of mesons made of a u or d and a $\bar{b}$, called B mesons, indicate that V_{cb} equals 0.041 ± 0.003, based on decay processes where the $\bar{b}$ turns into a $\bar{c}$. This gives $A = 0.84 \pm 0.06$. Measurements of V_{ub} (where the b turns into a u) and a number of other processes involving B mesons (mixing and CP violation) provide a measurement of

$$\rho = 0.17 \pm 0.07 \quad \text{and} \quad \eta = 0.36 \pm 0.04.$$

The parameters for the CKM matrix are recorded in Table 8.2. Later we will look at some of the processes used to determine these parameters.

Using the charged currents involving quarks and leptons, we can now write the low energy weak effective Hamiltonian density for virtual W exchange between two vertices:

$$\mathcal{H}_W = \frac{G_F}{\sqrt{2}} J_\mu J_\mu^\dagger.$$

TABLE 8.2 Parameters of the CKM Matrix.

Parameter	Value
λ	0.221 ± 0.002
A	0.84 ± 0.06
ρ	0.17 ± 0.07
η	0.36 ± 0.04

Here, J_μ contains both vector and axial vector terms as discussed earlier and is a weak isospin raising operator, while $J_\mu^\dagger$ involves the lowering operator. J_μ is given by the sum of three terms involving the three lepton generations and three terms involving the three quark generations summed over colors. Each quark term in J_μ is then itself made of three terms because of the CKM mixing. Which of the pieces of $\mathcal{H}_W$ contribute in a given situation is determined by the chosen initial and final states.

We consider next the general question of *CP* violation arising from the weak interaction currents, and return later to review the weak decays of hadrons systematically.

8.4.2 ■ CP Violation

We turn now to the behavior of the weak amplitudes under several special transformations. The amplitudes we examine involve matrix elements for states related to each other via C, P, and T transformations. These symmetries were introduced in Section 3.9. We start with the general vertex of the charged current for one quark generation chosen to be u, d. The diagrams for the processes that actually exist at the quark level are shown in Figure 8.4. We assume for simplicity that

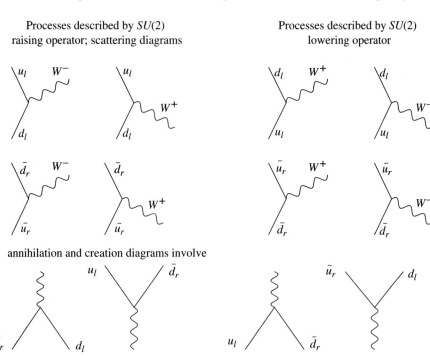

FIGURE 8.4 Space-time processes contained in the weak Hamiltonian. All couplings are shown for the scattering diagrams. Only half of the diagrams, showing the quark-antiquark participants, are shown for the annihilation and creation process. Time travels upward in all the diagrams.

8.4 Charged Weak Currents for Quarks

the momenta are high enough that we can ignore masses, so that all particles are left-handed and all antiparticles right-handed. We indicate the helicity by l(left) and r(right) for these two cases. We will focus on the scattering diagrams in Figure 8.4. The particle momenta and specific W helicities are not indicated in the figure, since the Hamiltonian density contains all choices.

We study a family of single vertex amplitudes for states related by the various transformations. It is sufficient to look at the first raising operator diagram in Figure 8.4. The other choices for related amplitudes work the same way. Thus we start with

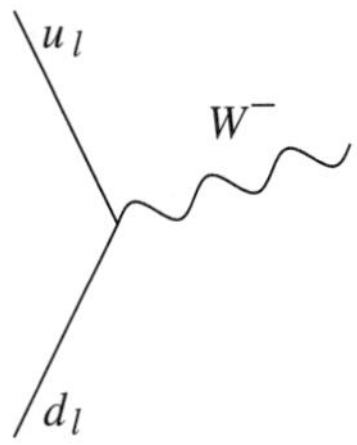

as the initial process. In Figure 8.5 we indicate the processes that are generated by using states related to the initial state by C, P, CP and CPT, whereas for CPT we choose states for which incoming and outgoing particles are interchanged.

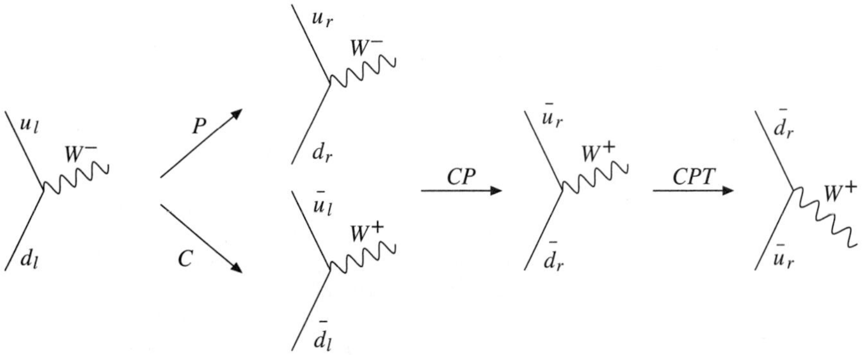

FIGURE 8.5 Family of processes generated by various transformations.

Comparing the processes in Figure 8.5 to those in Figure 8.4, we see that the processes generated by using states related to the initial state by C or P do not occur so that the weak interactions violate the C and P symmetries. It is also clear from the figures that QCD and QED, which contain diagrams with all helicity choices, have valid processes related by the C and P transformations.

The process generated by the CP transformation is a valid process that is contained in the terms given by the $SU(2)$ lowering operator, as indicated in Figure 8.4. *CP* symmetry is good, provided the couplings for the raising and lowering operator are identical. For the weak eigenstates this is guaranteed by the $SU(2)$ gauge symmetry—that is, the gauge theory has only one coupling constant. However, we always make measurements between mass eigenstates. In this case, the coupling for the raising operator is gV_{ud}, while for the lowering operator it is gV_{ud}^*. Therefore, *CP* can be violated if V_{ud} is complex. This is exactly what happens for some of the terms for three generations. Thus *CP* is violated in weak processes; however, the source is the mass generating sector, not the weak interactions per se. Measuring the *CP* violation is, however, not simple. Since rates depend on the squares of the couplings, a process given by one diagram will give the same rate as its *CP* partner for complex conjugate couplings. The manifestation of the *CP* violation therefore requires processes where several diagrams can interfere, and where the amplitudes for the several diagrams are not all just complex conjugates of each other.

Finally, note that the process related by the *CPT* symmetry in Figure 8.5 is contained in the same operator with which we started (the raising operator in Figure 8.4), so it automatically has the same coupling constant as the initial process. This depends only on the individual fields having both creation and annihilation operators. The *CPT* symmetry is therefore a good symmetry of the weak interactions. Embedding a weak vertex in a set of quantum events conserves this symmetry, since it is a symmetry of all of the other interactions as well. The necessity of a field containing both creation and annihilation processes that obey the *CPT* property is required by Lorentz invariance, as discussed in Section 3.9.

8.5 ■ CHARGED PION DECAY

We turn now to explicit weak hadronic decays. The simplest ones are those of π^+ and π^-. All the pion decays are interesting as a test of the ideas presented so far. The various decay branching ratios are given in Table 8.3. The decay $\pi^- \to \pi^0 e^- \bar{\nu}_e$ is very small, mainly because of the very small phase space for the decay. The smallness of $\pi^- \to e^- \bar{\nu}_e$ compared to $\pi^- \to \mu^- \bar{\nu}_\mu$ must be due to an interesting dynamical effect. This effect is called helicity suppression. The π^- lifetime is 2.603×10^{-8} sec, long enough that we can create beams of these particles.

We first consider the two decays in Table 8.3 that have only leptons in the final state. These are called leptonic decays. We indicate the charged lepton mass

TABLE 8.3 Pion Decay Branching Ratios.

Final state	Branching ratio
$\mu^- \bar{\nu}_\mu$	1.0
$e^- \bar{\nu}_e$	$1.230 \pm 0.004 \times 10^{-4}$
$\pi^0 e^- \bar{\nu}_e$	$1.025 \pm 0.034 \times 10^{-8}$

8.5 Charged Pion Decay

by m_l. The current matrix elements involved are $\langle 0| J_\mu^V + J_\mu^A |\pi^-\rangle$ for the π^-, and $\bar{u}_l \gamma_\mu (1 - \gamma_5) v_{\bar{\nu}}$ for the leptons. The matrix element is then

$$M_{fi} = \frac{G_F}{\sqrt{2}} [\bar{u}_l \gamma_\mu (1 - \gamma_5) v_{\bar{\nu}}] \langle 0| J_\mu^V + J_\mu^A |\pi^-\rangle V_{ud}. \tag{8.19}$$

The π^- matrix element involves an annihilation of the d and $\bar{u}$ in the π^-, yielding a W^-, which then makes a transition to the leptons. At the quark level, this process arises from the annihilation term contained in the raising operator of Figure 8.4. Note also the factor $V_{ud} = \cos\theta_c$ included for this doublet.

The pion current matrix element is dependent on the strong interactions, so it is not simple to calculate. It can, however, be written in terms of one constant. The vector current $\langle 0| J_\mu^V |\pi^-\rangle$ must be a momentum space Lorentz pseudovector since the pion is a pseudoscalar particle. However, to make a pseudovector requires using $\varepsilon_{\mu\nu\sigma\rho}$ and three independent vectors. Since there is only one, p_μ^π, there can be no contribution from the vector current. Thus, the only term contributing is the axial vector, which can be written in terms of one constant, f_π, as

$$\langle 0| J_\mu^A |\pi^-\rangle = f_\pi p_\mu^\pi. \tag{8.20}$$

We introduced this matrix element in Section 7.3.1. (The factor i appearing there is absorbed into f_π to simplify the notation).

We can now use the current matrix element to calculate the π decay rate. Substituting Eq. 8.20 into Eq. 8.19:

$$M_{fi} = \frac{G_F}{\sqrt{2}} f_\pi \cos\theta_c \bar{u}_l \not{p}_\pi (1 - \gamma_5) v_{\bar{\nu}}.$$

Writing $p_\mu^\pi = p_\mu^l + p_\mu^{\bar{\nu}}$, and using $\bar{u}_l \not{p}_l = m_l \bar{u}_l$, $\not{p}_\nu v_{\bar{\nu}} = 0$ for a massless neutrino, gives

$$M_{fi} = \frac{G_F}{\sqrt{2}} f_\pi m_l \cos\theta_c \bar{u}_l (1 - \gamma_5) v_{\bar{\nu}}. \tag{8.21}$$

Equation 8.21 indicates why the rate for $\pi^- \to \mu^- \bar{\nu}_\mu$ is so much greater than that for $\pi^- \to e^- \bar{\nu}_e$. The factor m_l in M_{fi} accounts for this, since $m_e \ll m_\mu$. For a current-current matrix element, any mix of vector and axial vector currents for the leptons would give this factor, whereas other operators (for example, a pseudoscalar-pseudoscalar interaction) would not. This is, therefore, a nice check that the Lorentz structure of the effective weak Hamiltonian is the current-current interaction. For vector or axial vector interactions of the leptons, the massless limit yields a particle and antiparticle of opposite helicity, which means an angular momentum along the direction of motion of the pair of one unit (which cannot come from a spinless initial state). Thus M_{fi} must vanish as $m_l \to 0$. For $m_l \neq 0$ the antineutrino remains right-handed with the charged lepton now forced to also be right-handed.

We can understand the m_l dependence in another way. The p_μ^π coming from the π^- current results in a matrix element that is proportional to the divergence of the lepton current. This divergence vanishes in the massless limit. The symmetry breaking, which generates the masses, also causes the current to be imperfectly conserved, and the nonconservation is proportional to m_l.

We now complete the rate calculation and also check what charged lepton helicity is allowed. Squaring

$$|M_{fi}|^2 = \frac{G_F^2 f_\pi^2 m_l^2}{2} \text{Tr}[v_{\bar{\nu}}\bar{v}_{\bar{\nu}}(1+\gamma_5)u_l\bar{u}_l(1-\gamma_5)]\cos^2\theta_c.$$

Summing over $\bar{\nu}$ spins (although only one helicity gives a nonzero result) and fixing the l helicity to be given by the spin vector s_l, we have

$$\sum_{\text{spins}} v_{\bar{\nu}}\bar{v}_{\bar{\nu}} = \slashed{p}_{\bar{\nu}} \quad \text{while} \quad u_l\bar{u}_l = (\slashed{p}_l + m_l)\left(\frac{1+\gamma_5\slashed{s}_l}{2}\right).$$

Inserting these and calculating the trace, the matrix element squared is

$$|M_{fi}|^2 = G_F^2 f_\pi^2 m_l^2 [p_{\bar{\nu}} \cdot p_l + m_l(p_{\bar{\nu}} \cdot s_l)]\cos^2\theta_c.$$

In the pion rest frame, for an l helicity ± 1, the various 4-vectors are

$$p_l = (E_l, |\vec{p}_l|\hat{n}), \quad p_{\bar{\nu}} = (1, -\hat{n})|\vec{p}_l|, \quad m_l s_l = (|\vec{p}_l|, E_l\hat{n})(\hat{s}\cdot\hat{n}).$$

Here $\hat{n}$ gives the direction of the charged lepton. The term $[p_{\bar{\nu}} \cdot p_l + m_l p_{\bar{\nu}} \cdot s_l] = p_{\bar{\nu}} \cdot p_l(1 + \hat{s}\cdot\hat{n})$. Thus we see that $\hat{s}\cdot\hat{n} = 1$ (right-handed l) is allowed, while the result for $\hat{s}\cdot\hat{n} = -1$ (left-handed l) vanishes. This is the one type of reaction where our expectation that all particles produced are left-handed, which is an $m_l = 0$ relativistic limit, is wrong. It does allow us to create fully polarized muon beams by selecting muons from π decay.

Using $p_{\bar{\nu}} \cdot p_l = \frac{1}{2}(m_\pi^2 - m_l^2)$, integrating over the phase space, and dividing by $2m_\pi$, we get

$$\Gamma_{\pi \to l\nu} = \frac{\cos^2\theta_c G_F^2 f_\pi^2 m_l^2 m_\pi}{8\pi}\left(1 - \frac{m_l^2}{m_\pi^2}\right)^2. \tag{8.22}$$

Since the other constants are known, we can measure $f_\pi = 0.94 m_\pi = 131$ MeV from the measured π^- lifetime and the branching ratio into $\mu^-\bar{\nu}_\mu$. Note that f_π has dimensions of mass.

The current-current form for the matrix element can be tested stringently by comparing the rates for $e^-\bar{\nu}_e$ to $\mu^-\bar{\nu}_\mu$. For an accurate comparison, radiative corrections should be included. The result of this calculation is

$$\frac{\Gamma_{\pi \to e\nu}}{\Gamma_{\pi \to \mu\nu}} = \left(\frac{m_e}{m_\mu}\right)^2 \left(\frac{m_\pi^2 - m_e^2}{m_\pi^2 - m_\mu^2}\right)^2 \left[1 - \frac{16.9\alpha}{\pi}\right].$$

8.5 Charged Pion Decay

The theoretical prediction for the ratio is 1.233×10^{-4}. The measured value is $1.230 \pm 0.004 \times 10^{-4}$.

8.5.1 ■ Conserved Vector Current

The dominant π^- decay rate to leptons was written in terms of a product of factors, $G_F^2 \cos^2 \theta_c f_\pi^2$. However, since f_π is not known a priori, we cannot easily use this expression to measure $\cos \theta_c$. We look next at the decay $\pi^- \to \pi^0 e^- \bar{\nu}_e$. This decay, called a semileptonic decay, will be seen to allow just such a measurement.

We now write the matrix element as

$$M_{fi} = \frac{G_F}{\sqrt{2}} \cos \theta_c \left\langle \pi^0 \right| J_\mu \left| \pi^- \right\rangle \bar{u}_e \gamma_\mu (1 - \gamma_5) v_{\bar{\nu}}.$$

J_μ is again $J_\mu^V + J_\mu^A$. However, the matrix element

$$\left\langle \pi^0 \right| J_\mu^A \left| \pi^- \right\rangle = 0,$$

since we do not have three independent momenta to dot into $\varepsilon_{\mu\nu\sigma\rho}$ to make an axial vector in momentum space. We define

$$\left\langle \pi^0 \right| J_\mu \left| \pi^- \right\rangle = \left\langle \pi^0 \right| J_\mu^V \left| \pi^- \right\rangle$$
$$= f_+^\pi (q^2) \left(p_\mu^{\pi^-} + p_\mu^{\pi^0} \right) + f_-^\pi (q^2) \left(p_\mu^{\pi^-} - p_\mu^{\pi^0} \right). \quad (8.23)$$

Here the momentum transfer q_μ is $p_\mu^{\pi^-} - p_\mu^{\pi^0}$. The expression above for the current, which is the most general vector we can write, involves two invariant functions called form factors. In the limit of a perfect isospin symmetry we can show that $f_-^\pi = 0$, and for $q^2 = 0$ calculate f_+^π, which is $\sqrt{2}$. Thus the matrix element is completely determined since q^2, which varies from $m_e^2 \leq q^2 \leq (m_{\pi^-} - m_{\pi^0})^2$, is always very small, allowing $f_+^\pi(q^2)$ to be taken equal to $f_+^\pi(0)$ to good approximation. We can then use the experimental rate to measure $\cos^2 \theta_c$, giving $\cos^2 \theta_c \simeq 0.95$.

Before turning to the rate calculation, we consider the constraints on the pion current matrix element. The vector current that contributes at the quark level is $\bar{\psi}_u \gamma_\mu \psi_d$, which is conserved in the isospin symmetry limit. Current conservation means that if we dot the momentum of the W into J_μ^V we get zero, that is,

$$\left(p_\mu^{\pi^-} - p_\mu^{\pi^0} \right) \left\langle \pi^0 \right| J_\mu^V \left| \pi^- \right\rangle = 0.$$

In the symmetry limit $(p_\mu^{\pi^-} - p_\mu^{\pi^0})(p_\mu^{\pi^-} + p_\mu^{\pi^0}) = m_{\pi^-}^2 - m_{\pi^0}^2 = 0$. Thus we must take $f_-^\pi = 0$, so that the second term in Eq. 8.23 does not spoil the current conservation.

We can, however, go further and predict f_+^π at $q^2 \simeq 0$. The two states π^- and π^0 are in the same isospin multiplet and the current operator is just the isospin raising operator, if we consider only isospin transformations. Applying the raising operator between the pion states, which are in an $I = 1$ multiplet, gives $\sqrt{2}$. This result is analogous to the situation where the matrix element for the electromagnetic current at $q^2 = 0$ is completely determined for a particle in terms of its electric charge, which equals the sum of the constituent charges for a composite object. The importance of having no momentum transfer is that the particle constituents are in the same spatial wave function before and after the W emission.

8.5.2 ■ Charge Operators

The value of the current matrix element at $q^2 = 0$ can be derived in a more formal way. The calculation illustrates more clearly what is meant by the particles carrying a non-Abelian "charge." To prove that $f_+^\pi(q^2 = 0) = \sqrt{2}$ we determine the current and charge operators. For a Lagrangian with a continuous symmetry, we can introduce a conserved current J_μ^a for each symmetry generator of index a, where

$$\frac{\partial}{\partial x_\mu} J_\mu^a = 0.$$

This implies that the associated charge $Q^a = \int d^3x\, J_0^a(x)$ is independent of time. For example, using the quark field operators,

$$Q^a = \int \psi^\dagger T_a \psi\, d^3x,$$

where ψ is a column vector of quark fields and T_a is a matrix representation of the generator. Using the commutation properties of the field operators we can show (although we will not prove this), that the Q^a operators are generators of the symmetry group $[Q^a, Q^b] = i f_{abc} Q^c$, where f_{abc} are the structure constants for the specific group. These relations, based on quark field operators, are true even if the space of states does not look like the free quark spectrum. For π^- decay, the charge operator related to the decay is the isospin raising operator Q_+. In general we know the effect of the raising operator on a state of given isospin (where we suppress the other labels for the state, which remain unchanged):

$$Q_+ |I, I_3\rangle = \sqrt{(I - I_3)(I + I_3 + 1)}\, |I, I_3 + 1\rangle.$$

This relation follows from the commutation relations for the charge operators. Taking a matrix element between a π^- of momentum $\vec{p}$ (for which $I = 1$, $I_3 = -1$, in the formula above) and a final π^0 with momentum $\vec{p}\,'$,

$$\left\langle \pi^0 \right| Q_+ \left| \pi^- \right\rangle = \sqrt{2} \left\langle \pi^0, \vec{p}\,' \,|\, \pi^0, \vec{p} \right\rangle.$$

In the matrix element calculation, we are as usual using a normalization of $2E$ particles per unit volume. Thus the corresponding state normalization is $\langle \pi^0, \vec{p}\,' \,|\, \pi^0, \vec{p} \rangle = 2EV$ for $\vec{p} = \vec{p}\,'$, 0 otherwise. We can write this normal-

8.5 Charged Pion Decay

ization in another way, making reference to the momenta only. Noting that for a large volume:

$$\int_V e^{i(\vec{p}'-\vec{p})\cdot x} d^3x = \begin{cases} V & \text{for } \vec{p}' = \vec{p}, \\ 0 & \text{otherwise,} \end{cases}$$

and that this integral can be written as $(2\pi)^3 \delta^3(\vec{p}' - \vec{p})$, we can write the state normalization as

$$\langle \pi^0, \vec{p}' | \pi^0, \vec{p} \rangle = 2E(2\pi)^3 \delta^3(\vec{p}' - \vec{p}).$$

We now look at the current matrix element directly:

$$\langle \pi^0 | J_0^V(x) | \pi^- \rangle = e^{i(p'-p)\cdot x} f_+^\pi(q^2)(E + E').$$

This is the generalization of Eq. 8.23, for which $x = 0$, to a general point in space-time. The exponential factor comes from writing the operator $J_\mu^V(x)$ as a translation applied to $J_\mu^V(0)$ and then applying the translation operators to the states. Integrating over d^3x to turn this into the charge matrix element gives

$$\langle \pi^0 | Q_+ | \pi^- \rangle = f_+^\pi(q^2)(E + E') \int e^{i(p'-p)\cdot x} d^3x.$$

In the isospin symmetry limit, Q_+ is independent of time so we can just take $t = 0$ in the integral. This gives

$$\langle \pi^0 | Q_+ | \pi^- \rangle = f_+^\pi(0) 2E (2\pi)^3 \delta^3(\vec{p}' - \vec{p}).$$

Comparing to the relation above for $\langle \pi^0 | Q_+ | \pi^- \rangle$, we have

$$f_+^\pi(q^2 = 0) = \sqrt{2}.$$

The determination of the current matrix element of the vector current (arising from the $d \to u$ transition) between states in an isospin multiplet in terms of the raising operator is quite general. For example, in neutron decay, $n \to p e^- \bar{\nu}_e$; we have for the vector current for q^2 near zero

$$\langle p | J_\mu^V | n \rangle = g_V(q^2) \bar{u}_p \gamma_\mu u_n.$$

The conserved vector current (called CVC for short) then implies immediately that $g_V(q^2 = 0) = 1$, the value of the matrix element of the isospin raising operator for the p and n doublet. This type of relation works also for nuclear beta decay linking nuclear states in an isospin multiplet. Such nuclear decays, in cases where only the vector current can contribute, provide the most accurate determination of $\cos\theta_c$.

8.5.3 ■ Rate for π^- Semileptonic Decay

Returning now to the $\pi^- \to \pi^0 e^- \bar{\nu}_e$ decay and using $f_+^\pi(q^2) = \sqrt{2}$ (that is, ignoring the very small q^2 variation of f_+) we can complete the rate calculation. The π^0 is very nonrelativistic in the π^- rest frame. Defining

$$\Delta = \text{maximum } e^- \text{ energy} = \frac{m_\pi^2 - m_{\pi^0}^2 + m_e^2}{2m_{\pi^-}} \quad \text{and} \quad x = \frac{m_e}{\Delta} \simeq \frac{1}{9},$$

it is reasonably straightforward to do the calculation in the limit where we ignore m_e. In this case the e^- energy spectrum and rate are given by

$$\frac{d\Gamma}{dE_{e^-}} = \frac{G_F^2 \cos^2 \theta_c}{\pi^3} E_{e^-}^2 (\Delta - E_{e^-})^2, \quad \Gamma = \frac{G_F^2 \cos^2 \theta_c \Delta^5}{30\pi^3}.$$

Unlike the decay spectrum for e^- in μ^- decay, the e^- spectrum here is symmetric about the average energy.

An exact calculation gives

$$\Gamma = \frac{G_F^2 \cos^2 \theta_c \Delta^5}{30\pi^3}$$

$$\times \left[\sqrt{1-x^2} \left(1 - \frac{9x^2}{2} - 4x^4 \right) + \frac{15}{2} x^4 \log\left(\frac{1+\sqrt{1-x^2}}{x} \right) \right]. \quad (8.24)$$

The factor in the brackets $= 0.94$. Using the measured value of the branching ratio, which is $1.025 \pm 0.034 \times 10^{-8}$, and the predicted rate above, we can extract $\cos \theta_c = 0.975$ with an error equal to 0.016. This is in good agreement with the value for $\cos \theta_c$ presented in Section 8.4.1.

8.6 ■ STRANGENESS CHANGING CURRENT OPERATOR AND KAON DECAY

We turn next to the kaon system. This system has a rich set of decay phenomena, many of which have historically generated key ideas or tests of the Standard Model. We might expect to find two pairs of related kaons: K^+ and K^- that have similar decays and lifetimes; and a second similar pair, K^0 and $\bar{K}^0$. The first surprise is that there are three kinds of kaons from the point of view of weak decays: K^+ and K^-, which decay similarly, and K_S^0 and K_L^0, that are linear combinations of K^0 and $\bar{K}^0$. The weak interactions, as a consequence of the CKM mixing matrix, allow K^0 and $\bar{K}^0$ to make transitions to the same final states. In quantum mechanics, initially degenerate states will mix completely even due to small perturbations, and this happens in the K^0 and $\bar{K}^0$ system. In the limit where CP is conserved we can immediately derive the K^0 and $\bar{K}^0$ mix that makes up the K_S^0 and K_L^0: Namely, the CP even and odd linear combinations of K^0 and $\bar{K}^0$ cannot make transitions into each other and so must be the eigenstates. We calculate the CP transformation based on the states defined in Section 5.2 and with the

8.6 Strangeness Changing Current Operator and Kaon Decay

same phase conventions as for the G parity discussion in Section 5.5.1. The CP relations are then, for a few states:

$$CP\,|K^0\rangle = |\bar{K}^0\rangle, \qquad CP\,|\pi^+\rangle = |\pi^-\rangle,$$
$$CP\,|K^+\rangle = -|K^-\rangle, \qquad CP\,|\pi^0\rangle = -|\pi^0\rangle. \tag{8.25}$$

It turns out that K_S^0 (short lifetime) is the (nearly) CP even state and K_L^0 (long lifetime) the (nearly) odd state. Thus:

$$|K_S^0\rangle \simeq \frac{|K^0\rangle + |\bar{K}^0\rangle}{\sqrt{2}} \quad \text{and} \quad |K_L^0\rangle \simeq \frac{|K^0\rangle - |\bar{K}^0\rangle}{\sqrt{2}}.$$

The lifetimes of the three types of kaons are:

$$\tau_{K^+} = \tau_{K^-} = 1.24 \times 10^{-8} \text{ sec}, \quad \tau_{K_S^0} = 0.894 \times 10^{-10} \text{ sec}, \quad \tau_{K_L^0} = 5.17 \times 10^{-8} \text{ sec}.$$

The K_L^0 and K_S^0 also have a slight mass difference: $m_{K_L^0} - m_{K_S^0} = 3.49 \times 10^{-12}$ MeV. This is a tiny number $\simeq \frac{1}{2}\Gamma_{K_S^0}$.

The mean flight distance for a decaying particle is $\beta\gamma\tau$, where

$$\beta = \text{velocity} = \frac{P}{E}, \quad \gamma = \frac{E}{m}.$$

For $\beta\gamma = 1$, the K_S^0 flies on average about 3 cm, while the other kaons fly a much longer distance. These distances are large enough to allow the creation of kaon beams and many detailed experimental measurements.

We list some of the interesting branching ratios for the various kaons in Table 8.4. First, notice the decays where the current-current matrix element involves a lepton pair for one current and a hadronic matrix element for the other current. At the quark level the hadronic current is

$$\left[\bar{\psi}_u \gamma_\mu (1 - \gamma_5)\psi_s\right] \sin\theta_c = \left(J_\mu^V + J_\mu^A\right) \sin\theta_c.$$

Within the $SU(3)$ flavor space the current transforms the $s \to u$. This corresponds to one of the $SU(3)$ raising operators (the V spin operator of Section 5.6). This weak isospin current operator is therefore an $SU(3)$ flavor current in the same octet as the current responsible for π^- decay. It generates analogous decays to those in the π^- system.

The simplest decay is given by the annihilation of the quarks in the K^-, which is an $|s, \bar{u}\rangle$ state. This is determined by the axial current, as in the π^- system. The rate for $K^- \to l\bar{\nu}_l$ can be obtained from the π^- decay rate formula by replacing m_π with m_K, and $f_\pi \cos\theta_c$ by $f_K \sin\theta_c$. We again expect that $e^-\bar{\nu}_e$ is heavily suppressed relative to $\mu^-\bar{\nu}_\mu$, in agreement with the data. Using the measured K^- branching ratio and the value of $\sin\theta_c = 0.221$ allows the extraction of the value

TABLE 8.4 Some Interesting K Decay Branching Ratios.

K Type	Mass (MeV)	Channel	Branching ratio
K^-	493.7	$\mu^- \bar{\nu}_\mu$	0.634
		$e^- \bar{\nu}_e$	1.55×10^{-5}
		$\pi^0 e^- \bar{\nu}_e$	0.049
		$\pi^0 \mu^- \bar{\nu}_\mu$	0.033
		$\pi^- \pi^0$	0.211
		$\pi^+ \pi^- \pi^-$	0.056
		$\pi^- \pi^0 \pi^0$	0.017
K_L^0	497.7	$\pi^\pm \mu^\mp \nu_\mu$	0.272
		$\pi^\pm e^\mp \nu_e$	0.388
		$\pi^+ \pi^-$	2.08×10^{-3}
		$\pi^0 \pi^0$	0.94×10^{-3}
		$\pi^+ \pi^- \pi^0$	0.126
		$\pi^0 \pi^0 \pi^0$	0.211
		$\gamma\gamma$	6.0×10^{-4}
		$\mu^+ \mu^-$	7.3×10^{-9}
K_S^0	497.7	$\pi^+ \pi^-$	0.686
		$\pi^0 \pi^0$	0.314
		$\pi^\pm e^\mp \nu_e$	7×10^{-4}
		$\pi^\pm \mu^\mp \nu_\mu$	5×10^{-4}
		$\gamma\gamma$	2.5×10^{-6}

for f_K, which is

$$\frac{f_K}{f_\pi} = 1.22, \quad \text{or } f_K = 160 \text{ MeV}. \tag{8.26}$$

For perfect $SU(3)$ symmetry, we expect the ratio in Eq. 8.26 to be $= 1$. We cannot use this decay rate to make a precise measurement of $\sin\theta_c$, unless the degree of $SU(3)$ violation can be calculated. Calculations using the lattice gauge technique, briefly described in Section 7.6, predict both meson decay constants with uncertainties of a few percent. As such calculations are improved, they may eventually lead to the most accurate value for $\sin\theta_c$.

The next simplest decays are the semileptonic decays analogous to the $\pi^- \to \pi^0 e^- \bar{\nu}_e$ decay. There are several such analogous decays, for example:

$$K^- \to \pi^0 e^- \bar{\nu}_e, \quad K^- \to \pi^0 \mu^- \bar{\nu}_\mu, \quad \bar{K}^0 \to \pi^+ e^- \bar{\nu}_e,$$
$$\bar{K}^0 \to \pi^+ \mu^- \bar{\nu}_\mu, \quad K^0 \to \pi^- e^+ \nu_e, \quad K^0 \to \pi^- \mu^+ \nu_\mu.$$

8.6 Strangeness Changing Current Operator and Kaon Decay

To be specific, we will focus on $K^- \to \pi^0 e^- \bar{\nu}_e$. For this decay,

$$M_{fi} = \frac{G_F}{\sqrt{2}} \sin\theta_c \langle \pi^0 | J_\mu^V | K^- \rangle \bar{u}_e \gamma_\mu (1-\gamma_5) v_{\bar{\nu}_e}. \tag{8.27}$$

The most general Lorentz Invariant result for the current matrix element is

$$\langle \pi^0 | J_\mu^V | K^- \rangle = f_+^{K^-}(q^2)(p_\mu^K + p_\mu^\pi) + f_-^{K^-}(q^2)(p_\mu^K - p_\mu^\pi).$$

We can predict $f_+^{K^-}(q^2 = 0)$ in the limit of exact $SU(3)$ symmetry. This matrix element of the appropriate $SU(3)$ raising operator can be evaluated, as was done for the analogous pion transitions in Sections 8.5.1 and 8.5.2. It is simple to figure out the value by just looking at what happens in the symmetry limit at $q^2 \to 0$. In this case the current just changes one quark type to another, leaving the state unchanged, so that the system emerges as a different meson in the same multiplet. The processes for $\pi^- \to \pi^0$, $K^- \to \pi^0$, and $\bar{K}^0 \to \pi^+$ are shown in Figure 8.6. In the $SU(3)$ symmetry limit, $f_+^{K^-}(q^2 = 0) = 1/\sqrt{2}$.

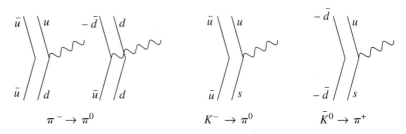

FIGURE 8.6 Analogous transitions responsible for meson semi-leptonic decays. Final states must be projected onto $\pi^0 = (|u,\bar{u}\rangle - |d,\bar{d}\rangle)/\sqrt{2}$ or $\pi^+ = -|u,\bar{d}\rangle$. Note that $\bar{K}^0 = -|s,\bar{d}\rangle$, and there is a minus sign in the $\bar{u} \to \bar{d}$ transition for $\pi^- \to \pi^0$ since the antiquark isospin states are $-|\bar{d}\rangle$ and $|\bar{u}\rangle$. Adding the two $\pi^- \to \pi^0$ terms coherently gives the $SU(3)$ symmetry predictions $f_+^\pi = \sqrt{2}$, $f_+^{K^-} = 1/\sqrt{2}$, $f_+^{\bar{K}^0} = 1$.

8.6.1 ■ Vector Dominance Model for Kaon to Pion Current Matrix Element

We can make a more detailed model for the $K^- \to \pi^0$ current matrix element using vector meson dominance. This model was used for the electromagnetic current coupling to charged pions at low q^2 in Chapter 6; we now extend it to the full $SU(3)$ multiplet of current transitions. The relevant diagram for the current matrix element is shown in Figure 8.7.

The ingredients to calculate the current matrix element are:

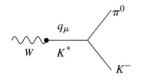

FIGURE 8.7 Vector dominance picture for current matrix element for $K^- \to \pi^0$.

1. W, K^* vertex, which is $g_{WK^*} e_\mu^{K^*} m_{K^*}^2$.
2. K^* propagator.
3. K^* vertex coupling for $K^- \to \pi^0$, which is $g_{K^*K^-\pi^0} e_\mu^{K^*}(p_\mu^K + p_\mu^\pi)$.
4. Sum over intermediate K^* spins.

Putting these together:

$$\langle \pi^0 | J_\mu^V | K^- \rangle = \frac{g_{WK^*} m_{K^*}^2 g_{K^*K^-\pi^0}}{q^2 - m_{K^*}^2} \left[-g_{\mu\nu} + \frac{q_\mu q_\nu}{m_{K^*}^2} \right] (p_\nu^K + p_\nu^\pi)$$

$$= g_{WK^*} g_{K^*K^-\pi^0} \left[\frac{m_{K^*}^2}{m_{K^*}^2 - q^2} \right]$$

$$\times \left((p_\mu^K + p_\mu^\pi) - \frac{(m_K^2 - m_\pi^2)}{m_{K^*}^2} (p_\mu^K - p_\mu^\pi) \right).$$

In the $SU(3)$ symmetry limit, at $q^2 = 0$, we expect from CVC that

$$g_{WK^*} g_{K^*K^-\pi^0} = \frac{1}{\sqrt{2}}.$$

We can check this relation if we assume that the couplings don't change from $q^2 = m_{K^*}^2$ to $q^2 = 0$. The couplings involved can then be measured separately using other processes. We can measure g_{WK^*} in the decay $\tau \to K^* \nu_\tau$ provided $\sin \theta_c$ is known separately. Using the analog of Eq. 8.13:

$$\Gamma_{\tau^- \to K^* \nu_\tau} = \frac{G_F^2 g_{WK^*}^2}{16\pi} \left(\frac{m_{K^*}}{m_\tau} \right)^2 (m_\tau^2 - m_{K^*}^2)^2 \left[m_\tau + \frac{2m_{K^*}^2}{m_\tau} \right] \sin^2 \theta_c.$$

The measured K^* branching ratio is $1.29 \pm 0.05\%$. Taking $\sin \theta_c = 0.221 \pm 0.002$ gives $g_{WK^*} = 0.24 \pm 0.01$. Using the values from Table 5.5, we can calculate $g_{K^*K^-\pi^0} = 3.17$. The product $g_{WK^*} g_{K^*K^-\pi^0} = 0.76 \pm 0.03$ is very close to $1/\sqrt{2}$, which is a nice consistency check. Note that the near equality of g_{WK^*} and $g_{W\rho}$ is an additional check on the value of $\sin \theta_c$, within the uncertainties of $SU(3)$ breaking.

The vector dominance model predicts the q^2 dependence of both form factors:

$$f_+^{K^-}(q^2) \simeq f_+^{K^-}(0) \left[1 + \frac{q^2}{m_{K^*}^2} \right], \quad \text{for } q^2 \ll m_{K^*}^2, \quad \text{and}$$

$$\frac{f_-^{K^-}(q^2)}{f_+^{K^-}(q^2)} = \frac{-(m_K^2 - m_\pi^2)}{m_{K^*}^2} = -0.28.$$

Both of these are in fairly good agreement with the experimentally measured parameters in the decay. Note that the predicted dependence on q^2 is quite weak, since q^2 is $\ll m_{K^*}^2$ over the whole decay phase space.

Given our specific model for the hadronic current matrix element, we now return to the full matrix element of Eq. 8.27. The term involving $f_-^{K^-}(q^2) q_\mu$ does not contribute in the limit where the leptonic current is conserved. For finite lepton

8.6 Strangeness Changing Current Operator and Kaon Decay

masses it gives a term proportional to m_l, which is completely negligible for the electron. Thus $f_-^{K^-}(q^2)$ is actually measured by looking at the $\mu \bar{\nu}_\mu$ final state.

Since $q^2 = m_K^2 + m_\pi^2 - 2m_K E_\pi$, it is convenient in calculating the rate to integrate over the electron and neutrino variables, leaving E_π (and therefore q^2) fixed. It is adequate to assume the electron is fully relativistic. Carrying out the integrals over the lepton variables gives

$$\frac{d\Gamma}{dE_\pi} = \frac{G_F^2 \sin^2 \theta_c}{6\pi^3} \left(\frac{m_K}{2}\right) |\vec{p}_\pi|^3 \left|f_+^{K^-}(q^2)\right|^2.$$

The q^2 dependence of $f_+^{K^-}$ can be measured from this expression using data. The vector dominance model, which we use below to calculate the rate, predicts (including the CVC constraint at $q^2 = 0$)

$$\left|f_+^{K^-}(q^2)\right|^2 = \frac{1}{2}\left(1 + \frac{2q^2}{m_{K^*}^2}\right),$$

to lowest order in q^2. We can get a rough estimate of the rate by assuming that the π is always relativistic, that is, taking

$$|\vec{p}_\pi| = E_\pi \quad \text{and} \quad E_\pi^{\max} = \frac{m_K}{2}.$$

With this approximation the integral over E_π is simple, giving

$$\Gamma = \frac{G_F^2 \sin^2 \theta_c}{48\pi^3} \left(\frac{m_K}{2}\right)^5 \left[1 + \frac{2}{5}\frac{m_K^2}{m_{K^*}^2}\right].$$

The second term in the brackets comes from the term proportional to q^2 in the form-factor and increases the rate by about 12%. Doing the integral over E_π exactly:

$$\Gamma_{K^- \to \pi^0 e^- \bar{\nu}_e} = 0.578 \frac{G_F^2 \sin^2 \theta_c}{48\pi^3} \left(\frac{m_K}{2}\right)^5 \left[1 + \frac{1.4}{5}\frac{m_K^2}{m_{K^*}^2}\right]$$

$$= 0.628 \frac{G_F^2 \sin^2 \theta_c}{48\pi^3} \left(\frac{m_K}{2}\right)^5. \qquad (8.28)$$

Using the measured K^- lifetime, the branching ratio into $\pi^0 e^- \bar{\nu}_e$, and the predicted rate in Eq. 8.28, gives the measured result

$$\sin^2 \theta_c = 0.0490 \pm 0.0007 \quad \text{and} \quad \sin \theta_c = 0.221 \pm 0.002.$$

To extract these numbers, an estimate of the radiative corrections and the small effect of $SU(3)$ violation on the rate is included.

We used the $SU(3)$ flavor symmetry in deriving the value of $f_+^{K^-}(q^2 = 0)$. To measure $\sin\theta_c$ in several other processes will give us more confidence in the method, which is based on the $SU(3)$ version of CVC. We can do this using the semileptonic decays of the strange baryons. We will discuss these very briefly.

If we consider a semileptonic baryon decay $B_1 \to B_2 + e^- + \bar{\nu}_e$, we can write the matrix element:

$$M_{fi} = \frac{G_F}{\sqrt{2}} \left\langle B_2 \left| J_\mu^V + J_\mu^A \right| B_1 \right\rangle \bar{u}_e \gamma_\mu (1 - \gamma_5) v_{\bar{\nu}_e} \begin{cases} \cos\theta_c, \\ \sin\theta_c. \end{cases}$$

Here $\cos\theta_c$ arises for $d \to u$ transitions and $\sin\theta_c$ for $s \to u$ transitions. Examples of the latter are $\Lambda^0 \to pe^-\bar{\nu}_e$ and $\Xi^0 \to \Sigma^+ e^- \bar{\nu}_e$. Leaving out terms proportional to $q_\mu = P_\mu^{B_1} - P_\mu^{B_2}$ in the hadronic current, since they yield terms which are proportional to m_e after dotting into the lepton current, and small magnetic moment type terms, we can write in general:

$$\left\langle B_2 \left| J_\mu^V + J_\mu^A \right| B_1 \right\rangle = \bar{u}_{B_2} \gamma_\mu (g_V - g_A \gamma_5) u_{B_1}.$$

The vector and axial form factors g_V and g_A are functions of q^2 and depend on the baryons B_1 and B_2. Since q^2 is reasonably small, we can approximate g_V and g_A as constants that depend on the baryon choices B_1 and B_2.

In the B_1 rest frame the B_2 is rather nonrelativistic. In this limit:

$$\Gamma_{B_1 \to B_2 e^- \bar{\nu}_e} = \frac{G_F^2 (m_{B_1} - m_{B_2})^5}{60\pi^3} \left[g_V^2 + 3g_A^2 \right] \begin{cases} \cos^2\theta_c, \\ \sin^2\theta_c. \end{cases}$$

The constant g_V can be predicted using CVC. It depends on the appropriate isospin or V spin raising operator matrix element within the octet of baryon states. Unfortunately the various g_A cannot be predicted. We can, however, use the Wigner–Eckart theorem to write the g_A in terms of Clebsch–Gordan coefficients and two constants. This comes about because J_μ^A transforms as an octet, and in the coupling octet $\to$ octet via an octet operator, two reduced matrix elements come from the two different octets that arise in the decomposition of $8 \otimes 8$ into irreducible $SU(3)$ representations. Since there are several semileptonic baryon decays, we can fit the rates in terms of the two octet couplings needed for the various g_A, and $\sin\theta_c$, the constant we want to measure. These decays give a $\sin\theta_c$ value in good agreement with the value extracted from kaon decay. The theoretical uncertainties in extracting $\sin\theta_c$ from the various measurements limits the comparison to about the 3% level.

8.6.2 ■ Operator for K Decay to All Hadronic Final States

In Table 8.4, we see that K mesons also decay into final states containing only hadrons. These arise through the piece of the weak effective Hamiltonian with current-current operators involving only quarks. Because of the CKM matrix, there are 9×9 such terms in $J_\mu J_\mu^\dagger$ after expanding in terms of individual quark

8.6 Strangeness Changing Current Operator and Kaon Decay

fields (to be summed also over colors). The current pairs relevant to a given physical process are determined by the change in flavor in the process, since the weak interactions are the only ones that change one quark flavor into another. However, in addition to the flavor change, we will have strong interactions between the constituents as the system evolves from the initial to the final state. This makes it difficult to actually calculate the rates.

For hadronic kaon decay we expect the relevant currents to change $s \to u$ (or $\bar{s} \to \bar{u}$; to be specific we look at $s \to u$) with the resulting W absorbed at a second vertex involving only quarks and antiquarks. The choices for the second vertex provide three types of diagrams:

1. W creates a new pair, $d\bar{u}$, as constrained by charge and energy conservation in the final state. The light antiquark in the kaon does not participate in the short distance W exchange. This diagram is called the spectator diagram and is analogous to the diagram for μ decay or $K^- \to \pi^0 e^- \bar{\nu}_e$.

2. The light antiquark within the kaon is involved in the short distance process, either by absorbing the W or annihilating the s quark. These are called weak scattering and annihilation diagrams, respectively. For the annihilation diagram, the W will produce a $d\bar{u}$ pair if we require only hadrons in the final state. This is analogous to the $K^- \to \mu^- \bar{\nu}_\mu$ process.

3. The W can reattach to the same quark that emitted it, with the quark making a transition to a new flavor. Since the quark states we are using are mass eigenstates, they propagate with no flavor change unless external interactions intervene. This diagram therefore only provides a real transition if we internally attach another interaction, for example, the exchange of a gluon with the light antiquark in the state. This diagram does not follow the simple expectation that only $s \to u$ contributes.

These three classes of diagrams are shown in Figure 8.8 for both K^- and $\bar{K}^0$ mesons. Also indicated are the final isospins that are allowed by each diagram. We can calculate the isospin by just adding together the isospin of the various quarks and antiquarks that appear in the final state. Diagrams (1) and (2) in Figure 8.8 come from the term in the weak effective Hamiltonian:

$$\frac{G_F}{\sqrt{2}} \left[\bar{\psi}_d \gamma_\mu (1 - \gamma_5) \psi_u \right] \left[\bar{\psi}_u \gamma_\mu (1 - \gamma_5) \psi_s \right] \sin\theta_c \cos\theta_c. \qquad (8.29)$$

We must add the effect of strong interactions to get the real initial and final states and the full interaction during the evolution of the system. Diagram (3) has some of the strong interactions already indicated; in this sense it is already a higher-order diagram. Since the gluon interactions are strong at the kaon mass scale, this type of diagram is not necessarily small. It is called a penguin diagram.

The hadronic final states produced in K decay are 2π or 3π final states. The discussion in Section 5.4.1 indicates that both types of decay can occur only if parity is violated by the decay interaction. By angular momentum conservation, the

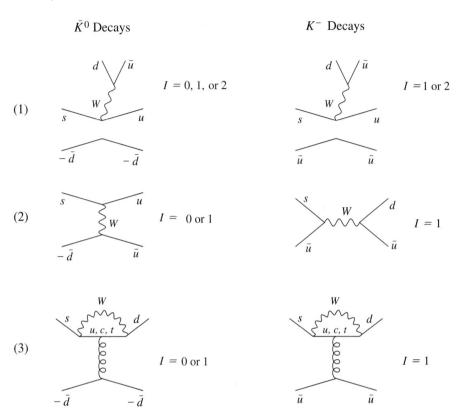

FIGURE 8.8 Types of diagrams at the quark level contributing to hadronic kaon decay. Diagrams differ by how the exchanged W is absorbed.

2π system has $J = 0$, which is a fully symmetric state. The isospin states that are possible for 2π are $I = 0, 1,$ or 2. Bose symmetry for the pions then tells us which isospin state can go with a particular spatial state. Since $J = 0$ is symmetric, we must choose symmetric isospin states, which for 2π are the states with $I = 0$ or 2 (the $I = 1$ state is antisymmetric). The specific decay channels are $K^- \to \pi^-\pi^0$, $\bar{K}^0 \to \pi^+\pi^-$ and $\pi^0\pi^0$. Since $\pi^-\pi^0$ has $I_3 = -1$, this final state cannot have $I = 0$ and it must be in an $I = 2$ state.

The decay to 3π must yield a $J = 0$ final state with very little available phase space. The form of the matrix element is similar to that of $\eta \to 3\pi$, but the phase space is even more limited. Therefore, we expect a fully symmetric s-wave final state to dominate the amplitude. In this case we can again find the final state isospin, since the isospin state needs to be fully symmetric. In addition, based on the diagrams in Figure 8.8, it must have $I \leq 2$. The only state of 3π satisfying this is an $I = 1$ state. The expectations for 2π and 3π decays are listed in Table 8.5. For 3π we assume the fully symmetric s-wave final state, which is a good approximation.

8.6 Strangeness Changing Current Operator and Kaon Decay

TABLE 8.5 Isospin Properties of Hadronic K Decay.

Decay type	Final isospin	Contributing operators
$\bar{K}^0 \to \pi\pi$	0, 2	$\Delta I = \frac{1}{2}, \frac{3}{2}$
$K^- \to \pi^-\pi^0$	2	$\Delta I = \frac{3}{2}$
$\bar{K}^0 \to 3\pi$	1	$\Delta I = \frac{1}{2}, \frac{3}{2}$
$K^- \to 3\pi$	1	$\Delta I = \frac{1}{2}, \frac{3}{2}$

Last, we can write the weak effective Hamiltonian as an operator with specific isospin transformation properties by adding the isospins of the two currents vectorially. The operator specified in Eq. 8.29 transforms as a mix of $\Delta I = \frac{1}{2}$, $\Delta I_3 = -\frac{1}{2}$ (that is, like an $I = \frac{1}{2}$, $I_3 = -\frac{1}{2}$ state under isospin rotations), and $\Delta I = \frac{3}{2}$, $\Delta I_3 = -\frac{1}{2}$ operators (given by the top two processes in Figure 8.8), while the penguin operator (the bottom process in Figure 8.8) transforms as $\Delta I = \frac{1}{2}$, $\Delta I_3 = -\frac{1}{2}$. Since strong interactions conserve isospin, the change in isospin that is possible in the process is governed by the weak operators. For a kaon doublet, for example K^- and $\bar{K}^0$, the same operators are responsible for the decay and we can use the Wigner–Eckart theorem to write the full amplitudes in terms of two reduced amplitudes (for $\Delta I = \frac{1}{2}$ and $\Delta I = \frac{3}{2}$) and Clebsch–Gordan coefficients for the 2π or the 3π channels. Denoting the operator as $O = O_{\Delta I = \frac{3}{2}, \Delta I_3 = -\frac{1}{2}} + O_{\Delta I = \frac{1}{2}, \Delta I_3 = -\frac{1}{2}}$, the Wigner–Eckart theorem gives:

$$O\left|K^-\right\rangle = O\left|I=\frac{1}{2}, I_3=-\frac{1}{2}\right\rangle$$

$$= a^{I=1}_{\Delta I=\frac{1}{2}}|1,-1\rangle + a^{I=2}_{\Delta I=\frac{3}{2}}\frac{\sqrt{3}}{2}|2,-1\rangle + \frac{a^{I=1}_{\Delta I=\frac{3}{2}}}{2}|1,-1\rangle$$

$$O\left|\bar{K}^0\right\rangle = O\left|I=\frac{1}{2}, I_3=\frac{1}{2}\right\rangle$$

$$= \frac{a^{I=1}_{\Delta I=\frac{1}{2}}}{\sqrt{2}}|1,0\rangle - \frac{a^{I=1}_{\Delta I=\frac{1}{2}}}{\sqrt{2}}|0,0\rangle + \frac{a^{I=2}_{\Delta I=\frac{3}{2}}}{\sqrt{2}}|2,0\rangle - \frac{a^{I=1}_{\Delta I=\frac{3}{2}}}{\sqrt{2}}|1,0\rangle . \quad (8.30)$$

Each a will be an amplitude when matrix elements are taken with given final states.

To project onto the 2π states, we use the isospin decomposition:

$$|2,-1\rangle = \frac{1}{\sqrt{2}}\left(\left|\pi^-,\pi^0\right\rangle + \left|\pi^0,\pi^-\right\rangle\right)$$

$$|2, 0\rangle = \tfrac{1}{\sqrt{6}}\left(\left|\pi^+, \pi^-\right\rangle + \left|\pi^-, \pi^+\right\rangle\right) + \sqrt{\tfrac{2}{3}}\left|\pi^0, \pi^0\right\rangle$$

$$|0, 0\rangle = \tfrac{1}{\sqrt{3}}\left(\left|\pi^+, \pi^-\right\rangle + \left|\pi^-, \pi^+\right\rangle\right) - \tfrac{1}{\sqrt{3}}\left|\pi^0, \pi^0\right\rangle.$$

Using these, we now write the kaon to 2π matrix elements in terms of two complex amplitudes (which we label by the isospin of the final state):

$$M_{K^- \to \pi^- \pi^0} = \sqrt{\tfrac{3}{2}} a_2$$

$$M_{\bar{K}^0 \to \pi^+ \pi^-} = \sqrt{\tfrac{1}{3}} a_2 + \sqrt{\tfrac{2}{3}} a_0$$

$$M_{\bar{K}^0 \to \pi^0 \pi^0} = \tfrac{2}{\sqrt{3}} a_2 - \sqrt{\tfrac{2}{3}} a_0. \qquad (8.31)$$

We take the conventional choice of signs for a_2, a_0 in this expression. The amplitudes are symmetrized for $\pi^0 \pi^0$ and, therefore, when integrating over the final phase space, we must include a factor of $\tfrac{1}{2}$ after integrating over all directions because of the identical particles in the final state. We look at the 3π decays in Section 8.9.

We will next look at a simple model for the 2π decays. It is what one might try, given knowledge of the decays involving leptons. It will not give the correct result, but provides an interesting example for comparison. The model assumes simple factorization of amplitudes for each current and that Eq. 8.29 provides the full weak effective Hamiltonian density, whose action to lowest order gives the decay matrix element. The diagrams are shown in Figure 8.9 for K^- decay.

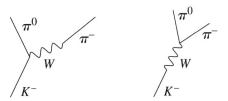

FIGURE 8.9 Simple factorization model for K^- decay. The second diagram vanishes by CVC in the isospin symmetry limit.

The only diagram that contributes, if we ignore the $\pi^- - \pi^0$ mass difference, is the diagram where $K^- \to \pi^0$ for one current matrix element and π^- is created from the vacuum by the other. For simplicity, since the model will not be adequate anyway, we will take $\langle \pi^0 | J_\mu | K^- \rangle \simeq f_+^{K^-}(q^2 = 0)(p_\mu^{K^-} + p_\mu^{\pi^0})$. Also, $\langle \pi^- | J_\mu | 0 \rangle = f_\pi p_\mu^{\pi^-}$. Using these current matrix elements in Eq. 8.29 gives:

$$M_{K^- \to \pi^- \pi^0} = \frac{G_F}{\sqrt{2}} \sin\theta_c \cos\theta_c f_\pi f_+^{K^-}(0) \left(m_K^2 - m_\pi^2\right).$$

8.6 Strangeness Changing Current Operator and Kaon Decay

An analogous calculation for $M_{\bar{K}^0 \to \pi^+\pi^-}$ would replace $f_+^{K^-}(0)$ by $f_+^{\bar{K}^0}(0)$. This model predicts that $M_{\bar{K}^0 \to \pi^0\pi^0} = 0$, which corresponds to a specific relation between the isospin amplitudes, namely $a_0 = \sqrt{2}a_2$ in Eq. 8.31. Making this substitution gives the relation

$$\frac{M_{\bar{K}^0 \to \pi^+\pi^-}}{M_{K^- \to \pi^-\pi^0}} = \sqrt{2},$$

which agrees with a direct calculation using the ratio

$$\frac{f_+^{\bar{K}^0}(0)}{f_+^{K^-}(0)} = \sqrt{2}.$$

Thus, this model contains both isospin changing operators with a fixed relationship.

Taking $m_K^2 - m_\pi^2 \simeq m_K^2$ and integrating over the phase space we get from this model:

$$\Gamma_{K^- \to \pi^-\pi^0} \simeq \frac{G_F^2 \sin^2\theta_c \cos^2\theta_c f_\pi^2 m_K^3}{64\pi}.$$

Using

$$\Gamma_{K^- \to \mu^-\bar{\nu}_\mu} \simeq \frac{G_F^2 f_K^2 \sin^2\theta_c m_K m_\mu^2}{8\pi},$$

we get the prediction:

$$\frac{\Gamma_{K^- \to \pi^-\pi^0}}{\Gamma_{K^- \to \mu^-\bar{\nu}_\mu}} \simeq \frac{1}{8}\left(\frac{f_\pi}{f_K}\right)^2 \left(\frac{m_K}{m_\mu}\right)^2 \cos^2\theta_c \simeq 1.5.$$

The real answer for this ratio is 0.33. The matrix element we are using for this $\Delta I = \frac{3}{2}$ process is too large by a little more than a factor of two. The real $\Delta I = \frac{3}{2}$ operator is therefore suppressed by the strong interactions relative to the model.

For the $\bar{K}^0$ decays, we see that the striking prediction of the model that $\bar{K}^0 \to \pi^0\pi^0$ is absent, is not correct. This prediction comes about through the cancellation of two real, in phase, $I = 0$ and $I = 2$ amplitudes. Including strong interactions, we expect that the amplitudes will not be real, which can spoil such a cancellation. In the present case this is, however, not the primary reason for the nonvanishing of the $\pi^0\pi^0$ amplitude. The model predicts that

$$\frac{\Gamma_{\bar{K}^0 \to \pi^+\pi^-}}{\Gamma_{K^- \to \pi^-\pi^0}} = 2.$$

The actual ratio is about 225, where we use the average of K_S^0 and K_L^0 to $\pi^+\pi^-$ to extract the $\bar{K}^0$ rate. This large ratio implies a large enhancement in the matrix element of the $\Delta I = \frac{1}{2}$ operator compared to naive expectations. From Eq. 8.31

and the measured rates, keeping only the $\Delta I = \frac{1}{2}$ part for $\bar{K}^0 \to \pi^+\pi^-$, we can estimate

$$\left|\frac{a_2}{a_0}\right| \simeq \frac{2}{3\sqrt{225}} = 0.044.$$

If we compare the measured amplitudes to the factorization calculation, the a_2 amplitude is suppressed by ~ 2.5 and the a_0 amplitude is enhanced by ~ 8.

Ignoring the a_2 amplitude, we predict that $\Gamma_{\bar{K}^0 \to \pi^0 \pi^0} = \frac{1}{2}\Gamma_{\bar{K}^0 \to \pi^+ \pi^-}$. This is quite close to the measured ratio of about 0.46 for both K_S^0 and K_L^0. The presence of the small a_2 amplitude reduces the $\pi^0\pi^0$ rate relative to the $\pi^+\pi^-$ rate by $\sim 10\%$.

A large enhancement in a given amplitude can arise as a result of a hadronic resonance. This is not the case here, since there is no low mass $I = 0$ resonance in the $\pi\pi$ system. Exactly how the strong interactions produce this enhancement is not understood quantitatively. It occurs in all the low energy hadronic weak decays involving the strange quark, not only kaon decay but also strange baryon decay, for example $\Lambda \to \pi^- p$ or $\pi^0 n$. It is called the $\Delta I = \frac{1}{2}$ rule since it is always tied to this operator.

Although we are unable to calculate the rates, we can deduce some interesting general constraints on the phases of the amplitudes. We turn to this next, considering a general weak decay of a pseudoscalar meson. We will return to the K decays later. Note that our discussion so far has been for the decays of the K^-, $\bar{K}^0$ isospin doublet; the more general discussion below will be needed to relate these to the K^+, K^0 doublet decays.

8.7 ■ GENERAL FRAMEWORK FOR WEAK DECAY OF PSEUDOSCALAR MESONS

We look at the general framework for describing the weak decay to hadrons of a pseudoscalar meson P in its rest frame. By angular momentum conservation and energy conservation the final state has $J = 0$ and $E = m_P$. The S matrix elements for the decay include strong interaction effects among the final state particles, resulting from the fact that we produce final state hadrons and not noninteracting quarks. We write generally:

$$S = S_0 - iM.$$

Here S_0 is the S matrix for the strong interactions, ignoring the weak interactions entirely. M is the contribution to S from the weak interactions acting to lowest order only, including, however, strong interactions acting to all orders. In the language of the iteration expansion for S of Chapter 2, S_0 is the sum of the terms containing the strong Hamiltonian to all orders, and M is the sum of all terms where the weak effective Hamiltonian appears only once. In cases where several weak diagrams contribute, as in Figure 8.8, each diagram makes a separate

8.7 General Framework for Weak Decay of Pseudoscalar Mesons

contribution to M. We have dropped terms in which the weak interactions appear more than once. Although this is a weak perturbation expansion, it is meant to be exact with regard to the strong interactions so that the outgoing particles interact significantly. Although we cannot solve the strong interaction problem, unitarity and the simplicity of the initial state provide constraints that allow some insight into the solution.

From the unitarity of S and S_0, we get to lowest order in M, by calculating $SS^\dagger$:

$$MS_0^\dagger = S_0 M^\dagger \quad \text{or} \quad M = S_0 M^\dagger S_0.$$

We next note that the pseudoscalar mesons that decay weakly (K, D, B mesons for the s, c, b quarks, respectively) are all eigenstates of the strong interactions—they would be stable in the absence of weak interactions. Thus, $\langle P|S_0|P\rangle = 1$ and $\langle n|S_0|P\rangle = 0$ for $|n\rangle \neq |P\rangle$. We can therefore ignore the last S_0 provided we are looking at an initial state $|P\rangle$. This gives $M = S_0 M^\dagger$. Note that M is not Hermitian (as it would be for weakly interacting particles only).

We next exploit angular momentum conservation for a final state basis of hadronic states with $J = 0$, $E = m_P$. We have not typically used angular momentum states, except for occasionally noting that the angular distribution in a decay could be understood from the point of view of angular momentum conservation. We specify the states produced in the decay with given particle content (for instance, some number of kaons and pions of given charges) and $J = 0$ as $|m_i\rangle$. These correspond to a spherically symmetric wave diverging from the origin. We normalize the flux of outgoing particles to correspond to the same event rate for each $|m_i\rangle$. We also consider a related set of states that correspond to a spherically symmetric wave starting in the distant past and converging on the origin. We write these as $|m_i, in\rangle$. The strong interaction S matrix $(S_0)_{ij} = \langle m_i | m_j, in\rangle = \langle m_i|S_0|m_j\rangle$. Thus S_0 is a unitary matrix that relates the two bases $|m_i\rangle$ and $|m_i, in\rangle$. If only elastic scattering is allowed, unitarity tells us that the flux in equals the flux out; therefore, $S_0 = e^{2i\delta}$, where we have considered only one state. This corresponds to the partial wave solution in nonrelativistic quantum mechanics, where δ is called the phase shift in the $J = 0$ angular momentum channel.

Relativistically, particles are destroyed and created so that the S_0 matrix will not be diagonal. However, by considering linear combinations of the states $|m_i\rangle$ we can find a new set $|n_i\rangle$, which diagonalize S_0. We call $|m_i\rangle$ the physical states (with particle content we detect) and $|n_i\rangle$ the strong scattering eigenstates (which contain mixtures with different particle content, however constrained, so that all $|m_i\rangle$ that mix into a given $|n_i\rangle$ have the same conserved strong quantum numbers). In the $|n_i\rangle$ basis, $\langle n_i|S_0|n_j\rangle = e^{2i\delta_{n_i}}\delta_{n_i n_j}$. Note that the eigenvalues of a unitary matrix are all just phase shifts, as above.

We now return to the weak scattering matrix and expand the relation between M and S_0 in the $|n_i\rangle$ basis. This gives

$$\langle n_i |M| P\rangle = e^{2i\delta_{n_i}} \langle n_i |M^\dagger| P\rangle = e^{2i\delta_{n_i}} \langle P |M| n_i\rangle^*.$$

We next use the *CPT* theorem, which M satisfies, giving (in the notation of Section 3.9)

$$\langle P|M|n_i\rangle^* = \langle CPTn_i|M|CPTP\rangle^*.$$

To simplify the notation, we write

$$|CPTP\rangle = |\bar{P}\rangle \quad \text{and} \quad |CPTn_i\rangle = |\bar{n}_i\rangle.$$

Since $|P\rangle$ is a pseudoscalar at rest, $|\bar{P}\rangle$ is its antiparticle at rest. $|\bar{n}_i\rangle$ is obtained from $|n_i\rangle$ by changing all particles to antiparticles and reversing all helicities. Thus,

$$\langle n_i|M|P\rangle = e^{2i\delta_{n_i}} \langle \bar{n}_i|M|\bar{P}\rangle^*.$$

We can repeat our entire analysis above starting with the state $|\bar{P}\rangle$. The *CPT* theorem applied to S_0 implies that $|\bar{n}_i\rangle$ are eigenstates with the same phase shifts as $|n_i\rangle$. This gives the complex conjugate of the equation above:

$$\langle \bar{n}_i|M|\bar{P}\rangle = e^{2i\delta_{n_i}} \langle n_i|M|P\rangle^*.$$

These equations imply that $|\langle \bar{n}_i|M|\bar{P}\rangle| = |\langle n_i|M|P\rangle|$, which we write as A_{n_i}. Using A_{n_i}, we can write in general:

$$\langle n_i|M|P\rangle = e^{i\delta_{n_i}} A_{n_i} e^{i\phi_{n_i}}$$
$$\langle \bar{n}_i|M|\bar{P}\rangle = e^{i\delta_{n_i}} A_{n_i} e^{-i\phi_{n_i}}. \tag{8.32}$$

The factor δ_{n_i} is called the strong phase, ϕ_{n_i} the weak phase. Having only an outgoing wave for the case of P decay, the strong phase shift is only half as great as for the strong scattering situation, which involves incoming and outgoing waves.

Although our expression above involves a number of unknown factors, its form allows some general conclusions:

1. We can show, from Eq. 8.32 above, that the lifetime of $|P\rangle$ and $|\bar{P}\rangle$ are the same. Normalizing $|A_{n_i}|^2$ to give the partial width for $|P\rangle \to |n_i\rangle$ directly, we have

$$\Gamma_P = \sum_i |\langle n_i|M|P\rangle|^2 = \sum_i \left|\langle \bar{n}_i|M|\bar{P}\rangle\right|^2 = \Gamma_{\bar{P}}.$$

Here $|n_i\rangle$ form a complete set, which we sum over to get the total rate. In reality the states we detect are $|m_i\rangle$. These are, however, related by a unitary transformation to the $|n_i\rangle$, so the result is the same if we sum over partial widths to $|m_i\rangle$.

2. The expressions given by Eq. 8.32 imply the same partial widths for $|P\rangle \to |n_i\rangle$ and $|\bar{P}\rangle \to |\bar{n}_i\rangle$. However, if we make a transformation to physical states, the

amplitudes for P or $\bar{P}$ will involve the same unitary transformations applied to the appropriate terms given in Eq. 8.32. The presence of the weak phase ϕ_{n_i} implies that the magnitude for sums involving amplitudes for $|P\rangle$ or $|\bar{P}\rangle$ will differ unless ϕ_{n_i} is the same, or δ_{n_i} is the same, for all channels summed over.

3. If CP is conserved, we can organize the decays for $|P\rangle$ or $|\bar{P}\rangle$ to corresponding physical states with identical partial widths or branching ratios. With CP conserved, we have directly for the weak amplitudes (for scattering states or physical states; we look at the former first):

$$\langle n_i | M | P \rangle = \langle CP n_i | M | CP\, P \rangle ,$$

and $|CP\, P\rangle = |\bar{P}\rangle$ for a single pseudoscalar at rest.[1] We write $|CP n_i\rangle = |\tilde{n}_i\rangle$. Therefore

$$\langle n_i | M | P \rangle = \langle \tilde{n}_i | M | \bar{P} \rangle .$$

In an experiment we detect the states

$$|m_i\rangle = \sum_j U_{ij} |n_j\rangle \quad \text{or} \quad |\tilde{m}_i\rangle = \sum_j U_{ij} |\tilde{n}_j\rangle .$$

The CP invariance of the strong interactions implies that the matrix relating $|m_i\rangle$ and $|n_i\rangle$ is the same as that for $|\tilde{m}_i\rangle$ and $|\tilde{n}_i\rangle$. These relations then imply

$$\langle m_i | M | P \rangle = \sum_j U_{ij} \langle n_j | M | P \rangle = \langle \tilde{m}_i | M | \bar{P} \rangle$$

and equal partial widths for $|P\rangle \to |m_i\rangle$ and $|\bar{P}\rangle \to |\tilde{m}_i\rangle$. The effect of the matrix U_{ij} is called strong rescattering, since it (combined with the phase shifts) amounts to a mixing between physical channels due to strong scattering.

A difference in partial widths for $|P\rangle \to |m_i\rangle$ and $|\bar{P}\rangle \to |\tilde{m}_i\rangle$ is called direct CP violation. It can only come from the weak phases ϕ_{n_i}. If CP violation arises from the CKM matrix only, then these phases are in principle determined from the angles and nonzero phase in this matrix.

8.8 ■ AMPLITUDES FOR KAON DECAY TO TWO PIONS

We can constrain the parameterization of the K decay matrix elements in light of the discussion on strong scattering and weak phases. The final states are simple, limited to 2π and 3π by energy conservation. Unfortunately, the actual values for the amplitudes still remain incalculable due to the complexities of the strong interactions.

[1] Note that $|\bar{P}\rangle$ can be $+$ or $-$ the physical state; for $|K^0\rangle$ it is $+$ for our phase choice. This leads to a $+$ or $-$ sign between amplitudes for physical states, which doesn't change the partial width relation.

The strong scattering eigenstates can be determined from the strong symmetries. These imply (ignoring the small isospin violation):

1. There are no transitions between 2π and 3π because of G parity conservation. Thus we can look at 2π individually.
2. For 2π with $J = 0$, the possible states have $I = 0$ or 2. By I spin conservation, these states scatter among themselves. Thus, for 2π there are strong phases δ_0 and δ_2 for the two isospin channels.
3. If we assume that weak phases come only from the CKM matrix, then the weak effective Hamiltonian in Eq. 8.29 provides no weak phase (this is true using the exact CKM matrix, Eq. 8.17). Thus, weak phases can arise only from the penguin diagram, mainly via the sequence of transitions $s \to t \to d$, where the $t \to d$ transition brings in the phase factor in our convention for the CKM matrix. This is a $\Delta I = \frac{1}{2}$ operator, so the weak phase occurs mainly in this term. This allows a rewriting of the matrix elements in Eq. 8.31 in terms of specific phases for each amplitude a_0 and a_2.
4. The CP properties for the states are determined by Eq. 8.25 for a $J = 0$ state.

Using these implications, we can parameterize the amplitudes for each kaon to decay into two pions. Based on Eq. 8.32 for each isospin channel and the sum over isospins in Eq. 8.31, the matrix elements are

$$M_{K^- \to \pi^- \pi^0} = M_{K^+ \to \pi^+ \pi^0} = \sqrt{\tfrac{3}{2}} e^{i\delta_2} A_2$$

$$M_{K^0 \to \pi^+ \pi^-} = \sqrt{\tfrac{1}{3}} e^{i\delta_2} A_2 + \sqrt{\tfrac{2}{3}} e^{i\delta_0} A_0 e^{-i\phi}$$

$$M_{K^0 \to \pi^0 \pi^0} = \tfrac{2}{\sqrt{3}} e^{i\delta_2} A_2 - \sqrt{\tfrac{2}{3}} e^{i\delta_0} A_0 e^{-i\phi}$$

$$M_{\bar{K}^0 \to \pi^+ \pi^-} = \sqrt{\tfrac{1}{3}} e^{i\delta_2} A_2 + \sqrt{\tfrac{2}{3}} e^{i\delta_0} A_0 e^{i\phi}$$

$$M_{\bar{K}^0 \to \pi^0 \pi^0} = \tfrac{2}{\sqrt{3}} e^{i\delta_2} A_2 - \sqrt{\tfrac{2}{3}} e^{i\delta_0} A_0 e^{i\phi} \quad (8.33)$$

A_0 and A_2 are real and δ_0, δ_2 have been independently measured to be about $35°$ and $-7°$, respectively, in scattering of pions on virtual pions emitted by nucleons. Note that the strong phases are indeed nonnegligible. The weak phase ϕ violates CP invariance, leading to direct CP violation. The phase ϕ has been measured to be $\sim 10^{-4}$, in rough agreement with estimates based on the CKM matrix and top quark penguin contributions. The direct CP violation is thus measured to be very small.

If the K^0 and $\bar{K}^0$ mesons had been the propagating mass eigenstates, we could directly measure ϕ by the difference in partial widths for decays to the $\pi^+ \pi^-$ or $\pi^0 \pi^0$ CP conjugate final states. Since the K^0 and $\bar{K}^0$ mix, we have to measure

8.9 ■ AMPLITUDES FOR *K* DECAY TO 3π

decay branching ratios for the eigenstates generated by mixing. We will therefore have to come back to these measurements after the mixing discussion in Chapter 9. Note, finally, that the rate for the sum of $\pi^+\pi^-$ and $\pi^0\pi^0$ is the same for K^0 and $\bar{K}^0$ (that is, both are independent of ϕ) as required by the *CPT* theorem for a sum over a complete set of states that scatter into each other.

8.9 ■ AMPLITUDES FOR *K* DECAY TO 3π

We look at all 3π decays within the following two good approximations:

1. Symmetric *s*-wave final state. This implies an $I = 1$ final state. From Eq. 8.30, keeping only the $I = 1$ terms, the K^- and $\bar{K}^0$ amplitudes are:

$$O\left|K^-\right\rangle = a^{I=1}_{\Delta I=\frac{1}{2}}|1,-1\rangle + \frac{a^{I=1}_{\Delta I=\frac{3}{2}}}{2}|1,-1\rangle$$

$$O\left|\bar{K}^0\right\rangle = \frac{a^{I=1}_{\Delta I=\frac{1}{2}}}{\sqrt{2}}|1,0\rangle + \frac{a^{I=1}_{\Delta I=\frac{3}{2}}}{\sqrt{2}}|1,0\rangle.$$

2. Dominance of the $\Delta I = \frac{1}{2}$ amplitude. This is implied by the data, as we will see below, and is a strong interaction effect. The data indicate that

$$\frac{a^{I=1}_{\Delta I=\frac{3}{2}}}{a^{I=1}_{\Delta I=\frac{1}{2}}} \sim 0.06.$$

Keeping only the $\Delta I = \frac{1}{2}$ amplitude, we have:

$$O\left|K^-\right\rangle = a^{I=1}_{\Delta I=\frac{1}{2}}|1,-1\rangle$$

$$O\left|\bar{K}^0\right\rangle = \frac{a^{I=1}_{\Delta I=\frac{1}{2}}}{\sqrt{2}}|1,0\rangle.$$

This implies that $\Gamma_{K^-\to 3\pi} = 2\Gamma_{\bar{K}^0\to 3\pi}$, where all combinations of 3π are summed over and we have ignored the mass difference between pions.

If *CP* is a good symmetry, then the amplitude $a^{I=1}_{\Delta I=\frac{1}{2}}$ has a phase $e^{i\delta_1}$ wherever it occurs. This phase has not been measured but drops out of the rate calculations, since only the one isospin state is involved. In principle, there is also a very small *CP* violating weak phase (as in the $\pi\pi$ case), which changes sign when we go from the amplitudes for the $K^-, \bar{K}^0$ doublet to the K^+, K^0 doublet. We ignore this since it has no effect on the rates for only one final isospin state. Finally, given the *CP* transformations in Eq. 8.25, there is an overall minus sign between analogous decays of K^+, K^- or K^0, $\bar{K}^0$, for example,

$$M_{K^-\to\pi^-\pi^-\pi^+} = -M_{K^+\to\pi^+\pi^+\pi^-}; \quad M_{\bar{K}^0\to 3\pi^0} = -M_{K^0\to 3\pi^0}.$$

We can calculate the rate into a given 3π channel in terms of one amplitude by using Clebsch–Gordan coefficients for the symmetric $I=1$ states. Table 8.6 summarizes the predictions and data. Included are the significant phase space corrections, which are large because $3m_\pi$ is not small compared to m_K and the pions do not all have the same mass. The K_L^0 rates are taken to be twice the $\bar{K}^0$ rates in order to use the data.

TABLE 8.6 Predictions and Data for $K \to 3\pi$.

Relative phase space factors are:

$K_L^0 \to 3\pi^0$	:	$K_L^0 \to \pi^+\pi^-\pi^0$	:	$K^- \to \pi^-\pi^0\pi^0$	:	$K^- \to \pi^-\pi^-\pi^+$
1.48	:	1.31	:	1.24	:	1.00

Prediction based on symmetric s-wave: Data:

$$\frac{\Gamma_{K^-\to\pi^-\pi^-\pi^+}}{\Gamma_{K^-\to\pi^-\pi^0\pi^0}} = 4\left(\frac{1}{1.24}\right) = 3.23 \qquad 3.23 \pm 0.04$$

$$\frac{\Gamma_{\bar{K}^0\to\pi^+\pi^-\pi^0}}{\Gamma_{\bar{K}^0\to 3\pi^0}} = \frac{2}{3}\left(\frac{1.31}{1.48}\right) = 0.590 \qquad 0.595 \pm 0.012$$

Prediction based also on $\Delta I = \frac{1}{2}$ rule: Data:

$$\frac{\Gamma_{K^-\to\pi^-\pi^-\pi^+} + \frac{1}{1.24}\Gamma_{K^-\to\pi^-\pi^0\pi^0}}{\frac{1}{1.31}\Gamma_{K_L^0\to\pi^+\pi^-\pi^0} + \frac{1}{1.48}\Gamma_{K_L^0\to 3\pi^0}} = 1 \qquad 0.82 \pm 0.02$$

The measured ratio not equal to 1 indicates the presence of a small $\Delta I = \frac{3}{2}$ amplitude.

8.10 ■ RARE K^0 DECAYS

Included in Table 8.4 are a few of the interesting rare decays for K_S^0 and K_L^0. The rare decays listed are to the $\gamma\gamma$ or $\mu^+\mu^-$ final states. We will not try to calculate the rates, but rather just state the physics behind the rates.

The electromagnetic processes leading to $\gamma\gamma$ or $\mu^+\mu^-$ are of higher order than the other decays in Table 8.4. As such, they illustrate the expansion of the time evolution operator into successive processes for small couplings. They serve to check that we have not left out any physics to the level that we understand the rates. Both K_S^0 and K_L^0 decay into $\gamma\gamma$, but through mechanisms that are quite different.

We begin with the diagrams in Figure 8.8 and consider the higher-order effects that can occur. The scattering and penguin diagrams for $\bar{K}^0$ yield, at the quark level, final states with $J=0$ and $u\bar{u}$ or $d\bar{d}$ quark content. The simplest such

8.10 Rare K^0 Decays

states at the hadron level are just the neutral pseudoscalars π^0, η^0, η'. The weak interactions will therefore provide a very tiny mixing for the neutral kaons with these pseudoscalars. Rather than considering K^0 and $\bar{K}^0$, we start with the states K_S^0 and K_L^0, which are very close to CP eigenstates. The light pseudoscalars are CP odd so that we expect a tiny bit of mixing of K_L^0 and π^0, η, η'. Since the latter decay to $\gamma\gamma$, we expect to find a small decay of K_L^0 into $\gamma\gamma$ in a final state that has $J^P = 0^-$. A calculation following these ideas gives a reasonable estimate of the branching ratio for $K_L^0 \to 2\gamma$.

The mixing with the π^0, η, η' should be absent for K_S^0 in the approximation that it is even under CP. The primary K_S^0 decay is to $\pi\pi$. This, however, provides another mechanism to produce $\gamma\gamma$, since the charged $\pi^+\pi^-$ will scatter to $\gamma\gamma$. The photons produced by this mechanism are in a state with $J^P = 0^+$, which is the state for the initial $\pi^+\pi^-$ system. A calculation of $K_S^0 \to \pi^+\pi^- \to \gamma\gamma$ gives a good description of this rate in terms of the measured rate to $\pi^+\pi^-$.

The rescattering mechanism can be taken one step further by considering $\gamma\gamma \to \mu^+\mu^-$ for the $J^P = 0^\pm$ states above. The branching ratio for $K_S^0 \to \mu^+\mu^-$ is too small to be measured, given the short K_S^0 lifetime. The $K_L^0 \to \mu^+\mu^-$ decay can be seen and the rate is as expected for $K_L^0 \to 2\gamma \to \mu^+\mu^-$. Note that in calculating such a rate, the imaginary part, which is expected to dominate the rate, can be calculated using unitarity in terms of real intermediate photons, whereas the real part can only be estimated.

The measured branching ratio for $K_L^0 \to \mu^+\mu^-$ is 7.3×10^{-9}. The smallness of this number is an impressive example that strangeness changing neutral currents do not occur in the weak Lagrangian directly. The rate for this decay is, however, so small that we should verify that it is consistent with expectations from second-order charged current weak processes. The second-order weak diagram is shown in Figure 8.10. This is called a box diagram. If we calculate this diagram for a virtual u quark alone, we get a result that is considerably larger than the measured branching ratio. The CKM coupling factor for the u quark is $V_{us}V_{ud}^*$. However, since the intermediate quark is virtual, we need to include intermediate states of larger mass, as also shown in Figure 8.10. The state with the largest coupling is the c quark. Its CKM factor is $V_{cs}V_{cd}^*$. In the two-generation approximation, $V_{us}V_{ud}^* + V_{cs}V_{cd}^* = \sin\theta_c \cos\theta_c - \cos\theta_c \sin\theta_c = 0$, by the unitarity of

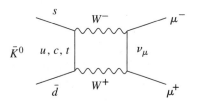

FIGURE 8.10 Box diagram contribution to $\bar{K}^0 \to \mu^+\mu^-$. A similar diagram contributes for the K^0. Virtual u, c, t interfere, with amplitudes for each quark dependent on quark masses and CKM factors at the two vertices.

the matrix. Thus at large invariant masses for the virtual quark ($\gg m_c$) the amplitude given by the box diagram is cut off by the cancellation between u and c, and in fact we get a result that is much smaller than the $\gamma\gamma$ contribution. Such a cancellation in the box diagram is called the GIM (Glashow–Iliopoulos–Maiani) mechanism, and was used to predict the existence and approximate mass of the c quark. In the three-generation case, the u, c cancellation at large masses is still very accurate. The t quark has a very small CKM factor, $V_{ts}V_{td}^*$, and therefore does not contribute much to the amplitude. For box diagrams involving initial b quarks instead of s quarks, the CKM factors are much larger for the virtual t quark contribution. As a result, virtual t quarks typically provide the dominant amplitude, compared to u, c exchange, for processes in the B meson system when box diagram contributions are important.

8.11 ■ WEAK DECAYS INVOLVING THE HEAVY QUARKS c, b, t

The weak interactions are the source of flavor changing decays for the lightest hadrons containing heavy quarks. We expect these hadrons to be pseudoscalar mesons and spin $\frac{1}{2}$ baryons. The decays are analogous to the kaon and strange baryon decays; however, many more final states are possible, given the greater energy released in the underlying quark transition. Summing over these final states, the total decay rate can be predicted more accurately as the heavy quark mass increases.

The most accurately-predicted decay is expected to be that of the t quark. Since it is heavier than the W, it decays via the weak single vertex interaction rather than the higher-order weak effective Hamiltonian. The diagram is shown in Figure 8.11. Given the values of the CKM matrix elements, the decay is completely dominated by the $t \to b$ transition. The emitted W is real and subsequently produces a mix of states with probabilities typical of W decay. Ignoring small mass and radiative corrections, and small CKM matrix elements, the W decays into (in the ratio shown):

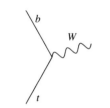

FIGURE 8.11 Top quark decay.

$$e^-\bar{\nu}_e, \quad \mu^-\bar{\nu}_\mu, \quad \tau^-\bar{\nu}_\tau, \quad \bar{u}d, \quad \bar{c}s.$$
$$1, \quad 1, \quad 1, \quad 3, \quad 3.$$

Including the small CKM matrix elements, $(\bar{u}, d)$ is replaced by

$$95\%(\bar{u}, d), \quad 5\%(\bar{u}, s)$$

and a very tiny amount of $(\bar{u}, b)$; $(\bar{c}, s)$ is replaced by $95\%(\bar{c}, s), 5\%(\bar{c}, d)$ and a very small fraction of decays into $(\bar{c}, b)$.

The t decay matrix element is

$$M_{fi} = \frac{g}{2\sqrt{2}} e_\mu^{W*} \bar{u}_b \gamma_\mu (1 - \gamma_5) u_t, \qquad (8.34)$$

8.11 Weak Decays Involving the Heavy Quarks c, b, t

allowing a straightforward calculation of the decay width. Summing over all spins and dividing by 2 for an unpolarized t quark:

$$\frac{1}{2}\sum_{spins}|M_{fi}| = \frac{g^2}{16}\left[\sum_{\lambda=1,2,3} e_\mu^W(\lambda)e_\nu^{W*}(\lambda)\right]$$
$$\text{Tr}\left[(\not{p}_b + m_b)\gamma_\mu(1-\gamma_5)(\not{p}_t + m_t)\gamma_\nu(1-\gamma_5)\right].$$

If we ignore m_b, keep only terms with an even number of γ matrices, and commute $1-\gamma_5$ through to the right:

$$\frac{1}{2}\sum_{spins}|M_{fi}|^2 = \frac{g^2}{16}\left[\sum_\lambda e_\mu^W(\lambda)e_\nu^{W*}(\lambda)\right]\text{Tr}\left[\not{p}_b\gamma_\mu\not{p}_t\gamma_\nu(1-\gamma_5)\right].$$

We see that

$$\sum_\lambda e_\mu^W(\lambda)e_\nu^{W*}(\lambda) = -g_{\mu\nu} + \frac{p_\mu^W p_\nu^W}{M_W^2}$$

is a symmetric tensor, so we need only keep the symmetric part of the trace, which means the γ_5 term does not contribute. Using $\text{Tr}[\not{p}_b\gamma_\mu\not{p}_t\gamma_\nu] = 4[p_\mu^b p_\nu^t + p_\mu^t p_\nu^b - g_{\mu\nu}(p_b \cdot p_t)]$ we get finally:

$$\frac{1}{2}\sum_{spins}|M_{fi}|^2 = \frac{g^2}{2}\left[(p_b \cdot p_t) + 2\frac{(p_W \cdot p_b)(p_W \cdot p_t)}{M_W^2}\right].$$

Using $p_W \cdot p_b = (p_t - p_b) \cdot p_b = p_t \cdot p_b - m_b^2 \simeq p_t \cdot p_b$,

$$\frac{1}{2}\sum_{spins}|M_{fi}|^2 = \frac{g^2}{2}(p_b \cdot p_t)\left[1 + \frac{2(p_W \cdot p_t)}{M_W^2}\right].$$

Completing the calculation in the t rest-frame:

$$p_b \cdot p_t = m_t E_b = \frac{m_t^2 - M_W^2}{2}, \quad p_t \cdot p_W = m_t E_W = \frac{m_t^2 + M_W^2}{2},$$

if we ignore m_b^2; therefore:

$$\frac{1}{2}\sum_{spins}|M_{fi}|^2 = \frac{g^2}{4}(m_t^2 - M_W^2)\frac{[2M_W^2 + m_t^2]}{M_W^2}.$$

Putting in the phase space factor:

$$\Gamma_t = \frac{g^2}{4\pi}\frac{(m_t^2 - M_W^2)^2(m_t^2 + 2M_W^2)}{16M_W^2 m_t^3}. \qquad (8.35)$$

Using

$$\frac{g^2}{4\pi} = \frac{1}{30}, \quad m_t = 175 \text{ GeV}, \quad M_W = 80.4 \text{ GeV},$$

gives $\Gamma_t = 1.5$ GeV.

This width is large, in fact so large that the t quark will decay before it really can bind into a hadronic system. Thus we are justified in talking about t-quark decay, rather than meson or baryon decay as for the other quark systems.

Given the very heavy t-quark mass, it has so far been produced only at the highest energy colliding beam machine, a 2 TeV proton-antiproton collider. It is studied in events containing t and $\bar{t}$ quark pairs, each of which decay as above. Each resulting b quark from the $t \to b$ transition really produces a hadronic jet over longer time scales, and each W decays into final states with probabilities given earlier.

8.11.1 ■ Weak Decay of Charm

We study next the case of charm. The c quark transition, like that in t decay, is dominated by the charged current term within one generation, in this case $c \to s$. However, the off-diagonal term $c \to d$ is not as suppressed as $t \to s$ or $t \to d$, so we have measurably small ($\sim 5\%$ of total) Cabibbo suppressed decays. The lightest charmed meson types are $D^+ = -|c, \bar{d}\rangle$, $D^0 = |c, \bar{u}\rangle$, and $D_s^+ = |c, \bar{s}\rangle$, along with their antiparticles. A rough estimate of the decay rate can be derived by treating the c as a freely decaying fermion (like the τ) and ignoring the masses of the lighter fermions. This gives

$$\Gamma_c = \Gamma_\tau \left(\frac{m_c}{m_\tau}\right)^5 \simeq \frac{1}{2}\Gamma_\tau,$$

assuming $m_c/m_\tau \simeq 0.9$. The lifetime of the charmed particles is therefore expected to be $\sim 6 \times 10^{-13}$ sec. Table 8.7 lists the lifetimes of the charmed mesons and one charmed baryon (Λ_c^+ made of u, c, d) for completeness. The naive estimate is therefore good to a factor ~ 2.

TABLE 8.7 Charmed Particle Lifetimes.

Particle	Mass (GeV)	Lifetime (sec)
D^+, D^-	1.869	$10.51 \pm 0.13 \times 10^{-13}$
$D^0, \bar{D}^0$	1.865	$4.12 \pm 0.03 \times 10^{-13}$
D_s^+, D_s^-	1.969	$4.90 \pm 0.09 \times 10^{-13}$
Λ_c^+	2.285	$2.00 \pm 0.06 \times 10^{-13}$

8.11 Weak Decays Involving the Heavy Quarks c, b, t

Unlike in the kaon system, the neutral meson mixing, which now involves D^0 and $\bar{D}^0$, does not result in strikingly interesting phenomena and the two neutral D lifetimes are equal to very good approximation. Neglecting CP violation, the mass eigenstates will be

$$\frac{|D^0\rangle \pm |\bar{D}^0\rangle}{\sqrt{2}}.$$

However, initially produced $|D^0\rangle$ or $|\bar{D}^0\rangle$ states will not mix noticeably while propagating if they decay before they can mix, and if the lifetimes of the two mass eigenstates are close to the same value. This is exactly the situation for D^0 and $\bar{D}^0$, as discussed in Chapter 9.

The free quark calculation for the charmed lifetime assumes that the charm version of the spectator diagram of Figure 8.8 dominates the hadronic decays, and, moreover, that the final quarks produced by the virtual W do not interact significantly with the remaining quarks. These assumptions would result in a single lifetime for all the weakly decaying charmed particles, which is seen not to be true.

The assumption that the spectator diagram is the dominant weak term in the time evolution of the system is likely to be a fairly good approximation. The penguin diagram is Cabibbo suppressed and the resulting decay rate is estimated to be very small compared to the spectator decay rate, which is Cabibbo allowed. Looking at the other diagrams, the D^0 has a Cabibbo allowed weak scattering diagram while the D_s^+ has a Cabibbo allowed weak annihilation diagram. If we consider the decaying charm quark mass as a variable, then these diagrams become less significant as the quark mass rises. This can be understood by looking at a leptonic decay (annihilation diagram) versus a semileptonic decay (spectator diagram). The rate for the latter, if we can ignore all masses except the heavy quark mass, is proportional to m_c^5, while the leptonic decay is proportional to $m_c f_D^2 m_l^2$ based on the analogous calculations for K and π decay (see, for example, Eqs. 8.24 and 8.22). The constant f_D, which depends on the details of the bound state, is not expected to change dramatically with the quark type and is expected to be $\sim f_K$. Therefore, the spectator decay rate is much larger than the annihilation decay rate for a large charm quark mass. Explicit estimates of rates indicate that the nonspectator diagrams contribute $\sim 10\%$ of the rate for the D^0 and D_s^+.

The assumption that the final quarks produced by the W in the spectator diagram do not interact significantly with the quarks from the initial charmed hadron, requires that the invariant mass between these two systems be large compared to λ_{QCD}. This is not the case for charm decays, but is a better approximation for the heavier b quark decay.

For D mesons, most of the charm decays are to 2-body systems composed of light pseudoscalar and/or vector mesons in low angular momentum states. Considering only the Cabibbo favored currents, the quark level spectator diagrams for D^0 and D^+ involve the following sets of quarks:

1. For D^0: $(c, \bar{u}) \to (s, \bar{u}) + W^+ \to (s, \bar{u}) + (u, \bar{d})$, which can make meson pairs with quark content $(s, \bar{u}) + (u, \bar{d})$ or $(s, \bar{d}) + (u, \bar{u})$ if the systems interact. Both combinations are found.

2. For D^+: $(c, \bar{d}) \to (s, \bar{d}) + W^+ \to (s, \bar{d}) + (u, \bar{d})$, which can make meson pairs with quark content $(s, \bar{d}) + (u, \bar{d})$ for both pairings of quarks with antiquarks. Thus, for D^+ we have not only the strong interactions but also interference between paths followed to the final state, since both pairings yield the same flavor choices. Destructive interference between the amplitudes for the two pairings is believed to be the major reason for the lifetime difference between the charged and neutral D.

The strong interactions spoil the naive expectation that all charmed particles have the same lifetime, as would happen for a spectator process with no final state interactions. The hadronic effects are, however, already much smaller than for the kaons, where

$$\frac{\Gamma_{K_S}}{\Gamma_{K^-}} = 139, \quad \text{while for charm} \quad \frac{\Gamma_{D^0}}{\Gamma_{D^+}} = 2.5.$$

The D meson hadronic decays also allow a nice check that the CKM structure applies to the currents that participate. We show in Figure 8.12 the full set of processes that take place through the spectator diagram once we go beyond just the Cabibbo favored flavor transitions. To be specific, the spectator quark is taken to be $\bar{u}$, but it could equally well be $\bar{d}$ or $\bar{s}$. There are four possibilities after writing out all the terms in the currents. The CKM factors, which can all be written to

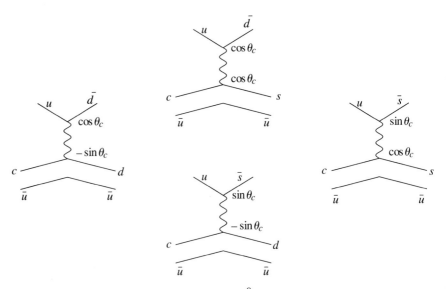

FIGURE 8.12 Set of spectator diagrams for D^0 decay. The CKM factor at each vertex is indicated.

good approximation in terms of $\cos\theta_c$ and $\sin\theta_c$, are indicated for each diagram.

The diagrams yield amplitudes that are Cabibbo favored ($\sim \cos^2\theta_c$, leading to strangeness $= -1$), singly Cabibbo suppressed ($\sim \cos\theta_c \sin\theta_c$, leading to strangeness $= 0$), or doubly Cabibbo suppressed ($\sim \sin^2\theta_c$, leading to strangeness $= 1$). Decays of each of these types have been observed with approximately the expected rates. Examples of final states for an initial D^0 are $K^-\pi^+$ (Cabibbo favored), K^+K^- and $\pi^+\pi^-$ (singly suppressed) and $K^+\pi^-$ (doubly suppressed). Unfortunately the strong interactions, combined with the violation of the $SU(3)$ flavor symmetry, preclude a careful quantitative comparison of the rates as a test of the theory.

While the rates for specific hadronic final states in D decays cannot be accurately calculated, relations between amplitudes for D^0 and D^+ can be calculated based on isospin symmetry, as was done for specific K^- and $\bar{K}^0$ decays such as $K \to 2\pi$. Within the framework of the spectator diagrams in Figure 8.12, we can also ask whether we expect CP violation when comparing amplitudes for the doublets D^+, D^0 to those for $D^-, \bar{D}^0$. Using the approximate form of the CKM matrix, Eq. 8.18, which provides the couplings in Figure 8.12, we get no weak phases. Only through the very small terms, which are included in the exact matrix in Eq. 8.17, can we get direct CP violating effects. This will happen only when two amplitudes with different strong and weak phases contribute to a given final state. This can occur for the two singly Cabibbo suppressed diagrams, which can lead to the same final state. This is also the channel to which the penguin diagram will contribute. To date, however, no CP violating effects have been seen in the D system. Finally, we mention that approximations such as factorization of the two currents into separate simple hadronic matrix elements, which worked poorly for the K^- decay in Section 8.6.2, give a better approximation for D decays. The factorization approximation will still not provide the hadronic scattering phase and hadronic rescattering phenomena for specific physical final states.

8.11.2 ■ Weak Decay of b Quark Systems

The final heavy flavored systems that decay weakly are the lightest mesons and baryons containing a b quark. The lightest mesons are $\bar{B}^0 = -|b,\bar{d}\rangle$, $B^- = |b,\bar{u}\rangle$, $\bar{B}^0_s = |b,\bar{s}\rangle$, and their antiparticles.[2] We also expect a heavier $B_c^- = |b,\bar{c}\rangle$, for which both b and c quark weak currents will contribute significantly to the decay. Table 8.8 lists the lightest particles with b quarks, including the lightest baryon, and their lifetimes. Note that the B mesons live longer than the D mesons, despite their larger mass. This comes about because of a much smaller CKM factor in the decay matrix element. The b transition involves either a change from the 3rd $\to$ 2nd or the 3rd $\to$ 1st generation, while the D decays are dominated by transitions within the 2nd generation.

For the b spectator diagrams, a wide variety of final states are produced, following the $b \to c$ or $b \to u$ transition, since energy conservation allows the W

---

[2]Note that the B meson is defined as the state with a $\bar{b}$ antiquark. This is analogous to the convention for the kaons. For the D meson, the D has a c quark.

TABLE 8.8 Bottom Particle Lifetimes.

Particle	Mass (GeV)	Lifetime (sec)
$B^0, \bar{B}^0$	5.279	$1.54 \pm 0.02 \times 10^{-12}$
B^+, B^-	5.279	$1.67 \pm 0.02 \times 10^{-12}$
$B_s^0, \bar{B}_s^0$	5.370	$1.46 \pm 0.06 \times 10^{-12}$
Λ_b^0	5.624	$1.23 \pm 0.08 \times 10^{-12}$
B_c^+, B_c^-	6.4	$0.46 \pm 0.16 \times 10^{-12}$

to produce $e^-\bar{\nu}_e$, $\mu^-\bar{\nu}_\mu$, $\tau^-\bar{\nu}_\tau$, $\bar{u}d$, $\bar{c}s$ (Cabibbo allowed) and $\bar{u}s$, $\bar{c}d$ (Cabibbo suppressed). These decay types can be experimentally separated based on the flavor content of the final state. In addition, the penguin diagrams for the b system, shown in Figure 8.13, are small but not negligible. In particular, the product of the $b \to t \to s$ CKM factors is the same as that for the dominant $b \to c$ transition. Note that the penguin diagram will lead to a different flavor content in the final state than that for most of the spectator diagrams.

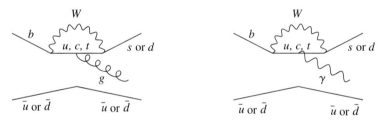

FIGURE 8.13 The penguin diagrams: (a) mediated by the strong interactions; (b) mediated by the electromagnetic interactions. Not shown is the diagram where W radiates the γ for (b).

Included in Figure 8.13 is the electromagnetic penguin, which yields a real photon and predominantly a strangeness -1 final state. This provides a rather distinct final state that has a branching ratio $\simeq 3 \times 10^{-4}$. For the hadronic penguin, the diagram is drawn with the gluon unattached to the other quark. In reality, the two quarks and gluon interact strongly, producing a range of multi-particle final states. Given the rate for the electromagnetic penguin, we can expect the hadronic penguin to have a branching ratio $\sim 1\%$, which is a small fraction of the total width.

The near equality of the B meson lifetimes indicates that the spectator diagram provides a good approximation for calculating the overall decay width, which involves a sum over all the many specific decay channels. Except for a few percent of the decays that occur through the $b \to u$ transition or the penguin diagram, the spectator transitions that account for the bulk of the width are given below.

8.12 Heavy Quark Effective Theory

Decay	$b \to ce^-\bar{\nu}_e$	$b \to c\mu^-\bar{\nu}_\mu$	$b \to c\tau^-\bar{\nu}_\tau$	$b \to c\bar{u}d$	$b \to c\bar{c}s$
Relative Rate	1	1	0.3	4	2

The expected relative rates are shown, where estimates have been made for the decrease in the branching ratio due to the phase space reduction for the heavier τ and $\bar{c}$. The calculations include an estimate of the *QCD* corrections that increase the hadronic branching ratio, for example, increasing the $b \to c\bar{u}d$ to $b \to ce^-\bar{\nu}_e$ ratio from the factor of 3 coming from counting colors alone. Including the Cabibbo suppressed processes, we should replace $(\bar{u}d) \to 95\%(\bar{u}d) + 5\%(\bar{u}s)$ and $(\bar{c}s) \to 95\%(\bar{c}s) + 5\%(\bar{c}d)$.

The factors above imply a branching ratio to $ce^-\bar{\nu}_e$ of about 12%, a little larger than the measured value, which is about 11%. The semileptonic decays provide a way to measure the CKM matrix elements $|V_{cb}|$ and $|V_{ub}|$. The semileptonic decay rate formula, except for the square of the CKM factor, is the same as for $\tau^- \to e^-\bar{\nu}_e\nu_\tau$, where we replace m_τ by m_b and m_{ν_τ} by m_c or m_u. In the rate, we sum over all final state hadronic systems. This is called an inclusive measurement and is expected to give an accurate result for quark mass differences in the transition $\gg \lambda_{QCD}$. The rate predictions are limited by the uncertainty in the b and c quark masses, which are about 4.9 GeV and 1.5 GeV, respectively, with an uncertainty of about 100 MeV for each. In addition, the $b \to u$ coupling is small enough that it has not been possible to make a very accurate measurement of the inclusive semileptonic decay rate for this quark transition. The measured semileptonic rates have led to the values

$$|V_{cb}| = 0.041 \pm 0.003,$$

$$|V_{ub}| = 0.0036 \pm 0.0005.$$

In terms of the CKM parameterization,

$$\frac{|V_{ub}|}{|V_{cb}|} = \sin\theta_c\sqrt{\rho^2 + \eta^2},$$

so the above values imply $\sqrt{\rho^2 + \eta^2} = 0.40 \pm 0.06$.

We have not mentioned neutral B meson mixing as yet. For the B system, two pairs can mix: $B^0 - \bar{B}^0$ and $B_s^0 - \bar{B}_s^0$. The mixing does not result in states with very different lifetimes, so we have not had to worry about mixing when discussing the lifetimes. The mixing effects are, however, large and interesting, and will be an important topic discussed in Chapter 9.

8.12 ■ HEAVY QUARK EFFECTIVE THEORY

The extraction of $|V_{cb}|$ using inclusive semileptonic decays suffers from the uncertainties in the quark masses. If we could find a method that involves the measured meson masses instead, we could reduce this uncertainty. This was done, using CVC, in the case of semileptonic K decay. The $b \to c$ transition involves heavy nonrelativistic quarks in the meson rest-frame, rather than the very relativistic $s \to u$ quarks for K decay. The spectator quark in both types of decay is a relativistic light quark.

The heavy quark limit for $b \to c$ allows a prediction of the semileptonic B decay amplitude to a pseudoscalar D or vector D^* meson in terms of $|V_{cb}|$ at a specific point in phase space for the decay, as we will see below. To understand this, consider a meson made of a very heavy quark of mass $m_Q \gg \lambda_{QCD}$ and a light antiquark of mass $m_q \ll \lambda_{QCD}$. The heavy quark acts as a static source of the color field, generating a bound state with the light antiquark. If we replace one heavy quark with another we generate a new bound state, for which the light antiquark can be expected to have the same wave function. But the overall mass for the bound state, determined mainly by the heavy quark mass, will change. An additional feature of the heavy quark interaction is the suppression of the spin interaction between Q and $\bar{q}$, as expected from the constituent quark model for the meson masses in Chapter 6. Thus, the 0^- and 1^- ground state mesons are much closer in mass than mesons in the light 0^- and 1^- $SU(3)$ multiplets, and the assumption that they have the same spatial wave function should be a good approximation.

For the weak decay of a B meson, consider the transition $b \to c + W$. If we start with a b at rest and emit a W that leaves the c at rest, then the spectator has precisely the correct wave function to be in the same bound state with a c as it was with the b. It is for this point in the phase space for the decay that the matrix element for the semileptonic decay can be calculated in terms of $|V_{cb}|$. This configuration corresponds to the maximum q^2 that the W can carry.

The hadronic matrix element we need in order to calculate the full amplitude is $\langle D|J_\mu^V + J_\mu^A|B\rangle$. For the case of no recoil we can ignore the light quarks and just calculate the matrix element between the heavy quarks at rest. Evaluating $\gamma_\mu(1 - \gamma_5)$ between spinors at rest gives $\sqrt{(2m_b)(2m_c)}$ for $\mu = 0$, and 0 for the three other indices. The mass factors come from the covariant normalization of the spinors. In the heavy quark limit, these are equal to $\sqrt{2m_B 2m_D}$, and when squared, will cancel the same $1/2m$ factors in the rate formula. Note that the $\gamma_\mu(1 - \gamma_5)$ also has a nonvanishing matrix element for a final spinor flipped in direction relative to the initial spinor. This leads to a predictable matrix element for a D^* final state when the D^* is produced at rest in the B rest-frame.

Returning to the D, we can write the matrix element calculated above in a covariant notation:

$$\left\langle D\left|J_\mu^V + J_\mu^A\right|B\right\rangle = \left\langle D\left|J_\mu^V\right|B\right\rangle$$
$$= \sqrt{m_B m_D}\left(\frac{p_\mu^B}{m_B} + \frac{p_\mu^D}{m_D}\right)h_+ \left(\frac{p_\mu^B}{m_B} \cdot \frac{p_\mu^D}{m_D}\right),$$

where $h_+(1) = 1$ from the foregoing discussion. This is not the most general current matrix element, which, from general considerations, can be written as

$$\sqrt{m_B m_D} \left[\left(\frac{p_\mu^B}{m_B} + \frac{p_\mu^D}{m_D} \right) h_+ \left(\frac{p_\mu^B}{m_B} \cdot \frac{p_\mu^D}{m_D} \right) \right.$$
$$\left. + \left(\frac{p_\mu^B}{m_B} - \frac{p_\mu^D}{m_D} \right) h_- \left(\frac{p_\mu^B}{m_B} \cdot \frac{p_\mu^D}{m_D} \right) \right],$$

where $h_+(1) = 1$ and $h_-(1) = 0$ in the heavy quark limit.[3] Analyses of the decay spectrum near the point where the D or D^* is at rest provide a good measurement of V_{cb}, yielding a value in agreement with that obtained from the inclusive semileptonic decay rate. Note that the decay to the D is proportional to $|\vec{p}_D|^3$ (like the analogous K decay in Section 8.6.1), so that the D^* decay, which is less suppressed for the kinematic region of interest, is actually the more practical choice for making the measurement.

This discussion can be put into a more general framework by organizing QCD with heavy quarks into terms of increasing powers of $1/m_Q$. This is called the Heavy Quark Effective Theory. Using this approach, we can show that, in the heavy quark limit, $h_- = 0$, and h_+ is a universal function that describes both the decays to D and D^* completely. All the relevant form factors can be written in terms of this one nonperturbative function, called the Isgur–Wise function.

8.13 ■ CONCLUSION

In this chapter we have studied a large body of data for the weak decays of leptons and quarks. In the next chapter, we will look at mixing phenomena for the same systems. The data involve patterns of flavor transformations that are predicted very nicely by the Standard Model assumptions, which are primarily:

1. Universal $SU(2)$ weak interaction involving left-handed fermions,
2. Symmetry breaking generating a large W mass and the experimentally determined fermion masses,
3. CKM matrix for three generations arising from a mismatch between weak and mass eigenstates of the fermions.

In many cases we have been able to test the self-consistency of the picture with reasonably high precision. Some of the precision semileptonic decay measurements were, however, needed to extract the parameters of the model. The large number of these parameters point to the need for a more comprehensive theory to explain the symmetry breaking.

---

[3]Note that this formula is just another way to write the general matrix element that we used for the light mesons, which is

$$\left\langle D \left| J_\mu^V \right| B \right\rangle = f_+(q^2)(p_\mu^B + p_\mu^D) + f_-(q^2)(p_\mu^B - p_\mu^D).$$

CHAPTER 8 HOMEWORK

8.1. What would the electron spectrum look like for $\mu^- \to e^- \bar{\nu}_e \nu_\mu$, for the case of a pure phase space decay? Compare to the real spectrum. In both cases ignore the electron mass and radiative corrections.

8.2. Consider a matrix element that has $(1 - i\gamma_5)$ appearing instead of $\gamma_\mu(1 - \gamma_5)$ for $\mu^- \to e^- \bar{\nu}_e \nu_\mu$. Calculate the e^- spectrum in the μ^- rest frame for this hypothesis for the decay matrix element. Ignore m_e.

8.3. Suppose that, for the τ doublet, the ν_τ were heavy and τ^- nearly massless. The decay would then be:

$$\nu_\tau \to \tau^- e^+ \nu_e.$$

(a) Assuming we can ignore m_τ and m_e compared to m_{ν_τ}, find the e^+ spectrum in the ν_τ rest frame.

(b) How does the shape of the spectrum compare to the e^+ spectrum in μ^+ decay?

(c) Calculate the ν_τ decay rate for this decay. Note that the e^+ spectrum provides a way for us to tell whether the transition involves particles or antiparticles for the ν_τ, τ^- doublet. The transition $c \to s$ is an example of an analogous situation to the one considered here.

8.4. Consider the reaction $\nu_\mu e^- \to \mu^- \nu_e$, where a ν_μ beam hits a stationary e^- in the laboratory. Show that the ν_μ laboratory energy must be

$$E_{\nu_\mu} > \frac{m_\mu^2 - m_e^2}{2m_e}$$

for the reaction to occur.

8.5. Several τ decays can be predicted from other measured quantities.

(a) Using the measured value of f_π, calculate the decay width for $\tau^- \to \pi^- \nu_\tau$. Using the τ lifetime, calculate the expected branching ratio for this decay. Compare to the measured branching ratio.

(b) Calculate the expected ratio of branching ratios for $\tau^- \to K^- \nu_\tau$ versus $\tau^- \to \pi^- \nu_\tau$.

8.6. If the τ mass had been only 400 MeV, the predominant hadronic decay mode would be $\tau^- \to \pi^- + \nu_\tau$. Compare the rate you would get for this decay for a 400 MeV τ to the rate expected for the decay into a noninteracting massless $u, \bar{d}$ quark doublet. Would the free-quark calculation for rates work even approximately as an estimate of the hadronic decay rate at this very low mass?

8.7. Consider the case of two families of quark doublets with the usual left-handed weak interactions and a new right-handed weak interaction with heavier gauge bosons. Assume the weak eigenstates, except for handedness, are the same for both interactions. When we diagonalize the general mass matrix, can we still get rid of all phase angles in the resulting 2×2 CKM matrices?

8.8. G parity is an important constraint for τ decays.

(a) Based on the G parity properties of the currents, namely

$$GJ_\mu^V G^{-1} = J_\mu^V$$
$$GJ_\mu^A G^{-1} = -J_\mu^A,$$

do we expect to see the decay:

$$\tau^- \to \eta \pi^- \nu_\tau ?$$

(b) Because of isospin breaking, we expect a small component of the η to have $I = 1$. Taking this small amplitude to be $a_{I=1}$, calculate the expected branching ratio for $\rho^- \to \pi^- \eta^0$ in terms of $|a_{I=1}|^2$. Using this branching ratio, calculate the expected branching ratio for $\tau^- \to \eta \pi^- \nu_\tau$ via the ρ^- channel. In Chapter 6, we estimated that $a_{I=1} = 1.8 \times 10^{-2}$ ($\eta - \pi^0$ mixing). Using this value, what is the expected value of the branching ratio?

8.9. Consider the ground state of a $\mu^+ e^-$ atom. Depending on the spin orientations, there are two nearly degenerate states with $J^P = 0^-$ and 1^-, both with orbital $\vec{L} = 0$.

(a) These states can decay by $\mu^+ \to e^+ \nu_e \bar{\nu}_\mu$ with the e^- a spectator, or by the weak annihilation process $\mu^+ e^- \to \bar{\nu}_\mu \nu_e$. Draw a diagram for the annihilation process. From general arguments show that the $J^P = 0^-$ state cannot annihilate to $\bar{\nu}_\mu \nu_e$ for massless neutrinos.

(b) For the $J^P = 1^-$ state, show that

$$\frac{\Gamma(\mu^+ e^- \to \bar{\nu}_\mu \nu_e)}{\Gamma(\mu^+ \to e^+ \nu_e \bar{\nu}_\mu)} = 64\pi \left(\frac{m_e}{m_\mu}\right)^3 \alpha^3.$$

(Note that you can use the nonrelativistic character of the initial state and the methods of Section 4.7).

8.10. Suppose hadrons were made up of very heavy quarks that are weakly bound (like the $\mu^+ e^-$ above). For this model, calculate in terms of particle masses and the wave function at the origin:

(a) f_π, π decay constant in $\pi \to \mu \nu$.

(b) $g_{W\rho}$, coupling of the ρ to the charged W.

8.11. The $\pi^- \to \pi^0 e^- \bar{\nu}_e$ allows a measurement of $\cos \theta_c$ using CVC. Writing

$$\langle \pi^0 | J_\mu | \pi^- \rangle = \sqrt{2}\left[\frac{f_+(q^2)}{f_+(0)} \left(p_\mu^{\pi^-} + p_\mu^{\pi^0}\right) + \frac{f_-(q^2)}{f_+(0)} \left(p_\mu^{\pi^-} - p_\mu^{\pi^0}\right)\right],$$

and using vector meson dominance through the ρ^-, what is

$$\frac{f_+(q^2)}{f_+(0)}$$

at the largest q^2 of the process? What is

$$\frac{f_-(q^2)}{f_+(q^2)}?$$

8.12. Consider the weak vector current matrix elements for $q^2 \to 0$ for states in the vector meson octet. Calculate $\langle \rho^0 | J_0^V | \rho^- \rangle$ and $\langle \omega^0 | J_0^V | \rho^- \rangle$ at $q^2 = 0$. Assume ideal mixing, that is,

$$\omega^0 = \frac{|u,\bar{u}\rangle + |d,\bar{d}\rangle}{\sqrt{2}}.$$

Since these mesons decay strongly, we cannot determine these experimentally.

8.13. In the approximation that the $\Delta I = \frac{1}{2}$ rule gives the whole decay amplitude, we can calculate the Clebsch–Gordan coefficients for K decay by using quark decay diagrams, starting with an initial weak transition that guarantees only $\Delta I = \frac{1}{2}$ operators. Starting with the penguin diagrams and assuming that two $q\bar{q}$ pairs are created, calculate the ratio of the various rates for $K \to 3\pi$. (Be careful to include an $n!$ in the amplitude due to exchange for identical bosons in the same s-wave state and $1/n!$ when you integrate over all final momenta). These coefficients were used to get the ratios in Table 8.6. Show, using the decay diagrams, that $K^- \to \pi^-\pi^0$ vanishes for the penguin diagram, as expected.

8.14. A simple way to find isospin states for states made only of pions is to use our knowledge of ordinary vectors. In particular, for two or three vectors we can make symmetric combinations:

$$\vec{I}_1 \cdot \vec{I}_2 \; (I=0) \quad \text{and} \quad \vec{I}_1(\vec{I}_2 \cdot \vec{I}_3) + \vec{I}_2(\vec{I}_1 \cdot \vec{I}_3) + \vec{I}_3(\vec{I}_1 \cdot \vec{I}_2) \; (I=1).$$

To turn these into pion states, we need the correspondence:

$$\pi^+ = -\left(\frac{I_x + iI_y}{\sqrt{2}}\right), \; \pi^- = \left(\frac{I_x - iI_y}{\sqrt{2}}\right), \; \pi^0 = I_z,$$

between the components of $\vec{I}$ and the pion states (these are the same as the relations between the Cartesian basis and the circular polarization states). Using this correspondence, find the $I = 0$ 2π combination used in $K \to 2\pi$ decays to arrive at Eq. 8.31. [Note that the antisymmetric final states in $\rho \to 2\pi$ and $\omega \to 3\pi$ can be calculated from $\vec{I}_1 \times \vec{I}_2$ and $\vec{I}_1 \cdot (\vec{I}_2 \times \vec{I}_3)$, which transform as $I = 1$ and $I = 0$, respectively].

8.15. Consider the decays $D_s^- \to \tau^- \bar{\nu}_\tau$ and $\mu^- \bar{\nu}_\mu$. These can be calculated in terms of one constant, f_{D_s}. Calculate the ratio of the two branching ratios. Using the measured D_s lifetime, calculate the expected branching ratio to $\tau^- \nu_\tau$, assuming $f_{D_s} = 280$ MeV.

8.16. Consider the D^+ meson. Which weak hadronic currents (that is, vector, axial, or both) are responsible for these decays?
(a) $D^+ \to \mu^+ \nu_\mu$
(b) $D^+ \to \bar{K}^0 \mu^+ \nu_\mu$
(c) $D^+ \to \bar{K}^{*0} \mu^+ \nu_\mu$

8.17. D mesons have both Cabibbo favored and Cabibbo suppressed semileptonic decays. Draw the quark-level Feynman diagrams for these two decays.

8.18. Draw the quark-level Feynman diagrams for hadronic D decays using
(a) Annihilation diagrams,
(b) Scattering diagrams,
(c) Penguin diagrams.

CHAPTER 9

Weak Mixing Phenomena

9.1 ■ INTERPLAY OF PRODUCTION, PROPAGATION, AND DETECTION

In this chapter we look at a number of quantum mechanical mixing phenomena that arise ultimately from a mismatch between the quark and lepton weak interaction eigenstates and mass eigenstates. The mixing phenomena occur in neutrino processes and also for neutral mesons that decay weakly. They provide beautiful examples of quantum mechanics in action in the particle world. In the case of neutrinos, production and detection involve weak interactions, while propagation involves mass eigenstates. The weakly produced neutrino state is generally a mix of mass eigenstates. Coherent interference between the propagating neutrino mass eigenstates affects the position dependence of the final weak detection of a given neutrino type and is called neutrino oscillation. Hadrons, such as K, D, and B mesons, result from particle collisions that yield flavor eigenstate quarks that bind into hadrons. The lightest charged mesons of a given flavor (or any of the lightest baryons containing heavy quarks) propagate as mass eigenstates, subsequently decaying weakly if no interactions occur with intervening matter. For neutral mesons, the weak interactions contribute to create the true mass eigenstates, which are mixes of particle and antiparticle mesons. These propagate with fixed masses and lifetimes. An initial neutral meson state, produced, for example, in a hadronic collision, is a mixture of these true mass eigenstates.

The mass pattern for the neutrino mass eigenstates (analogous to the quark masses) is not understood, but a crucial feature for observable oscillations is that the mass splittings are sufficiently small. For the mixed neutral mesons, the mass splittings are very tiny, arising from weak interactions. The magnitude of these splittings can, however, be understood once the quark masses and CKM matrix are specified. These very small splittings produce very large mixing of states because of the complete degeneracy of the initial particle and antiparticle meson states.

The existence of observable interference and oscillations in the propagation of the states requires an uncertainty in energy and momentum of the source of the particles, related to the localization in space and time of the particle production. Thus, the uncertainty principle, which has not played much of a role in earlier discussions (for example, we typically used perfectly defined momentum eigenstates for all the particles in a process), is a necessary ingredient for oscillations to occur. To understand this, we look at the case of two neutrino mass eigenstates that can mix in the case of π^+ decay to $\mu^+ \nu_\mu$. We specify the mass eigenstates as $|\nu_1\rangle$ and $|\nu_2\rangle$ so that the decay involves $\pi \to \mu\nu_1$, $\pi \to \mu\nu_2$ in the mass eigenstate basis. We do not indicate helicities, since only one helicity is produced to very good approximation. For a π at rest we can calculate the possible neutrino energies in terms of the particle masses:

$$E_{\nu_1} = \frac{m_\pi^2 + m_{\nu_1}^2 - m_\mu^2}{2m_\pi}, \quad E_{\nu_2} = \frac{m_\pi^2 + m_{\nu_2}^2 - m_\mu^2}{2m_\pi}.$$

If m_{ν_1}, m_{ν_2} are sufficiently large and different from each other, we can tell which one was produced and there are no oscillations. In general:

$$E_{\nu_1} - E_{\nu_2} = \frac{m_{\nu_1}^2 - m_{\nu_2}^2}{2m_\pi}.$$

For the measured neutrino parameters discussed later in this chapter, this is approximately 10^{-11} eV, and an initial smearing uncertainty (large compared to this) leads to coherent interference between the amplitudes for the two neutrino types at distances and times large compared to the uncertainty in the space-time location of the production point. For comparison to the energy difference above, the π^+ width is 2.5×10^{-8} eV.

Writing $|\nu_\mu\rangle = |\nu_1\rangle \cos\theta + |\nu_2\rangle \sin\theta$, the state produced from an individual π decay is given by

$$|\psi\rangle = \sum_{\vec{p}_\mu, \vec{p}_{\nu_1}, \vec{p}_{\nu_2}} A_1(\vec{p}_{\nu_1}, \vec{p}_\mu) |\vec{p}_{\nu_1}\rangle |\vec{p}_\mu\rangle \cos\theta + A_2(\vec{p}_{\nu_2}, \vec{p}_\mu) |\vec{p}_{\nu_2}\rangle |\vec{p}_\mu\rangle \sin\theta.$$

The amplitudes A_1 and A_2 carry the information specifying the wave-packet properties of the correlated ν and μ state. In general, predictions for subsequent measurements on the neutrino state, where a measurement is a weak interaction in a material, are specified by a density matrix obtained by projecting over the muon degrees of freedom. Thus the density matrix ρ is given by

$$\rho = \sum_{\vec{p}_\mu} \langle \vec{p}_\mu | \psi \rangle \langle \psi | \vec{p}_\mu \rangle.$$

If $A_1(\vec{p}_{\nu_1}, \vec{p}_\mu)$ and $A_2(\vec{p}_{\nu_2}, \vec{p}_\mu)$ are large only at different values of $\vec{p}_{\nu_1}$ and $\vec{p}_{\nu_2}$, for a given $\vec{p}_\mu$, then the neutrinos behave incoherently; that is, the interference term in ρ between ν_1 and ν_2 can be ignored. If the slight difference in the mean neutrino momentum for A_1 and A_2 is negligible compared to the width in the neutrino momentum variable for a given $\vec{p}_\mu$, then we are in the fully coherent limit where we can take $A_1 = A_2$. The latter is true for the mass differences of states that we are looking at in this chapter. We can then discuss the oscillating states themselves, without reference to the other particles in the event that created the neutrino.

Although the uncertainties in energy and momentum play an important role in providing the fully coherent condition, they have essentially no effect on the actual final result. For example, if the width of A_1 and A_2 in the neutrino momentum is

9.2 ■ MIXING FOR WEAKLY DECAYING PSEUDOSCALAR MESONS

small compared to the mean neutrino momentum, smearing over the tiny width has almost no effect on any measurements far away from the particle source. The primary smearing will come from the event-to-event variation in the mean momentum and flight path, which comes from summing data over many decays.

9.2 ■ MIXING FOR WEAKLY DECAYING PSEUDOSCALAR MESONS

Consider first the mixing for mesons. We indicate the neutral meson states as $|P^0\rangle$ and $|\bar{P}^0\rangle$. These are each eigenstates of the strong and electromagnetic interactions. While propagating, these mesons experience weak interactions that can create transitions to other states, such as the decays discussed in Chapter 8, but also transitions between the two can occur. The amplitude matrix describing the transitions between the two is $\pi_{ij} = M_{ij}$, where M_{ij} is calculated in principle in the usual way using Feynman diagrams. Here the indices i and j specify the neutral meson or its antiparticle.

Including the interactions during the propagation gives for the propagator the perturbation sum drawn in Figure 9.1. We discussed this previously in Sections 5.3 and 6.4.3 and will study the mixing phenomenon more thoroughly here. The propagating mass eigenstates are those that diagonalize π_{ij}. We can sum the set of diagrams, giving the corrected propagator:

$$\frac{i}{p^2 - m^2 \delta_{ij} - \pi_{ij}(p^2)}.$$

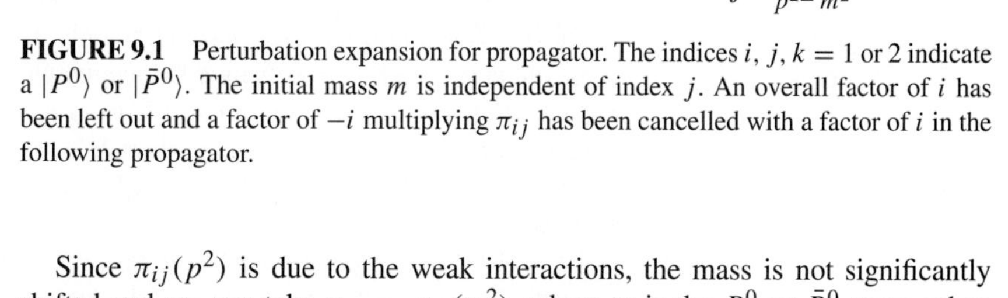

FIGURE 9.1 Perturbation expansion for propagator. The indices $i, j, k = 1$ or 2 indicate a $|P^0\rangle$ or $|\bar{P}^0\rangle$. The initial mass m is independent of index j. An overall factor of i has been left out and a factor of $-i$ multiplying π_{ij} has been cancelled with a factor of i in the following propagator.

Since $\pi_{ij}(p^2)$ is due to the weak interactions, the mass is not significantly shifted and we can take $\pi_{ij} = \pi_{ij}(m^2)$, where m is the P^0 or $\bar{P}^0$ mass, when calculating the eigenstates. In addition, since $|\pi_{ij}| \ll m^2$,

$$p^2 - m^2 \delta_{ij} - \pi_{ij} \simeq p^2 - \left(m\delta_{ij} + \frac{\pi_{ij}}{2m}\right)^2.$$

We can write a general matrix in terms of two Hermitian matrices. Defining:

$$\pi = \frac{(\pi + \pi^\dagger)}{2} - i\left[i\frac{(\pi - \pi^\dagger)}{2}\right] = \pi^{(+)} - i\pi^{(-)},$$

$\pi^{(+)}$ and $\pi^{(-)}$ are each Hermitian. Dividing by $2m$, we get the matrix that gets added to $m\delta_{ij}$ in the propagator:

$$\Delta m - \frac{i\Gamma}{2} = \frac{\pi^{(+)}}{2m} - \frac{i\pi^{(-)}}{2m}.$$

The matrices Δm and Γ are called the mass and width matrices, respectively. They are each Hermitian by construction.

The matrix π gets contributions from virtual and real intermediate states that can link $|P^0\rangle$ and $|\bar{P}^0\rangle$. The term $\pi^{(+)}$ is due to virtual states and $\pi^{(-)}$ is due to real states. The latter follows from unitarity, as we show next.

The S matrix, $S_{fi} = \delta_{fi} - i(2\pi)^4 \delta^4(p_f - p_i) M_{fi}$, satisfies $S^\dagger S = \delta_{fi}$, which implies that:

$$i(M_{fi} - M^*_{if}) = i(M_{fi} - M^\dagger_{fi}) = (2\pi)^4 \sum_n \delta^4(p_n - p_i) M^*_{nf} M_{ni}.$$

Within the subspace of states spanned by the two pseudoscalar mesons, M is the matrix π, so that we have:

$$2\pi^{(-)}_{fi} = (2\pi)^4 \sum_n \delta^4(p_n - p_i) M^*_{nf} M_{ni}$$

and

$$\Gamma_{fi} = \frac{1}{2m} \sum_n M^*_{nf} M_{ni} (2\pi)^4 \delta^4(p_n - p_i).$$

Here i is the initial pseudoscalar meson, and f is the final pseudoscalar meson. If there were no mixing (that is, Δm and Γ diagonal), the formulas would give the usual mass renormalization and total width that characterize each individual decaying state. With mixing, the off-diagonal terms in Γ get contributions from real decays (since $p_n = p_i$ due to the δ-function) that both $|P^0\rangle$ and $|\bar{P}^0\rangle$ share in common (that is, for which both M_{ni} and M^*_{nf} are nonzero). Note that for $i = f$, all states n contribute a positive term to Γ_{ii}; for $i \neq f$ the terms can contribute with different signs.

The form of the mixing matrix can be simplified using *CPT* symmetry, where the states related by the symmetry transformation were discussed in Section 3.9. For the present case, this relates the matrix elements $\pi_{ij} = \pi_{CPTj,CPTi}$. For $i = j = 1$ ($|P^0\rangle$ meson) and $i = j = 2$ ($|\bar{P}^0\rangle$ meson) this gives $\pi_{11} = \pi_{22}$. For $i = 1, j = 2$ or $i = 2, j = 1$, it gives no extra information. Thus we can write in general:

9.2 Mixing for Weakly Decaying Pseudoscalar Mesons

$$\Delta m - \frac{i\Gamma}{2} = \begin{pmatrix} \Delta m_{11} - \frac{i\Gamma_{11}}{2} & \Delta m_{12} - \frac{i\Gamma_{12}}{2} \\ \Delta m_{12}^* - \frac{i\Gamma_{12}^*}{2} & \Delta m_{11} - \frac{i\Gamma_{11}}{2} \end{pmatrix}. \quad (9.1)$$

Here Δm_{11} and Γ_{11} are real.

We have an additional relation for the case of CP invariance, which is $\pi_{ij} = \pi_{CPi,CPj}$. Now taking $i = 1(|P^0\rangle)$ and $j = 2(|\bar{P}^0\rangle)$ we get $\pi_{12} = \pi_{21}$, that is, π is a symmetric matrix. In this case $\Delta m_{12} = \Delta m_{12}^*$ and $\Gamma_{12} = \Gamma_{12}^*$. The eigenstates of Eq. 9.1 are then automatically

$$\frac{|P^0\rangle \pm |\bar{P}^0\rangle}{\sqrt{2}}.$$

Note that the CP symmetry expectation is spoiled by phases in either Δm_{12} (from virtual states) or Γ_{12} (from real transitions), as we expect to get from the three-generation CKM matrix.

The general solution for the eigenstates, as we show below, is

$$|P_1\rangle = p\left|P^0\right\rangle + q\left|\bar{P}^0\right\rangle$$
$$|P_2\rangle = p\left|P^0\right\rangle - q\left|\bar{P}^0\right\rangle. \quad (9.2)$$

The complex constants p and q satisfy $|p|^2 + |q|^2 = 1$. Note that $|P_1\rangle$ and $|P_2\rangle$ are not automatically orthogonal, that is, $|p|^2 \neq |q|^2$ in all cases. This comes about because the combination $\Delta m - (i\Gamma/2)$ is not Hermitian.

The matrix in Eq. 9.1 is of the form

$$\begin{pmatrix} A & B \\ C & A \end{pmatrix},$$

which has eigenvalues $\lambda_{1,2} = A \pm \sqrt{BC}$, with eigenvectors

$$\frac{1}{\sqrt{|B|^2 + |C|^2}} \begin{pmatrix} \sqrt{B} \\ \sqrt{C} \end{pmatrix} \quad \text{and} \quad \frac{1}{\sqrt{|B|^2 + |C|^2}} \begin{pmatrix} \sqrt{B} \\ -\sqrt{C} \end{pmatrix}.$$

The ratio

$$\frac{q}{p} = \sqrt{\frac{C}{B}} = \sqrt{\frac{\Delta m_{12}^* - i\Gamma_{12}^*/2}{\Delta m_{12} - i\Gamma_{12}/2}}. \quad (9.3)$$

This ratio will be 1 if CP is conserved. It will be a phase factor $e^{i\psi}$, that is, have magnitude 1, if $|\Delta m_{12}| \gg |\Gamma_{12}|$ or $|\Gamma_{12}| \gg |\Delta m_{12}|$.

The states $|P_1\rangle$ and $|P_2\rangle$ will propagate in space-time in their rest-frame as

$$|P_1(t)\rangle = e^{-im_1 t - \Gamma_1 t/2} |P_1(0)\rangle \quad \text{or} \quad |P_2(t)\rangle = e^{-im_2 t - \Gamma_2 t/2} |P_2(0)\rangle, \quad (9.4)$$

which gives a probability of finding the state at time t equal to $e^{-\Gamma_1 t}$ or $e^{-\Gamma_2 t}$, respectively, if we are initially in a mass eigenstate. The eigenvalues are (with

1, 2 corresponding to the upper and lower signs, respectively):

$$m_{1,2} = m + \Delta m_{11} \pm \text{Re}\sqrt{\left(\Delta m_{12} - \frac{i\Gamma_{12}}{2}\right)\left(\Delta m_{12}^* - \frac{i\Gamma_{12}^*}{2}\right)}$$

$$\frac{\Gamma_{1,2}}{2} = \frac{\Gamma_{11}}{2} \mp \text{Im}\sqrt{\left(\Delta m_{12} - \frac{i\Gamma_{12}}{2}\right)\left(\Delta m_{12}^* - \frac{i\Gamma_{12}^*}{2}\right)} \qquad (9.5)$$

If we can ignore *CP* violation, these simplify to:

$$m_{1,2} = m + \Delta m_{11} \pm \Delta m_{12},$$

$$\Gamma_{1,2} = \Gamma_{11} \pm \Gamma_{12}. \qquad (9.6)$$

9.3 ■ $K^0, \bar{K}^0$ SYSTEM

Although the mixing formalism is the same for each pseudoscalar meson pair, the visible results are quite particular to each flavor. We will look first at the K^0 and $\bar{K}^0$ system, which we discuss in the most detail. *CP* violation is a small effect for this system, so we will look initially at what happens in the limit of *CP* symmetry. We discuss the widths first, which are easiest to understand, involving real decay states.

For *CP* symmetry, the eigenstates (sometimes called K_1 and K_2) are:

$$\left|K_S^0\right\rangle = \frac{\left|K^0\right\rangle + \left|\bar{K}^0\right\rangle}{\sqrt{2}}, \quad \left|K_L^0\right\rangle = \frac{\left|K^0\right\rangle - \left|\bar{K}^0\right\rangle}{\sqrt{2}}.$$

Of the predominant decays, both K^0 and $\bar{K}^0$ decay to 2π and 3π states. These final states are *CP* even and *CP* odd (in the fully symmetric *s*-wave approximation for 3π), respectively. Therefore, we have for the individual amplitudes, as discussed in Chapter 8:

$$M_{K^0 \to 2\pi} = M_{\bar{K}^0 \to 2\pi}, \quad M_{K^0 \to 3\pi} = -M_{\bar{K}^0 \to 3\pi}.$$

The semileptonic decays are not common decays for $\bar{K}^0$ and K^0, since in one case we get $\pi^+ e^- \bar{\nu}_e$ and in the other $\pi^- e^+ \nu_e$. Thus we can write, ignoring the very rare decays,

$$\Gamma_{12} = \Gamma_{2\pi} - \Gamma_{3\pi}.$$

Here $\Gamma_{2\pi}$ and $\Gamma_{3\pi}$ are the K^0 or $\bar{K}^0$ widths to the given final state. Using Eq. 9.6, we get to good approximation:

$$\Gamma_{K_S^0} = 2\Gamma_{2\pi} + \Gamma_{\text{semileptonic}}, \quad \Gamma_{K_L^0} = 2\Gamma_{3\pi} + \Gamma_{\text{semileptonic}}. \qquad (9.7)$$

9.3 K^0, $\bar{K}^0$ System

We thus see that $\Gamma_{K_S^0} \gg \Gamma_{K_L^0}$ because $\Gamma_{2\pi} \gg \Gamma_{3\pi}$. This is, however, due merely to the much greater phase space for the 2π decay. We would not expect such a large difference in decay rates between CP even and odd final states for the pseudoscalar mesons made of heavier quarks.

The mass difference between K_S^0 and K_L^0 is not as easily predicted, stemming from all possible virtual transitions that link K^0 and $\bar{K}^0$. We define

$$\Delta m_K = m_{K_L^0} - m_{K_S^0} = -2\Delta m_{12}. \qquad (9.8)$$

Δm_K is measured to be positive and equal to $(0.474 \pm 0.001)\Gamma_S$.

If we think in terms of hadrons (called long-distance contributions), some contributing terms to Δm_K are

$$K_L^0 \to \pi^0, \eta, \eta' \to K_L^0 \quad (CP \text{ odd contribution}).$$
$$K_S^0 \to 2\pi_{\text{virtual}} \to K_S^0 \quad (CP \text{ even contribution}).$$
$$K_L^0 \to 3\pi_{\text{virtual}} \to K_L^0 \quad (CP \text{ odd for small virtual mass}).$$

We do not, however, know how to calculate such hadronic amplitudes. We can also look at the mixing in terms of quarks (called short-distance contributions, since this is expected to be a correct description at distance scales much smaller than the hadronic size). In this case, the mixing is due to box diagrams, shown in Figure 9.2.

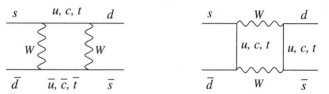

FIGURE 9.2 Box diagrams involved in K^0, $\bar{K}^0$ mixing. We can also have penguin diagrams with two loops linking the K^0 and $\bar{K}^0$ states; these are very small.

For these diagrams, the GIM mechanism (the cancellation at large virtual masses between diagrams with propagating u and c quarks, discussed in Section 8.10) and the small CKM elements for the t quark contribution limit the value of the box diagrams. An estimate of their contribution to Δm_K is

$$\Delta m_K \simeq \frac{G_F^2}{6\pi^2} f_K^2 m_K m_c^2 \cos^2\theta_c \sin^2\theta_c. \qquad (9.9)$$

Using $m_c = 1.5$ GeV, Eq. 9.9 gives a value for Δm_K that is about 15% smaller than the measured value. This approximate agreement indicates that we understand, at least qualitatively, the value of Δm_K.

9.3.1 ■ Kaon Oscillations

In the approximation of *CP* conservation, the K_S^0 and K_L^0 decay patterns are simple for final states that are *CP* eigenstates. If we look at decays to 2π we see only one amplitude, that of the rapidly decaying K_S^0. Similarly, if we observe 3π we see only the slowly decaying K_L^0. However, there are also decays, primarily the semileptonic decays, to which both K_S^0 and K_L^0 contribute. In this case, the decay at a given time exhibits an interference, containing contributions from both propagating eigenstates. We consider this situation next.

To keep track of the K_S^0 and K_L^0 components over time, we need to know the initial state. This state is determined by the production process for the kaons. A K^0 state can be produced via low energy $\pi^- + p$ scattering into states containing a strange baryon. Thus, for example $\pi^- + p \to \Lambda^0 + K^0$ produces a K^0. To produce a $\bar{K}^0$ requires more final state particles, which is disfavored at low energies and forbidden by energy conservation at sufficiently low energy. Using a production process that fixes the pseudoscalar's flavor at birth is called flavor tagging in production. A final K decay that can only come from one of K^0 or $\bar{K}^0$ is called flavor tagging in the decay.

To look at kaon oscillations for a flavor-tagged initial state, we assume that at $t = 0$ we have a state that is $|K^0\rangle$. For simplicity we assume *CP* conservation. Therefore,

$$|K(t=0)\rangle = \left|K^0\right\rangle = \frac{\left|K_S^0\right\rangle + \left|K_L^0\right\rangle}{\sqrt{2}}.$$

This state, if undisturbed by collisions, will propagate in its rest-frame as

$$\begin{aligned}|K(t)\rangle &= \frac{\left|K_S^0\right\rangle e^{-\left(\frac{\Gamma_S}{2}+im_S\right)t} + \left|K_L^0\right\rangle e^{-\left(\frac{\Gamma_L}{2}+im_L\right)t}}{\sqrt{2}} \\ &= \frac{\left[e^{-\left(\frac{\Gamma_S}{2}+im_S\right)t} + e^{-\left(\frac{\Gamma_L}{2}+im_L\right)t}\right]}{2}\left|K^0\right\rangle \\ &\quad + \frac{\left[e^{-\left(\frac{\Gamma_S}{2}+im_S\right)t} - e^{-\left(\frac{\Gamma_L}{2}+im_L\right)t}\right]}{2}\left|\bar{K}^0\right\rangle \end{aligned} \quad (9.10)$$

This gives for the intensities for the two kaon components:

$$I(K^0) = \left|\left\langle K^0 | K(t)\right\rangle\right|^2 = \frac{1}{4}\left[e^{-\Gamma_S t} + e^{-\Gamma_L t} + 2e^{\frac{-(\Gamma_S+\Gamma_L)}{2}t}\cos(\Delta m_K t)\right]$$

$$I(\bar{K}^0) = \left|\left\langle \bar{K}^0 | K(t)\right\rangle\right|^2 = \frac{1}{4}\left[e^{-\Gamma_S t} + e^{-\Gamma_L t} - 2e^{\frac{-(\Gamma_S+\Gamma_L)}{2}t}\cos(\Delta m_K t)\right].$$

$$(9.11)$$

9.3 $K^0, \bar{K}^0$ System

Tagging decays, indicating the presence of K^0 or $\bar{K}^0$, are observed to occur with a pattern in time given by the appropriate intensity above. If we start with a number of $K^0 = N_{K^0}$ at $t = 0$, and the width to a given tagging decay is Γ_{tag}, then the number of decays to the tagging final state is, for example,

$$\frac{dN_{\text{tag}}}{dt} = \Gamma_{\text{tag}} N_{K^0} I(K^0)$$

for a final K^0. This is translated into a pattern in space if the K system is moving in the laboratory. The $\cos(\Delta m_K t)$ term makes the pattern oscillatory. Since the state decays, the oscillations will be noticeable only if Δm_K is not small compared to $(\Gamma_S + \Gamma_L)/2$. This is the situation for the K system.

Since we started with a K^0, the degree of mixing is specified by how often we see a $\bar{K}^0$ decaying rather than a K^0. We can quantify this, over the whole time history, by the ratio

$$r = \frac{\int I(\bar{K}^0) dt}{\int I(K^0) dt} = \frac{\left(\frac{\Gamma_S - \Gamma_L}{\Gamma_S + \Gamma_L}\right)^2 + 4\left(\frac{\Delta m_K}{\Gamma_S + \Gamma_L}\right)^2}{2 - \left(\frac{\Gamma_S - \Gamma_L}{\Gamma_S + \Gamma_L}\right)^2 + 4\left(\frac{\Delta m_K}{\Gamma_S + \Gamma_L}\right)^2}. \quad (9.12)$$

For the K system, $\Gamma_S \gg \Gamma_L$; in this limit $r \simeq 1$, independent of Δm_K.

9.3.2 ■ CP Violation in the Kaon System

With *CP* not conserved, we must use Eq. 9.2 to express the K^0_S and K^0_L eigenstates in terms of K^0 and $\bar{K}^0$. For a tagged initial state (specifically K^0) we can repeat the algebra leading to Eq. 9.10. This gives the general propagation equation:

$$|K(t)\rangle = \left[\frac{e^{-\left(\frac{\Gamma_S}{2} + im_S\right)t} + e^{-\left(\frac{\Gamma_L}{2} + im_L\right)t}}{2}\right] |K^0\rangle$$

$$+ \frac{q}{p}\left[\frac{e^{-\left(\frac{\Gamma_S}{2} + im_S\right)t} - e^{-\left(\frac{\Gamma_L}{2} + im_L\right)t}}{2}\right] |\bar{K}^0\rangle. \quad (9.13)$$

Thus the only change due to *CP* violation is the factor q/p in the second term. If we had started with a $\bar{K}^0$, we would get the expression for this case by exchanging $|K^0\rangle \leftrightarrow |\bar{K}^0\rangle$ and $q/p \leftrightarrow p/q$ in Eq. 9.13.

The intensity functions are now the same as for no *CP* violation, except for a factor $|q/p|^2$ multiplying the expression for $I(\bar{K}^0)$ in Eq. 9.11. The combined effect of *CP* violation and kaon oscillations is often plotted by calculating the asymmetry:

$$A(t) = \frac{I(K^0) - I(\bar{K}^0)}{I(K^0) + I(\bar{K}^0)}. \quad (9.14)$$

The intensities are measured via the time distribution of tagging decays for K^0 and $\bar{K}^0$, for example, decays to $\pi^- e^+ \nu_e$ and $\pi^+ e^- \bar{\nu}_e$, respectively. The very beautiful data for $A(t)$ are shown in Figure 9.3. From these data we can extract $|\Delta m_K|$.

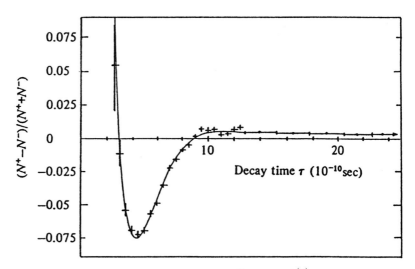

FIGURE 9.3 Charge asymmetry in the decay $K^0 \to \pi^{\mp} e^{\pm} \overset{(-)}{\nu}$ as a function of proper time. Both the K_L, K_S interference and the CP violating offset are clearly shown. [From S. Gjesdal et al., *Phys. Lett.* 52B, 113(1974).]

The result for $A(t)$ is particularly simple for $t \gg \tau_S$. Here K_S^0 has decayed away and we observe the K_L^0 part only. In this regime, $A(t)$ is independent of time and is

$$A = A(t \gg \tau_S) = \frac{1 - \left|\frac{q}{p}\right|^2}{1 + \left|\frac{q}{p}\right|^2} = \frac{|p|^2 - |q|^2}{|p|^2 + |q|^2} = |p|^2 - |q|^2. \quad (9.15)$$

For the kaon system, the CP violation is quite small. We can therefore, to first order, write:

$$p = \frac{1+\varepsilon}{\sqrt{2}}, \quad q = \frac{1-\varepsilon}{\sqrt{2}}.$$

With this notation, $A = 2\text{Re}(\varepsilon)$ to first order in ε. This provides a direct measurement of $\text{Re}(\varepsilon)$. The result is

$$\text{Re}(\varepsilon) = (1.64 \pm 0.06) \times 10^{-3}. \quad (9.16)$$

CP violation was discovered, however, not through observing the tagged decays, but rather through decays to the CP eigenstate $\pi\pi$ final state. Any nonzero

9.3 K^0, $\bar{K}^0$ System

rate of K_L^0 to $\pi\pi$ is a signal for *CP* violation. In the limit where *CP* violation stems only from virtual states, that is, from the mass matrix, the *CP* violating amplitudes are easy to calculate in terms of ε. In terms of the CP^+ and CP^- kaon states $(K^0 \pm \bar{K}^0)/\sqrt{2}$, we can write to first order:

$$\left|K_S^0\right\rangle = \left|K_{CP+}^0\right\rangle + \varepsilon \left|K_{CP-}^0\right\rangle$$
$$\left|K_L^0\right\rangle = \left|K_{CP-}^0\right\rangle + \varepsilon \left|K_{CP+}^0\right\rangle. \tag{9.17}$$

With no *CP* violation in the decay amplitudes $|K_{CP+}^0\rangle \to CP$ even states, $|K_{CP-}^0\rangle \to CP$ odd states and therefore for any given *CP* even or odd final states, we get amplitude ratios

$$\frac{A_{K_L^0 \to CP^+ \text{state}}}{A_{K_S^0 \to CP^+ \text{state}}} = \frac{A_{K_S^0 \to CP^- \text{state}}}{A_{K_L^0 \to CP^- \text{state}}} = \varepsilon.$$

This gives a good, but not exact, description of the rates, as we will see below. Note that the two-family CKM matrix, which is sufficient for most decays, gives no *CP* violation in the decays; ε, which comes from virtual states (with contributions from all three families), violates the *CP* symmetry.

Focusing on the 2π system, we define more generally, that is, also including *CP* violation in the decay amplitudes:

$$\eta_{+-} = \frac{A_{K_L^0 \to \pi^+\pi^-}}{A_{K_S^0 \to \pi^+\pi^-}}, \quad \eta_{00} = \frac{A_{K_L^0 \to \pi^0\pi^0}}{A_{K_S^0 \to \pi^0\pi^0}}. \tag{9.18}$$

We can measure the magnitudes of these using the measured K_L^0, K_S^0 lifetimes and their respective branching ratios to $\pi\pi$. For example,

$$|\eta_{+-}|^2 = \frac{B\left(K_L^0 \to \pi^+\pi^-\right)}{B\left(K_S^0 \to \pi^+\pi^-\right)} \left(\frac{\tau_S}{\tau_L}\right),$$

with a similar relation for $\pi^0\pi^0$. From these we find

$$|\eta_{00}| \simeq |\eta_{+-}| = (2.28 \pm 0.02) \times 10^{-3}. \tag{9.19}$$

Using Eq. 9.17, we have for the amplitude ratios, including direct *CP* violation in the decay amplitudes for the *CP* odd kaon state:

$$\eta_{+-} = \frac{A_{CP- \to \pi^+\pi^-} + \varepsilon A_{CP+ \to \pi^+\pi^-}}{A_{CP+ \to \pi^+\pi^-} + \varepsilon A_{CP- \to \pi^+\pi^-}}, \quad \eta_{00} = \frac{A_{CP- \to \pi^0\pi^0} + \varepsilon A_{CP+ \to \pi^0\pi^0}}{A_{CP+ \to \pi^0\pi^0} + \varepsilon A_{CP- \to \pi^0\pi^0}}.$$

Since the *CP* violation in the decay will be seen to be small, the product $\varepsilon A_{CP- \to \pi\pi}$ is very small and can be ignored. This gives

$$\eta_{+-} = \varepsilon + \frac{A_{CP^- \to \pi^+\pi^-}}{A_{CP^+ \to \pi^+\pi^-}}, \quad \eta_{00} = \varepsilon + \frac{A_{CP^- \to \pi^0\pi^0}}{A_{CP^+ \to \pi^0\pi^0}}.$$

We next construct the various amplitudes using the formulas for the K decays from Chapter 8, Eq. 8.33:

$$A_{CP^- \to \pi^+\pi^-} = \sqrt{\tfrac{1}{3}} e^{i\delta_0} A_0 \left(e^{-i\phi} - e^{i\phi} \right)$$

$$A_{CP^- \to \pi^0\pi^0} = -\sqrt{\tfrac{1}{3}} e^{i\delta_0} A_0 \left(e^{-i\phi} - e^{i\phi} \right).$$

$$A_{CP^+ \to \pi^+\pi^-} = \sqrt{\tfrac{2}{3}} e^{i\delta_2} A_2 + \sqrt{\tfrac{1}{3}} e^{i\delta_0} A_0 \left(e^{i\phi} - e^{-i\phi} \right)$$

$$A_{CP^+ \to \pi^0\pi^0} = 2\sqrt{\tfrac{2}{3}} e^{i\delta_2} A_2 - \sqrt{\tfrac{1}{3}} e^{i\delta_0} A_0 \left(e^{i\phi} - e^{-i\phi} \right).$$

Taking $e^{-i\phi} - e^{i\phi} \simeq -2i\phi$, $e^{i\phi} + e^{-i\phi} \simeq 2$, and expanding to first order in the small quantity A_2/A_0, which equals 0.044, gives the result:

$$\eta_{+-} = \varepsilon - i\phi + \varepsilon', \quad \eta_{00} = \varepsilon - i\phi - 2\varepsilon', \quad \text{where} \quad \varepsilon' = i\phi \frac{A_2}{\sqrt{2} A_0} e^{i(\delta_2 - \delta_0)}. \quad (9.20)$$

By very carefully measuring the relative rates for K_L^0 to $\pi^0\pi^0$ versus $\pi^+\pi^-$, and assuming that ε' is approximately in phase with η_{+-}, it is possible to determine experimentally that:

$$\left| \frac{\varepsilon'}{\varepsilon - i\phi} \right| \simeq 2 \times 10^{-3}.$$

This gives for ϕ:

$$\phi \simeq \frac{2 \times 10^{-3} \times |\eta_{+-}|}{\frac{A_2}{\sqrt{2} A_0}} = 1.4 \times 10^{-4}. \quad (9.21)$$

Thus to good approximation $\eta_{+-} = \eta_{00} = \varepsilon$. Note that the phase of ε' is $90° + \delta_2 - \delta_0 \simeq 48°$. This is rather close to the measured phase of η_{+-}, about 43.5°, which justifies the assumption above of nearly equal phases.

The phases of η_{+-} and η_{00} are measured by looking at the detailed interference pattern between K_S^0 and K_L^0 as they evolve coherently in time and decay to $\pi^+\pi^-$ or $\pi^0\pi^0$. The initial state for such an experiment can be a flavor-tagged state such as K^0. These measurements also yield the sign and magnitude of Δm_K. We look at the data after discussing the expectation for the phases.

Within the CKM framework we can accurately predict the phase of η_{+-} and η_{00}. To do this we look first at ε. Since it is small, we can conclude from Eq. 9.3 that $\text{Im}(\Delta m_{12}) \ll \text{Re}(\Delta m_{12})$ and $\text{Im}(\Gamma_{12}) \ll \text{Re}(\Gamma_{12})$. With the imag-

9.3 $K^0, \bar{K}^0$ System

inary parts small we can solve for ε from Eq. 9.3, keeping only the first order terms:

$$\varepsilon = \frac{\left[i\text{Im}(\Delta m_{12}) + \text{Im}\frac{\Gamma_{12}}{2}\right]}{2\left[\text{Re}(\Delta m_{12}) - i\text{Re}\frac{\Gamma_{12}}{2}\right]}. \quad (9.22)$$

Using this formula for ε and ignoring the very small ε' term in Eq. 9.20, gives

$$\eta_{+-} = \frac{i\text{Im}(\Delta m_{12}) + \text{Im}\frac{\Gamma_{12}}{2}}{2\left[\text{Re}(\Delta m_{12}) - i\text{Re}\frac{\Gamma_{12}}{2}\right]} - i\phi$$

$$= \frac{i\left[\text{Im}(\Delta m_{12}) - 2\text{Re}(\Delta m_{12})\phi\right] + \left[\text{Im}\frac{\Gamma_{12}}{2} - 2\phi\text{Re}\frac{\Gamma_{12}}{2}\right]}{2\left(\text{Re}(\Delta m_{12}) - i\text{Re}\frac{\Gamma_{12}}{2}\right)}.$$

In the very good approximation that the 2π decay mode gives the entire value of $\text{Im}\Gamma_{12}$, and $\text{Re}\Gamma_{12}$ and that we can ignore the $\Delta I = \frac{3}{2}$ contribution to $\text{Re}\Gamma_{12}$, we have from Eq. 8.33:

$$\text{Im}\Gamma_{12} = 2\phi\text{Re}\Gamma_{12},$$

so that the second term in brackets vanishes. This gives then:

$$\eta_{+-} = \frac{i\left[\frac{\text{Im}(\Delta m_{12})}{\text{Re}(\Delta m_{12})} - 2\phi\right]}{2\left[1 - \frac{i\text{Re}\Gamma_{12}}{2\text{Re}(\Delta m_{12})}\right]}.$$

Using Eq. 9.6 to replace $\text{Re}\Gamma_{12}$ by $(\Gamma_S - \Gamma_L)/2$ and $2\text{Re}(\Delta m_{12})$ by $-\Delta m_K$ in the denominator gives

$$\eta_{+-} = \frac{i\left[\frac{\text{Im}(\Delta m_{12})}{2\text{Re}(\Delta m_{12})} - \phi\right]}{1 + \frac{i(\Gamma_S - \Gamma_L)}{2\Delta m_K}} = \frac{\left[\frac{(\Gamma_S - \Gamma_L)}{2\Delta m_K} + i\right]\left[\frac{\text{Im}(\Delta m_{12})}{2\text{Re}(\Delta m_{12})} - \phi\right]}{1 + \left(\frac{\Gamma_S - \Gamma_L}{2\Delta m_K}\right)^2}. \quad (9.23)$$

We see that the phase of η_{+-}, ϕ_{+-}, can be accurately predicted from the first term in brackets in Eq. 9.23:

$$\tan\phi_{+-} = \left(\frac{2\Delta m_K}{\Gamma_S - \Gamma_L}\right).$$

It involves only directly measured quantities. The result predicted, $\phi_{+-} = 43.5°$, is in excellent agreement with the measured value of $43.51 \pm 0.06°$. We can also use $|\eta_{+-}|$ to calculate $\text{Im}(\Delta m_{12})$. Since ϕ is small, we neglect it, also

$$\frac{\Gamma_S - \Gamma_L}{2\Delta m_K} \simeq 1,$$

so that from Eq. 9.23:

$$\frac{\mathrm{Im}(\Delta m_{12})}{\mathrm{Re}(\Delta m_{12})} \simeq 2\sqrt{2}\,|\eta_{+-}| \simeq 6 \times 10^{-3}. \qquad (9.24)$$

As a final topic, we investigate the expected time distribution for decay into $\pi^+\pi^-$ starting from a tagged K^0 initial state. This is used to measure ϕ_{+-}. At $t = 0$ we have

$$|K(t=0)\rangle = |K^0\rangle = \frac{1}{2p}\left[|K_S^0\rangle + |K_L^0\rangle\right].$$

$|K_S^0\rangle$ and $|K_L^0\rangle$ evolve as mass eigenstates. Thus the amplitude to go to $\pi^+\pi^-$ is

$$A_{K(t)\to\pi^+\pi^-} = \frac{1}{2p} A_{K_S^0 \to \pi^+\pi^-}\left[e^{-(\frac{\Gamma_S}{2}+im_S)t} + \eta_{+-}\,e^{-(\frac{\Gamma_L}{2}+im_L)t}\right].$$

Squaring, we get the time dependent intensity:

$$I_{\pi^+\pi^-}(t) = I_0\left[e^{-\Gamma_S t} + |\eta_{+-}|^2 e^{-\Gamma_L t}\right.$$
$$\left. + 2e^{-\left(\frac{\Gamma_S+\Gamma_L}{2}\right)t}|\eta_{+-}|\cos(\Delta m_K t - \phi_{+-})\right]. \qquad (9.25)$$

Figure 9.4 shows the very nice data. At small times, we have dominantly K_S^0 decay. At times such that $e^{-\Gamma_S t/2} \sim |\eta_{+-}|$, the oscillatory term is also significant, while at long times we have essentially only K_L^0 decay left.

The mixing phenomena for the kaon system are characterized by a number of observables: masses and widths of eigenstates, asymmetry in tagging decays, and interference in common decays. The measurable parameters are $\Gamma_S, \Gamma_L, \Delta m_K, |p|^2 - |q|^2, \eta_{+-}, \eta_{00}$. We can also consider other parameters such as the phase of p and q, which are not directly measurable. With a specific choice for the CKM matrix, these phases are calculable. Since other choices are possible, we can ask what would happen with another parameterization of the CKM matrix. The answer is that the unmeasured phases, such as those of p and q, are changed, but all measurable parameters are left unchanged. To get correct results we have to stick to a given choice for the CKM matrix for all parts of a calculation. We will always use the parameterization introduced in Chapter 8. Within this parameterization, we will treat all phases as meaningful—even if some are intermediate to arriving at the measurables—in order to be sure that our calculations are fully consistent. Since the *CP* violation in our CKM convention arises primarily from transitions from the $3 \to 1$ generation, we conclude that for the kaon system, the phase ϕ arises mainly from the t quark in the penguin diagram and $\mathrm{Im}(\Delta m_{12})$ from the t quark in the box diagram. These are the diagrams in Figures 8.8 and 9.2 containing the t quark. The CKM element that contains the phase in both of these processes is V_{td}. It will occur again in $B^0, \bar{B}^0$ mixing, where the t quark dominates the box diagram calculation.

9.4 ■ $D^0, \bar{D}^0$ System

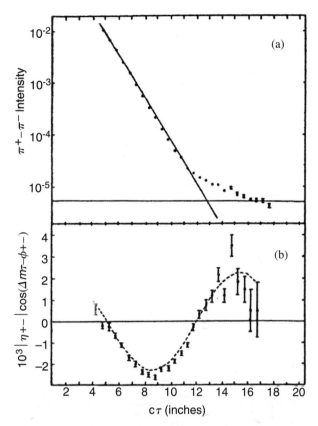

FIGURE 9.4 Intensity of $\pi^+\pi^-$ as a function of proper time τ. Curves in the top figure show the two exponential contributions. These are modified by a sinusoidal term shown in the lower figure, where the curve gives the best fit to $|\eta|$ and ϕ. [From J. H. Christenson et al., *Phys. Rev. Lett.* 43, 1212(1979).]

9.4 ■ $D^0, \bar{D}^0$ SYSTEM

For meson pairs other than $K^0, \bar{K}^0$, the mass eigenstates have nearly identical widths and the lifetimes are rather short. In addition, as we will see, we can take $|p| = |q|$ to good approximation. For short lifetimes, the simplest observable is the time integrated mixing parameter r in Eq. 9.12. What is usually quoted as the degree of mixing is the fraction of tagged decays of opposite flavor from the initially tagged meson to the total number of meson decays tagged as $|P\rangle$ or $|\bar{P}\rangle$. For $|q/p| = 1$ this is

$$\chi = \frac{r}{1+r}.$$

Chapter 9 Weak Mixing Phenomena

Defining the mean lifetime $\bar{\Gamma}$ as

$$\frac{\Gamma_S + \Gamma_L}{2}$$

and the lifetime difference $\Delta\Gamma = \Gamma_S - \Gamma_L$, we have that:

$$\chi = \frac{\left(\frac{\Delta m}{\bar{\Gamma}}\right)^2 + \left(\frac{\Delta\Gamma}{2\bar{\Gamma}}\right)^2}{2\left[1 + \left(\frac{\Delta m}{\bar{\Gamma}}\right)^2\right]}. \qquad (9.26)$$

This is quadratic in the various ratios, so small values of $\Delta m/\bar{\Gamma}$ (often called x) and $\Delta\Gamma/2\bar{\Gamma}$ (often called y) will give very small mixing fractions.

The parameters $\Delta m/\bar{\Gamma}$ and $\Delta\Gamma/2\bar{\Gamma}$ for the D system are expected to be small and, in fact, no mixing (or CP violation) has yet been observed for this system. The present experimental limits are approximately

$$\frac{\Delta m_D}{\bar{\Gamma}_D} \quad \text{and} \quad \frac{\Delta\Gamma_D}{2\bar{\Gamma}_D} < 0.02,$$

although expectations are that they are each $\ll 0.01$. The mixing parameters for the D^0, $\bar{D}^0$ system are expected to be small for three reasons, as follows:

1. Cabibbo Suppression: For $\Delta\Gamma_D$, mixing must involve common final states of D^0 and $\bar{D}^0$. The diagrams from Figure 8.12, combined with the analogous $\bar{D}^0$ diagrams, provide four such mixing contributions at the quark level:

(a) A Cabibbo favored D^0 amplitude $\times$ Doubly Cabibbo suppressed $\bar{D}^0$ amplitude;
(b) A Doubly Cabibbo suppressed D^0 amplitude $\times$ Cabibbo favored $\bar{D}^0$ amplitude;
(c) Two types of singly suppressed amplitudes for D^0 and $\bar{D}^0$.

These mixing contributions to Γ_{12} all contribute with a CKM factor, which is $\sin^2\theta_c \cos^2\theta_c \simeq 0.05$, whereas $\bar{\Gamma}$ does not have such a suppression factor.

Evaluating the box diagram for the short distance part of Δm_D, we also get the same CKM factor of $\sin^2\theta_c \cos^2\theta_c$.

2. GIM Suppression: In the box diagram we again have the GIM cancellation as for the kaons, but now between the nearly degenerate d and s quarks. In addition, the virtual quarks in the box d, s, b (instead of u, c, t for the K or B systems) give a small numerical result given the specific quark masses.

However, we now also have a GIM cancellation for the real final states, since d and s can both be produced in the final state, by the heavier c quark. The diagrams in Figure 8.12 have very specific signs for the CKM factors. In the limit of perfect $SU(3)$ symmetry, the contribution to $\Delta\Gamma_D$ from amplitudes obtained from diagrams of type (a) and (b) cancel the contribution from the two types of diagrams of type (c) that were described in (1) above.

3. Phase Cancellation: Within a diagram type, for example type (c), the terms making up Γ_{12} have plus and minus signs when real hadronic final states are considered (analogous to $\Gamma_{12} = \Gamma_{2\pi} - \Gamma_{3\pi}$ for the kaon system) and will tend to cancel. The final states for diagrams of type (c) can in fact be arranged into CP even and odd states, which give opposite sign terms in Γ_{12}. Such a cancellation also occurs in the long distance part of Δm_D.

Finally, we note that the CP violation in the D system should also be very small. The direct CP violation, which can occur for both charged and neutral D meson decay, is very small because the CKM elements for all major decay diagrams are very close to being real. In the mixing, the box diagrams containing the CKM phase are also small, involving the virtual b quark (instead of the t quark in the K system) whose contribution is small because of the small value of $V_{ub}V_{cb}^*$ and the small value of m_b (compared to m_t).

9.5 ■ $B^0, \bar{B}^0$ AND $B_s^0, \bar{B}_s^0$ SYSTEMS

We examine the expectation for the widths of the mass eigenstates first. For the $\bar{B}^0$ meson, all decays that involve the dominant $b \to c$ transition yield quark final states that differ in flavor from those obtained in B^0 decays via $\bar{b} \to \bar{c}$, provided the W makes a transition into a Cabibbo favored quark-antiquark pair. Mixing will result from the Cabibbo suppressed $b \to c\bar{c}d$ decay sequence, which yields a $c\bar{c}d\bar{d}$ final state after including the spectator quark. Arising mainly from Cabibbo suppressed decays, we expect the mixing due to real common decays of B^0 and $\bar{B}^0$ to be small, implying that $\Delta\Gamma_B \ll \bar{\Gamma}_B$. For the $B_s^0, \bar{B}_s^0$ system, common decays can arise from the larger Cabibbo favored $b \to c\bar{c}s$ decay sequence, which at the quark level yields a final state $c\bar{c}s\bar{s}$ after including the spectator $\bar{s}$ quark. This decay sequence corresponds to about 20% of the decays, based on the relative rates in Section 8.11. We can, however, divide $c\bar{c}s\bar{s} \to$ hadrons into CP even and odd final states, whose contributions will tend to cancel in $\Delta\Gamma_{B_s}$. Thus we expect $\Delta\Gamma_{B_s}/\bar{\Gamma}_{B_s}$ to be less than 10%.

In contrast to the width term, the mass mixing term is large for both meson pairs. Some of the diagrams for B^0 and B_s^0 mixing are shown in Figure 9.5. The dominance of the box diagram involving only t quarks makes the B system ideal for studying the physics responsible for both mixing and CP violation. We said earlier that high mass physics is difficult to see because it involves short-distance interactions. This is clear for the case of a propagator and also for the running coupling, as given for example in Eq. 7.14, after renormalization. In the broken gauge theory of the weak interactions, higher-order dimensional quantities, such as the value of the box diagram (apart from CKM factors) and the W mass shift discussed in Chapter 10, are largest for the heaviest quarks that contribute. The sensitivity to high-mass objects means that these higher-order quantities provide constraints on the existence of new physics. Any extra constituents must con-

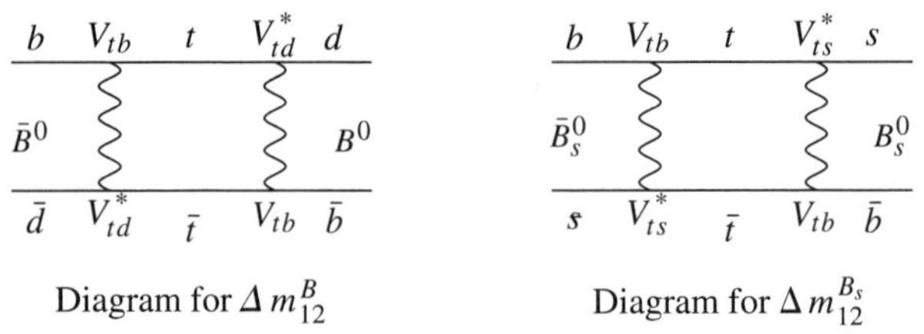

FIGURE 9.5 One of two contributing mixing box diagrams for the B^0 and B_s^0 systems involving top quarks. The CKM factor for each vertex is indicated.

tribute in a way that does not spoil the agreement (within the errors) of the measured quantities and the Standard Model calculation of these quantities, which to date gives a good description of all of the data.

For the B^0 box diagram in Figure 9.5, the product of CKM factors appearing in Δm_{12}^B is $(V_{td}^*)^2$, since $V_{tb} \simeq 1$. V_{td} involves the 3rd $\rightarrow$ 1st generation transition and brings in the phase factor from the CKM matrix. This will have important consequences for CP violation, discussed in the next section. For the B^0, $\bar{B}^0$, mixing is measured both through χ_B in Eq. 9.26 and by observing B oscillations in time. These measurements imply that

$$\frac{\Delta m_B}{\bar{\Gamma}_B} = 0.755 \pm 0.015. \tag{9.27}$$

The value of Δm_B, which for the B system $\simeq 2|\Delta m_{12}^B|$, can be calculated through the box diagrams involving t quarks, with about 30% uncertainties coming from QCD corrections and uncertainties regarding the B meson wave function. The measured value for the B^0 system allows a calculation of $|V_{td}|$ and is used as one of the inputs for determining ρ and η in the CKM matrix.

For the B_s^0, the CKM factors in the box diagram are $(V_{ts}^*)^2$, for $V_{tb} \simeq 1$. In the approximation of $SU(3)$ invariance, we can predict the ratio of mass differences within the two B systems in terms of CKM factors, since the rest of the calculation for Δm is identical. This gives the prediction (using Eq. 9.5 with $|\Gamma_{12}| \ll |\Delta m_{12}|$ in the calculation of Δm):

$$\frac{\Delta m_{B_s}}{\Delta m_B} = \frac{|V_{ts}|^2}{|V_{td}|^2} = \frac{1}{\sin^2 \theta_c \left[(1-\rho)^2 + \eta^2\right]}. \tag{9.28}$$

The ratio is predicted to be about 25. This gives, using the measured value of $\Delta m_B \bar{\Gamma}_{B_s}$,

$$\frac{\Delta m_{B_s}}{\bar{\Gamma}_{B_s}} = 26, \tag{9.29}$$

9.5 $B^0, \bar{B}^0$ and $B_s^0, \bar{B}_s^0$ Systems

after including a 40% increase of the ratio in Eq. 9.28, expected from $SU(3)$ violation. The mixing for the B_s^0 system is a very large effect, in particular, to very good approximation, $\chi_{B_s} = \frac{1}{2}$, the value for complete mixing.

The predicted value in Eq. 9.29 has an uncertainty of about 20% coming primarily from the error on the CKM parameter ρ. A good measurement of this ratio would therefore improve our knowledge of ρ. The exact value for the ratio in Eq. 9.29 has not been determined, but the measured lower limit is close to the expected value.

Finally, note that Δm_{12} for B_s is expected to be nearly real since V_{ts}, which appears in the box diagram, involves a 3rd→ 2nd generation transition. Therefore we expect that CP violation due to mixing is very small for the B_s system.

9.5.1 ■ *CP Violation in the $B^0, \bar{B}^0$ System*

For the B system, we expect CP violation whenever the phases from V_{td} or V_{ub} appear in the amplitude for a process; although observable direct CP violation requires in addition that the given process occurs through two amplitudes with different strong and weak phases, as discussed in Section 8.7. The phase from V_{ub} can affect the small fraction of decays occurring through the $b \to u$ transition. The mixing through the box diagram introduces a significant phase, which comes from the $b \to t \to d$ transition sequence making up part of the box diagram. To measure this, we have to find an observable in which the phase remains, in order to have a tangible CP violating effect. To find such an observable, we first examine the mass eigenstates that determine the space-time propagation.

For the $B^0, \bar{B}^0$ system, we can relate the coefficients q and p of the mass eigenstates through the relation in Eq. 9.3. Since $\Gamma_{12} \ll \Delta m_{12}$, $|q/p| = 1$, and we get no CP violation in the rates for tagged final states if we start with a tagged initial state. This can be verified by using the analog of Eq. 9.13 for the B system. If we define the phase of Δm_{12} by $\Delta m_{12} = -|\Delta m_{12}|e^{2i\beta}$, then from Eq. 9.3,

$$\frac{q}{p} = \sqrt{\frac{\Delta m_{12}^*}{\Delta m_{12}}} = \frac{V_{td}}{V_{td}^*} = e^{-2i\beta}.$$

We can thus take $p = \frac{1}{\sqrt{2}}e^{i\beta}$, $q = \frac{1}{\sqrt{2}}e^{-i\beta}$. Using $V_{td} = A\lambda^3(1 - \rho - i\eta)$ from Eq. 8.18,

$$\tan \beta = \left(\frac{\eta}{1 - \rho}\right), \qquad (9.30)$$

which implies that $\tan \beta \simeq 0.43$.

To find an observable sensitive to β, we can look at common decays for B^0 and $\bar{B}^0$. Specifically, we assume that we start with a decay initially tagged as B^0. Let f be a final state to which both B^0 and $\bar{B}^0$ can decay. An example of such a final state at the quark level is $c\bar{c}d\bar{d}$, which at the hadron level can yield the D^+D^- final state. Another example of a final state is $c\bar{c}s\bar{d}$, which does not mix with $c\bar{c}\bar{s}d$ at the quark level, but does at the hadron level because of kaon mixing. Examples are the final states $J/\psi K_S^0$ and $J/\psi K_L^0$, which are produced from either B^0 or $\bar{B}^0$. The final states mentioned above are all CP eigenstates (ignoring the small parameter ε for kaon mixing). We define the two amplitudes A_f and $\bar{A}_f$ as the amplitudes for $B^0 \to f$ and $\bar{B}^0 \to f$, respectively. In the case of the CP eigenstates above, no extra weak phases arise in the decay, since the transitions involve CKM elements for $b \to c$ or $\bar{b} \to \bar{c}$, then $A_f = \pm \bar{A}_f$ with the sign given by the CP of the final state.

Using Eq. 9.13 for the B system, with the assumption that both mass eigenstates have nearly the same lifetime and that the mass difference has the same sign as in the kaon system, results in the amplitude for the projection onto the state f:

$$A_{B(t) \to f} = e^{-\left(\frac{\bar{\Gamma}}{2} + i\bar{m}\right)t} A_f \left[\left(\frac{e^{\frac{-i\Delta m_B t}{2}} + e^{\frac{i\Delta m_B t}{2}}}{2}\right)\right]$$

$$+ \frac{q}{p} \frac{\bar{A}_f}{A_f} \left[\left(\frac{e^{\frac{-i\Delta m_B t}{2}} - e^{\frac{i\Delta m_B t}{2}}}{2}\right)\right]$$

$$= e^{-\left(\frac{\bar{\Gamma}}{2} + i\bar{m}\right)t} A_f \left[\cos\left(\frac{\Delta m_B t}{2}\right) - \frac{q}{p} \frac{\bar{A}_f}{A_f} i \sin\left(\frac{\Delta m_B t}{2}\right)\right]. \quad (9.31)$$

For example, for decays tagged in the final state as B^0 or $\bar{B}^0$, respectively, it gives predicted rates proportional to

$$|A_f|^2 e^{-\bar{\Gamma} t} \cos^2\left(\frac{\Delta m_B t}{2}\right) \quad \text{or} \quad |\bar{A}_f|^2 e^{-\bar{\Gamma} t} \sin^2\left(\frac{\Delta m_B t}{2}\right), \quad (9.32)$$

which exhibits the B^0, $\bar{B}^0$ oscillations but cannot be used to measure CP violation. On the other hand, taking a CP even or odd common decay final state for which no extra weak phases arise in the decay, we get

$$\left|A_{B(t) \to f}\right|^2 = |A_f|^2 e^{-\bar{\Gamma} t} \left(1 \mp \sin(2\beta) \sin(\Delta m_B t)\right),$$

where the sign in the second term is determined by the CP of the final state. We can repeat the analysis starting with a tagged $\bar{B}^0$ initial state at $t = 0$. The result is

$$\left|A_{\bar{B}(t) \to f}\right|^2 = |\bar{A}_f|^2 e^{-\bar{\Gamma} t} \left(1 \pm \sin(2\beta) \sin(\Delta m_B t)\right).$$

This leads to a time-dependent asymmetry that is independent of $|A_f|^2$ or $\bar{\Gamma}$:

$$A = \frac{\left|A_{B(t) \to f}\right|^2 - \left|A_{\bar{B}(t) \to f}\right|^2}{\left|A_{B(t) \to f}\right|^2 + \left|A_{\bar{B}(t) \to f}\right|^2} = \mp \sin(2\beta) \sin(\Delta m_B t), \quad (9.33)$$

which allows a nice measurement of the *CP* violating parameter β. This is a direct measurement of the parameter β, as opposed to extracting it from the difficult-to-calculate parameter ε from the kaon system. Note, however, that within the uncertainties in the calculation of ε, both systems give consistent results for the phase in the CKM matrix. The experimentally-measured value for $\sin 2\beta$, 0.73 ± 0.05, agrees very well with the value expected from other determinations of the CKM parameters.

An experimentally convenient method for studying neutral pseudoscalar mesons is to produce $|P^0, \bar{P}^0\rangle$ nearly at rest using the e^+e^- annihilation reaction. This requires finding an e^+e^- center of mass energy so that the production occurs through the decay of a convenient $J^P = 1^-$ resonance. This gives a high rate, with the resonant state produced in a significant fraction of all the e^+e^- annihilation events. The fact that we start from a very specific two-body state for the P^0 and $\bar{P}^0$ provides constraints on the final state decay pattern. This follows from the fact that the e^+e^- reaction in the center of mass yields a *p*-wave, two-body system, which is antisymmetric under exchange. Thus, if the first decay tags the parent as $|P^0\rangle$ we know that the other particle starts as $|\bar{P}^0\rangle$ at that time. It then evolves in time via the formulas we have discussed, where t = time difference from the first decay. In this way, the first decay serves to tag the second meson, which can then be studied. For the B system, the experiments are done at a center of mass energy of 10.58 GeV, which is the mass of the lightest resonance above the threshold for production of B and $\bar{B}$. We anticipate further checks of the CKM framework for B meson *CP* violation in the next few years. This will include measurements involving final states produced through a variety of decay diagrams, for example, penguin diagrams, or the $b \to u$ transition that brings in an extra phase. Examples of interesting final states are ϕK_S^0 and $\pi^+\pi^-$.

9.6 ■ NEUTRINO OSCILLATIONS

We turn next to neutrino oscillations. In the Standard Model the neutrino lifetimes are very long, so only the mass differences and mixing matrix contribute to create oscillations. The masses and terms in the mixing matrix are not theoretically calculable at present, so the goal is to extract them from data. This is an area where rapid progress is being made through a variety of experiments.

Neutrino production and detection using charged current interactions always involves tagging the neutrino as one of the three lepton flavor eigenstates. Neutral current interactions are the same for all flavors and do not allow tagging of the final state neutrino. Propagation between interactions occurs through the mix of mass eigenstates corresponding to the initially tagged state. Because of the very small masses, the neutrinos are generally very relativistic in the laboratory. Whereas the pseudoscalar meson mixing was conveniently calculated in the meson rest-frame, neutrino oscillations are most easily calculated in the laboratory in the ultra-relativistic limit.

We indicate the lepton flavor eigenstates as $|\nu_\alpha\rangle$ and the mass eigenstates as $|\nu_i\rangle$ with masses m_i. At $t = 0$ we assume that we produce the state $|\nu_\alpha\rangle$, which we expand in terms of $|\nu_i\rangle$:

$$|\nu_\alpha\rangle = \sum_i U_{\alpha i} |\nu_i\rangle . \qquad (9.34)$$

$U_{\alpha i}$ is the neutrino analogue of the CKM matrix. This gives at a later time, at a position $\vec{x}$:

$$|\nu(t)\rangle = \sum_i U_{\alpha i} e^{-i E_i t} e^{i \vec{p} \cdot \vec{x}} |\nu_i\rangle , \qquad (9.35)$$

where we assume we can ignore the details of the wave-packet and use the centroid of momentum $\vec{p}$. We assume the ν state interacts weakly at time t and that we tag the state as the lepton flavor eigenstate $|\nu_\beta\rangle$. This process has an intensity:

$$I_{\beta\alpha} = |\langle \nu_\beta | \nu(t)\rangle|^2 = \left| \sum_i U_{\alpha i} U_{\beta i}^* e^{-i E_i t} \right|^2 . \qquad (9.36)$$

In the ultra-relativistic limit:

$$E_i = p + \frac{m_i^2}{2p},$$

so that

$$I_{\beta\alpha} = \left| \sum_i U_{\alpha i} U_{\beta i}^* e^{-i \frac{m_i^2 t}{2p}} \right|^2 . \qquad (9.37)$$

For m_i^2 all equal, or all masses sufficiently small so that all the exponential factors are near unity, the unitarity of the U matrix gives $I_{\beta\alpha} = \delta_{\beta\alpha}$, that is, no apparent mixing or oscillations. Eq. 9.37 is the basic mixing equation that we use here. For the case of propagation in a material, for example within the sun, this equation will be modified as discussed in Section 9.6.3.

The discussion has so far been general. We now assume for simplicity that only two generations participate. Then:

$$U = \begin{pmatrix} \cos\theta & \sin\theta \\ -\sin\theta & \cos\theta \end{pmatrix}.$$

Starting with the initial state specified by $\alpha = 1$, there are two intensities, for which $\beta = 1$ or 2. Using U above, these are:

$$I_{11} = 1 - I_{21},$$

9.6 Neutrino Oscillations

$$I_{21} = 4\cos^2\theta \sin^2\theta \sin^2\left[\frac{(m_1^2 - m_2^2)t}{4p}\right]$$

$$= \sin^2 2\theta \sin^2\left[\frac{(m_1^2 - m_2^2)t}{4p}\right]. \tag{9.38}$$

These equations have three regimes:

1. $\dfrac{|m_1^2 - m_2^2|t}{4p} \ll 1$. No oscillations are noticeable. This is the situation close to the source.

2. $\dfrac{|m_1^2 - m_2^2|t}{4p} \sim 1$. An oscillation pattern is noticeable as t (which for a relativistic particle is the length from the source divided by the velocity of light) or p vary, allowing a determination of $|m_1^2 - m_2^2|$.

3. $\dfrac{|m_1^2 - m_2^2|t}{4p} \gg 1$. Very rapid oscillations occur with an experiment typically averaging over them. Since

$$\sin^2\left[\frac{(m_1^2 - m_2^2)t}{4p}\right]$$

averages to $\frac{1}{2}$, we get:

$$\bar{I}_{21} = \tfrac{1}{2}\sin^2(2\theta), \quad \bar{I}_{11} = \tfrac{1}{2}(1 + \cos^2(2\theta)).$$

A nonzero value for $\bar{I}_{21}$ provides evidence for oscillations but does not allow a measurement of $|m_1^2 - m_2^2|$.

Data using atmospheric neutrinos, which contain a mix of ν_μ and ν_e (and their antiparticles) created in cosmic ray showers through the decays $\pi \to \mu + \nu_\mu$ and $\mu \to e + \bar{\nu}_e + \nu_\mu$, indicate that ν_μ and ν_τ mix copiously, while ν_e does not mix very significantly for distances and momenta sensitive to $|m_1^2 - m_2^2| \gtrsim 10^{-4}$ eV^2. The data for the mixing of ν_μ and ν_τ are in fact described well by choosing:

$$\left|m_1^2 - m_2^2\right| \simeq 2.5 \times 10^{-3} \text{ eV}^2 \quad \text{and} \quad \sin^2(2\theta) \simeq 1, \quad \text{or} \quad \theta \simeq 45°.$$

A very large, deep underground, water Cherenkov detector is used to measure the flux of atmospheric neutrinos. The earth above the detector absorbs nearly all particles other than neutrinos. Using water as the interaction medium provides a minimum cost experiment. In this detector, the Cherenkov rings produced by relativisitic charged particles travelling through the water are detected. Muons produced by ν_μ are registered as particles that slowly lose energy through ionization only. Electrons produced by ν_e generate electromagnetic showers in the water. The incident neutrino direction and energy can be determined approximately from the Cherenkov rings for the particles produced in the event. To reduce systematic errors, the data are plotted as a ratio of downward-going neutrinos that traverse the atmosphere and a small amount of Earth (short baseline) to upward-going neutrinos that traverse the atmosphere and most of the Earth (long baseline). The data for mixing of neutrinos produced in the atmosphere are shown in Figure 9.6. Fortuitously, the diameter of the Earth is exactly right for observing large oscillations of ν_μ to ν_τ for momenta of a few GeV.

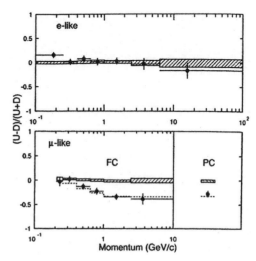

FIGURE 9.6 Up-down asymmetry calculated from the ratio of detected μ (from ν_μ) and e (from ν_e) events produced by upward- and downward-going neutrinos. Muon events are divided into those fully contained in the detector, allowing an energy measurement, and those only partially contained, which typically have energies above 10 GeV. [From Y. Fukuda et al., *Phys. Rev. Lett.* 81, 1562(1998).]

9.6.1 ■ Three Neutrino Generations

Studies of ν_e oscillations use not only the data from atmospherically produced neutrinos shown in Figure 9.6, but also ν_e from nuclear reactions in the sun and $\bar{\nu}_e$ from nuclear reactors. For three neutrino generations, the data for oscillations can be described in terms of four angles, one of which is a *CP* violating phase, analogous to the CKM angles, and three differences of masses squared: $\Delta m_{12}^2 = m_1^2 - m_2^2$, $\Delta m_{13}^2 = m_1^2 - m_3^2$, and $\Delta m_{23}^2 = m_2^2 - m_3^2$, where only two are independent. The data for ν_e oscillations using atmospheric neutrinos from cosmic ray showers, shown in Figure 9.6, as well as data from nuclear reactors, indicate that two of the neutrino mass eigenstates are much more degenerate with each other than they are with the third. In this case we have $|\Delta m_{13}^2| \simeq |\Delta m_{23}^2| \gg |\Delta m_{12}^2|$. For this situation the oscillation Eq. 9.37 sim-

9.6 Neutrino Oscillations

plifies, for $\alpha \neq \beta$, to

$$I_{\beta\alpha} = 4\left|U_{\alpha 3} U^*_{\beta 3}\right|^2 \sin^2\left(\frac{\Delta m^2_{23} t}{4p}\right). \tag{9.39}$$

To arrive at this equation, we use the unitarity of the three generation U matrix and

$$\frac{\Delta m^2_{12} t}{4p}$$

is assumed small, as appears true for the atmospheric neutrino oscillation measurements.

Using the notation of the CKM matrix, given in Eq. 8.17, we can rewrite Eq. 9.39 as

$$I(\nu_e \to \nu_\mu) = \sin^2\theta_{23} \sin^2 2\theta_{13} \sin^2\left(\frac{\Delta m^2_{23} t}{4p}\right)$$

$$I(\nu_e \to \nu_\tau) = \cos^2\theta_{23} \sin^2 2\theta_{13} \sin^2\left(\frac{\Delta m^2_{23} t}{4p}\right)$$

$$I(\nu_\mu \to \nu_\tau) = \cos^4\theta_{13} \sin^2 2\theta_{23} \sin^2\left(\frac{\Delta m^2_{23} t}{4p}\right). \tag{9.40}$$

The data from Figure 9.6 indicate that

$$\left|\Delta m^2_{23}\right| \simeq 2.5 \times 10^{-3} \text{ eV}^2, \theta_{23} \simeq 45°,$$

and θ_{13} is small. These values for the neutrino parameters lead to a suppression of ν_e oscillations to both ν_μ and ν_τ and allow ν_μ to ν_τ mixing at a rate given by the two-generation formula of the previous section. Unlike the situation for the quarks, where an approximate two-generation mixing formula requires one angle (the Cabibbo angle) to be much larger than the other two, for the neutrinos the third angle θ_{12} is not constrained. Both small or large values are possible for the mixing angle θ_{12}. We will see that this angle is in fact large.

The formulas in Eq. 9.40 apply to oscillations of atmospheric or accelerator produced neutrinos, where the energies are typically greater than a few hundred MeV. Reactor antineutrinos have energies that are typically 100 to 1000 times smaller and can therefore show measurable oscillations due to Δm^2_{12} as small as $\sim 10^{-5}$ eV^2, for distances between the source and detector (baseline) on the order of a few hundred km. For $\theta_{12} \gg \theta_{13}$, a deficit of electron antineutrinos, relative to expectations for no mixing, is given by an oscillation formula that can be derived from Eq. 9.37:

$$I_{11} = I(\bar{\nu}_e \to \bar{\nu}_e) = 1 - \sin^2 2\theta_{12} \sin^2 \frac{\Delta m_{12}^2 t}{4p}. \tag{9.41}$$

If θ_{12} is also small, then no appreciable oscillations can be seen.

A number of experiments looking at reactor produced $\bar{\nu}_e$ have been performed. The $\bar{\nu}_e$ typically have energies of a few MeV and are detected through the reaction $\bar{\nu}_e + p \to e^+ + n$. Experiments that are located within 10 km of the reactor $\bar{\nu}_e$ source see no deficit of $\bar{\nu}_e$ relative to the numbers expected from a calculation based on the power generated by the reactor. A very recent experiment, observing $\bar{\nu}_e$ from a number of reactors in Japan, has an average baseline of 180 km. This experiment found a reduction in detected $\bar{\nu}_e$ to about 61% of the calculated rate for no oscillations. Allowing for oscillations, a fit to the rate and $\bar{\nu}_e$ spectrum gives a value of

$$\left|\Delta m_{12}^2\right| \simeq 6.9 \times 10^{-5} \text{ eV}^2$$

and the constraint that $\sin^2(2\theta_{12})$ is larger than about 0.5. The θ_{12} angle is presently better measured using the remarkable data for the detected ν_e from the sun, which we discuss in the next sections. More reactor data are being collected and will yield results of better precision over time.

These neutrino oscillation results involve only differences of masses squared and do not give us a value for any individual masses. In fact, it is not known if the mass hierarchy (that is, which is heaviest and lightest) is the same as in the charged lepton case. The neutrino masses have not been measured; only upper limits have been determined. The most stringent limit comes from measuring the endpoint of the tritium beta decay spectrum, which indicates that the mass of the neutrino accompanying the electron is less than a few eV.

9.6.2 ■ Oscillations of Neutrinos from the Sun

The nuclear processes that produce radiant energy from the sun all involve the charged current weak interaction, and result in the creation of electron neutrinos. Thus solar neutrinos are tagged at production to be ν_e. The expected neutrino spectrum is shown in Figure 9.7, based on a detailed model of the nuclear processes taking place in the sun. The neutrinos are typically produced near the center of the sun and therefore have to traverse the sun as well as the intervening space to reach detectors on earth. Unfortunately, detecting low energy neutrinos is difficult and experiments cannot make measurements over the entire spectrum.

Solar neutrino experiments are of two types. Both types are performed deep underground to minimize backgrounds. The first type of experiment uses a very large water Cherenkov detector, also used to study atmospheric neutrinos. These water Cherenkov detector experiments are sensitive to neutrinos of energy larger than about 5 MeV. They therefore mainly detect the neutrinos coming from ^{8}B reactions in Figure 9.7. These experiments have the virtue that the neutrino energy can be determined and the events allow a rough determination of the incident neu-

9.6 Neutrino Oscillations

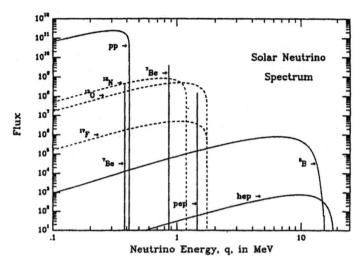

FIGURE 9.7 The solar neutrino energy spectrum predicted by the Standard Solar Model. Curves are labeled by nuclei involved in neutrino production processes. Fluxes from continuum sources are given in units of number per cm^2 per sec per MeV at a distance of one astronomical unit; line sources are integrated over energy. [From J. N. Bahcall and R. K. Ulrich, *Rev. of Modern Phys.* 60, 297(1988).]

trino direction, useful for separation of signal and background. Figure 9.8 shows the angular correlation of the electron neutrino signal with the direction of the sun, indicating very clearly that the sun is radiating neutrinos.

The reaction detected in ordinary water is the elastic scattering on electrons, which has charged and neutral current contributions from ν_e and neutral current contributions from ν_μ and ν_τ. By using heavy water containing deuterium, additional reactions can be observed, which are used to isolate the event fraction due to ν_e. An example is the charged current interaction, which is sensitive to ν_e only:

$$\nu_e + d \to p + p + e^-.$$

In addition, the neutral current interaction

$$\nu + d \to p + n + \nu$$

has the same cross section for each of the three neutrino types. It therefore provides an absolute normalization of the ν flux and also a check that the solar model is correct. Using these reactions on the deuteron, the flux ratio of electron neutrinos to all neutrinos from ^{8}B is determined to be 0.34 ± 0.03. The neutral current data also indicate that the solar model provides a good description of the incident neutrino flux.

The data on neutrino elastic scattering on electrons in an ordinary water target further indicate the existence of oscillations. This measurement yields a rate that is 0.45 ± 0.02 of the expectation from the solar model without oscilla-

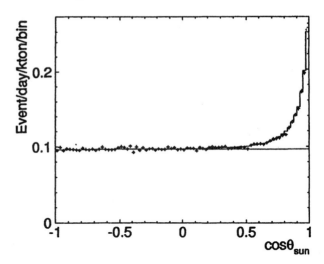

FIGURE 9.8 Distribution of the cosine of the event angle relative to the direction of the sun. The points are data, the histogram is a best fit to a solar signal, and the horizontal line is the estimated background. Data are restricted to energies between 5.0 and 20.0 MeV. The number of solar neutrinos detected is about 18,000. [From S. Fukuda et al., *Phys. Rev. Lett.* 86, 5651(2001).]

tions. Calculating expectations from the dominant ν_e cross section (based on 0.34 ν_e after oscillations) and the smaller cross sections for ν_μ or ν_τ (based on 0.66 for the sum of these after oscillations) provides a very consistent cross check between the deuteron and electron scattering measurements. This provides good evidence that a significant fraction of the ν_e have turned into ν_μ or ν_τ between production and detection. The measured neutrino spectrum using elastic electron scattering is shown in Figure 9.9 as a ratio to the standard solar model without oscillations. The ratio is very constant, showing that we see the spectral shape from Figure 9.7 with an overall reduction, consistent with a constant factor over the whole energy range. This spectral shape is an important constraint on the oscillation parameters, as we will see in the next section.

The second method to measure the flux of solar neutrinos uses large volumes of material containing nuclei that can undergo nuclear transformations when hit by ν_e. These reactions are sensitive only to the ν_e component and require a final state that can be separated reliably from the initial volume and also subsequently detected. The initial neutrino detection experiment used the conversion of ^{37}Cl to ^{37}Ar and provided the first evidence for neutrino oscillations. These experiments do not determine the initial neutrino energy, and integrate over the product of the incident spectrum and the cross section for the nuclear reaction. The ^{37}Cl reaction is primarily sensitive to the ^{8}B neutrinos and gives a result that is consistent with the water Cherenkov experiments.

A more recent set of experiments uses the transformation of ^{71}Ga to ^{71}Ge. This reaction is primarily sensitive to the more copiously produced lower-energy

9.6 Neutrino Oscillations

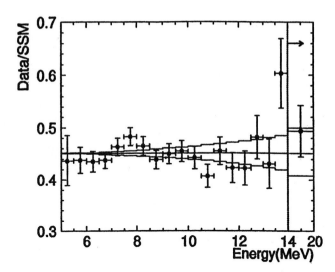

FIGURE 9.9 The measured neutrino spectrum normalized to the Standard Solar Model. The data from 14 MeV to 20 MeV are combined into a single bin. The solid line shows the average, while the band around this line indicates the energy correlated uncertainty. [From S. Fukuda et al., *Phys. Rev. Lett.* 86, 5651(2001).]

neutrinos. The reaction threshold is about 0.23 MeV for the neutrino energy. The data measured in the ^{71}Ga experiments, normalized to the solar model without oscillations, gives a flux ratio of 0.60 ± 0.06. This is significantly larger than the value for the higher energy ^{8}B neutrinos. The consistency of all the neutrino results requires a change in the oscillation pattern somewhere between about 1 MeV and 5 MeV. Over this range, the fraction of initially produced ν_e detected as ν_e changes from about 0.60 to about 0.34.

9.6.3 ■ Matter-Induced Oscillations in the Sun

The sun is so far away that we expect the distance to the Earth to correspond to many neutrino oscillation lengths, unless the mass difference of the oscillating pair is extremely small, for example, of order 10^{-10} eV^2. We will consider a different possibility, which is that the matter in the sun changes the oscillation probability. This is called the MSW (Mikheyev, Smirnov, and Wolfenstein) effect. It corresponds to the same phenomenon responsible for the index of refraction for light in a clear dielectric. The data on solar neutrinos indicate that this effect plays an important role in the oscillation phenomena.

Consider the creation of an electron neutrino of given energy in the sun. In vacuum, the amplitude for other neutrino components would build up during propagation if there is mixing, as we discussed before. The neutrino cross section is sufficiently small that the likelihood of the neutrino being absorbed due to the weak interaction with the solar material is very small, so we might imagine that the result for neutrino propagation in the sun is the same as in vacuum. However, the elastic coherent scattering in the forward direction from the solar matter modifies the propagating wave. This is the origin of the index of refraction for light in a dielectric. Taking the direction of motion to be the z axis, the amplitude for a given momentum p is modified by the material so that it varies as e^{inpz}, where the index of refraction is

$$n = 1 + \frac{2\pi N f(0)}{p^2}.$$

Here $f(0)$ is the forward scattering amplitude in the notation of nonrelativistic quantum mechanics, and N is the number of scattering centers per unit volume. Using the optical theorem, this formula gives the usual result for the attenuation of the number of particles due to the total cross section as they move along z.

In the neutrino case, the elastic forward scattering amplitude for the neutral current is the same for each type of neutrino. The neutral current contribution to the index of refraction produces an overall phase shift that is the same for all neutrino types. Therefore, the neutral current process does not create any mixing. However, one additional process has a finite elastic forward amplitude. This is the charged current process for $\nu_e e^-$ scattering on the electrons in the sun. This process exists only for the ν_e component. The relevant Feynman diagram is shown in Figure 1.5 and the amplitude is given by Eq. 8.5. One subtlety in calculating M_{ii} arises from the exchange property for fermions, and leads to an overall extra factor of -1. For the forward amplitude, we want to calculate the transition for $|\nu_e, e^-\rangle$ going to $|\nu_e, e^-\rangle$. The amplitude in Eq. 8.5 is that for a transition from $|\nu_e, e^-\rangle$ to $|e^-, \nu_e\rangle$, requiring an extra exchange of particles in the final state to arrive at the desired amplitude. The relativistic amplitude for forward scattering can be related to $f(0)$ by the formula

$$f(0) = -\frac{M_{ii}}{8\pi m_e}. \tag{9.42}$$

The relation in Eq. 9.42 follows from equating the expression for the forward cross section in the nonrelativistic and relativistic formalisms, with the correct sign convention given in Section 2.13. M_{ii} for the charged current process is $(2m_e)(2E)\sqrt{2}G_F$, with the energy factors just resulting from the normalization of the spinors. This gives, for the charged current contribution to the index of refraction,

$$\frac{2\pi N_e f(0)}{p^2} = \frac{-\sqrt{2}G_F N_e}{p}. \tag{9.43}$$

The density N_e is that for electrons only and E is set equal to p for the neutrino.

If there were no mixing and the value of N_e was constant, we would expect for the electron neutrino an amplitude of the form (we leave out the neutral current contribution, which is flavor-independent):

$$e^{i(pz - \sqrt{2}G_F N_e z - Et)}. \tag{9.44}$$

9.6 Neutrino Oscillations

Including the neutrino mass to lowest order, this is

$$e^{iE(z-t)}e^{-i\left(\frac{m^2}{2p}+\sqrt{2}G_F N_e\right)z}. \tag{9.45}$$

With mixing, however, ν_e is not a mass eigenstate. Propagation without mixing occurs for the eigenstates resulting from the combined effect of the index of refraction and the original mass matrix. To find these eigenstates we have to diagonalize the matrix that appears in the phase in Eq. 9.45. After multiplying by $2p$, the matrix is

$$U\begin{pmatrix} m_1^2 & 0 \\ 0 & m_2^2 \end{pmatrix} U^\dagger + \begin{pmatrix} m_0^2 & 0 \\ 0 & 0 \end{pmatrix}, \tag{9.46}$$

where for simplicity we assume mixing primarily between two neutrinos only. The basis chosen is the lepton flavor eigenstate basis, not the vacuum mass eigenstate basis. The masses m_1 and m_2 are the two mass eigenvalues in vacuum and U is the two-flavor mixing matrix introduced before. It is specified by one angle θ. The presence of matter provides the extra term for ν_e only, where

$$m_0^2 = 2\sqrt{2} G_F N_e p. \tag{9.47}$$

Using a value of $p = 10$ MeV and electron densities expected near the center of the sun, $m_0^2 \approx 10^{-4} \text{eV}^2$. For $|m_2^2 - m_1^2|$ less than or of the order of this number, the matter in the sun will have a significant effect on the mixing. This is true even though the deviation from 1 of the index of refraction, given in Eq. 9.43, is very small. Since m_0^2 is momentum dependent, the effect due to the matter in the sun will vary over the neutrino spectrum.

We can diagonalize the matrix in Eq 9.46 in terms of a unitary matrix U_m, which equals U in vacuum. We can calculate the mixing angle in U_m, resulting in the expression

$$\sin^2(2\theta_m) = \frac{\sin^2(2\theta)}{\left[\left(\frac{m_0^2}{(m_2^2-m_1^2)} - \cos(2\theta)\right)^2 + \sin^2(2\theta)\right]}.$$

For $m_0^2 \gg |m_2^2 - m_1^2|$, $\theta_m \to 0$ and ν_e is a mass eigenstate; for $m_0^2 \ll |m_2^2 - m_1^2|$, $\theta_m \to \theta$, the vacuum solution.

We have assumed above that N_e is a constant, which is not correct as the neutrino moves through the sun. Therefore an exact solution requires propagating each energy component through the solar medium. We will look at a simple case that agrees with the data and does not require any complicated calculations. We assume that $|m_2^2 - m_1^2|$ is a few times 10^{-5} eV^2. This was the value from the reactor data in the previous section. In this case the material in the sun has little effect on electron neutrinos produced with $p \leq 1$ MeV. For these neutrinos we can just use the solution in vacuum to good approximation. The fraction of neutrinos that are detected as ν_e in vacuum was given in Section 9.6, and is:

$$I_{11} = \tfrac{1}{2}\left(1 + \cos^2(2\theta)\right).$$

Using the result from the ^{71}Ga experiments, we can estimate that $I_{11} \simeq 0.60$, and $\theta \simeq 32°$. If we consider three generations, then the present situation is the same as the reactor experiments, given by Eq. 9.41, with an extremely long baseline. The formula for I_{11} is the same as for the two-generation case we have been considering, with the angle θ_{12} appearing in place of θ. The angle θ above is therefore θ_{12} in this case.

For the ^{8}B neutrinos, the solar material dominates the mass matrix so that ν_e is approximately an eigenstate near the center of the sun in this momentum range. For the solar density varying slowly in space compared to the wavelength corresponding to the neutrino momentum, the neutrino remains in the one eigenstate that is continuously connected to the initial eigenstate as it traverses the sun. This is called the adiabatic approximation. Since ν_e is the heavier eigenstate, it will emerge from the sun in the heavier of the two eigenstates, which we assume is ν_2. Note that the state ν_3 is much heavier (or lighter, depending on the mass hierarchy) than the other two states and plays no role. Since ν_2 is a mass eigenstate, the neutrino remains in this state until it is detected. Projecting this state onto ν_e, which is the species measured directly in the experiments, and squaring, gives a fraction of ν_e detected equal to $\sin^2 \theta$. For θ_{13} small, the mixing matrix for three generations gives the same result with $\theta = \theta_{12}$. Equating $\sin^2 \theta_{12}$ to 0.34, based on the water Cherenkov experiments, gives the result $\theta_{12} = 35°$. The two measured values for θ_{12}, using low momentum and high momentum solar neutrinos, are very consistent. Note that this result also resolves part of the mass hierarchy, namely that m_2 is larger than m_1.

Before concluding this section, we want to mention that the phenomenon analogous to the neutrino index of refraction also exists for kaons and has been used in experiments. If we allow a beam of neutral kaons to travel a long distance, the K_S^0 component decays, leaving a pure K_L^0 beam. If we then pass this beam through a piece of material, the beam exiting the material contains a K_S^0 component. This is called kaon regeneration and comes from the different forward scattering amplitude of the K^0 and $\bar{K}^0$. While the amplitudes in the neutrino case are essentially real, the kaon amplitudes contain a real and an imaginary part. The real parts produce phase shift as in the neutrino case, while the imaginary parts attenuate the K^0 and $\bar{K}^0$ components by different amounts. After passing through the regenerating material, the K_S^0 and K_L^0 parts of the beam interfere as they propagate, yielding a decay pattern characteristic of the interference.

9.6.4 ■ CP Violation in Neutrino Mixing

The data discussed so far indicate that we can approximate the neutrino mixing matrix as

9.6 Neutrino Oscillations

$$U = \begin{pmatrix} \cos\theta_{12} & \sin\theta_{12} & \theta_{13}e^{i\delta} \\ -\sin\theta_{12}\cos\theta_{23} & \cos\theta_{12}\cos\theta_{23} & \sin\theta_{23} \\ \sin\theta_{12}\sin\theta_{23} & -\cos\theta_{12}\sin\theta_{23} & \cos\theta_{23} \end{pmatrix}.$$

Here $\theta_{12} \simeq 35°$, $\theta_{23} \simeq 45°$, $\theta_{13} < 10°$ based on analysis of all of the various types of data, and δ is the *CP* violating phase. This is the CKM matrix of Eq. 8.17 with the angle θ_{13} taken to be small. We have also chosen the opposite sign for the phase angle δ. None of the experiments to date have been sensitive to δ, which will be difficult to measure because of the small value of θ_{13}. The neutrino matrix is very different from the quark-sector CKM matrix. None of the neutrino lepton flavor states are even approximately mass eigenstates. As in the quark sector, the results indicate that the Standard Model is incomplete. Still missing is the physics that determines the masses and mixing.

To measure *CP* violation will require terrestrial experiments able to measure the small terms left out of the oscillation Eq. 9.40. These experiments will probably require the controlled conditions of accelerator produced neutrinos and antineutrinos, where the difference in oscillations between the two indicates *CP* violation. The angle θ_{13} will need to be either simultaneously measured or measured in earlier experiments, if it is sufficiently large. Whether *CP* violation can be measured depends on having a sufficiently large value for θ_{13}. To perform these experiments will likely require building neutrino sources of very high intensity and using long baselines for the detectors.

Expanding to second order in small quantities, the expression in Eq. 9.40 is modified to be

$$I(\nu_e \to \nu_\mu) = \sin^2\theta_{23}(2\theta_{13})^2 \sin^2\left(\frac{\Delta m_{23}^2 t}{4p}\right)$$

$$+ \cos^2\theta_{23} \sin^2 2\theta_{12} \left(\frac{\Delta m_{12}^2 t}{4p}\right)^2$$

$$+ \sin 2\theta_{12} \sin 2\theta_{23} 2\theta_{13} \left(\frac{\Delta m_{12}^2 t}{4p}\right)$$

$$\times \sin\left(\frac{\Delta m_{23}^2 t}{4p}\right) \cos\left(\delta - \frac{\Delta m_{23}^2 t}{4p}\right).$$

The oscillation process involving antineutrinos has the same expression with δ replaced by $-\delta$. This difference, if it can be measured experimentally, would indicate *CP* violation. We note finally that, for a sufficiently large energy, the effect of the index of refraction due to the matter of the Earth will need to be taken into account. The forward scattering amplitude for $\bar{\nu}_e$ and ν_e have an opposite sign, so the Earth will create a difference between the propagation of ν_e and $\bar{\nu}_e$.

CHAPTER 9 HOMEWORK

9.1. Consider the kaon system in the limit of the two light generations only. Suppose we use for the CKM matrix the general unitary matrix:

$$\begin{pmatrix} e^{i\phi_1} \cos\theta_c & e^{i(\phi_1+\phi_2)} \sin\theta_c \\ -e^{i\phi_3} \sin\theta_c & e^{i(\phi_2+\phi_3)} \cos\theta_c \end{pmatrix}.$$

(a) What is the overall CKM factor in the various neutral K decays and in the various terms contributing to the box diagram?

(b) Show that for $\Gamma_S, \Gamma_L, \Delta m_K, |p|^2 - |q|^2$, the phase angles disappear and only θ_c is relevant, when these measurables are calculated using the matrix above.

9.2. Consider the box diagram with only virtual t quarks. We can write the contribution of this diagram to $K, \bar{K}$ and $B, \bar{B}$ mixing through Δm_{12}, which is given by:

$$\text{For } K, \bar{K}: \quad \frac{G_F^2}{6\pi^2} f_K^2 m_K V_{ts}^2 \left(V_{td}^*\right)^2 \times \text{box calculation}$$

$$\text{For } B, \bar{B}: \quad \frac{G_F^2}{6\pi^2} f_B^2 m_B V_{tb}^2 \left(V_{td}^*\right)^2 \times \text{box calculation}$$

Assuming that the box calculation, which gives a real constant stemming from short-distance physics, is the same for both systems and that $f_B \simeq 1.3 f_K$, calculate

$$\frac{\text{Im}\left(\Delta m_{12}^K\right)}{\text{Re}\left(\Delta m_{12}^K\right)}$$

for the $K^0, \bar{K}^0$ system in terms of the measured values of Δm_B and Δm_K. Note that we assume virtual t quarks (through the box diagram) dominate the value of $\text{Im}(\Delta m_{12}^K)$ and are negligible in $\text{Re}(\Delta m_{12}^K)$, while for $B^0, \bar{B}^0$ they dominate both real and imaginary parts of Δm_{12}^B.

9.3. The box diagram involving t quarks gives mixing terms between

$$B^0, \bar{B}^0, \quad B_s^0, \bar{B}_s^0, \quad \text{and} \quad B^0, \bar{B}_s^0.$$

(a) Draw the box diagrams for these three processes, indicating the CKM factors at each vertex.

(b) Is the diagram for $B^0, \bar{B}_s^0$ larger or smaller in magnitude than for $B^0, \bar{B}^0$?

(c) If larger, why isn't $B^0, \bar{B}_s^0$ mixing a physically-observed process?

9.4. Consider the set of strong scattering eigenstates in the mixing calculation.

(a) Can states of opposite CP be part of the same strong scattering eigenstate?

(b) For the case where $\langle n|$ and $\langle \bar{n}|$ are the same state, show that the strong phases drop out, while the weak phases remain in a calculation of Γ_{12}. Show that $\Gamma_{21} = \Gamma_{12}^*$ for the contribution of these states to the mixing.

9.5. Consider the two vector mesons K^{*0} and $\bar{K}^{*0}$.

(a) Is the weak transition between the two nonzero?

(b) Ignoring *CP* violation, what are the mass eigenstates?

(c) Why is it unnecessary to ever use these mass eigenstates?

9.6. Consider the CKM matrix given in Eq. 8.17.

(a) Show that a good approximation for V_{ts} that keeps both real and imaginary parts is

$$V_{ts} = -A\lambda^2(1 + i\lambda^2\eta).$$

(b) The *CP* violating angle β, arising from mixing, was defined for the B^0 system in Eq. 9.30. The analogous angle for the B_s^0 system is called χ. Show that

$$\tan \chi = -\lambda^2\eta.$$

How large is χ?

(c) *CP* violation from mixing in the B_s^0 system is given in terms of χ. Give an example of a decay of B_s^0 where the decay asymmetry, Eq. 9.33, can be used to measure $\sin 2\chi$. This requires that no additional weak phases occur in the decay amplitude and that the final state is a *CP* eigenstate.

9.7. The penguin diagrams for *B* decay were shown in Figure 8.13. Assume that the t quark is the dominant contributor to these diagrams.

(a) Are there any phases in the CKM factors for the transition $b \to s + g$? What do you expect for

$$\frac{q}{p}\frac{\bar{A}_f}{A_f}$$

in Eq. 9.31 for the final state ϕK_S^0 produced through this penguin diagram?

(b) Are there any phases in the CKM factors for the transition $b \to d + g$? What do you expect for

$$\frac{q}{p}\frac{\bar{A}_f}{A_f}$$

in Eq. 9.31 for the final state $\pi^+\pi^-$ produced through this penguin diagram?

(c) Are there other quark-level diagrams that should lead to the same final state as in (b) above? Do these have the same phase in the CKM factors?

9.8. For $m_2^2 - m_1^2 = 3 \times 10^{-3} \text{eV}^2$ and $p = 1$ GeV, find the distance at which the oscillation parameter

$$\frac{(m_2^2 - m_1^2)t}{4p} = 1.$$

9.9. For the assumptions stated in the text, derive Eqs. 9.40 and 9.41.

9.10. For solar neutrinos arriving at the earth with an energy of 10 MeV, what fraction of the neutrinos are ν_μ and what fraction ν_τ?

CHAPTER 10

The Electroweak Gauge Theory and Symmetry Breaking

10.1 ■ WEAK NEUTRAL CURRENT

In the previous two chapters we explored the wide range of interesting phenomena arising from the charged current coupling of fermions to W^+ and W^-. From these we might expect additional phenomena calculable from the coupling of quarks and leptons to the neutral member of the W triplet. The neutral current terms in the interaction were given in Chapter 8; for each doublet we get a term

$$\frac{g}{2} \left[\bar{\psi}_L^U \gamma_\mu \psi_L^U - \bar{\psi}_L^D \gamma_\mu \psi_L^D \right] W_\mu^3. \tag{10.1}$$

This interaction gives quite specific predictions for the coupling and for the helicity of the fermions that interact. The exchanged neutral W might also have the same mass as that of the charged W, provided the mechanism that generates the W mass is the same for all members of the triplet. None of these expectations turn out to agree with what is observed for the weak neutral current interaction, which occurs through the exchange of the Z^0 whose mass is about 10 GeV greater than the W mass. What is remarkable about the deviations from the naive expectations is that the electromagnetic interactions are intimately related to the creation of these deviations.

10.2 ■ NEUTRAL CURRENT MIXING

To find the correct weak neutral current interaction, we have to introduce an additional interaction mediated by a gauge boson called the B^0.[1] Including this interaction, we have for the weak interaction Hamiltonian density,

$$\mathcal{H} = g J_\mu^i W_\mu^i + g' \frac{J_\mu^Y}{2} B_\mu. \tag{10.2}$$

The first term is the weak interaction from the last two chapters, so that

$$J_\mu^i = \bar{\psi}_L \gamma_\mu T_i \psi_L$$

for each doublet. The neutral current part is given by Eq. 10.1. The second term is an Abelian gauge interaction and has a new coupling constant g'. The factor of $\frac{1}{2}$

---
[1] This should not be confused with the pseudoscalar B mesons containing a b quark or $\bar{b}$ antiquark.

10.2 Neutral Current Mixing

in this term is a convention. We do not yet know the handedness of the fermions for the current J_μ^Y, since we have not yet connected the B^0 interaction to any physical effects. The quantum number Y is called the weak hypercharge. This charge is associated with an Abelian gauge interaction and can therefore take on different values for the various fermions. The B^0 and W^0 can mix significantly when the gauge boson masses are generated, since they are initially degenerate. The number of gauge bosons matches what is needed for the weak and electromagnetic interactions. Requiring that we get the known electromagnetic and charged current weak interaction turns out to be enough to define the remaining neutral current interaction completely, after mixing, in terms of one mixing angle. We will go as far as we can here without introducing an explicit symmetry breaking mechanism.

For the new interaction, we have to specify the value of the Y quantum number for each of the left- and right-handed fermion fields, which are separate fields prior to mass generation. We specify the states as $|T, T_3, Y\rangle$, where T, T_3 specify the weak isospin quantum numbers and Y specifies the hypercharge. At this point, there is no electromagnetism, so we do not specify electric charges separately. We suppress the color label, which exists for the quarks, as well as the helicity (which is important) and momentum labels. The interactions in Eq. 10.2 have a gauge symmetry, which is,

$$SU(2) \times U(1),$$

involving the product of the weak isospin ($SU(2)$) and weak hypercharge ($U(1)$). We assume that the gauge transformations for the two groups commute. The W bosons carry no hypercharge and the B boson carries no weak isospin. If we include the $SU(3)$ color group, the symmetry transformations are of the type

$$SU(3) \times SU(2) \times U(1)$$

acting on left- and right-handed fermions, which are independent states prior to mass generation.

From J_μ^Y we can define a charge operator Q^Y by integrating $J_0^Y(x)$ over all space. This was discussed in Section 8.5.1 for the weak isospin, when the conserved vector current idea was introduced. The hypercharge of a state is given by

$$\langle T', T_3', Y' | Q^Y | T, T_3, Y\rangle = Y \langle T', T_3', Y' | T, T_3, Y\rangle.$$

A very important consequence of assuming commuting transformation groups is that the value of Y is the same for all members of a weak isospin doublet (or more generally, any weak isospin multiplet). This follows from $[Q^Y, R] = 0$ for any weak isospin rotation (or isospin generator) R. Applying the commutator to the state $|T = \frac{1}{2}, T_3 = \frac{1}{2}, Y\rangle$ and choosing R as the rotation, which takes the state with $T_3 = \frac{1}{2}$ to that with $T_3 = -\frac{1}{2}$, we get

$$\left[Y_{T_3=1/2} - Y_{T_3=-1/2}\right] \left| T = \frac{1}{2}, T_3 = -\frac{1}{2}, Y\right\rangle = 0.$$

FIGURE 10.1 Generic propagator coupling through amplitude π, which is a matrix in the W^0, B^0 space.

Thus Y is the same for both states. Note that this property is not true for the electric charge, since the doublet members do not have the same charge; the weak isospin generators and the electric charge operator do not commute.

We will now introduce mixing for B^0 and W^0 in a generic fashion, independent of the exact mechanism. We assume that after symmetry breaking the propagator contains an interaction that couples the W^0 and B^0, as shown in Figure 10.1. The interaction must involve particles that have both nonzero weak hypercharge and weak isospin, in order to couple to the two gauge bosons. Since the B^0 and W^0 are degenerate (both are initially massless), any common interaction will mix them, as was true for the K^0 and $\bar{K}^0$ systems. After the mixing, we write the mass eigenstate fields as

$$A_\mu = B_\mu \cos\theta_W + W_\mu^3 \sin\theta_W$$
$$Z_\mu = -B_\mu \sin\theta_W + W_\mu^3 \cos\theta_W, \tag{10.3}$$

with inverses

$$W_\mu^3 = \sin\theta_W A_\mu + \cos\theta_W Z_\mu$$
$$B_\mu^3 = \cos\theta_W A_\mu - \sin\theta_W Z_\mu. \tag{10.4}$$

The mixing angle is called the weak mixing angle. It is measured to be $\sin^2\theta_W = 0.2312$, as we discuss in the next section. We choose A_μ to remain a massless field, which we identify with the photon, providing the only unbroken gauge symmetry. This requires a coupling of A_μ to the vector current only, since this is the only conserved current that occurs for massive fermions. The mixing will be constrained to give the correct current interaction for electrodynamics, in order to agree with what is seen experimentally.

Rewriting the neutral current interaction in terms of A_μ and Z_μ gives:

$$g J_\mu^3 W_\mu^3 + g' \frac{J_\mu^Y}{2} B_\mu = \left[g \sin\theta_W J_\mu^3 + g' \cos\theta_W \frac{J_\mu^Y}{2} \right] A_\mu$$
$$+ \left[g \cos\theta_W J_\mu^3 - g' \sin\theta_W \frac{J_\mu^Y}{2} \right] Z_\mu. \tag{10.5}$$

However, we know how the electromagnetic interaction behaves and the structure of the electromagnetic current, so we can use this to constrain the terms above. In particular, we must take

$$e J_\mu^{EM} = \left[g \sin\theta_W J_\mu^3 + g' \cos\theta_W \frac{J_\mu^Y}{2} \right], \tag{10.6}$$

where the coupling e has been factored out of the electromagnetic current. J_μ^3 has left-handed fields only, so J_μ^Y must add the correct amount of left- and right-handed fields to make a pure vector current for J_μ^{EM}.

10.2 Neutral Current Mixing

We can use Eq. 10.6 to derive a consistency check of the approach. We use the fact that Y, for a weak isospin doublet, is the same for both doublet members. Writing the relation in Eq. 10.6 for the charge operators and subtracting the values for the $T_3 = \frac{1}{2}$ and $-\frac{1}{2}$ left-handed states implies that

$$e\left[Q^{EM}_{T_3=1/2} - Q^{EM}_{T_3=-1/2}\right] = \left[\frac{1}{2} - \left(\frac{-1}{2}\right)\right]g\sin\theta_W = g\sin\theta_W.$$

This says that the electric charge difference in all doublets is the same, which agrees with what is seen. For example, the charge difference is one unit for

$$\begin{pmatrix}\nu_e \\ e^-\end{pmatrix} \quad \text{and} \quad \begin{pmatrix}u \\ d\end{pmatrix}.$$

It thus explains the pattern of charge differences for the U and D doublet members. In addition, using the measured charge values in units of e, gives the relation

$$e = g\sin\theta_W. \tag{10.7}$$

Note that for this to work, we must have $e < g$, as seen experimentally.

Turning again to Eq. 10.6, we can relate the hypercharge and the electric charge directly for right-handed particles (for which the weak isospin is zero), and for doublets by summing the expression for the two doublet members. However, the value of Y and g' for the particles cannot be independently fixed in the absence of additional physics that relates the weak isospin to the hypercharge. We choose, by convention,

$$e = g\sin\theta_W = g'\cos\theta_W, \quad \text{which implies} \quad J^{EM}_\mu = J^3_\mu + \tfrac{1}{2}J^Y_\mu. \tag{10.8}$$

This fixes $Y = -1$ for the left-handed electron, $Y = -2$ for the right-handed electron, and also determines the Y values for the various other particles.

We also find an interesting relation from Eq. 10.7:

$$\frac{G_F}{\sqrt{2}} = \frac{g^2}{8M_W^2} = \frac{e^2}{8M_W^2\sin^2\theta_W}.$$

Since G_F and $e^2 = 4\pi\alpha$ are very well measured, a measurement of either M_W or $\sin^2\theta_W$ constrains the other. Taking $\sin^2\theta_W = 0.2312$ gives the prediction

$$M_W = 77.6 \text{ GeV}.$$

This is wrong by a few percent. What should we conclude from this discrepancy? It stems from small higher-order corrections, the largest coming from our use of $\alpha = 1/137$. The coupling is not independent of scale, and we should use α evaluated at $q^2 = M_W^2$. Choosing this scale, $\alpha(M_W^2) = 1/128.9$. Using this value, we get the predictions

$$\frac{g^2}{4\pi} = \frac{1}{128.9(\sin^2\theta_W)} = \frac{1}{29.8} \quad \text{and} \quad M_W = 80 \text{ GeV},$$

which are very close to the measured value. We will return to relations involving the mass later, since these offer a high-precision test of the Standard Model.

What about the other neutral current interaction? This term is

$$\left[g \cos\theta_W J_\mu^3 - g' \sin\theta_W \frac{J_\mu^Y}{2}\right] Z_\mu.$$

Replacing

$$\frac{J_\mu^Y}{2} \quad \text{with} \quad J_\mu^{EM} - J_\mu^3$$

and

$$g' \sin\theta_W \quad \text{with} \quad \frac{g \sin^2\theta_W}{\cos\theta_W}$$

gives

$$\frac{g}{\cos\theta_W}\left[J_\mu^3 - \sin^2\theta_W J_\mu^{EM}\right] Z_\mu. \tag{10.9}$$

The term multiplying Z_μ gives the coupling to the neutral weak Z boson. We have managed to define the Z^0 interaction without knowing the details of the electroweak symmetry breaking. Both the helicity structure and magnitude of the coupling are modified from the weak $SU(2)$ neutral current term alone, but completely determined from our knowledge of weak isospin and electromagnetism. Note that the mixing model explains the coupling of the photon to the charged W as being due to the non-Abelian gauge coupling of the W_μ^3 piece of A_μ.

10.3 ■ Z^0 PHENOMENOLOGY

Equation 10.9 provides the coupling of Z^0 to the set of leptons and quarks. Combined with the Z^0 mass, it defines the low-energy theory for Z^0 mediated processes, analogous to the W mediated processes in the previous two chapters. The measured Z^0 mass is 91.19 GeV, which cannot be obtained from the general discussion in the previous section; it requires specifying the symmetry breaking mechanism.

The Z^0 couplings to fermions do not have the universal behavior for quarks and leptons that the W couplings have. For a given fermion flavor, f, we have:

$$J_\mu^3 = \bar\psi_L^f \gamma_\mu T_3 \psi_L^f = \frac{T_3^f}{2} \bar\psi_f \gamma_\mu (1-\gamma_5) \psi_f \quad \text{and} \quad J_\mu^{EM} = Q^f \bar\psi_f \gamma_\mu \psi_f.$$

Therefore, the coupling term in Eq. 10.9 is

$$\frac{g}{2\cos\theta_W}\left[\bar\psi_f\left(\left(T_3^f - 2\sin^2\theta_W Q^f\right)\gamma_\mu - T_3^f \gamma_\mu \gamma_5\right)\psi_f\right]. \tag{10.10}$$

10.3 Z^0 Phenomenology

Defining $C_V^f = T_3^f - 2\sin\theta_W Q^f$, $C_A^f = T_3^f$, we have for the coupling:

$$\frac{g}{2\cos\theta_W}\bar{\psi}_f \gamma_\mu \left(C_V^f - C_A^f \gamma_5\right)\psi_f,$$

which is a general mix of vector and axial vector currents. We can now form a general second-order weak effective Hamiltonian density describing low-energy scattering of two flavors, shown in Figure 10.2, analogous to the W interaction in Figure 8.1. Writing this as a matrix element analogous to Eq. 8.5, as would describe, for example, $\nu_\mu e^- \to \nu_\mu e^-$, gives

$$M_{fi} = \frac{g^2}{4\cos^2\theta_W m_Z^2}\left[\bar{\psi}_1 \gamma_\mu (C_V^{f_1} - C_A^{f_1}\gamma_5)\psi_1\right]\left[\bar{\psi}_2 \gamma_\mu (C_V^{f_2} - C_A^{f_2}\gamma_5)\psi_2\right]. \tag{10.11}$$

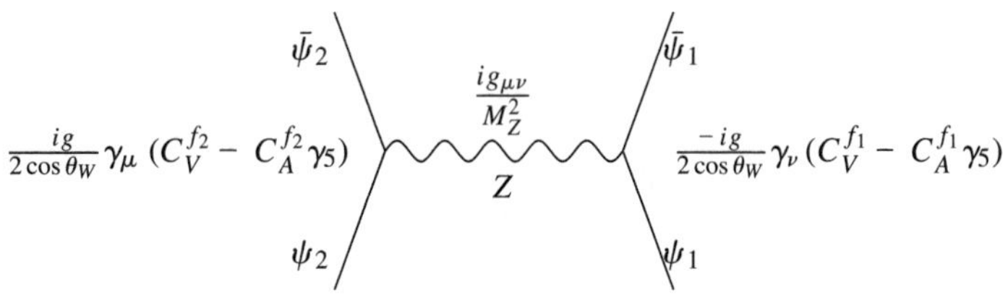

FIGURE 10.2 Second-order weak neutral current process for $q^2 \ll M_Z^2$.

In Table 10.1 we list the various coupling factors C_V^f and C_A^f. The numbers in the table are based on the value $\sin^2\theta_W = 0.2312$. In reality, a careful measurement of the couplings allows us to determine $\sin^2\theta_W$, which has now been measured with an uncertainty of about 0.0002. A test of the theory is then provided by the consistency of a large body of data with this one value. These data include ν neutral current scattering on e^- and on nucleons, very small weak neutral current effects that violate parity in e^- scattering on nucleons, and very detailed measurements on the final states produced from the Z^0 itself in the process

TABLE 10.1 Weak Neutral Current Couplings.

Flavor	Q^f	$T_3^f = C_A^f$	$C_V^f = T_3^f - 2\sin^2\theta_W Q^f$
ν_e, ν_μ, ν_τ	0	$\frac{1}{2}$	$\frac{1}{2}$
e^-, μ^-, τ^-	-1	$-\frac{1}{2}$	-0.037
u, c, t	$\frac{2}{3}$	$\frac{1}{2}$	0.191
d, s, b	$-\frac{1}{3}$	$-\frac{1}{2}$	-0.345

$e^+e^- \to Z^0 \to f\bar{f}$. The comparison of theory to the most precise measurements requires higher-order corrections and a careful definition of parameters, such as $\sin^2\theta_W$, and their relation to a given renormalization scheme.

From Eq. 10.8 we can also determine the hypercharge values for the various fermion types. These are given in Table 10.2. The ν_R has no interactions of any kind (ignoring gravity), other than whatever is responsible for generating neutrino masses. The electric charge is the only quantum number in common for the left- and right-handed members, which get joined into a mass eigenstate. Note also that the question of why the quark charges are fractional is transferred to the question of why Y is fractional for quarks.

TABLE 10.2 $SU(2) \times U(1)$ Quantum Numbers.

Flavor	T	T_3	Q	Y
ν_L	$\frac{1}{2}$	$\frac{1}{2}$	0	-1
ν_R	0	0	0	0
e_L^-	$\frac{1}{2}$	$-\frac{1}{2}$	-1	-1
e_R^-	0	0	-1	-2
u_L	$\frac{1}{2}$	$\frac{1}{2}$	$\frac{2}{3}$	$\frac{1}{3}$
u_R	0	0	$\frac{2}{3}$	$\frac{4}{3}$
d_L	$\frac{1}{2}$	$-\frac{1}{2}$	$-\frac{1}{3}$	$\frac{1}{3}$
d_R	0	0	$-\frac{1}{3}$	$-\frac{2}{3}$

The Z^0 couplings in Table 10.1 allow a direct calculation of the Z^0 width. The decay matrix element to a given flavor-pair is:

$$M_{fi} = \frac{g}{2\cos\theta_W} e_\mu^Z \bar{u}_f \gamma_\mu (C_V^f - C_A^f \gamma_5) v_{\bar{f}}.$$

Ignoring the fermion masses relative to M_Z, this gives for the partial width

$$\Gamma_{Z^0 \to f\bar{f}} = \frac{g^2 M_Z}{48\pi \cos^2\theta_W} \left[\left|C_V^f\right|^2 + \left|C_A^f\right|^2 \right]. \tag{10.12}$$

Using Table 10.1 and summing over three neutrino generations, three charged lepton generations, three down-type quark generations × three colors, and two up-type generations × three colors gives:

$$\Gamma_Z = 2.4 \text{ GeV}.$$

10.3 Z^0 Phenomenology

Higher-order *QCD* corrections increase this prediction to about 2.5 GeV. The measured value, which is 2.495 ± 0.002, allows a measurement of $\alpha_s(M_Z^2) = 0.12 \pm 0.02$ based on the size of the correction, which is a factor of

$$\left[1 + \frac{\alpha_s(M_Z^2)}{\pi}\right]$$

for each quark decay.

We can also use Eq. 10.12 to evaluate the Z^0 branching ratios to the various fermion types. The smallest branching ratios (3.37%) are to each of the charged leptons, the largest are to each of the down-type $q\bar{q}$ pairs (15.2%).

An analogous calculation for Γ_W gives:

$$\Gamma_W = \frac{3}{4}\left(\frac{g^2}{4\pi}\right) M_W = 2.0 \text{ GeV}.$$

Higher-order *QCD* corrections again increase the quark decay terms by

$$\left[1 + \frac{\alpha_s(M_W^2)}{\pi}\right]$$

for the W, so that the predicted width is 2.1 GeV. As in the Z^0 case, this is in excellent agreement with direct measurements, which are, however, not as precise as in the Z^0 case.

High-precision studies of the Z^0 parameters have been carried out at e^+e^- colliders running at center of mass energies near M_Z. A number of very interesting measurements can be made with high statistics, since the cross section is large because of the resonance. The process for production of an $f\bar{f}$ pair is shown in Figure 10.3. The matrix element is given by Eq. 10.11, except that we need to keep the full Z^0 propagator rather than taking the $q^2 \ll M_Z^2$ limit. This gives

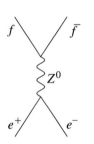

FIGURE 10.3 Feynman diagram for $e^+e^- \to f\bar{f}$ via an intermediate Z^0 resonance.

$$M_{fi} = \frac{-g^2}{4\cos^2\theta_W} \frac{\left[\bar{u}_f \gamma_\mu \left(C_V^f - C_A^f \gamma_5\right) v_{\bar{f}}\right]\left[\bar{v}_{e^+}\gamma_\mu \left(C_V^e - C_A^e \gamma_5\right) u_{e^-}\right]}{\left[q^2 - M_Z^2 + iq^2\frac{\Gamma_Z}{M_Z}\right]},$$

(10.13)

where the variation of the width with q^2 has been included in the denominator. The small nonresonant photon contribution has been left out. This matrix element, including its helicity dependence, serves as the basis for predicting the many measurements that can be made. These include the following.

Measurement of M_Z, Γ_Z and the cross section by measuring rates versus q^2. Individual branching ratios will also determine $|C_V^f|^2 + |C_A^f|^2$. The predicted total cross section for unpolarized e^+e^-, prior to radiative corrections and excluding the photon exchange contribution, is

$$\sigma(q^2) = 12\pi \left(\frac{q^2}{M_Z^2}\right) \frac{\Gamma_{Z\to e^+e^-}\Gamma_Z}{(q^2 - M_Z^2)^2 + \frac{q^4\Gamma_Z^2}{M_Z^2}}. \tag{10.14}$$

If we look at a single final state $f\bar{f}$, the Γ_Z in the numerator should be replaced by $\Gamma_{Z\to f\bar{f}}$. At $q^2 = M_Z^2$ the total cross section is

$$\sigma(M_Z^2) = 12\pi \frac{\Gamma_{Z\to e^+e^-}}{M_Z^2 \Gamma_Z}.$$

The enhancement in the cross section due to the Z^0 is a factor $\sim 10^3$ compared to photon exchange with no Z^0.

The observed total cross section is actually smaller than the result given in Eq. 10.14, since the events where $Z^0 \to \nu_f \bar{\nu}_f$ produce undetectable final states. This is about 20% of the events for three neutrino generations. The determination of the invisible cross section is the basis for saying that there are only three light neutrino generations. The invisible cross section still contributes to the width in the denominator in Eq. 10.14, which can be measured by scanning in q^2. A scan of the cross section into hadronic final states is shown in Figure 10.4. The data are beautifully described by the Standard Model with three neutrino species.

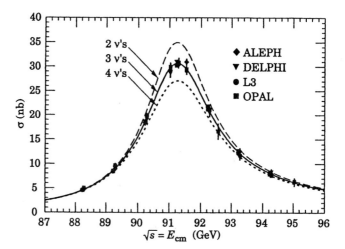

FIGURE 10.4 Cross section for e^+e^- to hadronic final states as a function of the center of mass energy. Curves are the Standard Model predictions, after radiative corrections, for two, three, or four light neutrino species. [Data compilation from the Particle Data Group, K. Hagiwara et al., *Phys. Rev.* D66, 010001(2002)].

10.3 Z^0 Phenomenology

The unpolarized cross section comes from averaging over cross sections for left-handed electrons, which annihilate with right-handed positrons, and right-handed electrons, which annihilate left-handed positrons. The pairing between electron-positron helicities is true for any mix of vector and axial currents in the relativistic limit. These two cross sections are, however, not equal because the weak coupling contains unequal left- and right-handed components. These two components are, for a general flavor,

$$C_L^f = \frac{C_V^f + C_A^f}{2} \quad \text{and} \quad C_R^f = \frac{C_V^f - C_A^f}{2},$$

respectively. We can exploit this feature to measure $\sin^2 \theta_W$, which appears in the couplings, by using polarized electrons and measuring the cross section difference for left-handed and right-handed electrons. Defining an asymmetry (called the left-right asymmetry):

$$A_{LR} = \frac{\sigma_L - \sigma_R}{\sigma_L + \sigma_R},$$

$$A_{LR} = \frac{|C_L^e|^2 - |C_R^e|^2}{|C_L^e|^2 + |C_R^e|^2} = \frac{2 C_V^e C_A^e}{|C_V^e|^2 + |C_A^e|^2}.$$

This term measures the degree of parity violation, since it requires both vector and axial terms in the same current. We define in general for any flavor,

$$A_f = \frac{2 C_V^f C_A^f}{\left|C_V^f\right|^2 + \left|C_A^f\right|^2} \tag{10.15}$$

which depends on $\sin^2 \theta_W$ through C_V^f. The measurement of A_{LR} has provided the single most accurate value for $\sin^2 \theta_W$.

The final state fermions are polarized for the same reason that A_{LR} is nonzero. In practice, the polarization can only be measured for τ by using its decay to analyze the degree of polarization. Taking τ^- (τ^+ has the opposite polarization), we define the polarization in terms of the difference of the number of left- and right-handed τ^- produced:

$$P_\tau = \frac{N_L - N_R}{N_L + N_R}.$$

Using Eq. 10.13 this can be calculated to be, for an unpolarized initial $e^+ e^-$:

$$P_\tau = \frac{A_\tau (1 + \cos^2 \theta) + 2 A_e \cos \theta}{(1 + \cos^2 \theta) + 2 A_\tau A_e \cos \theta},$$

where θ is the production angle relative to the incident e^- direction. The $A_\tau A_e$ term is small, so if we sum events over all directions:

$$\bar{P}_\tau \simeq A_\tau, \quad \text{which should equal} \quad A_{LR} = A_e.$$

The measurement of $\bar{P}_\tau$ is much more difficult than A_{LR} and it yields a lower precision measurement of $\sin^2\theta_W$, although it agrees excellently with the expected value from A_{LR}.

The parity violation in the coupling implies that a correlation exists between the direction of the final state particle f and the initial e^- (and the $\bar{f}$ with e^+). Using the matrix element in Eq. 10.13, the angular distribution for f can be calculated to be, for an unpolarized e^+e^- initial state:

$$\frac{d\sigma^{f\bar{f}}}{d\cos\theta} = \frac{3}{8}\sigma_{f\bar{f}}\left[(1+\cos^2\theta) + 2A_e A_f \cos\theta\right].$$

Counting the number of events where f has $\cos\theta \geq 0$ (forward production) and those where $\cos\theta \leq 0$ (backward production), we can define an asymmetry, called the forward-backward asymmetry:

$$A_{FB}^f = \frac{N_F - N_B}{N_F + N_B} = \frac{3}{4}A_e A_f.$$

This can be evaluated for final charged leptons directly and gives an accurate measurement of $\sin^2\theta_W$. Since the quarks produce jets rather than being detected directly, a measurement of A_{FB}^f for quarks requires using the average jet directions to define the quark direction and also tagging the jet type to know whether the parent was f or $\bar{f}$. This can be done best for b jets and c jets. Despite these uncertainties for the quarks, we can still make a reasonably accurate measurement.

The measurements discussed above can all be described in terms of g^2 (which is related to e^2 and $\sin^2\theta_W$) and two input parameters, M_Z and $\sin^2\theta_W$, arising from electroweak symmetry breaking. These two have now been measured with fractional errors of about 10^{-4} and 10^{-3}, respectively, allowing an excellent description of all the Z^0 data, often with fractional errors of 10^{-3}.

10.4 ■ INTERACTIONS AMONG THE GAUGE BOSONS THEMSELVES

A key distinction between an Abelian and a non-Abelian gauge theory is that the gauge particles interact with each other in the non-Abelian theory. In addition, this interaction is completely fixed in character by the gauge symmetry. The weak gauge particle couplings result from a Lagrangian analogous to the QCD Lagrangian discussed in Chapter 7, but now for $SU(2)$. Recent data using e^+e^- annihilation at energies up to a little more than 200 GeV allow a high-precision test of these predictions of the electroweak theory. The reactions here

10.4 Interactions Among the Gauge Bosons Themselves

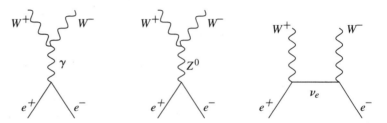

FIGURE 10.5 Lowest-order electroweak diagrams for $e^+e^- \to W^+W^-$.

are $e^+e^- \to W^+W^-$ and $e^+e^- \to Z^0Z^0$. The lowest-order diagrams for the W^+W^- case are shown in Figure 10.5.

The data for $e^+e^- \to W^+W^-$ are shown in Figure 10.6. These data provide a very nice verification of the full gauge theory structure. Also shown are calculations of the cross section using only subsets of the diagrams, indicating that all diagrams are needed to get the correct cross section. Figure 10.7 shows the data for $e^+e^- \to Z^0Z^0$. These are also nicely described by the electroweak gauge theory.

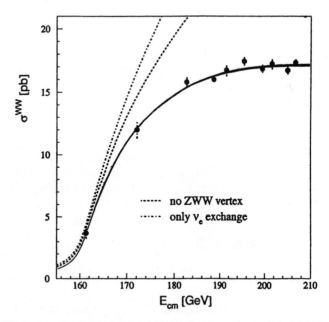

FIGURE 10.6 W pair production cross section compared to the Standard Model prediction, and, for comparison, cross section calculated with only a subset of contributing diagrams. [Reprinted with permission of the Electroweak Working Group, CERN, Geneva, Switzerland.]

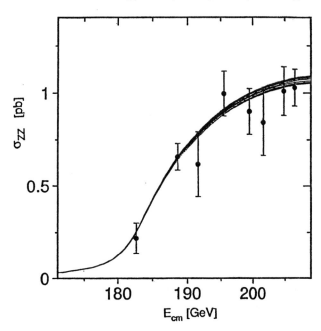

FIGURE 10.7 Observed Z pair production cross section compared to the Standard Model prediction. [Reprinted with permission of the Electroweak Working Group, CERN, Geneva, Switzerland.]

10.5 ■ HIGGS MECHANISM

We turn next to the symmetry breaking mechanism, which is the only element still lacking from the rather detailed discussion of the electroweak physics presented earlier. Symmetry breaking requires the addition of scalar fields that interact with both the fermions and gauge bosons, generating their masses. Since the interactions are local and the masses are always present during free propagation, we can conclude that the scalar must also always be locally present, that is, it is everywhere. Thus the vacuum is actually uniformly filled with the scalar field. This is called a nonvanishing vacuum expectation value. This mechanism is called the Higgs mechanism, after Peter Higgs, who worked out this idea.

The fields and their interactions for the Higgs mechanism provide the following set of renormalizable terms that we can have in a Lagrangian:

1. Spin zero fields with interaction terms up to 4th order in the scalars. The scalar fields therefore have mass terms and self interactions.
2. Spin 1 gauge fields. These are initially massless for a renormalizable theory. They interact with all objects that carry the "charge" of the gauge transformation.

10.5 Higgs Mechanism

3. Massless left- and right-handed fermion fields. These interact with both the scalars (called a Yukawa interaction) and gauge fields.

Except for the scalar terms, we have discussed (2) and (3) extensively. We therefore focus on (1) and the scalar-gauge field interactions. We leave the scalar-fermion issues for later. We look at a simplified situation first in order to develop some familiarity with how the Higgs mechanism works. The simplest situation is one charged scalar field interacting with one Abelian gauge field. We write the scalar field $\phi(x)$ and the gauge field as $A_\mu(x)$. We use the terminology of electromagnetism because of its familiarity. The real situation involves a multiplet of scalars interacting with W_μ^i and B_μ.

Since the scalar is charged, it is described by a complex field

$$\phi(x) = \frac{\phi_1(x) + i\phi_2(x)}{\sqrt{2}}.$$

We assume the charge of the scalar is $-e$ and that of its antiparticle is e. The Lagrangian for the scalars and gauge field is

$$\mathcal{L} = \left[\left(\frac{-i\partial}{\partial x_\mu} + eA_\mu\right)\phi^*\right]\left[\left(\frac{i\partial}{\partial x_\mu} + eA_\mu\right)\phi\right]$$
$$- \mu^2 \phi^* \phi - \lambda(\phi^*\phi)^2 - \frac{1}{4}F_{\mu\nu}F_{\mu\nu}. \tag{10.16}$$

Without the gauge interaction, $\mathcal{L}$ describes a scalar theory, discussed briefly in Chapter 2, with a conserved current reflecting the global invariance under an overall phase change for ϕ. The mass would be μ and the term involving λ, in a perturbation expansion, gives a charge conserving interaction involving four scalar particles at a vertex.

The gauge interaction was introduced through the prescription

$$\left[\frac{i\partial}{\partial x_\mu}\right]\phi \to \left[\frac{i\partial}{\partial x_\mu} + eA_\mu\right]\phi,$$

and the free gauge field included in $\mathcal{L}$ through the field tensor term. With the gauge field present, the electromagnetic current is modified from

$$-e\left(i\phi^*\frac{\partial\phi}{\partial x_\mu} - i\phi\frac{\partial\phi^*}{\partial x_\mu}\right),$$

to

$$J_\mu^{EM} = -e\left[\phi^*\left(\frac{i\partial}{\partial x_\mu} + eA_\mu\right)\phi - \phi\left(\frac{i\partial}{\partial x_\mu} - eA_\mu\right)\phi^*\right]$$
$$= -e\left[i\phi^*\frac{\partial\phi}{\partial x_\mu} - i\phi\frac{\partial\phi^*}{\partial x_\mu}\right] - 2e^2\phi^*\phi A_\mu.$$

The field equation for A_μ, including interactions in the Lorentz gauge, is then

$$\frac{\partial^2}{\partial t^2} A_\mu - \nabla^2 A_\mu = J_\mu^{EM}.$$

The generation of mass for A_μ amounts to noticing that, if $\phi^*\phi \neq 0$, then the propagation of A_μ is modified even if there is no dynamic scalar field present (that is, even if $\partial \phi / \partial x_\mu = 0$, so no ϕ particles are physically propagating). A constant value of $\phi^*\phi$, written $\langle \phi^*\phi \rangle$, is called a nonzero vacuum expectation value for the field. Ignoring the dynamic term in J_μ^{EM}, which involves propagating ϕ particles, and focusing on the second term, we have the field equation for A_μ:

$$\frac{\partial^2}{\partial t^2} A_\mu - \nabla^2 A_\mu = -2e^2 \langle \phi^*\phi \rangle A_\mu. \tag{10.17}$$

Looking at a plane wave solution, which represents a propagating gauge particle, we try

$$A_\mu \sim e^{-ik \cdot x} = e^{i(\vec{k} \cdot \vec{x} - \omega t)}.$$

Substituting the plane wave into Eq. 10.17 then gives

$$\omega^2 = \vec{k}^2 + 2e^2 \langle \phi^*\phi \rangle.$$

In this scalar-filled medium, the gauge field propagates as if it had a mass squared, given by $2e^2 \langle \phi^*\phi \rangle = m_A^2$.

We are used to the propagation of electromagnetic fields being modified in media. For example, in a dielectric, the velocity of light is changed. The propagation of electromagnetic waves with an apparent nonzero mass occurs in a plasma, for example, in the ionosphere. These media, however, have other material properties. Since we are not otherwise aware of the presence of the scalar field, postulating such a mechanism for gauge particle masses is perhaps a surprise and, in fact, not all of its ramifications are understood (for example, what happens to gravitational effects and the geometry of space-time for the scalar-filled vacuum).

The mass generation through the nonzero vacuum expectation value requires that the minimum energy for the ϕ field occurs when $\langle \phi^*\phi \rangle \neq 0$. This has to be true independent of any excitations of the field, so it is a property of the term in the Hamiltonian density involving the fields only and no derivatives. This term is called the potential energy term $V(\phi)$. It is, for Eq. 10.16,

$$V(\phi) = \mu^2 \phi^*\phi + \lambda (\phi^*\phi)^2. \tag{10.18}$$

We normally expect μ^2 to be the scalar mass squared and to be > 0. If, however, $\mu^2 < 0$, then $\phi = 0$ is not the minimum. This is the mechanism envisioned for moving the minimum away from $\phi = 0$. It is the simplest $V(\phi)$ giving a nonzero vacuum expectation value, but not very well motivated by physics (other than the need to generate the particle masses).

10.5 Higgs Mechanism

With $\mu < 0$, $\lambda > 0$, the minimum occurs for

$$\phi^*\phi = \frac{-\mu^2}{2\lambda}.$$

If we know the parameters $-\mu^2$ and λ, we can calculate

$$m_A^2 = \frac{-e^2\mu^2}{\lambda} = e^2 v^2, \tag{10.19}$$

where we define $v^2 = -\mu^2/\lambda$. The parameter v, determined by the vacuum expectation value of the field, has the dimensions of a mass.

If we write ϕ in terms of the two real fields, the global phase invariance is just a rotational invariance among the two fields ϕ_1 and ϕ_2. The minimum then corresponds to

$$\phi^*\phi = \frac{1}{2}\left(\phi_1^2 + \phi_2^2\right) = \frac{v^2}{2}. \tag{10.20}$$

The actual ground state of the system can be in any direction in the ϕ_1, ϕ_2 space and this direction is unpredictable a priori. The choice nature takes in a given case creates "spontaneous symmetry breaking," since once the choice is made, the directions in the space are no longer equivalent. In this case the ϕ_1, ϕ_2 rotational symmetry is still a feature of $\mathcal{L}$, but not of the physics, which occurs relative to this ground state or apparent vacuum state.

We will now put together all of the terms in $\mathcal{L}$ for the case where $v \neq 0$. We already know what happens to A_μ; it has a mass m_A. The question is, what happens to ϕ; what physics does it give, and can we see the physics? Prior to taking this step, however, we consider the scalar theory, without the gauge particle, to see what this simpler situation would be like in the case of a spontaneously broken global symmetry. In this case, we define the ϕ_1 and ϕ_2 directions so that v is along ϕ_1. In addition, we reference the dynamic components to v, so that we can write

$$\phi = \frac{1}{\sqrt{2}}\left(v + h(x) + i\theta(x)\right). \tag{10.21}$$

Substituting this into $\mathcal{L}$ for the scalar fields alone, we get an equation in terms of the new real fields $h(x)$ and $\theta(x)$:

$$\mathcal{L} = \text{Const.} + \frac{1}{2}\left(\frac{\partial \theta(x)}{\partial x_\mu}\right)^2 + \frac{1}{2}\left(\frac{\partial h(x)}{\partial x_\mu}\right)^2 + \mu^2 h^2(x)$$

$$- \lambda v h(x) \left[h^2(x) + \theta^2(x)\right] - \frac{\lambda}{4}\left(h^2(x) + \theta^2(x)\right)^2. \tag{10.22}$$

There is no quadratic term in $\theta(x)$ when we expand around the minimum. This $\mathcal{L}$ now describes two particles that are no longer degenerate. The $h(x)$ particle has a mass $\sqrt{-2\mu^2} = \sqrt{2\lambda v^2}$ (which is now positive) and the $\theta(x)$ particle is massless. The massless particle is called a Goldstone boson. There are still two types of excitations, as was the case for a symmetric vacuum, but they are not degenerate. The cubic and quartic terms represent interactions between the two types of particles. Note that the number of particles at a vertex now does not have to be four. The cubic term involves the vacuum value v as the fourth field, so that three physical particles can interact at a point. For the cubic term, the vacuum does not bring in or take away any energy or momentum. It acts as a dimensional coupling factor.

We now return to the full, locally gauge invariant Lagrangian. We will see that, for this theory, the massive scalar remains as a physical excitation, but the Goldstone boson does not. The local gauge symmetry implies that we can transform ϕ and A_μ via

$$\phi(x) \to e^{i\theta(x)}\phi(x)$$
$$A_\mu(x) \to A_\mu(x) - \frac{1}{e}\frac{\partial \theta(x)}{\partial x_\mu}, \qquad (10.23)$$

resulting in the same Lagrangian. We will make use of the gauge freedom to make an explicit gauge choice. This means that we give up the freedom to make further gauge transformations in favor of an explicit Lagrangian where the physical particle degrees of freedom are most evident. The gauge choice we make is to remove the imaginary part of $\phi(x)$. That is, we write

$$\phi(x) = \frac{1}{\sqrt{2}}(v + H(x)). \qquad (10.24)$$

The physics of the imaginary part will thus be ascribed entirely to the presence of an $A_\mu(x)$. We can thus rewrite $\mathcal{L}$ in this gauge as

$$\mathcal{L} = \frac{1}{2}\left[\left(\frac{2H}{\partial x_\mu}\right)^2 - 2\lambda v^2 H^2\right] - \lambda v H^3 - \frac{\lambda}{4}H^4$$
$$- \left(\frac{1}{4}F_{\mu\nu}F_{\mu\nu} - \frac{1}{2}e^2 v^2 A_\mu^2\right) + \frac{1}{2}e^2 H^2 A_\mu^2 + v e^2 H A_\mu^2. \qquad (10.25)$$

The terms include:

1. A massive scalar particle H, called the Higgs particle. Its mass is $m_H = \sqrt{2\lambda v^2}$.
2. A massive spin 1 A particle with mass $m_A = ev$. Note that the mass of the A particle cannot be predicted from that of H, since different constants enter. Both masses are, however, proportional to v, which is the primary mass scale in the problem.
3. Interactions of H with itself, involving cubic (three H particles at a vertex) and quartic (four H particles at a vertex) terms.
4. Interactions of A with H, which involve two A particles and one or two H particles. Note that the charge conservation pattern of the symmetric theory

is destroyed by the vacuum in the spontaneously broken case. For example, A_μ can radiate a single H.

This theory thus contains significant new physics, particularly if we can reach an energy where the Higgs particle can actually be produced through its coupling to A_μ. Taking the interaction term with one H field, which allows H production at the lowest energy, H is radiated from A_μ with a well-defined coupling ve^2. This coupling is determined by m_A itself and is therefore fully predictable once the mass of A is measured. In this sense, the primary uncertainty in predicting phenomena involving H are the particle masses.

We comment on a few issues before turning to the $SU(2) \times U(1)$ version of the Higgs phenomenon. The first concerns the number of physical degrees of freedom. In the symmetric version of the theory, which describes the physics for particle eigenstates when $v = 0$, the degrees of freedom are the ϕ particle, its antiparticle, and the two transverse helicity states of the massless A. In the case of the spontaneously broken symmetry, the number of degrees of freedom for the particle states are the same, with the particle content being H and the three helicity states of the massive A particle. Both realizations of the theory have been shown to provide a renormalizable quantum theory. It is also possible to work in an arbitrary gauge, rather than our choice, but at the cost of greater complexity.

10.6 ■ THE THEORY OF WEINBERG AND SALAM

We are now ready to combine the $SU(2) \times U(1)$ gauge theory with the Higgs mechanism. This is the theory of Weinberg and Salam, which provides the simplest model for electroweak symmetry breaking. We focus again on the scalar terms and the scalar-gauge boson interactions, which are the crucial elements for the Higgs mechanism. We return to the other terms in the Lagrangian after working out the gauge boson mass pattern. The part of $\mathcal{L}$ that generates the gauge boson masses is

$$\mathcal{L}_{\text{scalar}} = \left|\left(i\frac{\partial}{\partial x_\mu} - gT_i W^i_\mu - g'\frac{Y}{2}B_\mu\right)\phi\right|^2 - V(\phi), \quad \text{where}$$

$$V(\phi) = \mu^2 \phi^\dagger \phi + \lambda \left(\phi^\dagger \phi\right)^2. \tag{10.26}$$

Quadratic terms in the gauge boson fields with constant coefficients correspond to mass terms when combined with the quadratic field tensor terms, which have to be added to Eq. 10.26. The field ϕ has both weak isospin and hypercharge, since we want it to mix W^3_μ and B_μ after spontaneous symmetry breaking. The form for $\mathcal{L}$ is obtained by replacing $i(\partial/\partial x_\mu)$ with the gauge invariant derivative. Under gauge transformations, ϕ is assumed to transform as $\phi \to e^{i\alpha_i T_i}\phi$ for $SU(2)$ transformations and $\phi \to e^{iY\beta}\phi$ for the hypercharge transformations. The type of $SU(2)$ multiplet will determine the dimensionality and form for the matrix representation of the T_i.

The simplest choice for ϕ is to take a doublet of charged scalars, where charged now refers to hypercharge. Since Y is the same for the whole multiplet, we need to specify the single Y value. The electric charge of the members of the doublet will be:

$$Q_{T_3=1/2} = \frac{1}{2} + \frac{Y}{2}, \quad Q_{T_3=-1/2} = -\frac{1}{2} + \frac{Y}{2}.$$

Choosing $Y = 1$, we get the upper member of the doublet with electric charge $= 1$, the lower with electric charge $= 0$ (these are for the particles; the antiparticles have the opposite charge). Since we want electromagnetism to remain unbroken, a member of the doublet must be electrically neutral. This member gets a non-vanishing vacuum expectation value after symmetry breaking. Nature has told us which direction in the space we need to assign the nonzero v.

We will write

$$\phi = \frac{1}{\sqrt{2}} \begin{pmatrix} \phi_1 + i\phi_2 \\ \phi_3 + i\phi_4 \end{pmatrix}, \quad \phi^\dagger \phi = \frac{1}{2}\left(\phi_1^2 + \phi_2^2 + \phi_3^2 + \phi_4^2\right).$$

For $\mu^2 < 0$ we again get spontaneous symmetry breaking, with the minimum given by

$$\phi^\dagger \phi = \frac{v^2}{2}, \quad v^2 = -\frac{\mu^2}{\lambda}.$$

Consider now a general $\phi(x)$ whose length is referenced to v. It can be brought to an arbitrary direction in the isospin space by a weak isospin rotation. Thus we can write in general:

$$\phi(x) = e^{i\theta_i(x)T_i} \begin{pmatrix} 0 \\ \dfrac{v + H(x)}{\sqrt{2}} \end{pmatrix}.$$

The four parameters $\theta_1(x), \theta_2(x), \theta_3(x), H(x)$ determine $\phi(x)$ at a point x in space-time.

We can, however, eliminate the weak rotation by using the gauge freedom. This will tie us to a specific gauge choice, but one where the propagating particle degrees of freedom are the explicit fields. In addition, we have chosen to put v into the component of $\phi(x)$, which is electrically neutral. Thus the vacuum will not be able to contribute electric charge in a given scattering process, and electric charge will be conserved among the physically propagating particles.

To figure out the W, Z masses we need only take for ϕ the vacuum expectation term, since the gauge boson masses are present even with no dynamically propagating scalar particles. Thus, choosing

10.6 The Theory of Weinberg and Salam

$$\phi(x) = \begin{pmatrix} 0 \\ \frac{v}{\sqrt{2}} \end{pmatrix}, \qquad (10.27)$$

and substituting into $\mathcal{L}_{\text{scalar}}$ of Eq. 10.26, we get for the term involving the gauge field and the ϕ field,

$$\left| \left(g T_i W_\mu^i + \frac{g' B_\mu}{2} \right) \begin{pmatrix} 0 \\ \frac{v}{\sqrt{2}} \end{pmatrix} \right|^2.$$

Using

$$T_i = \frac{\sigma_i}{2}$$

gives:

$$\frac{1}{8} \left| \begin{pmatrix} g W_\mu^3 + g' B_\mu & g \left(W_\mu^1 - i W_\mu^2 \right) \\ g \left(W_\mu^1 + i W_\mu^2 \right) & -g W_\mu^3 + g' B_\mu \end{pmatrix} \begin{pmatrix} 0 \\ v \end{pmatrix} \right|^2$$

$$= \frac{1}{8} v^2 \left| \begin{matrix} g \left(W_\mu^1 - i W_\mu^2 \right) \\ -g W_\mu^3 + g' B_\mu \end{matrix} \right|^2$$

$$= \frac{g^2 v^2}{8} \left[\left(W_\mu^1 \right)^2 + \left(W_\mu^2 \right)^2 \right] + \frac{v^2}{8} \left[\left(g W_\mu^3 - g' B_\mu \right) \right]^2. \quad (10.28)$$

The interaction term maintains a symmetry between W_μ^1 and W_μ^2, which we can rewrite in terms of the charged W_μ fields $W_\mu^{(+)}$ and $W_\mu^{(-)}$ of Eq. 8.2. Namely, we can rewrite the first term in brackets in Eq. 10.28 as

$$\frac{1}{2} \left(\frac{gv}{2} \right)^2 \left[\left| W_\mu^{(+)} \right|^2 + \left| W_\mu^{(-)} \right|^2 \right].$$

Thus the $W_\mu^{(+)}$ and $W_\mu^{(-)}$ fields have a mass term equal to

$$\frac{gv}{2} = M_W.$$

Therefore,

$$M_W = \sqrt{4\pi \left(\frac{g^2}{4\pi} \right)} \frac{v}{2},$$

which implies that $v = 247$ GeV, if we use the measured values of M_W and $g^2/4\pi$. This new mass scale characterizes the electroweak spontaneous symmetry breaking.

The second term in brackets in Eq. 10.28 represents a neutral current mass term. To write it explicitly as a mass term, we need to define new normalized fields:

$$Z_\mu = \frac{g W_\mu^3 - g' B_\mu}{\sqrt{g^2 + g'^2}},$$

with the orthogonal field being

$$A_\mu = \frac{g B_\mu + g' W_\mu^3}{\sqrt{g^2 + g'^2}}.$$

The field A_μ has no mass term; it remains an unbroken gauge symmetry. The field Z_μ has a mass term, which is

$$\sqrt{g^2 + g'^2}\,\frac{v}{2} = M_Z.$$

Our formulas for the fields are the same as written earlier in Eq. 10.3, after using $g \sin \theta_W = g' \cos \theta_W$ from Eq. 10.8. Note that the symmetry breaking mechanism provides no insight into the value of $\sin^2 \theta_W$, since g and g' are put in as parameters separately. However, it provides a relation for M_Z and M_W, namely

$$M_Z = M_W \frac{\sqrt{g^2 + g'^2}}{g} = \frac{M_W}{\cos \theta_W}.$$

This is often written as

$$\sin^2 \theta_W = 1 - \left(\frac{M_W}{M_Z}\right)^2, \tag{10.29}$$

a key prediction of the model! If we use the measured masses, we get from this formula

$$\sin^2 \theta_W = 0.222.$$

This is off by about 4% from the value measured using the couplings on the Z^0. The discrepancy is due to higher-order corrections, which we will examine in the next section.

The mass formulas depend only on gauge field propagation in the Higgs ground state. We can also observe interaction terms between the scalar H and the various gauge bosons. We get these by using, in the Lagrangian of Eq. 10.26,

$$\phi(x) = \begin{pmatrix} 0 \\ \dfrac{v + H(x)}{\sqrt{2}} \end{pmatrix}. \tag{10.30}$$

10.6 The Theory of Weinberg and Salam

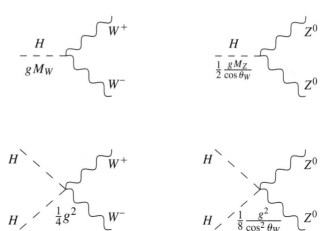

FIGURE 10.8 Feynman diagrams for H, W, Z interactions.

The interaction vertices and coupling factors are shown in Figure 10.8. The terms in $\mathcal{L}_{\text{scalar}}$ are

$$\frac{vg^2}{2} H(x) W^{(+)}_\mu W^{(-)}_\mu + \frac{vg^2}{4\cos^2\theta_W} H(x) Z^0_\mu Z^0_\mu$$

$$+ \frac{g^2}{4} H(x)^2 W^{(+)}_\mu W^{(-)}_\mu + \frac{g^2}{8\cos^2\theta_W} H(x)^2 Z^0_\mu Z^0_\mu.$$

Note that we can rewrite

$$\frac{vg^2}{2} = gM_W \quad \text{and} \quad \frac{vg^2}{4\cos^2\theta_W} = \frac{gM_Z}{2\cos\theta_W}$$

so that the coupling factor for the emission of an H from a W or Z is determined in terms of known constants and the W and Z masses. The main undetermined factor is the value of m_H, which sets the energy scale needed to produce the H.

We can also write the scalar self-interaction terms arising from $\mathcal{L}$. These are identical to the scalar terms for the Abelian case given in Eq. 10.25. Thus $m_H = \sqrt{2\lambda v^2}$, which means that we cannot predict the Higgs scalar mass from the known values of M_W and M_Z.

A few final points regarding the scalar-gauge boson terms:

1. There are no H interactions with A_μ. The H particle therefore has no electrical charge and is electrically neutral.

2. The number of degrees of freedom in the symmetric theory or in the spontaneously broken theory is 12. In the former case, it is four scalars and eight transverse helicity states for the massless gauge bosons; in the latter case it is

one scalar, two helicity states for the photon, and nine helicity states for the three massive gauge bosons.

3. The theory we have written has one doublet of scalars. We can get mass generation from a multiple number of scalar doublets, with the same quantum numbers for each doublet. But all vacuum expectation values must occur in the neutrally charged components of the doublets. In such a Lagrangian, the vacuum expectation values for the doublets are $v_1^2, \ldots, v_n^2$, for n doublets. The mass formulas for the W and Z^0 are the same as for one doublet, provided we replace v^2 with $v_1^2 + v_2^2 + \cdots + v_n^2$. The theory can also be extended to doublets with $Y = -1$. In this case, the field with no electric charge is the upper component. A vacuum expectation value for this component again gives the same result as a $Y = 1$ doublet. Since we have used up the gauge freedom for one doublet, we cannot remove the extra scalar components in the other doublets in a model with extra doublets. Thus we would get more propagating scalar particles in this case. For example, for two doublets, we get in addition to H an extra neutral pair of scalars and a charged pair H^+ and H^-. The success of the mass relation between M_W and M_Z cannot tell us how many doublets to expect.

10.7 ■ CORRECTIONS TO THE *W* AND *Z* MASSES

The calculations in this chapter have typically involved the lowest-order matrix elements, although we stated what the *QCD* corrections are for Γ_Z and Γ_W. To implement the calculations, we need three electroweak parameters: g, g' (or equivalently $e, \sin^2\theta_W$), and v (or one of M_W and M_Z). The data have now reached a sufficient level of precision that higher-order corrections, including a renormalization scheme, must be included in order to describe the physics. The resulting parameters of the renormalized theory are extracted from measured quantities. In practice, the three best-measured quantities are $\alpha = e^2/4\pi$, M_Z, and G_F (which, like α, is a $q^2 \simeq 0$ measurable). The three best-measured quantities, after including calculable higher-order effects, determine the other electroweak parameters, which provide predictions for various other measurements.

The most reasonable scale at which to renormalize the theory is $q^2 \simeq M_Z^2$. At this scale $\alpha(M_Z^2)$, which is larger than at $q^2 = 0$, is the relevant coupling parameter. Running down in q^2 from this point toward $q^2 = 0$, $\alpha(M_Z^2)$ goes to its low-energy value, which involves about a 6% decrease. The interactions of W and Z involve, after symmetry breaking, a theory with massive gauge bosons for which the running is not described by a running coupling constant. Rather, the corrections involve vertex corrections and changes in the denominator of the gauge boson propagators. These changes turn out to be much smaller than the change in $\alpha(q^2)$, resulting typically in corrections $\sim 1\%$.

The present status of the electroweak data is that a number of measurements with fractional errors at the 10^{-3} level have been made, besides the very accurate measurements of α, G_F, and M_Z. These include Γ_Z, $\Gamma_{Z \to \ell^+\ell^-}$, M_W, the

10.7 Corrections to the W and Z Masses

value of $\sin^2 \theta_W$ extracted from the asymmetries in Z^0 decay, and a number of less-accurate measurements in neutrino scattering. All these measurements are in agreement with expectations, once we include higher-order corrections needed to provide predictions accurate to the 10^{-3} level.

The higher-order predictions contain some contributions from aspects of the physics that are not constrained by the $SU(2) \times U(1)$ framework alone. The measurements in this case serve as a test or constraint on these extra elements of the physics. We will look at one very interesting example, the relation between the gauge boson masses.

The mass relation in Eq. 10.29,

$$\sin^2 \theta_W = 1 - \left(\frac{M_W}{M_Z}\right)^2,$$

is given by the Lagrangian after spontaneous symmetry breaking, but prior to considering the full quantum theory. In the quantum theory, the W and Z^0 are coupled to the other particles; this renormalizes the gauge boson propagators due to virtual states, as we have seen for other particle states. The process is shown generically in Figure 10.9 for the quark and lepton loops that we know exist.

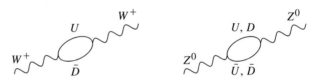

FIGURE 10.9 Fermion loops contributing to renormalization of the W and Z^0 propagators.

The intermediate states provide amplitudes π_{WW} and π_{ZZ} that modify the propagator, giving, for the denominators in the propagators,

$$q^2 - M_W^2 - \pi_{WW} \quad \text{and} \quad q^2 - M_Z^2 - \pi_{ZZ},$$

respectively. Near the pole, the imaginary parts are determined by the widths Γ_W and Γ_Z. The widths in turn are determined by the decays to the light quarks and leptons, as discussed in Section 10.3. The W and Z^0 masses are measured experimentally and are the masses from the Lagrangian modified by the real part of the virtual state corrections. They therefore are sensitive to the physics for all particles that couple, including those that might be too heavy to be directly produced. In fact, calculating the loop contributions for the three fermion families, we find that the mass shifts are completely dominated by the t quark loops. For the Z^0, these are $t\bar{t}$ virtual states; for W^+, they are $t\bar{b}$ virtual states. The large difference in mass between t and b makes the fractional mass shift different for the Z^0 and W bosons. In Figure 10.10 we show the other lowest-order loop diagrams expected for a propagating gauge boson.

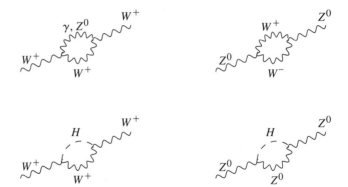

FIGURE 10.10 Gauge boson and Higgs particle loops.

We can relate the measured gauge boson masses, the loop corrections, and the masses in the Lagrangian prior to the loop corrections (which we indicate as M_W^0, M_Z^0) via

$$\left(\frac{M_W^0}{M_Z^0}\right)^2 = \left(\frac{M_W}{M_Z}\right)^2 \frac{1}{1+\delta\rho}.$$

This serves to define $\delta\rho$. The value of $\delta\rho$ is $\sim 1\%$. Both M_W^0 and M_Z^0 are reduced by the loop corrections, with M_Z^0 reduced by approximately twice as much as M_W^0 because of the larger value of the $t\bar{t}$ loop than the $t\bar{b}$ loop.

Including the $\delta\rho$ term, we can write the relation between the renormalized $\sin^2\theta_W$ and the physical gauge boson masses as

$$\sin^2\theta_W = 1 - \left(\frac{M_W}{M_Z}\right)^2 \frac{1}{1+\delta\rho}. \tag{10.31}$$

The loop calculations for the t quark contributions result in

$$\delta\rho_t = \left(\frac{G_F}{\sqrt{2}}\right)\left(\frac{3m_t^2}{8\pi^2}\right) = 0.0096,$$

for $m_t = 175$ GeV. To get an uncertainty in $\delta\rho_t < 10^{-3}$ requires a top mass measured to an accuracy of a few percent. Adding in the gauge boson loop terms we get a prediction, for $m_t = 175$ GeV,

$$\delta\rho = 0.0106 \pm 0.0006.$$

Using $\sin^2\theta_W = 0.2312 \pm 0.0002$ and the measured values of $M_W = 80.448 \pm 0.031$, $M_Z = 91.188 \pm 0.002$ gives a measured value $\delta\rho = 0.012 \pm 0.001$, from Eq. 10.31. We see there is very little room for any contributions to $\delta\rho$ from physics other than the known fermions and gauge bosons. Note that these considerations were used to determine the top quark mass before it was directly measured through the production of real top quarks in high energy hadronic collisions.

In the calculation of $\delta\rho$, we left out the term $\delta\rho_H$ coming from a Higgs boson in the loop. This contribution can be calculated to give

$$\delta\rho_H = \left(\frac{G_F M_W^2}{\sqrt{2}}\right)\left(\frac{-11}{12\pi^2}\frac{\sin^2\theta_W}{\cos^2\theta_W}\right)\log\left(\frac{m_H}{M_W}\right) = -0.0015\log\left(\frac{m_H}{M_W}\right).$$

If no extra physics contributes to $\delta\rho$, other than the terms above, we can conclude that $\delta\rho_H \simeq 0$, or $m_H \simeq M_W$. This tells us that H is not too heavy! Note that the logarithmic sensitivity to m_H results in a large error ($\sim M_W$) on m_H, given the present data. What the experiments will eventually find for m_H and the symmetry breaking mechanism is of enormous interest. These questions should be resolved in the next few years using the next generation of accelerators, which can search for H with a mass as large as several hundred GeV.

10.8 ■ GENERATION OF FERMION MASSES

The final piece of the theory is the generation of the fermion masses by spontaneous symmetry breaking. This must be accomplished through gauge invariant couplings in the initial symmetric Lagrangian. Thus the terms involving the fermion-scalar interaction must transform as zero hypercharge, weak isospin singlets under the $SU(2) \times U(1)$ gauge transformations.

We will use mass eigenstates, as discussed in Section 8.4, and look at one lepton generation first. The other generations provide an exact copy of the procedure, which unfortunately does not relate or constrain the various masses. Indeed, the various fermion masses do not reveal much of a pattern, although the very large value of m_t versus the other masses, and the very small neutrino masses suggest that there is more physics we do not yet understand.

Looking at the electron generation, we first list in Table 10.3 the relevant $SU(2) \times U(1)$ quantum numbers of the particles we wish to couple. To simplify the notation,

$$\phi = \frac{1}{\sqrt{2}}\begin{pmatrix}\phi_1 + i\phi_2 \\ \phi_3 + i\phi_4\end{pmatrix} \quad \text{is written as} \quad \begin{pmatrix}\phi_U \\ \phi_D\end{pmatrix},$$

indicating the T_3 assignments. Based on the quantum numbers in the table, an interaction of the type

$$e_R + \begin{pmatrix}\phi_U \\ \phi_D\end{pmatrix} \to \begin{pmatrix}\nu_e \\ e\end{pmatrix}_L \quad \text{and} \quad \begin{pmatrix}\nu_e \\ e\end{pmatrix}_L \to e_R + \begin{pmatrix}\phi_U \\ \phi_D\end{pmatrix} \quad (10.32)$$

conserves all the $SU(2) \times U(1)$ quantum numbers (and therefore also electric charge). The second term comes from the Hermitian conjugate of the first term if we write an equation involving the fields. Including an unknown coupling con-

TABLE 10.3
Quantum Numbers for Fermion-Scalar Coupling.

	T	Y
$\begin{pmatrix}\nu_e \\ e\end{pmatrix}_L$	$\frac{1}{2}$	-1
ν_{e_R}	0	0
e_R	0	-2
$\begin{pmatrix}\phi_U \\ \phi_D\end{pmatrix}$	$\frac{1}{2}$	1

stant for the fermion-scalar interaction, and using a notation that can be extended to the other lepton and quark mass eigenstate generations directly, the fermion-scalar part of the Lagrangian is

$$\mathcal{L} = -g_D \left[\left(\bar{\psi}_L^U, \bar{\psi}_L^D \right) \begin{pmatrix} \phi_U \\ \phi_D \end{pmatrix} \psi_R^D + \bar{\psi}_R^D \left(\phi_U^\dagger, \phi_D^\dagger \right) \begin{pmatrix} \psi_L^U \\ \psi_L^D \end{pmatrix} \right]. \quad (10.33)$$

We next assume that the scalar terms spontaneously break the symmetry, as discussed in the previous sections. Thus, the scalar fields for the gauge we are using are

$$\phi_U = 0, \quad \phi_D = \left(\frac{v + H(x)}{\sqrt{2}} \right).$$

Substituting ϕ into Eq. 10.33 gives

$$\mathcal{L} = -\frac{g_D v}{\sqrt{2}} \left[\bar{\psi}_L^D \psi_R^D + \bar{\psi}_R^D \psi_L^D \right] - \frac{g_D H}{\sqrt{2}} \left[\bar{\psi}_L^D \psi_R^D + \bar{\psi}_R^D \psi_L^D \right].$$

The term in the first brackets is just $\bar{\psi}^D \psi^D$, so that this is just a mass term. It gives

$$m_D = \frac{g_D v}{\sqrt{2}}, \quad (10.34)$$

allowing m_D to be expressed in terms of the a priori unknown coupling constant g_D. We can rewrite this part of the Hamiltonian density in terms of m_D as

$$\mathcal{H} = -\mathcal{L} = m_D \bar{\psi}^D \psi^D + \frac{m_D}{v} H \bar{\psi}^D \psi^D. \quad (10.35)$$

The second term is an interaction term that gives the amplitude for the D fermion to emit a physical Higgs particle. The coupling is completely determined, as was the coupling for the gauge particle to emit a scalar. In the fermion case, the coupling is m_D/v. Thus the light fermions have very small couplings for $v = 247$ GeV.

The masses for the up fermions can be generated in an analogous way. In this case we need a doublet with $Y = -1$, which can be obtained by using the charge conjugate doublet formed from ϕ. We can understand this doublet by analogy with the antiparticle doublet derived from the strong isospin (u, d) quark pair. For the quarks, the antiquark doublet is $(-\bar{d}, \bar{u})$. For the ϕ field we use

$$\begin{pmatrix} \phi_U \\ \phi_D \end{pmatrix} \rightarrow \begin{pmatrix} \phi_D^\dagger \\ -\phi_U^\dagger \end{pmatrix} \rightarrow \begin{pmatrix} \frac{v + H(x)}{\sqrt{2}} \\ 0 \end{pmatrix}$$

after spontaneous symmetry breaking. Our mass generating interaction is now

$$\nu_R + \begin{pmatrix} \phi_D^\dagger \\ -\phi_U^\dagger \end{pmatrix} \rightarrow \begin{pmatrix} \nu_e \\ e \end{pmatrix}_L \quad \text{and} \quad \begin{pmatrix} \nu_e \\ e \end{pmatrix}_L \rightarrow \nu_R + \begin{pmatrix} \phi_D^\dagger \\ -\phi_U^\dagger \end{pmatrix}.$$

This gives in $\mathcal{L}$ a term, using again the generation independent notation,

$$\mathcal{L} = -g_U \left[\left(\bar{\psi}_L^U, \bar{\psi}_L^D\right) \begin{pmatrix} \phi_D^\dagger \\ -\phi_U^\dagger \end{pmatrix} \psi_R^U + \bar{\psi}_R^U \left(\phi_D, -\phi_U\right) \begin{pmatrix} \psi_L^U \\ \psi_L^D \end{pmatrix} \right]. \quad (10.36)$$

Replacing the ϕ field with its form after spontaneous symmetry breaking gives a term in the Hamiltonian density:

$$\mathcal{H} = m_U \bar{\psi}^U \psi^U + \frac{m_U}{v} H \bar{\psi}^U \psi^U.$$

Again, the amplitude to emit an H is specified in terms of the U mass and v, and is small for a light fermion. For the top quark, however, this ratio is only somewhat smaller than 1.

An interesting variant of the above is a scheme where there are two different doublets with $Y = 1$ and -1, respectively. The $Y = 1$ doublet can then generate D masses and the $Y = -1$ doublet U masses, via vacuum expectation values v_D and v_U, respectively. To get the correct W and Z masses requires $v_U^2 + v_D^2 = v^2$. In this scheme, the U and D fermions will couple to different physical scalar particles.

10.9 ■ MAJORANA NEUTRINOS

The neutrinos have no electromagnetic or color quantum numbers, which allows an entirely different mass scheme that is consistent with the conserved quantum numbers at low energies. The weakly interacting neutrino species are the left-handed neutrino and the right-handed antineutrino. These can be coupled together through a mass term, yielding a four-component spinor that mixes particles and antiparticles. Such a field is called a Majorana fermion. The right-handed neutrino and left-handed antineutrino can also be coupled together through a separate mass term. The masses of the two Majorana neutrinos do not have to be related. The field involving the right-handed neutrino has no electroweak interactions at all and need not exist. It is called a sterile neutrino. This scheme can be generalized, with the neutrinos having both Majorana and Dirac mass terms and mixing between the different species.

We will not work out the Majorana spinor formalism. The neutrino mixing results in Chapter 9 are the same for Majorana or Dirac neutrinos. An important difference between the two schemes is that a Majorana neutrino that is emitted can subsequently behave like an antineutrino when it is absorbed. In the massless limit, this effect vanishes because the emitted particle is exactly left-handed, whereas the absorbed antiparticle would have to be right-handed to get a nonzero amplitude. Thus these lepton violating two-step processes are very small for highly relativistic neutrinos. The most promising method for observing this effect is in the neutrinoless double-beta decay process in some nuclei. In this process, forbidden for a Dirac neutrino, two neutrons turn into two protons and two electrons. The neutrino emitted in one neutron decay is absorbed in the second decay. It is thus virtual. Explicit decay rate estimates indicate that the rates are very small even for the most favorable choice for the nucleus. The experiments must look for a few decays in many tons of material. Experiments have not yet reached a sensitivity where we might expect to see an effect.

CHAPTER 10 HOMEWORK

10.1. Calculate the total cross section for

$$\nu_\mu e^- \to \nu_\mu e^- \quad \text{and} \quad \bar{\nu}_\mu e^- \to \bar{\nu}_\mu e^-.$$

Show that a comparison of the two cross sections allows a measurement of $\sin^2 \theta_W$. You may assume in the calculation that $E_\nu \gg m_e$, so that you can ignore the electron mass. The initial electron is unpolarized.

10.2. Calculate Γ_Z and Γ_W to lowest order. Ignore fermion masses for the energetically allowed decays.

10.3. For $e^+ e^- \to Z^0 \to f\bar{f}$, ignoring all fermion masses, calculate

(a)
$$A_{LR} = \frac{\sigma_L - \sigma_R}{\sigma_L + \sigma_R},$$

using the left- and right-handed e^- cross sections.

(b) For an unpolarized $e^+ e^-$ interaction, calculate

$$\frac{d\sigma_{f\bar{f}}}{d(\cos\theta)} \quad \text{and} \quad \sigma_{f\bar{f}},$$

where these are the cross sections to a given flavor pair at the Z^0 mass.

(c) Show that the q^2 dependence of the width in Eq. 10.13 is correct, assuming that we can ignore all fermion masses in the decay of the virtual Z, which is a good approximation for a heavy Z.

10.4. Draw the lowest-order Standard Model diagram for $e^+ e^- \to Z^0 Z^0$.

10.5. Consider the $SU(2) \times U(1)$ gauge theory coupled to a triplet of real scalar fields. The $SU(2)$ matrices that are the generators for rotations of the scalar field triplet are $(T_i)_{jk} = -i\varepsilon_{ijk}$.

(a) What is Y for the real triplet?

(b) Suppose the symmetry is spontaneously broken. What are the resulting gauge boson masses? What are the degrees of freedom for the physical particles?

10.6. Consider the $SU(2) \times U(1)$ gauge theory coupled to a single Higgs doublet, which was discussed in the text as the simplest version of the Standard Model. Work out the physical theory we would get if the vacuum expectation value were

$$\frac{1}{\sqrt{2}} \begin{pmatrix} v \\ 0 \end{pmatrix} \quad \text{instead of} \quad \frac{1}{\sqrt{2}} \begin{pmatrix} 0 \\ v \end{pmatrix}.$$

10.7. The width to lepton pairs, $\Gamma_{Z \to e^+ e^-}$, has been measured to be 84.0 ± 0.1 MeV. From this measured value and the measured values for M_Z and $\sin^2 \theta_W$, calculate $g^2/4\pi$. Compare this value to the value obtained from using

$$\frac{g^2}{4\pi} = \frac{1}{\sin^2 \theta_W} \alpha,$$

where α is evaluated at the weak scale. (The small difference found is due to higher-order electroweak corrections).

10.8. Calculate $g^2/4\pi$ using the relation

$$\frac{G_F}{\sqrt{2}} = \frac{g^2}{8M_W^2}$$

and the measured values of G_F and M_W^2. (The small difference from the value in problem 10.7 is again due to small electroweak corrections).

10.9. Why is the contribution of the diagram $e^+ e^- \to H \to W^+ W^-$ too small to be seen, when compared to the other diagrams in the electroweak theory?

CHAPTER 11

Large Cross Section Processes

11.1 ■ TYPES OF PROCESSES

The interactions between particles provide two types of processes with large cross sections that do not decrease as the overall energy increases. The first large cross sections are for the strong interactions of hadrons. Given the confinement of the color degrees of freedom, the interaction is short range. The hadrons interact strongly only when they overlap physically, which gives a cross section that is roughly related to the size of the particles. For example, at a center of mass energy of about 30 GeV, some of the measured total hadronic cross sections are

$$\sigma(pp) = 40 \text{ mb},$$
$$\sigma(\pi p) = 26 \text{ mb},$$
$$\sigma(Kp) = 21 \text{ mb}.$$

These cross sections grow slowly with energy. As an example, $\sigma(pp)$ doubles when going from a center of mass energy of 30 GeV to 2000 GeV. The total pp and $\bar{p}p$ cross sections are shown in Figure 11.1 over a large range in center of mass energy.

The particles emerging from such hadronic collisions (which are predominantly hadrons made of the light quarks), share the large initial energy in a characteristic manner. The initial particle directions define a collision or beam direction. The typical momenta transverse to the beam direction for the final state particles are governed by λ_{QCD}, the primary dimensional parameter for the nonperturbative regime of the strong interactions of the light quarks. The large longitudinal momentum of the initial particles is not turned into transverse momentum, and the particle motions are therefore mostly along the direction of the initial momentum. None of these features are calculable from first principles, but fit qualitatively with the idea of color confinement. The small fraction of the cross section where large transverse momentum jets are produced in the final state can be understood using perturbation theory, and will be discussed in Chapter 12.

The strong cross sections listed above imply that a hadron passing through a solid medium will interact in a distance that is not too large. This feature provides our criterion for a "large cross section." As an example, taking a medium with 10^{24} scattering centers (nucleons) per cubic centimeter and a cross section of 30 mb per nucleon, gives a mean interaction length of 33 cm.

11.1 Types of Processes

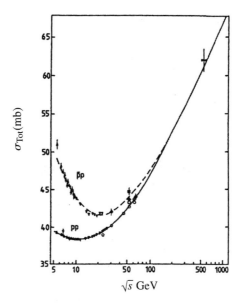

FIGURE 11.1 The pp and $\bar{p}p$ cross sections versus center of mass energy. [From A. Donnachie and P. V. Landshoff, *Nucl. Phys.* B267, 690 (1986). See also *Pomeron Physics and QCD* by A. Donnachie, G. Dosch, P. Landshoff, and O. Nachtmann (Cambridge University Press, 2002).]

The second set of processes that yield large cross sections are due to electromagnetic interactions. Although the coupling is much smaller than for the strong interactions (yielding, for example, an extra factor of α^2 in cross sections involving two electromagnetic vertices), the range is very large so that cross sections can be large. They will be large whenever a process can occur with very small momentum transfer. For example, a charged particle will elastically scatter off atoms in a material, in which case the range is given approximately by the size of the atom, beyond which the atom acts as a neutral object. This can be compared to the range for the strong interactions, which is $\sim 1/\lambda_{QCD}$ for color neutral hadrons. As a result, a high-energy charged hadron passing through a material has an electromagnetic cross section larger than its strong cross section. The infinite range of the electromagnetic interaction also allows coherent interactions of charged particles with an entire medium, where the idea of a cross section involving only two particles at a time does not apply. Examples are Cherenkov radiation, for particles traveling faster than the velocity of light in the medium, and transition radiation, when a particle goes from one medium into another.

The large cross section processes provide methods to detect the presence of particles using a particle detector. Thus, colliding beam experiments are usually arranged to create rare collisions involving the beam particles, detecting the emerging collision products through large cross section processes in surrounding detector elements. Particles such as neutrinos, which have no large cross section processes, escape the detector volume undetected.

The typical hadronic interaction length in a solid medium, discussed above, indicates that hadrons will have only a small probability to interact in a gaseous medium, which has a density typically about 10^{-3} that of a solid, unless they travel long distances in the medium. Thus a gaseous medium where particles travel, for example, a distance of 1 meter, is primarily sensitive to the electromagnetic interactions of charged particles. From the Coulomb scattering calculations in Chapter 3, we get an approximate cross section for a high momentum particle of charge e to scatter electrons in the medium:

$$\sigma = \frac{e^4}{8\pi m_e E_{\min}} = \frac{250 \text{ mb}}{E_{\min} \text{ (MeV)}}. \tag{11.1}$$

Here $E_{\min}$ is measured in MeV, the electron in the medium has been treated as a free particle initially at rest, and the cross section is the total cross section for final electron kinetic energies greater than $E_{\min}$. To get a true total cross section, we must add the contribution from collisions that create atomic excitations and account for the range of binding energies for a real complex atom. The dominant part of the cross section will, however, involve the outermost, weakly bound electrons of an atom. We can very roughly estimate a typical cross section by taking $E_{\min} = 5$ eV and a density of 3×10^{20} electrons per cubic centimeter (a gas with 10 easily-ionized electrons per molecule). This gives 15 electrons liberated per centimeter of gas traversed. These electrons typically all have small energies, so that the high energy particle is not greatly disturbed by the collisions. By measuring the positions of the electrons in the ionization trail we can determine the trajectory of the high momentum particle. In addition, the excitations produced will lead to the emission of light as atoms return to their ground states. This is the basis for the functioning of thin solid scintillators, which register the passage of a charged particle through the resulting light emitted in the material.

11.2 ■ MULTIPLE COULOMB SCATTERING

The simplest example of a large electromagnetic cross section is elastic Coulomb scattering of a fast charged particle in a medium. To get a large cross section, the momentum transfer must be small, so that the high-momentum particle is barely deflected. For simplicity we assume the medium is made of one type of atom.

We look at the kinematics first. We assume an incident fast particle of energy E_A and momentum along the z direction equal to p_A. The target particle is an atom with mass m_B and is initially at rest. After the scatter, the atom has a small momentum p_B and is scattered at an angle θ_B. It therefore has an approximate energy

$$m_B + \frac{p_B^2}{2m_B}$$

11.2 Multiple Coulomb Scattering

and momentum along z equal to $p_B \cos\theta_B$. After the collision, the incident particle will have an energy

$$E_A - \frac{p_B^2}{2m_B},$$

momentum along z equal to $p_A - p_B \cos\theta_B$, and transverse momentum of magnitude $p_B \sin\theta_B$. Requiring that the energy squared minus the momentum squared for the incident particle is equal before and after the collision, we can solve for $\cos\theta_B$, which is approximately

$$\cos\theta_B = \frac{p_B}{2m_B}.$$

Thus for a small momentum transfer ($\ll m_B$) the target particle is scattered at approximately $90°$ to the incident particle direction and the momentum transfer squared is approximately $q^2 = -p_B^2$.

To calculate the rate, we use the second-order matrix element given by Eq. 3.5 in terms of two individual current matrix elements:

$$M_{fi} = \frac{-J_\mu^A J_\mu^B}{q^2},$$

where A is the incident particle and B the target particle. Since the target particle is barely scattered, only the charge density is large for J_μ^B. Taking $q^2 \simeq -|\vec{q}|^2$, we therefore have, to good approximation,

$$M_{fi} = \frac{J_0^A J_0^B}{|\vec{q}|^2}.$$

Taking an incident particle with charge e and spin $\frac{1}{2}$ (the result is the same for spin zero) $J_0^A = eu_f^\dagger u_i = e(2E_A)$ for very small deflection. Here the final and initial spins are equal. For the target, we factor out of the density a charge e and $2m_B$ for the nonrelativistic case, and write $J_0^B = e2m_B \rho_B(\vec{q})$. The factor $\rho_B(\vec{q})$ is the Fourier transform of the target charge density. Therefore

$$|M_{fi}|^2 = \frac{e^4(2E_A)^2}{q^4}(2m_B)^2 |\rho_B(\vec{q})|^2, \qquad (11.2)$$

and the cross section is

$$d\sigma = \frac{e^4(2m_B)}{q^4 \beta} |\rho_B(\vec{q})|^2 \frac{d^3 p_A}{(2\pi)^2} \frac{d^3 p_B}{2E_B} \delta^4(p_f - p_i). \qquad (11.3)$$

The relative velocity is the incident particle velocity β, and we ignore the energy change for this particle; E_B is the final energy of the target particle. Integrating over $d^3 p_A$ eliminates the three spatial momentum δ-functions.

We next write $d^3 p_B = p_B E_B dE_B d\Omega_B$. The integral over the azimuthal angle gives a factor of 2π; the integral over $d(\cos\theta_B)$ will eliminate the last δ-function. This gives

$$d\sigma = \frac{e^4}{4\pi q^4 \beta} |\rho_B(\vec{q})|^2 \frac{p_B 2m_B dE_B}{\left|\frac{dE_A}{d\cos\theta_B}\right|}.$$

Since $E_A = \sqrt{(\vec{p}_A - \vec{p}_B)^2 + m_A^2} = \sqrt{p_A^2 + p_B^2 - 2p_A p_B \cos\theta_B + m_A^2}$,

$$\left|\frac{dE_A}{d\cos\theta_B}\right| = \frac{p_A p_B}{E_A} = \beta p_B.$$

Therefore:

$$d\sigma = \frac{e^4}{4\pi q^4 \beta^2} |\rho_B(\vec{q})|^2 2m_B dE_B.$$

Noticing that we can write $q^2 = -2m_B(E_B - m_B)$, a result which follows from squaring the difference of the initial and final target 4-momenta, we have finally, for a relativistic incident particle with $\beta \simeq 1$:

$$\frac{d\sigma}{d|\vec{q}|^2} = \frac{e^4}{4\pi} \frac{|\rho_B(\vec{q})|^2}{|\vec{q}|^4} = 4\pi\alpha^2 \frac{|\rho_B(\vec{q})|^2}{|\vec{q}|^4}. \quad (11.4)$$

The total cross section is determined mainly by the region of low $|\vec{q}|^2$.

11.2.1 ■ Multiple Scattering Angle

A fast charged particle usually experiences many Coulomb scattering collisions in passing through a material, since the cross section is large. This is called multiple Coulomb scattering. The primary effect of each elastic collision with an atom in the material is a slight deflection of the incident particle. We can calculate the most likely deflection for each collision by finding the average momentum transfer:

$$\langle |\vec{q}|^2 \rangle = \int \frac{|\vec{q}|^2 \frac{d\sigma}{d|\vec{q}|^2}}{\sigma} d|\vec{q}|^2.$$

For a small angle, $|\vec{q}|^2 = p_A^2 \theta^2$; therefore,

$$\langle \theta^2 \rangle = \frac{1}{p_A^2} \int \frac{|\vec{q}|^2 \frac{d\sigma}{d|\vec{q}|^2}}{\sigma} d|\vec{q}|^2. \quad (11.5)$$

Since the successive collisions are independent uncorrelated events, the mean square value for θ^2 after n collisions will be $n\langle\theta^2\rangle$. This applies provided that $n\langle\theta^2\rangle$ is still a small number, even though n is large. The number of collisions

11.2 Multiple Coulomb Scattering

$n = \sigma N t$, where N = number of scattering centers per unit volume and t is the thickness of material traversed. This gives finally a mean square value from many collisions:

$$\overline{\theta^2} = \frac{1}{p_A^2} \int |\vec{q}|^2 \frac{d\sigma}{d|\vec{q}|^2} d|\vec{q}|^2 \, Nt = \frac{4\pi \alpha^2}{p_A^2} \int \frac{|\rho(\vec{q})|^2}{|\vec{q}|^2} d|\vec{q}|^2 \, Nt. \tag{11.6}$$

We can separate these factors into a term that depends on the incident particle, a term that depends on the type of material, and the thickness traversed. An analogous material-dependent term occurs in other electromagnetic processes, such as pair production by a high-energy photon (shown in Figure 6.3) or photon radiation by a high-energy electron. These processes have an extra power of α in the rate compared to the Coulomb scattering calculation. We will use the normalization most suited to these higher-order processes and define a radiation length X_0 as

$$\frac{1}{X_0} = N\alpha \left(\frac{\alpha}{m_e}\right)^2 \int \frac{|\rho(\vec{q})|^2}{|\vec{q}|^2} d|\vec{q}|^2. \tag{11.7}$$

This gives then, for a relativistic particle,

$$\overline{\theta^2} = \frac{\pi}{\alpha} \left(\frac{2m_e}{p_A}\right)^2 \left(\frac{t}{X_0}\right). \tag{11.8}$$

Measuring p_A in GeV implies a value for $\overline{\theta^2}$ of about

$$\overline{\theta^2} = (2 \times 10^{-2})^2 \left(\frac{1}{p_A(\text{GeV})}\right)^2 \left(\frac{t}{X_0}\right). \tag{11.9}$$

The typical scattering angle is indeed small for a relativistic particle passing through a material for which t/X_0 is not too large. The multiple scattering provides a limit to how well we can determine the trajectory of a fast particle using a gas, since the material changes the trajectory. The other limit is usually given by how well we can determine the location of the ionization trail of electrons left by the particle in the gas. These limits are generally much more restrictive than the quantum limitation, which follows from having to represent a particle by a wave-packet rather than a state with a perfectly defined momentum.

11.2.2 ■ Radiation Length

Evaluating the radiation length requires an integral over the square of the Fourier transform of the spatial charge density,

$$\rho(\vec{q}) = \int \rho(\vec{x}) e^{i\vec{q}\cdot\vec{x}} d^3x. \tag{11.10}$$

The value of the form factor $\rho(\vec{q})$ can be estimated for a range of $\vec{q}$ values. For

$$|\vec{q}| \ll \frac{1}{r_{\text{atom}}},$$

where r_{atom} is the outer radius of the atom, the whole charge of the system contributes coherently. Since an atom is electrically neutral, $\rho(\vec{q}) \simeq 0$ in this regime. For

$$\frac{1}{r_{\text{atom}}} \ll |\vec{q}| \leq \frac{1}{r_{\text{nucleus}}}$$

only the nuclear charge contributes, giving $\rho(\vec{q}) = Z$. For

$$\frac{1}{r_{\text{nucleus}}} \ll |\vec{q}|,$$

the nuclear charge no longer contributes coherently and $\rho(\vec{q})$ approaches zero. Therefore, approximating $\rho(\vec{q}) = Z$ for

$$\frac{1}{r_{\text{atom}}} \leq |\vec{q}| \leq \frac{1}{r_{\text{nucleus}}},$$

and zero otherwise, gives for the radiation length in the multiple scattering formula:

$$\frac{1}{X_0} = 2N\alpha \left(\frac{\alpha}{m_e}\right)^2 Z^2 \log\left(\frac{r_{\text{atom}}}{r_{\text{nucleus}}}\right).$$

This expression can be calculated more accurately by using a cutoff Coulomb potential (that is, a Yukawa potential, whose parameters are typically estimated using the Thomas–Fermi model for the atom) to evaluate the integral at small $|\vec{q}|$, and a uniform charge density for the nucleus for the cutoff at high $|\vec{q}|$. These result in fixing the parameters, to good approximation (except for very low Z atoms where somewhat larger corrections are needed), to values:

$$r_{\text{atom}} = \frac{1.4 a_0}{Z^{1/3}},$$

where the Bohr radius

$$a_0 = \frac{1}{m_e \alpha}; \quad r_{\text{nucleus}} = 1.3 \times 10^{-13} \text{ cm} \quad A^{1/3} \simeq 1.7 \times 10^{-13} \text{ cm} \quad Z^{1/3}.$$

Using the classical radius of the electron,

$$\frac{\alpha}{m_e} = 2.8 \times 10^{-13} \text{ cm},$$

11.2 Multiple Coulomb Scattering

we can write

$$r_{\text{nucleus}} = 0.6 \frac{\alpha}{m_e} Z^{1/3}.$$

The ratio

$$\frac{r_{\text{atom}}}{r_{\text{nucleus}}} \simeq (r_{\text{atom}} m_e)^2 = \left(\frac{1.4}{\alpha Z^{1/3}}\right)^2.$$

With these approximations,

$$\frac{1}{X_0} = 4N\alpha \left(\frac{\alpha}{m_e}\right)^2 Z^2 \log\left(\frac{190}{Z^{1/3}}\right). \tag{11.11}$$

This expression is, to good approximation, the same as that for the radiation length in pair production (which we look at in Section 11.3.1) so that we typically speak of only one such quantity. The near equality of these quantities is, however, an accident stemming from the value for the classical radius of the electron and the fact that the logarithm of a large number is not very sensitive to small changes in the number. The most important material-dependent factor in Eq. 11.11, for a given material type (solid or gas), is the factor of Z^2. This makes X_0 much smaller for high Z materials than for low Z materials. Some examples of the radiation length for a number of materials are given in Table 11.1. The value for hydrogen involves an explicit calculation and the small contribution from incoherent scattering on the electrons in the material has been included to arrive at the numbers.

TABLE 11.1 Radiation Length for a Few Materials.

Material	Z	X_0
H_2 gas	1	7.3×10^3 m
H_2 liquid	1	8.7 m
C	6	18.8 cm
Fe	26	1.8 cm
Pb	82	0.56 cm

11.2.3 ■ Energy Loss

The coherent elastic scattering off atoms, discussed in Section 11.2.1, results in very little energy loss by the incident particle. The transfer of energy is dominated by processes where the target is left in an excited state. These processes include excitation of atomic electrons into higher energy atomic levels and, for greater energy transfer, the ejection of individual electrons from the atomic target. We investigate this latter process, but still for a nonrelativistic final state electron. In this case the momentum transfer is small, resulting in a large cross section. The calculation of atomic excitations can be done by using explicit wave functions for the atomic electrons.

For the case where the electron is ejected with energy $\gg$ ionization energy, we consider the initial electron to be at rest. We calculate the momentum transfer using the difference of the initial and final target four-momentum. This gives

$$q^2 = -2m_e(E_e - m_e), \tag{11.12}$$

where $E_e - m_e$ is the final kinetic energy of the target electron assumed initially to be at rest. Using energy conservation, it is also the energy lost by the incident relativistic beam particle. For a given q^2, the energy loss $\sim 1/m_e$, indicating why the scattering of electrons, as opposed to the coherent scattering of the whole atom, is the dominant mechanism by which the incident particle loses energy.

We can use Eq. 11.4 for nonrelativistic electron scattering, with $\rho(\vec{q}) = 1$ for the point-like electron. Substituting for q^2 from Eq. 11.12 gives finally

$$\frac{d\sigma}{dE} = \frac{e^4}{8\pi m_e E^2}, \tag{11.13}$$

where $E = E_e - m_e$ is the energy loss of the initial fast beam particle. Integrating Eq. 11.13 over the energy loss gives the cross section to produce a free ionized electron, given earlier in Eq. 11.1. This single electron cross section should be summed over the different electrons in an atom to get the total cross section for an atomic system.

To calculate the average energy loss per unit length of material traversed, written as dE/dx, we have to average E over the total cross section and multiply by the number of collisions per unit length. This gives, where N is the density of the material,

$$\frac{dE}{dx} = N \int E \frac{d\sigma}{dE} dE. \tag{11.14}$$

An exact calculation requires an understanding of the detailed atomic properties of the material. We will do an approximate calculation, which allows the material properties to be summarized in terms of one parameter, an effective ionization energy.

Since q^2 is proportional to the energy transferred to the target, provided it is initially at rest, we can rewrite Eq. 11.14 as

$$\frac{dE}{dx} = N \int |q^2| \frac{d\sigma}{dq^2} \frac{dq^2}{2m_e}. \tag{11.15}$$

Integrating over a region of q^2 for a given electron in the atom gives a contribution of the given electron to dE/dx:

$$\frac{4\pi \alpha^2 N}{m_e} \frac{1}{2} \log\left(\frac{q^2_{\max}}{q^2_{\min}}\right).$$

11.2 Multiple Coulomb Scattering

Summing this over the Z electrons, q_{max}^2 is nearly the same for each electron, while q_{min}^2 depends on the initial electron state. Defining an average q_{min}^2 by averaging the $\log(q_{min}^2)$ over the Z electrons, gives

$$\frac{dE}{dx} = \frac{4\pi\alpha^2 NZ}{m_e} \frac{1}{2} \log\left(\frac{q_{max}^2}{q_{min}^2}\right). \tag{11.16}$$

To calculate the maximum of $|q^2| = 2m_e(E_e - m_e)$, we look at the maximum energy that can be transferred to the electron. In terms of the momentum and energy of the initial relativistic particle, a calculation of the maximum energy transferred in a forward collision gives:

$$E_e^{max} - m_e = \frac{2m_e p_A^2}{2m_e E_A + m_A^2 + m_e^2}. \tag{11.17}$$

Since the electron is much lighter than all other charged particles,

$$E_e^{max} - m_e = 2m_e \left(\frac{p_A}{m_A}\right)^2$$

over the wide range of energies for which

$$E_A \ll \left(\frac{m_A}{m_e}\right) m_A$$

(the exception to this is the case of an incident electron for which $m_A = m_e$). We assume that this is the case and use

$$\left|q_{max}^2\right| = 2m_e(E_e^{max} - m_e) = (2m_e)^2 \gamma^2, \tag{11.18}$$

where p_A is taken to be equal to E_A in the relativistic limit.

Eq. 11.16 is based on the assumption of a nonrelativistic final electron. This is clearly not correct for q_{max}^2 given by Eq. 11.18. For a point-like spin $\frac{1}{2}$ beam particle incident on the target, the cross section can be calculated more accurately using the matrix element squared in Section 3.2.3. This gives a small correction to Eq. 11.16, whose calculation we leave to a homework problem. We will ignore this issue and use Eq. 11.16 for the energy loss, where $|q_{max}^2|$ is given by Eq. 11.18.

For a free electron, q_{min}^2 can be zero. For a bound electron, the energy transfer is quantized and has a minimum value, implying a nonzero q^2. Taking I to be the energy transfer between given atomic states, we can explicitly calculate q^2 for the case of no transverse momentum exchanged, which gives the minimum magnitude for the momentum transfer squared in such a transition. In this case, the spatial momentum exchanged is absorbed by the excited atomic system, whose final direction is the same as that of the incident particle. Note that this is different from the case of coherent elastic scattering where the target recoiled at 90° to the beam. We can calculate the minimum q^2 in terms of the energy change of the incident particle:

$$q^2 = (\Delta E_A)^2 - (\Delta p_A)^2 = (\Delta E_A)^2 \left[1 - \left(\frac{dp_A}{dE_A}\right)^2\right]$$

$$= (\Delta E_A)^2 \left[1 - \left(\frac{E_A}{p_A}\right)^2\right].$$

For $\Delta E_A = I$ and a very relativistic incident particle,

$$q^2 = -\frac{I^2}{\gamma^2}.$$

This gives then the smallest magnitude for the momentum transfer:

$$\left|q^2_{\min}\right| = \frac{I^2}{\gamma^2}. \tag{11.19}$$

Using Eqs. 11.18 and 11.19, we get finally from Eq. 11.16:

$$\frac{dE}{dx} = \frac{4\pi\alpha^2 NZ}{m_e} \log\left(\frac{2m_e\gamma^2}{I}\right). \tag{11.20}$$

The parameter I provides a compact way to summarize the atomic properties relevant to the energy transfer process.

The generalization of the dE/dx formula to the case where the incident particle is not necessarily relativistic is left to a homework problem. The resulting generalization to Eq. 11.20 is:

$$\frac{dE}{dx} = \frac{4\pi\alpha^2 NZ}{m_e\beta^2} \left[\log\left(\frac{2m_e\beta^2\gamma^2}{I}\right) - \beta^2\right].$$

This equation shows that the energy loss is greater for a slow particle than for a moderately relativistic particle. It has a broad minimum near $\beta\gamma = 3$, and then rises slowly because of the factor of γ^2 in the logarithm. The scattering of atomic electrons creates ion pairs in the medium. Table 11.2 provides some examples of the number of ion pairs created in various gases at standard temperature and pressure at the minimum of the ionization loss ($\beta\gamma$ of about 3 for the incident particle). Note that the ionized electrons of the material will subsequently produce additional ions when they lose their kinetic energy. This typically increases the number of ion pairs by a factor of two to four.

The calculation of energy loss has a long history. Bohr developed a classical theory in 1915; Bethe derived the quantum calculation in 1930; and Fermi worked out the reduction to the energy loss for highly relativistic particles due to the index of refraction of the medium (called the density effect, which we will not calculate) in 1940. Measurement of dE/dx for a particle traversing a gas volume is one of the techniques used (in conjunction with an independent momentum measurement) to identify the particle type, since it depends on β of the particle.

TABLE 11.2 Ion Pairs Per cm Traversed.

Gas	Ion Pairs/cm
H_2	5.2
He	5.9
O_2	22
Ar	29
Xe	44

11.3 ■ RADIATIVE PROCESSES FOR ELECTRONS AND PHOTONS

The Coulomb scattering process yields very large cross sections, but the energy loss per unit length for a relativistic charged particle is typically rather small. As an example, consider an "average" solid with $NZ = 10^{24}$ electrons/cm^3. So the first term in Eq. 11.20 is

$$\frac{4\pi \alpha^2 N Z}{m_e} = 0.52 \text{ MeV/cm}.$$

Assuming that the logarithmic factor in Eq. 11.20 is approximately 20 gives a typical energy loss of 10 MeV/cm, or 1 GeV/m. In a gas the number would be smaller by a factor of about 1000.

For relativistic muons, the dominant mechanism for energy loss up to very high energies is the Coulomb scattering of electrons in the medium, as calculated in Section 11.2.3. This allows the unique identification of muons as charged particles able to penetrate large thicknesses of absorber without hadronic interactions. Electrons also lose energy by Coulomb scattering and do not have hadronic interactions. However, they generate electromagnetic showers in a material, a phenomenon we look at next. These involve higher-order processes than Coulomb scattering, and therefore have smaller cross sections; however, they degrade the energy much more rapidly. This allows the identification of electrons using thick absorbers. However, Coulomb scattering remains the dominant process for electrons traversing moderate thicknesses of a gas (where "moderate" means small compared to the length scale given by the radiation length for the material, as we will see below).

The creation of electromagnetic showers involves both the emission of photons by electrons and positrons (called bremsstrahlung) and the conversion of photons into e^+e^- pairs in the field of an atom in the material. These processes must involve very small momentum transfer to the atom in the material in order for the cross section to be large. This in turn requires that the participating electrons and positrons are very relativistic. We will examine the pair production process in this section and then summarize briefly some of the results for bremsstrahlung. The diagrams for pair production were given in Figure 6.3. Those for bremsstrahlung are obtained by changing the outgoing e^+ to an incoming e^- and the incoming γ to an outgoing γ in this figure.

We start by looking at the kinematics for pair production. Assuming that the recoil atom absorbs very little momentum, implies that we can ignore the energy transfer to the atom. We specify the incoming photon four-momentum by $k = (\omega, \vec{k})$ and the outgoing positron and electron four-momenta as p_+ and p_-, respectively. Energy conservation then implies that $\omega = E_+ + E_-$. The minimum momentum transfer to the atom for given E_+ and E_- values is derived from a configuration where the e^+ and e^- both travel along the initial photon direction. This gives

$$q_{\min}^2 = -(\omega - |\vec{p}_+| - |\vec{p}_-|)^2.$$

Expanding the momenta as

$$E - \frac{m_e^2}{2E}$$

gives the relation

$$q_{min}^2 = -\left(\frac{m_e^2}{2E_+} + \frac{m_e^2}{2E_-}\right)^2 = \frac{-m_e^2}{4}\left(\frac{1}{\gamma_+} + \frac{1}{\gamma_-}\right)^2. \qquad (11.21)$$

For $|q_{min}^2|$ to be small, we indeed see that γ_+ and γ_- have to be large.

We know that the charge form factor of the atom is large only for a range of q^2 values. In particular, for a maximum cross section we want

$$|q_{min}^2| \leq \frac{1}{r_{atom}^2},$$

which then allows q^2 to range over all values where the form factor is large. Taking

$$\gamma_+ \simeq \gamma_- = \gamma \quad \text{and} \quad \frac{1}{r_{atom}} \sim \frac{\alpha m_e Z^{1/3}}{1.4},$$

as in the Thomas–Fermi model of the atom, the cross section will be large provided that

$$\gamma \geq \frac{1.4}{\alpha Z^{1/3}}.$$

This is true in all materials for photon energies in the 100 MeV range and up.

Equation 11.21 can also be used to determine when muons will have significant radiative processes. Taking hydrogen as an example, we can replace m_e with m_μ in Eq. 11.21 to get the requirement, based on the atomic form factor:

$$\gamma \geq \left(\frac{m_\mu}{m_e}\right)\frac{1}{\alpha} \simeq 2.9 \times 10^4. \qquad (11.22)$$

This requires energies $\simeq 3$ TeV for the muons, a number that decreases, like $1/Z^{1/3}$, for high Z atoms.

We will see shortly that the electron pair production cross section is proportional to $1/m_e^2$ as a result of the virtual e^+ or e^- propagator appearing in the matrix element. This feature does not appear in Coulomb scattering, which depends mostly on β of the beam particle. For pair production of muons,

$$\frac{1}{m_e^2} \quad \text{is replaced by} \quad \frac{1}{m_\mu^2}.$$

As a result, pair production to e^+e^- always dominates the photon cross section at high energies. Electromagnetic showers involve, to good approximation, only

11.3 Radiative Processes for Electrons and Photons

electrons, positrons, and photons. But if we look at an incoming muon instead of a photon, the formula for γ in Eq. 11.22 indicates correctly the energy at which a muon will significantly lose energy via radiative processes. However, once radiated, photons will generate showers involving e^+e^-, if they have large energies.

11.3.1 ■ Rate Calculation for Pair Production

The matrix element for pair production on atoms in a material is:

$$M_{fi} = \frac{e^3 J_0(\vec{q})}{|\vec{q}|^2} \bar{u}(p_-) \left[\gamma_0 \frac{(\slashed{k} - \slashed{p}_+ + m)\slashed{\epsilon}}{(k - p_+)^2 - m^2} + \frac{\slashed{\epsilon}(\slashed{k} - \slashed{p}_- + m)}{(k - p_-)^2 - m^2} \gamma_0 \right] v(p_+). \tag{11.23}$$

The first and second terms in M_{fi} correspond to the following diagrams.

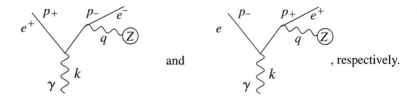

and , respectively.

The atom acts as a static source described by the Fourier transform of the charge density $J_0(\vec{q})$, where the charge e has been factored out. The photon spin is e_μ and can be chosen transverse to the photon direction. Taking the photon direction to be along z, the choices for the spin vector are $\hat{e}_x$ and $\hat{e}_y$.

As in the Coulomb scattering calculation, $J_0(\vec{q}) = 2m_{\text{atom}} \rho(\vec{q})$. This gives for the cross section, for small q^2:

$$d\sigma = \frac{(4\pi)^3 \alpha^3}{2\omega} \sum_{e^+, e^- \text{ spins}} |M|^2 \frac{d^3 p_+}{2E_+ (2\pi)^3} \frac{d^3 p_-}{2E_- (2\pi)^3}$$

$$\times (2\pi) \delta(\omega - E_+ - E_-) \frac{|\rho(\vec{q})|^2}{|\vec{q}|^4} \tag{11.24}$$

where M is the piece of M_{fi} left after taking out the factor

$$\frac{e^3 J_0(\vec{q})}{|\vec{q}|^2}.$$

Averaging also over all photon spins:

$$d\sigma = \frac{1}{(2\pi)^2} \frac{\alpha^3}{\omega} \frac{1}{2} \sum_{e^+, e^-, \gamma \text{ spins}} |M|^2 \frac{d^3 p_+}{E_+} \frac{d^3 p_-}{E_-} \delta(\omega - E_+ - E_-) \frac{|\rho(\vec{q})|^2}{|\vec{q}|^4}.$$

We will do the calculation at high energies, much larger than the electron mass m_e. In this case the process is quasi-one-dimensional. The transverse momenta are small compared to the longitudinal momenta over nearly the entire phase space. This allows a number of simplifications. Indicating the longitudinal and transverse components for each particle of the pair as $p_z^+, p_z^-, \vec{p}_t^{\,+}, \vec{p}_t^{\,-}$, we have for each:

$$p_z \simeq E - \frac{m_e^2 + p_t^2}{2E},$$

so that

$$E - p_z = \frac{m_e^2 + p_t^2}{2E}.$$

The combination $m_e^2 + p_t^2$ is called the transverse-mass squared. Defining $\vec{q}_t = -(\vec{p}_t^{\,+} + \vec{p}_t^{\,-})$, we can calculate, for negligible energy transfer to the target atom:

$$q^2 = -|\vec{q}\,|^2 = -\left[\left(\frac{m_e^2 + |\vec{p}_t^{\,-}|^2}{2E_-} + \frac{m_e^2 + |\vec{p}_t^{\,+}|^2}{2E_+}\right)^2 + |\vec{q}_t|^2\right]. \quad (11.25)$$

At very high energies, we can therefore approximate q^2 as $-|\vec{q}_t|^2$, so q^2 will depend only on the sum of the transverse momenta. This will always be a good approximation when the first term in brackets, coming from the longitudinal momentum transfer, is small compared to the q^2 cut-off determined by the size of the atom. This condition is called complete screening.

When integrating over the phase space, we also ignore the coupling of the longitudinal and transverse momenta, that is, we assume that to good approximation the integration limits for the energy or longitudinal momenta are independent of the values of the transverse momenta, for the small transverse momenta where the cross section is large.

The matrix element factor

$$\frac{1}{2}\sum_{e^+,e^-\text{ spins}} |M|^2$$

can be calculated using the trace techniques developed earlier. The result is the expression

$$\frac{|\vec{q}_t|^2 \omega^2}{(p_-\cdot k)(p_+\cdot k)} - \left[2E_+\frac{(\vec{p}_-\cdot \hat{e})}{(p_-\cdot k)} + 2E_-\frac{(\vec{p}_+\cdot \hat{e})}{(p_+\cdot k)}\right]^2$$
$$- q^2\left[\frac{(\vec{p}_-\cdot \hat{e})}{(p_-\cdot k)} - \frac{(\vec{p}_+\cdot \hat{e})}{(p_+\cdot k)}\right]^2.$$

The last term is higher order in the small quantities $\vec{p}_t^{\,+}, \vec{p}_t^{\,-}, q^2$, than the first two terms, since $\hat{e}$ is transverse by gauge invariance. We will therefore drop this term.

11.3 Radiative Processes for Electrons and Photons

We will also use

$$(p_+ \cdot k) = \omega(E_+ - p_z^+) = \omega\left(\frac{m_e^2 + |\vec{p}_t^{\,+}|^2}{2E_+}\right)$$

and the analogous expression for $(p_- \cdot k)$. Note that the matrix element vanishes for $\vec{p}_t^{\,+} = \vec{p}_t^{\,-} = 0$, because of the absence of a longitudinal photon polarization. With the above approximations:

$$\frac{1}{2E_+E_-}\sum_{e^+,e^-,\gamma \text{ spins}} |M|^2 = \frac{8|\vec{q}_t|^2}{(m_e^2 + |\vec{p}_t^{\,+}|^2)(m_e^2 + |\vec{p}_t^{\,-}|^2)}$$
$$- \frac{16 E_+ E_-}{\omega^2}\left(\frac{\vec{p}_t^{\,+}}{m_e^2 + |\vec{p}_t^{\,+}|^2} + \frac{\vec{p}_t^{\,-}}{m_e^2 + |\vec{p}_t^{\,-}|^2}\right)^2. \quad (11.26)$$

A convenient choice of coordinates for integrating over the phase space are cylindrical coordinates in momentum space, which separate the longitudinal and transverse momenta. We further approximate for both e^+ and e^-:

$$d^3 p = dp_z\, d^2 p_t \simeq dE\, d^2 p_t.$$

This gives, for the differential cross section,

$$d\sigma = \frac{\alpha^3}{(2\pi)^2} I\left(|\vec{q}_t|^2, \vec{p}_t^{\,+}, \vec{p}_t^{\,-}, E_+, E_-\right) \frac{|\rho(\vec{q}_t)|^2}{|\vec{q}_t|^4} d^2 p_t^+\, d^2 p_t^-\, \frac{dE_+}{\omega}, \quad (11.27)$$

where I is the expression given in Eq. 11.26.

To isolate the dependence on $|\vec{q}|^2$ in Eq. 11.27, we define the new variables

$$\vec{p}_t' = \frac{\vec{p}_t^{\,+} - \vec{p}_t^{\,-}}{2} \quad \text{and} \quad \vec{q}_t' = \frac{\vec{p}_t^{\,+} + \vec{p}_t^{\,-}}{2}.$$

These are analogous to the use of the center of mass and relative coordinates in space. Note, $\vec{q}_t' = -\vec{q}_t/2$, and we will use each variable below, based on whichever makes the notation simpler. We can rewrite $d^2 p_t^+ d^2 p_t^- = d^2 p_t' d^2 q_t$, which allows the integration over $\vec{q}_t$ after the integration over the independent transverse momentum $\vec{p}_t'$. Making the change of variables in Eq. 11.26 gives the expression:

$$\frac{8|\vec{q}_t|^2}{\left(m_e^2 + |\vec{q}_t' - \vec{p}_t'|^2\right)\left(m_e^2 + |\vec{q}_t' + \vec{p}_t'|^2\right)}$$
$$- \frac{16 E_+ E_-}{\omega^2}\left(\frac{(\vec{q}_t' + \vec{p}_t')}{m_e^2 + |\vec{q}_t' + \vec{p}_t'|^2} + \frac{(\vec{q}_t' - \vec{p}_t')}{m_e^2 + |\vec{q}_t' - \vec{p}_t'|^2}\right)^2. \quad (11.28)$$

The expression in Eq. 11.28 indicates that $\vec{p}_t'$ has a magnitude of order m_e. The variable $\vec{q}_t'$ is typically much smaller because of the extra factor of $1/|\vec{q}_t|^4$

appearing in Eq. 11.27. We will therefore make the approximation that $|\vec{q}_t'^2| \ll m_e^2 + |\vec{p}_t'|^2$ when integrating over $d^2 p_t'$. With this approximation, the first term in Eq. 11.28 is, to lowest order in $|\vec{q}_t'|^2$,

$$\frac{8 |\vec{q}_t|^2}{\left[m_e^2 + |\vec{p}_t'|^2\right]^2},$$

whose integral over $d^2 p_t'$ is

$$\frac{8\pi}{m_e^2} |\vec{q}_t|^2.$$

The second term in Eq. 11.28 is approximately

$$-\frac{16 E_+ E_-}{\omega^2} \left[4|\vec{q}_t'|^2 - \frac{16(\vec{q}_t' \cdot \vec{p}_t')^2}{(m_e^2 + |\vec{p}_t'|^2)} + \frac{16 |\vec{p}_t'|^2 (\vec{p}_t' \cdot \vec{q}_t')^2}{(m_e^2 + |\vec{p}_t'|^2)} \right] \frac{1}{\left[m_e^2 + |\vec{p}_t'|^2\right]^2},$$

whose integral over $d^2 p_t'$ is

$$-\frac{4}{3} \frac{E_+ E_-}{\omega^2} \left(\frac{8\pi}{m_e^2} |\vec{q}_t|^2 \right).$$

Integrating over all angles of the $d^2 q_t$ integral leaves finally the expression:

$$d\sigma = \left[2\alpha \left(\frac{\alpha}{m_e}\right)^2 \frac{|\rho(\vec{q}_t)|^2}{|\vec{q}_t|^2} d |\vec{q}_t|^2 \right] \left[\left(1 - \frac{4}{3}\frac{E_+ E_-}{\omega^2}\right) \frac{dE_+}{\omega} \right]. \quad (11.29)$$

The cross section expression is a product of two factors. The first term in brackets, after integration over $|\vec{q}_t|^2$ and multiplication by the density of scatterers, N, is defined to be the reciprocal of the radiation length for electromagnetic showers. In evaluating this integral, the lower cutoff is provided by the atomic radius. The upper cutoff is provided by taking into account the exact expression in Eq. 11.28. Our approximations, valid for low $|\vec{q}_t|^2$, are invalid for $|\vec{q}_t|^2 \gtrsim m_e^2$, where the terms in Eq. 11.28 are falling instead of rising. A more accurate evaluation of the integrals over $d^2 p_t'$ indicates that m_e^2 is the correct cutoff for the $|\vec{q}_t|^2$ integral. This gives for the radiation length:

$$\frac{1}{X_0} = 2N\alpha \left(\frac{\alpha}{m_e}\right)^2 \int_{q_{\min}^2}^{m_e^2} \frac{|\rho(\vec{q}_t)|^2}{|\vec{q}_t|^2} d|\vec{q}_t|^2$$

$$= 4N\alpha \left(\frac{\alpha}{m_e}\right)^2 Z^2 \log(m_e r_{\text{atom}}). \quad (11.30)$$

We used for the form-factor the constant value $\rho(\vec{q}_t) = Z$ over the range of the integral. The result in Eq. 11.30 is the same expression that we found in the multiple scattering calculation in Section 11.2.2.

The second factor in brackets in Eq. 11.29 gives an energy integral that can be rewritten in terms of fractional energies

$$x_+ = \frac{E_+}{\omega}, \quad x_- = \frac{E_-}{\omega},$$

giving an expression $[1 - \frac{4}{3}x_+(1-x_+)]\,dx_+$. The distributions for the fractional energies (and longitudinal momentum fractions at high energies) are independent of the overall energy scale. These are scale invariant quantities at very high overall energy. The integral over dx_+ gives a factor of $\frac{7}{9}$. Thus the number of high energy photons in a beam will be attenuated in a material by pair production, in a distance z, according to the formula:

$$N = N_0 e^{-(7/9)(z/X_0)},$$

creating electromagnetic showers in the material.

The calculation for an e^+ or e^- propagating in a material can be done analogously. These particles can radiate many photons, including soft photons for which the cross section diverges as the energy goes to zero. The cumulative effect of these photons can be summed in a calculation of the rate of energy loss. In this case, we can write an expression for the remaining energy of the e^+ or e^- as it penetrates the material. The energy attenuation is given by the formula:

$$E = E_0 e^{-z/X_0}.$$

We see that the radiation length is again the length-scale parameter defining the shower properties.

The above calculations assume that the production of electromagnetic showers is the dominant process for photons, which is correct for very high energies, particularly for materials with large Z. At lower energies, other processes are dominant, such as Compton scattering and the photoelectric effect for photons, and the dE/dx energy loss for e^+ and e^-. Note that the radiation length is smallest for high Z materials. These materials provide the best detector choice for distinguishing electrons, which shower, from hadrons, which interact strongly.

11.4 ■ INCLUSIVE DISTRIBUTIONS IN HADRONIC SCATTERING

We have looked at two types of electromagnetic processes, elastic Coulomb scattering with a differential cross section dependent on the momentum transfer, and pair production, an inelastic process where the differential cross section depends on the transverse momentum and the fractional energies of the final state leptons. These processes are quasi-one-dimensional at high energy, in the kinematic region where the cross section is large.

The large cross section hadronic processes also break up into elastic and inelastic processes that are quasi-one-dimensional at high energies. For simplicity we consider the scattering of a hadron on another hadron, rather than on a nucleus, as would be the case in a material.

The hadronic cross section is governed by the size of λ_{QCD} and involves non-perturbative physics. The elastic part of the cross section is typically about 20% of the total cross section and is dominated by small momentum transfers. The distribution in $|q^2|$ for the scatters is approximately exponential. This result contrasts with elastic Coulomb scattering, which has a power law dependence governed by the photon propagator. The hadronic differential elastic cross section $d\sigma/d|q^2|$ approaches a constant as $q^2 \to 0$, indicating that there is no long-range interaction. The confinement of the color degrees of freedom implies that we should not think in terms of individual gluons existing at long distances from colorless hadrons. An example of an elastic cross section, for high energy $\bar{p}p$ scattering, is shown in Figure 11.2.

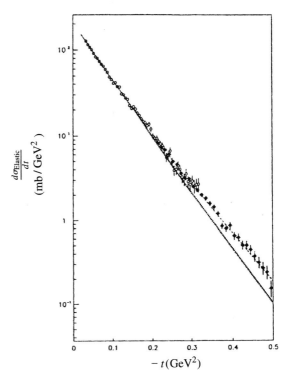

FIGURE 11.2 Differential elastic $\bar{p}, p$ cross section at a center of mass energy of 546 GeV. [From M. Bozzo et al., *Phys. Lett.* 147B, 385 (1984).]

A process related to elastic scattering, called diffraction dissociation, occurs with a cross section roughly comparable to elastic scattering. This process is somewhat similar to Coulomb scattering on an atomic system, where the atomic

11.4 Inclusive Distributions in Hadronic Scattering

system is left in an excited state. Since the hadrons are composite, they can scatter with little energy and momentum transfer through resonance production, where either one or both of the emerging hadrons are in internally excited states of the initial hadrons. As in the case of elastic scattering, no internal quantum numbers are exchanged between the beam and target systems for this process.

The remaining part of the cross section at high energies results in inelastic events with many final state particles. The vast majority of these events are quasi-one-dimensional. To quantify the distribution of particles in these events, an inclusive distribution function in the invariant phase space is formed:

$$\frac{E\,dN}{d^3p},$$

where the integral of this function over d^3p/E gives the average number of particles produced in an event. This distribution can further be broken down into contributions from different particle types, for example pions, kaons, and nucleons of various charges. These inclusive distribution functions, which depend on the initial colliding hadrons and the center of mass energy, exhibit a number of regularities.

The first regularity is a limited transverse momentum, yielding the quasi-one-dimensional behavior. Considering the various particle types, the distribution function is roughly an exponential function of the transverse mass, which favors the production of light hadrons over heavy hadrons. The particles produced most copiously are pions. Their transverse momentum is on average $\simeq \lambda_{QCD}$.

With regard to the longitudinal momentum, we can divide the phase space roughly into three regions in the center of mass. The first region is where the final state particles carry a significant fraction of the incident particle energy and follow its direction, called the beam fragmentation region. The internal quantum numbers of the particles in this region are strongly affected by those of the incident particle. The second region is the analogous region of phase space relative to the target particle. The particles in these two regions are called leading particles. The third region, called the central region, corresponds to particles with relatively low energies in the center of mass. In this region, the internal quantum numbers of the initial colliding particles have little affect, so that the particles are on average neutral with regard to internal additive quantum numbers. Most of the particles are produced in the central region, since energy conservation limits the number of leading particles.

A convenient variable for studying leading particle behavior is the fractional energy or fractional longitudinal momentum in the center of mass. In this frame, the maximum value of the energy that a particle can have is $\sqrt{s}/2$. Define

$$x = \frac{2p_z}{\sqrt{s}};$$

it ranges between $+1$ (leading particles for the beam direction) and -1 (leading particles for the target direction) if we ignore particle masses. Figure 11.3 shows

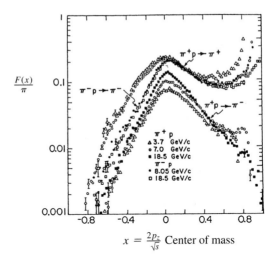

FIGURE 11.3 Inclusive distribution functions normalized to $F(x)/\pi$, where $F(x) = \frac{2E}{\sqrt{s}}\frac{dN}{dx}$. Data are for pions incident on protons. The incident pion beam momentum is indicated in the figure. [From M. Alston-Garnjost et al., *Phys. Lett.* B39, 402 (1972).]

the inclusive distributions, integrated over transverse momentum, for π^+ and π^- in $\pi^+ p$ and $\pi^- p$ interactions at several energies. The experiments are performed by colliding pions of various momenta on stationary proton targets. The distributions show that the leading particles in the beam direction are primarily those pions identical in charge to the incident pion. In the backward direction, the distribution functions are suppressed, since a leading particle should carry the baryon number of the target proton, which neither π^+ nor π^- do.

Figure 11.3 also suggests that the inclusive distribution is approaching a well-defined function of x alone at high center of mass energy, analogous to the behavior of the distribution function for the final state electron energy in pair production. Such behavior is called scaling. The distributions in hadronic reactions scale approximately, with slow logarithmic violations. These violations reduce the distributions at large x, resulting in more particles in the central region, as the overall energy increases. This is one of a number of quantities that show slow logarithmic changes with energy; another is the total cross section, which grows slowly with energy.

The inclusive distributions, integrated over transverse momentum, also illuminate how the energy is partitioned, assuming a quasi-one-dimensional process. Writing

$$F(x) = E\frac{dN}{dp_z} = \frac{2E}{\sqrt{s}}\frac{dN}{dx} \simeq x\frac{dN}{dx},$$

we expect that $\int_0^1 F(x)\,dx \simeq \int_{-1}^0 F(x)\,dx \simeq 1$, if we sum over all particle types. This is just a statement of energy conservation, with the energy assumed to be split

11.4 Inclusive Distributions in Hadronic Scattering

equally between forward- and backward-moving particles. By splitting $F(x)$ into contributions from different particle types, we can see the fraction of the energy each type carries.

If we wish to count the number of particles, rather than the energy carried, then we are interested in the function:

$$\frac{dN}{dx} = \frac{1}{x} F(x).$$

Since $F(x)$ is a constant at $x = 0$, dN/dx is large in the central region, indicating that most particles are produced in this region. This is, however, the region where our approximation that $p_z = E$ does not hold. As a result, the central region distributions are better displayed using different variables. The typical choice is to use a variable y, defined so that

$$\frac{dp_z}{E} d^2 p_t = dy\, d^2 p_t.$$

The variable y is called the rapidity, and is given by

$$y = \frac{1}{2} \log\left(\frac{E + p_z}{E - p_z}\right). \tag{11.31}$$

The data indicate that the inclusive distribution in y and p_t^2, for a given final state particle type, approximately factorizes into a distribution in y and one in terms of the transverse mass. If the dynamics populate the longitudinal phase space uniformly, the distribution in y will be constant. In practice, this is approximately true for $|y| \lesssim 3$ for center of mass energies ~ 100 GeV. This region is called the rapidity plateau. The height of the plateau and its width grow slowly with energy, providing a multiplicity of final particles that grows somewhat faster than logarithmically with the total center of mass energy. The average number of charged particles produced, which is easier to measure than the number of neutrals, is about 20 at a center of mass energy of 100 GeV. The choice of a central region in the center of mass system corresponds to a region in angle for the produced particles. Taking $p_z \simeq E \cos\theta$ gives the approximate formula

$$y \simeq \frac{1}{2} \log\left(\frac{1 + \cos\theta}{1 - \cos\theta}\right). \tag{11.32}$$

The region $|y| \leq 3$ then corresponds to a scattering angle $\geq 6°$ to the beam direction.

The total rapidity range for a given energy is determined by the logarithm of the total center of mass energy, $\sqrt{s}$. For large energy and forward-moving particles,

$$y = \frac{1}{2} \log\left(\frac{E + p_z}{E - p_z}\right) \simeq \log\left(\frac{2E}{\sqrt{m^2 + p_t^2}}\right).$$

Assuming a typical transverse mass ~ 1 GeV and a maximum value of $E = \sqrt{s}/2$, gives a maximum value of the rapidity,

$$y_{max} = \log\left(\sqrt{s}(\text{GeV})\right).$$

Including the region of negative y gives a total rapidity range of

$$\Delta y = 2\log\left(\sqrt{s}(\text{GeV})\right),$$

which grows logarithmically with the total center of mass energy.

An interesting property of the rapidity is that it shifts by a constant under a Lorentz transformation along z. As a result, rapidity distributions are the same in the center of mass and in a target rest frame, except for a constant overall shift.

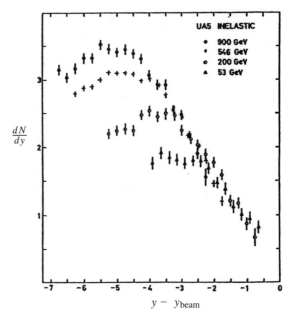

FIGURE 11.4 Charged particle rapidity distributions in $\bar{p}, p$ scattering at a range of center of mass energies. Points shown correspond to forward-moving particles in the center of mass. [From G. J. Alner et al., *Z. Phys.* C33, 1 (1986).]

The data in Figure 11.4 show the rapidity distribution for forward-moving charged particles in $\bar{p}p$ scattering at a range of center of mass energies. The rapidity is calculated using the approximation of Eq. 11.32, and the value of y for the incident beam particle has been subtracted so that the distributions all end at $y - y_{\text{beam}} = 0$. The data points correspond to all forward-moving particles in the center of mass. The distribution for backward-moving particles is the same as that for forward-moving particles with the correspondence $y \to -y$, when the distribution is plotted in the center of mass. The data show:

1. A nearly universal distribution for rapidities near y_{beam}, independent of the center of mass energy. This is equivalent to approximate scaling of the inclusive distribution function in the beam fragmentation region.
2. A logarithmic increase in the total rapidity range with center of mass energy.
3. An increase with center of mass energy in the width of the central region (the region over which the distribution is roughly constant).
4. A slow rise with center of mass energy of the height of the distribution in the central region.

CHAPTER 11 HOMEWORK

11.1. For small momentum transfer scattering of a fast particle from a heavy target initially at rest, show that the scattering angle for the heavy target is approximately given by:

$$\cos\theta_B = \frac{p_B}{2m_B}.$$

The heavy target mass is m_B, $q^2 = -p_B^2$ to good approximation.

11.2. Consider n Coulomb scatterings, each through small angles, giving a total spatial momentum transfer of

$$\vec{q} = \vec{q}_1 + \vec{q}_2 + \cdots + \vec{q}_n,$$

where each $\vec{q}_i$ is approximately transverse to the initial large momentum.

(a) Assuming uncorrelated scatters, show that the expectation values satisfy:

$$\langle |\vec{q}|^2 \rangle = n \langle |\vec{q}_i|^2 \rangle, \quad \text{assuming} \quad \langle |\vec{q}_i|^2 \rangle = \langle |\vec{q}_j|^2 \rangle$$

for all i and j.

(b) From $|\vec{q}|^2$ we can calculate a scattering angle θ^2. It is sometimes convenient to consider separately the two components of the transverse momentum, q_x and q_y. In spherical coordinates:

$$q_x = p_A \sin\theta \cos\phi \simeq p_A \theta \cos\phi = p_A \theta_x,$$
$$q_y = p_A \sin\theta \sin\phi \simeq p_A \theta \sin\phi = p_A \theta_y,$$

defines the projected angles θ_x and θ_y. Show that $\langle \theta_x^2 \rangle = \frac{1}{2} \langle \theta^2 \rangle$.

(c) After traversing a distance l of scattering material, the particle will suffer a deflection transverse to its initial direction. Assuming a small net deflection angle and many scatters, show that

$$\langle x^2 \rangle = \frac{l^2 \langle \theta_x^2 \rangle}{3}.$$

11.3. The scattering of a fast particle on a heavy target particle can be thought of as scattering on a fixed source of the electromagnetic field. Using the Hamiltonian density for the electromagnetic interaction, we get a contribution to the S matrix, to lowest order:

$$S_{fi} = -i \int \langle f| J_\mu(x) A_\mu(x) |i\rangle \, d^4x.$$

(a) For the target particle at rest, the fixed potential is just the scalar potential $V(\vec{x})$. The matrix element of J_μ is between states of the incident fast particle with momentum p_i and p_f. Show that this gives

$$S_{fi} = -i \int -e\bar{u}(p_f)\gamma_0 u(p_i) e^{i(p_f - p_i)\cdot x} V(\vec{x}) \, d^4x$$

(The incident particle has charge $-e$ and spin $\tfrac{1}{2}$). Integrating over the time coordinate gives

$$S_{fi} = i2\pi \, \delta(E_f - E_i) e\bar{u}(p_f)\gamma_0 u(p_i) \int e^{i(\vec{p}_f - \vec{p}_i)\cdot \vec{x}} V(\vec{x}) \, d^3x.$$

(b) The relation between the potential and charge density of the heavy target is

$$\nabla^2 V = -\rho(\vec{x}).$$

Show that this leads to

$$\int e^{i\vec{q}\cdot \vec{x}} V(\vec{x}) \, d^3x = \frac{1}{|\vec{q}^{\,2}|} \int e^{i\vec{q}\cdot \vec{x}} \rho(\vec{x}) \, d^3x = \frac{1}{|\vec{q}^{\,2}|} \rho(\vec{q}),$$

giving

$$S_{fi} = i2\pi \delta(E_f - E_i) e\bar{u}(p_f)\gamma_0 u(p_i) \frac{\rho(\vec{q})}{|\vec{q}^{\,2}|}.$$

This is an alternative way to get the cross section calculated in the text.

(c) Suppose we want to use elastic scattering of electrons to study the charge distribution of the nucleus. What typical value of q^2 is needed for such a study?

11.4. Consider elastic multiple Coulomb scattering of a fast charged particle in Helium gas. Assuming that the density of scattering centers is given by Avogadro's number of atoms in 22.4 liters of gas (ideal gas law for standard temperature and pressure) and that $\rho(\vec{q}) \simeq 2$ for $q^2 \geq 1/(\tfrac{1}{2}\text{Å})^2$, 0 otherwise, find the number of scatters per centimeter of gas traversed.

11.5. We can modify the dE/dx formula in Eq. 11.20 to be more general.

(a) If we keep the velocity β of the incident particle (that is, take $p_A = \beta E_A$ rather than $p_A = E_A$), show that the calculation in the text can be generalized to

$$\frac{dE}{dx} = \frac{4\pi \alpha^2 NZ}{m_e \beta^2} \log \frac{2m_e \beta^2 \gamma^2}{I}.$$

In the logarithm, the minimum momentum transfer is taken to be

$$q^2_{\min} = \frac{-I^2}{\beta^2 \gamma^2},$$

where we still assume that $E_A \gg I$.

(b) Assuming a point-like spin $\frac{1}{2}$ particle incident on a free electron, we can calculate the matrix element squared without assuming the scattered electron remains nonrelativistic. Show that the matrix element squared, averaged over initial spins and summed over final spins, in the frame where the electron is initially at rest, is

$$|\overline{M_{fi}}|^2 = \frac{8e^4}{q^4}\left[m_e^2(E_A)^2 + m_e^2(E_A')^2 + \frac{(m_e^2 + m_A^2)}{2}q^2\right].$$

E_A is the initial energy, E_A' is the final energy of the incident particle. For energies such that

$$E_A \ll \left(\frac{m_A}{m_e}\right)m_A, \quad E_A' \simeq E_A$$

and we have

$$|\overline{M_{fi}}|^2 = 16m_e^2 E_A^2 \frac{e^4}{q^4}\left[1 + \frac{m_A^2 q^2}{4m_e^2 E_A^2}\right].$$

(This can be compared to Eq. 11.2.) The expression in brackets is approximately

$$1 + \frac{\beta^2 q^2}{|q_{max}^2|}.$$

Show that the second term will give a correction to the energy loss formula through its contribution at large q^2, giving the result:

$$\frac{dE}{dx} = \frac{4\pi\alpha^2 NZ}{m_e\beta^2}\left[\log\left(\frac{2m_e\beta^2\gamma^2}{I}\right) - \frac{\beta^2}{2}\right].$$

A more correct treatment of the atomic transitions results in another $-\beta^2/2$ appearing inside the brackets, resulting in the Bethe–Bloch energy loss formula given in the text.

11.6. Consider a 100 GeV photon incident on a hydrogen atom.

  (a) Calculate the pair-production cross section using Eqs. 11.29 and 11.30.

  (b) Calculate the cross section for Compton scattering on the atomic electron. This was discussed in Chapter 3.

  (c) Calculate the cross section to produce hadrons by absorption of the photon by the proton. You can use the Vector Dominance model of Chapter 6 and assume that the meson-proton cross section is 30 mb.

11.7. Consider a Lorentz transformation along z, given by parameters β and γ. Calculate the rapidity of a particle in the new frame in terms of the Lorentz transformation parameters and the initial value of y.

CHAPTER 12

Scattering with Large Momentum Transfer

12.1 ■ TYPES OF PROCESSES

In Chapter 11 we discussed the large cross section processes that are the basis for detection of long-lived particles; here we study the rare large momentum transfer processes that are the basis for discovering new physics at short-distance and high-mass scales. These large momentum transfer processes played an important role in establishing the quark and gluon structure of the proton (discovered at distance scales that are small compared to the size of the proton) and in the discovery of all the heavier, and hence short-lived, particle types. The actual scale by which we define "large momentum transfer" or "large mass" has evolved with time, as new discoveries have been assimilated into the Standard Model, raising the scale to the next potential discovery region. At present, the next region expected to yield new discoveries and insights is the mass range of electroweak symmetry breaking, somewhere in the region of 100 GeV to 1 TeV.

The largest center of mass energies at accelerators occur for colliding beams. The individual beams can be e^+, e^-, p, or $\bar{p}$. The highest energies presently available in the center of mass are approximately 200 GeV for e^+e^-, 300 GeV for ep, and 2 TeV for $p\bar{p}$ scattering. A new pp collider, called the Large Hadron Collider, with center of mass energy of 14 TeV, is under construction. Reactions between the particles in the beams can proceed via exchange of a particle, for example a photon, in which case the large momentum transfer process appears as the tail of a distribution containing a wide range of momentum transfers. Large momentum transfer processes can also proceed via annihilation of the initial state particles. The simplest example of this is e^+e^- annihilation. This process, as a probe of high mass physics, has the advantage that the entire initial state energy is available for the production of new particles. For energies well below the Z^0 mass, e^+e^- annihilation is predominantly an electromagnetic process. Above the Z^0 mass scale, we have the full range of electroweak interactions. Calculations for large momentum transfer processes for initial and final state leptons were discussed in Chapter 3 (below the Z^0 mass); Chapter 10 generalized the results to the full electroweak theory.

The e^+e^- annihilation reaction does not directly couple to color and therefore involves small coupling constants. We could achieve higher rates if we could directly collide a quark and an antiquark or two gluons, since the coupling would then involve $\alpha_s(q^2)$, if the particles exchanged carry color. We are not able to do

12.2 ■ e^+e^- ANNIHILATION TO HADRONS

The e^+e^- annihilation reaction contributed greatly to our understanding of the Standard Model. The change with energy in cross sections and final states, when the full set of electroweak processes begin to contribute, is a nice example of the opening of "new" physics as the energy increases.

12.2.1 ■ Energies below Z^0, the Electromagnetic Regime

Figure 12.1 shows the data, accumulated over approximately three decades, for the cross section for $e^+e^- \to$ hadrons divided by $e^+e^- \to \mu^+\mu^-$. We focus below on the energy region below Z^0. The data show the very large resonance cross sections for vector meson production near threshold for the creation of a new

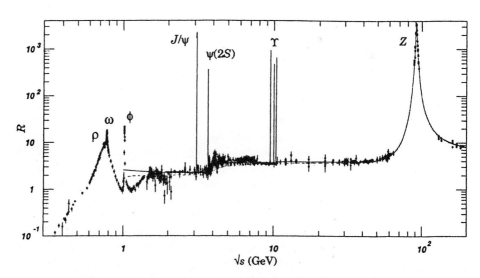

FIGURE 12.1 Data for $e^+e^- \to$ hadrons divided by $e^+e^- \to \mu^+\mu^-$, from a number of experiments. The sharp peaks correspond to the production of $J^P = 1^-$ resonances near flavor thresholds. The solid curve is the Standard Model prediction, which applies far from flavor thresholds. [Data compilation from the Particle Data Group, K. Hagiwara et al., *Phys. Rev.* D66, 010001 (2002)].

quark-antiquark pair. These include ρ, ω, ϕ for the light flavors, and the J/ψ and Υ systems for the charm and bottom flavors. Data collected at the resonances have provided much information on the binding potential, as well as heavy meson decay properties. Away from all thresholds, the cross section ratio R is nearly constant. In the region around 30 GeV energy in the center of mass, it is approximately equal to 3.85. The lowest-order expectation for R is

$$R = 3 \sum_i e_{q_i}^2 = 3\frac{2}{3},$$

based on the production of five flavors of nearly free quarks and antiquarks at short distances, including the factor of 3 from color. The lowest-order QCD correction, corresponding to the radiation of a gluon, shows up in the final state through the production of three jet events, where one jet is a gluon jet. This gives the more accurate prediction

$$R = 3 \sum_i e_{q_i}^2 \left[1 + \frac{\alpha_s(s)}{\pi}\right]. \tag{12.1}$$

In the regime where $s = (30 \text{ GeV})^2$, the data indicate a gluonic correction of about 5%, in agreement with QCD expectations, providing a good measurement of α_s at the given s value.

12.2.2 ■ Quark Fragmentation Functions

The idea of nearly free quarks and gluons at short distances allows a calculation of the e^+e^- total cross section, given by Eq. 12.1, as a ratio to the point-like cross section to $\mu^+\mu^-$. The quarks and gluons, however, fly apart, and at longer distances eventually produce a cascade of colorless particles in jet structures. The particles with a large fraction of the initial quark energy and momentum (called leading particles), can be expected on average to have internal quantum numbers correlated with those of the quark. Lower momentum particles have on average no net internal quantum numbers, but are required to neutralize the color fields that stretched between the initial rapidly separating quark and antiquark (for the case of no additional gluon jets). These lower momentum particles form a central region rapidity plateau along the $q, \bar{q}$ axis, analogous to that seen in inelastic hadron-hadron scattering discussed in Chapter 11.

The process leading to the final jets is called quark fragmentation and is described in terms of a number of functions that are matched to the data. These functions describe the probability of finding a hadron of a given type in a jet initiated by a specific quark (or antiquark, or gluon), where the hadron carries a given fraction of the large initial quark energy. The functions describe the leading particles within each jet in terms of the initiator of the jet and its energy, independent of the number of other jets. The particles in the central region cannot be described this simply, and their distributions depend somewhat on the types and interactions

12.2 e^+e^- Annihilation to Hadrons

of the full complement of jets. This simple picture is approximate even for the leading particle distributions, which show very slow logarithmic changes, analogous to those seen in the beam and target fragmentation distributions in inelastic hadron-hadron collisions. The quark fragmentation functions are, however, very useful, allowing nearly universal predictions of the hadrons produced in a jet, which is initiated by a quark far away from all other colored particles in momentum space, independent of the initial overall scattering process.

The transverse momentum distributions for the hadrons in a jet, relative to the jet axis, are rather similar to those in inelastic hadron scattering. These momentum components are exponentially limited, with the scale for the average transverse momentum $\sim \lambda_{QCD}$. A parameterization in terms of the transverse mass provides a good description across particle types.

The fractional energy carried by a hadron of type h in a high energy jet initiated by a quark of type q is defined to be

$$z = \frac{E_h}{E_q}. \tag{12.2}$$

The quark fragmentation function is then defined as

$$\frac{dN^h}{dz} = D_q^h(z); \tag{12.3}$$

it gives the distribution function for the number of hadrons of type h in the jet. Energy conservation gives the constraint

$$\sum_h \int_0^1 z \, D_q^h(z) \, dz = 1. \tag{12.4}$$

Charge conjugation and isospin symmetry provide relations between a number of the fragmentation functions. For example:

$$D_u^{\pi^+} = D_{\bar{u}}^{\pi^-} = D_d^{\pi^-} = D_{\bar{d}}^{\pi^+},$$

$$D_u^{\pi^-} = D_{\bar{u}}^{\pi^+} = D_d^{\pi^+} = D_{\bar{d}}^{\pi^-},$$

$$D_s^{\pi^+} = D_s^{\pi^-}.$$

For the light quarks, the fragmentation functions can be approximately parameterized as

$$D_q^h(z) = N \frac{(1-z)^n}{z}, \tag{12.5}$$

where N and n depend on q and h. In general, if h contains the quark q, n is smaller than if h does not. Thus, for example, $D_u^{\pi^+} \gg D_u^{\pi^-}$ for z near 1. If we imagine forming hadrons through a cascade of quarks and antiquarks created from the vacuum, along the colored fields linking the initially produced u and $\bar{u}$, then π^+ can be created from the initial high-energy u quark combining with an antiquark created from the vacuum, whereas π^- is made entirely of the quarks and antiquarks created from the vacuum. As a result, for particles moving along the initial u quark direction, the average π^+ energy is larger than the average π^- energy. For large z, $n \simeq 2$ for π^+ and $n \simeq 3$ for π^- in Eq. 12.5.

We expect the distribution for small z to be universal for a given type of hadron, independent of the light quark initiating the jet. Thus we have the general result that $D_q^h(z)$ is independent of q (or $\bar{q}$) for $q = u, d,$ or s, as $z \to 0$. By isospin symmetry, we expect these distributions to be equal for $h = \pi^+, \pi^0,$ or π^-. The distribution for $z \to 0$ is, however, significantly smaller for kaons than for pions, because hadronic resonances created in the cascade produce more pions than kaons when they subsequently decay, and also because the creation of $s\bar{s}$ from the vacuum is significantly suppressed relative to $u\bar{u}$ and $d\bar{d}$. The suppression factor is about 3, as determined from the data.

Figure 12.2 shows the inclusive fractional momentum distributions for a number of hadron species seen in e^+e^- annihilation at a range of center of mass energies. These represent a sum of the fragmentation functions over the quark, antiquark, and gluon jets produced. They illustrate the approximate scaling behavior and the large preponderance of pions among the hadron types.

The fragmentation function for the heavy quarks, c or b, to produce heavy mesons is very different from the light quark functions in Eq. 12.5, for two reasons. The first is that the cascade of quarks and antiquarks created from the vacuum contains only light quarks, to very good approximation. As already mentioned, even the $s\bar{s}$ component is significantly suppressed. Therefore, heavy mesons contain the initially produced heavy quark, but do not participate in the hadron cascade near $z = 0$. This cascade is responsible for the $1/z$ factor in the light quark to light hadron fragmentation function, but is missing in the heavy quark to heavy meson case. The second reason is that the primary meson containing the heavy quark has a distribution favoring a value of z near one. This distribution is very different from the function in Eq. 12.5. To understand this, consider the light quark (or antiquark) created from the vacuum that ends up in the primary heavy meson. If we assume that the particle creation process from the vacuum results in comparable energies for this quark and the energy left for the remainder of the cascade, then the remaining cascade will carry very little energy, which can be seen as follows. Since the light quark in a heavy meson carries a fraction of the meson energy $\simeq m_q/m_Q$ (where q refers to the light quark or antiquark and Q refers to the heavy quark or antiquark), it will carry at most this fraction of the total jet energy. Taking this to be comparable to the energy in the rest of the cascade, and using energy conservation, we predict that the average z for the heavy meson is

$$\langle z \rangle \simeq 1 - \frac{\lambda_{QCD}}{m_Q}.$$

12.2 e^+e^- Annihilation to Hadrons

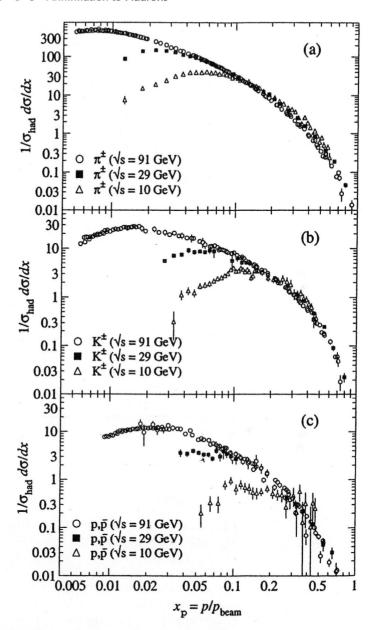

FIGURE 12.2 Scaled momentum spectra of (a) $\pi^{\pm}$, (b) $K^{\pm}$, and (c) p or $\bar{p}$ at $\sqrt{s} = 10$, 29, and 91 GeV. [Data compilation from the Particle Data Group, K. Hagiwara et al., *Phys. Rev.* D66, 010001 (2002)].

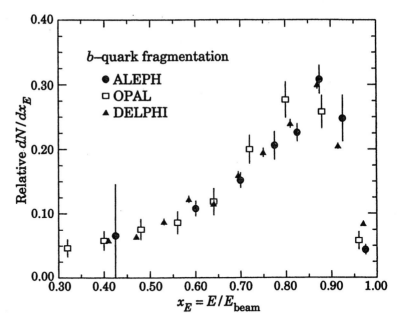

FIGURE 12.3 Measured e^+e^- fragmentation function of b quarks into B hadrons at $\sqrt{s} = 91$ GeV. [Data compilation from the Particle Data Group, K. Hagiwara et al., *Phys. Rev.* D66, 010001 (2002)].

In this formula, the light quark parameter mass is $\sim \lambda_{QCD}$. The measured averages are $\langle z \rangle \simeq 0.6$ for charm and $\langle z \rangle \simeq 0.75$ for bottom mesons, with the most likely values even larger. Figure 12.3 shows the fragmentation function for the b quark to produce B hadrons. The data come from the decays of the Z^0, but look the same as at lower energy.

12.2.3 ■ e^+e^- Annihilation in the Electroweak Regime

Figure 12.4 shows the measured e^+e^- cross section to hadrons versus center of mass energy, along with the calculated value based on the full electroweak contributions in the Standard Model. At the Z^0 resonance, the cross section is enhanced enormously; it then falls again as the energy increases beyond this resonance. The mix of fermion types and the angular distributions at Z^0 were discussed extensively in Chapter 10; they differ significantly from the lower energy values where the production is proportional to the electric charge squared of the fermion-antifermion pair. The quark fragmentation functions are, however, the same at Z^0 as for virtual photons. Other than at the Z^0 resonance, the cross section falls with increasing energy, requiring accelerators with larger luminosity as the energy rises, in order to provide enough interesting events to study.

The cross section in Figure 12.4 has various contributing electroweak processes. However, those due to the Higgs boson have not yet been seen. The direct production process $e^+e^- \to H^0$ is too small to be seen, since the Higgs couples

12.2 e^+e^- Annihilation to Hadrons

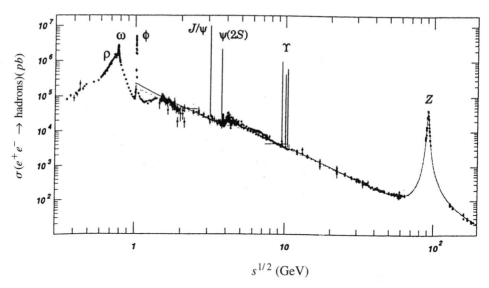

FIGURE 12.4 Measured cross section for $e^+e^- \to$ hadrons, including the region of Z^0. Solid curve is the Standard Model prediction. [Data compilation from the Particle Data Group, K. Hagiwara et al., *Phys. Rev.* D66, 010001 (2002)].

FIGURE 12.5 The dominant H^0 production process for e^+e^- annihilation in the Standard Model.

to mass and the electron is very light. The dominant process for Higgs production in e^+e^- annihilation is shown in Figure 12.5. The Higgs coupling is completely predicted in terms of the Z^0 mass in the Standard Model, which has just one single scalar excitation.

The Higgs boson produced via the process in Figure 12.5 should be easy to detect once the total energy is above $M_Z + m_H$. For example, it can be seen in a recoil mass distribution opposite Z^0, where Z^0 can be detected most cleanly through its decay to e^+e^- or $\mu^+\mu^-$. The present set of data, using the highest energy e^+e^- collider, imply that the mass of H^0 is greater than about 110 GeV, if it exists. Therefore, it will require the construction of a new, higher energy, e^+e^- accelerator to study this process.

Other interesting H^0 production processes that are large at high energies are shown in Figure 12.6. These again test the couplings to gauge bosons, which

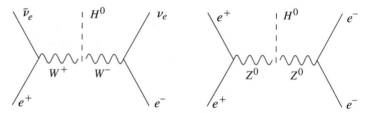

FIGURE 12.6 Additional interesting electroweak processes leading to an H^0 in e^+e^- annihilation.

are completely predicted in the Standard Model. Checking these couplings, as well as those responsible for H^0 decay to various final states, would constitute crucial tests of the symmetry breaking scheme. Since H^0 has yet to be found, we might find an entirely different mechanism. Whatever the answer is, these types of searches are at the forefront of our quest to advance the understanding of electroweak symmetry breaking.

12.2.4 ■ Energy Scale for Production of the Higgs Particle

We said at the beginning of this chapter that the source of electroweak symmetry breaking, the H^0 or whatever plays its role, should be found somewhere between 100 GeV and 1 TeV. The low end of the range is favored by the precision electroweak data, for example, the ratio of the W and Z masses, which favor a value of $m_H \sim 100$ GeV. The upper limit is provided by considering the diagrams of Figure 12.6 more carefully. These diagrams include gauge boson-gauge boson fusion, where the H^0 appears as a resonance. Allowing the H^0 to decay back into gauge boson pairs, we can consider this as a contributor to gauge boson-gauge boson scattering. There are, however, other diagrams involving only gauge bosons that contribute to gauge boson-gauge boson scattering, since the bosons interact with each other. For energies too small to produce the Higgs boson, these other diagrams are the only significant contributors. They grow with energy since the gauge boson longitudinal polarization terms by themselves would lead to a nonrenormalizable theory. At what energy would these terms violate unitarity for angular momentum $J = 0$, the channel where H^0 contributes, when the amplitudes are calculated perturbatively? The answer is about 1 TeV. Hence, before this energy, either the H^0 must contribute, restoring unitarity, or the interaction becomes strongly interacting so that the perturbative calculation breaks down. In either case, new physics will have to show up in the gauge boson-gauge boson scattering channel by 1 TeV.

12.3 ■ HADRON STRUCTURE AND SHORT DISTANCE SCATTERING

Rare, large-momentum transfer collisions are seen in pp and $p\bar{p}$ interactions at very high energies. The large transverse momenta are typically carried by jets of particles, with the properties of each jet rather similar to the light quark and gluon jets seen in e^+e^- annihilation to hadrons. In addition, heavy quark and antiquark pairs are produced with small cross sections. For the case of a large exchange of energy and momentum, the constituents of the proton or antiproton cannot coherently participate in the scattering process and we expect that we are seeing incoherent scattering of the individual point-like constituents of the parent hadrons. The interaction between these constituents is specified by QCD at a large q^2 scale. Since $\alpha_s(q^2)$ decreases at large q^2, this is the regime where perturbation theory should apply. In fact, the successful description of these scattering processes has been a major triumph of QCD.

12.3 Hadron Structure and Short Distance Scattering

To ensure a reasonably small coupling constant, the momentum transferred should be $\gtrsim 10\,\text{GeV}$. This is the perturbative regime where the simplest diagrams provide a rather good description of the physics. Thus, for example, the very rare process $gg \to t\bar{t}$ for top quark production is a very high q^2 process, and a rather good calculation of the rate in very high energy $p\bar{p}$ or pp collisions should be possible in terms of the probabilities for finding gluons of various momenta in the proton. This process should be the dominant way to produce $t\bar{t}$ in future pp colliders at center of mass energies larger than 10 TeV.

To calculate the dominant large-momentum transfer processes in $p\bar{p}$ or pp collisions, such as $gg \to q\bar{q}$ or $gg, gq \to gq, qq \to qq, q\bar{q} \to q\bar{q}$, or gg, which are analogues of the electromagnetic processes of Chapter 3 (with the important addition of the coupling between gluons), we have to specify how we end up in a color neutral system. For example, in a typical scattering process with large momentum transfer, the final state consists initially of colored remnants of the beam and target hadrons plus the colored scattered constituents. These are each widely separated in phase space and give rise to cascades of hadrons. The assumption that the longer time-scale hadronization process does not modify the overall rate works well. Thus the appropriate treatment of color, when calculating the scattering cross section, is to average the constituent cross section over colors for the initial constituents, reflecting the colorless initial hadrons, and sum over all colors for the final quarks and gluons.

An interesting check of how color should be treated can be made by comparing a strong and electromagnetic process. Figure 12.7 shows the diagrams for $q\bar{q}$ annihilation to $t\bar{t}$ (the dominant process for $t\bar{t}$ production at 2 TeV) and to large invariant mass e^+e^- pairs. The latter is called the Drell–Yan process. In both cases, the q and $\bar{q}$ must have the same flavor. For the e^+e^- final state, they must also have the same color, since the photon is neutral. For $t\bar{t}$, all final state colors contribute incoherently to the cross section. A consequence of color is that the factor of 3 in R for color in e^+e^- annihilation is replaced by a factor of $\frac{1}{3}$, compared to a calculation in a colorless theory, for color annihilation in the Drell–Yan cross section. These considerations are all verified by calculations that agree with the data.

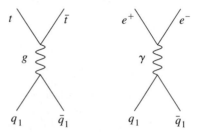

FIGURE 12.7 Diagram for $q_1\bar{q}_1$ annihilation through a gluon and through a photon. For the gluon, all colors of q_1 and $\bar{q}_1$ contribute; for the photon, only the colorless combination contributes.

The lowest-order two-body scattering cross sections can be calculated from the appropriate Feynman diagrams, where initial spins and colors are averaged and final spins and colors are summed over. As an example, the diagrams corresponding to gq and gg elastic scattering are shown in Figures 12.8 and 12.9, respectively.

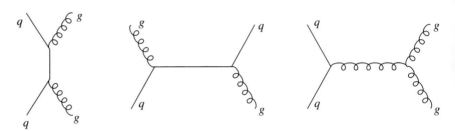

FIGURE 12.8 Lowest-order diagrams for quark-gluon scattering. The first two diagrams are the same as for Compton scattering; the third comes about because gluons couple to each other.

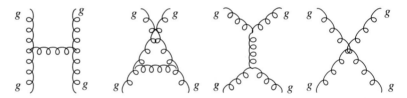

FIGURE 12.9 Lowest-order diagrams for gluon-gluon scattering. Time flows upward in the figure.

The differential cross section for each constituent pair, considering only light constituents, can be written as

$$\frac{d\sigma}{dt} = \frac{\pi \alpha_s^2}{s^2} F(s, t, u).$$

Here s, t, u are the Mandelstam variables for the constituent pair scattering process and all constituent masses have been ignored, as appropriate for high energies and light constituents. The function F is dimensionless, and is tabulated for the various scattering processes in Table 12.1. In order to get a feeling for the size of the various cross sections, the value of F for 90° scattering, where $t = u = -s/2$, is included in the table. Note that gg scattering has by far the largest cross section. The labels 1 and 2 indicate different light flavors.

12.3 Hadron Structure and Short Distance Scattering

TABLE 12.1 Cross Section Factors for Light Quarks and Gluons.

Process	$F(s, t, u)$	$F\left(s, -\dfrac{s}{2}, -\dfrac{s}{2}\right)$
$q_1 q_2 \to q_1 q_2$	$\dfrac{4}{9}\left[\dfrac{s^2+u^2}{t^2}\right]$	2.22
$q_1 \bar{q}_2 \to q_1 \bar{q}_2$	$\dfrac{4}{9}\left[\dfrac{s^2+u^2}{t^2}\right]$	2.22
$q_1 q_1 \to q_1 q_1$	$\dfrac{4}{9}\left[\dfrac{s^2+u^2}{t^2}+\dfrac{s^2+t^2}{u^2}\right]-\dfrac{8}{27}\dfrac{s^2}{ut}$	3.26
$q_1 \bar{q}_1 \to q_2 \bar{q}_2$	$\dfrac{4}{9}\left[\dfrac{t^2+u^2}{s^2}\right]$	0.22
$q_1 \bar{q}_1 \to q_1 \bar{q}_1$	$\dfrac{4}{9}\left[\dfrac{s^2+u^2}{t^2}+\dfrac{t^2+u^2}{s^2}\right]-\dfrac{8}{27}\dfrac{u^2}{st}$	2.59
$q_1 \bar{q}_1 \to gg$	$\dfrac{32}{27}\left[\dfrac{u^2+t^2}{ut}\right]-\dfrac{8}{3}\left[\dfrac{u^2+t^2}{s^2}\right]$	1.04
$gg \to q_1 \bar{q}_1$	$\left(\dfrac{3}{8}\right)^2 [q_1 \bar{q}_1 \to gg]$	0.15
$qg \to qg$	$\dfrac{u^2+s^2}{t^2}-\dfrac{4}{9}\left[\dfrac{u^2+s^2}{us}\right]$	6.11
$gg \to gg$	$\dfrac{9}{2}\left(3-\dfrac{ut}{s^2}-\dfrac{us}{t^2}-\dfrac{st}{u^2}\right)$	30.4

12.3.1 ■ Momentum Spectrum for Constituents

To use the *QCD* quark and gluon cross sections, we must specify the momentum spectrum for the incident constituents. We will take up this issue next. To be specific, we take the parent hadron to be a proton. If the proton were a nonrelativistic weakly bound system, each constituent would carry a fixed fraction of the proton 4-momentum (given by the ratio of the constituent mass to the total mass) in any frame. For a relativistic system it is more complicated; however, we will be able to make use of the limited momentum in the bound state to simplify the result at high energies.

Consider a constituent that at some moment in time has the 4-momentum $E, p_z, \vec{p}_t$ in the proton rest frame. The momentum components will change over time due to interactions, but with typical values $\sim \lambda_{QCD}$. Suppose we boost the

4-momentum to a high-momentum frame, for example, the center of mass of a subsequent collision. At high momentum, $\beta \simeq 1$, and we have for the transformed constituent 4-momentum:

$$E' = \gamma(E + p_z), \quad p'_z = \gamma(E + p_z), \quad \vec{p}'_t = \vec{p}_t.$$

Ignoring $\vec{p}'_t$ at large longitudinal momentum, the constituent carries a fraction of the total 4-momentum of the proton given by

$$x = \frac{E + p_z}{m_p}. \tag{12.6}$$

For a weakly bound nonrelativistic system, this gives the expected unique value for x. For the relativistic proton constituents, we expect a continuous spectrum of values for x, since p_z and E can be comparable. Thus, the constituent momenta are specified by a distribution function of the variable x for each type of constituent. We shall write these functions as $f_i(x)$, where i specifies the type of light quark or gluon. Energy conservation then requires

$$\int dx \sum_i x f_i(x) = 1. \tag{12.7}$$

We usually sum the probabilities over the color types, which must all be equally likely. Therefore, for a quark, $f_i(x)$ is the distribution for a flavor of type i with any color choice; for a gluon, the sum goes over the eight color choices, giving the probability to find a gluon of any color. The functions $f_i(x)$ depend on the details of the bound state and thus cannot be calculated from first principles in the absence of a detailed theory of the proton. They will have to be extracted from data and can then be used for a wide range of processes.

Consider the scattering of constituent number 1 from the beam particle on constituent number 2 in the target. We can calculate this in terms of the probabilities to find given fractional momentum constituents and their scattering cross sections. For fractional momenta x_1 and x_2, the invariant mass squared for the scattering system is $\hat{s} = x_1 x_2 s$, where s is the analogous quantity for the parent system. In this calculation we assume that all particle masses can be ignored. For a 2-jet final state, the constituent scattering differential cross section can be written as a function of the constituent momentum transfer $\hat{t}$ for a given $\hat{s}$. Therefore, we can write the differential cross section as

$$\frac{d\sigma}{dx_1 \, dx_2 \, d\hat{t}} = f_1(x_1) f_2(x_2) \frac{d\sigma}{d\hat{t}}(\hat{s}, \hat{t}, \hat{u}). \tag{12.8}$$

In this expression, $d\sigma/d\hat{t}$ must have the appropriate average over initial colors and sum over final colors. In addition, we typically do not separate events by jet type; so the expression has to be summed over all participating subprocesses, that is, over all choices for the constituent types. The various cross section factors were given in Table 12.1. What we need, in order to do a calculation, are the $f_i(x)$. We turn to this issue next.

12.4 ■ DEEP INELASTIC LEPTON-PROTON SCATTERING

To unravel the constituent structure of the proton, we need a probe whose structure is simple and well understood. Such a probe is provided by choosing a lepton of high energy and looking at scattering with $|q^2| \gg \lambda_{QCD}^2$. The interaction with the lepton involves the electroweak interactions, for which the lowest-order term is a good approximation because of the smallness of the coupling. We show in Figure 12.10 a few of the possible processes. We examine only $e^- p$ scattering at a momentum transfer such that $\lambda_{QCD}^2 \ll |q^2| \ll M_Z^2$; in this case, we need worry only about photon exchange. The other processes can be treated analogously to the $e^- p$ case. These reactions are called deep inelastic scattering, indicating that the momentum transfer is large. Deep inelastic $e^- p$ scattering is of great historical importance, having established for the first time that the proton has light constituents that interact in a point-like fashion. These constituents were subsequently identified as the light quarks and gluons, the objects needed to understand the spectrum of light hadronic states.

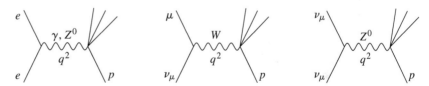

FIGURE 12.10 Several scattering processes using leptonic probes. The final hadronic state can contain many particles.

The final states produced at high energy in a deep inelastic ep event can contain many particles. Denoting a specific such state as $|n\rangle$, the matrix element for the process is, in the single photon exchange approximation:

$$M_{fi} = e^2 \bar{u}(p')\gamma_\mu u(p) \langle n| J_\mu |P\rangle / q^2. \tag{12.9}$$

Here the initial and final electron momenta are p and p', respectively; the proton momentum is P; and the charge e has been factored out of the matrix element of the hadronic current operator. However, focusing on a given final state $|n\rangle$ will obscure the basic physics, since the rate to a given state depends both on the basic constituent scattering and on the details of how the constituents re-form into hadrons. Summing over all final hadronic states, for a given value of the kinematic parameters of the scattered electron, allows a separation of the short-distance physics from the long-distance hadronization process. This is called an inclusive measurement.

Prior to looking at the expression for the inclusive cross section, consider the kinematics of the process. The scattering of an electron of a given initial energy on a proton can be described in terms of the electron scattering angle and its energy loss. These can be combined to calculate the 4-momentum transfer q. The invariant $\sqrt{-q^2}$ can be thought of as roughly the inverse of the distance scale probed by the virtual photon. For scales corresponding to 10^{-16} cm, the present experimental limit, the quarks and gluons appear to be point-like, whereas the proton structure is apparent at distances small compared to 10^{-13} cm.

From q and the initial proton momentum P, we can calculate two independent invariants for the virtual photon-proton system. The first is q^2. We often take $Q^2 = -q^2$, in order to use a positive quantity. The second independent invariant is $P \cdot q = m_p(E - E')$ in a frame where the proton is initially at rest. Thus the energy transferred in this frame, defined to be the variable $\nu = E - E'$, is also an invariant.

If we consider the scattering as due to the proton constituents, we can ask, what kinematic quantities describe a constituent in a given scatter? In a frame where the proton is initially moving at high momentum, we can solve for the final momentum of the constituent via the 4-momentum conservation relation:

$$p_{\text{constituent}}^{\text{final}} + p' = p_{\text{constituent}}^{\text{initial}} + p.$$

Using Eq. 12.6 and the momentum transfer q, this gives

$$p_{\text{constituent}}^{\text{final}} = xP + q.$$

Squaring this and assuming that we can ignore the constituent mass squared and the proton mass squared compared to the large quantity Q^2, gives the relation

$$x = \frac{-q^2}{2P \cdot q} = \frac{Q^2}{2P \cdot q} = \frac{Q^2}{2m_p\nu}. \tag{12.10}$$

The dimensionless variable x is called the Bjorken scaling variable. Scattering at a given Bjorken x provides a microscope with which to look for proton constituents carrying a fraction x of the momentum of the rapidly moving proton. The cross section will depend on the distribution function for finding such constituents.

Finally, we would like to ask, what invariant mass hadronic final states are produced in a given reaction? The 4-momentum of the entire hadronic final state is given by $P + q$. Squaring this, we get the hadronic final state invariant mass squared

$$m^2 = m_p^2 + 2P \cdot q + q^2 = m_p^2 + Q^2 \left(\frac{1}{x} - 1\right). \tag{12.11}$$

We see from this that $x = 1$ corresponds to elastic scattering and that $0 \le x \le 1$ in general, as expected from the interpretation as a fractional momentum for a constituent. We require Q^2 and m^2 to be $\gg m_p^2$ for the constituent picture to be valid.

12.4.1 ■ Cross Section for Deep Inelastic Scattering

We turn next to a calculation of the inclusive cross section. We do this first by an entirely general formulation based only on Lorentz invariance, parity conservation, and gauge invariance for the single-photon exchange term. We will then make the connection of the general result to the specific formulation from the constituent point of view.

We perform the calculation in the proton rest frame, the appropriate frame for the initial experiments that were performed. We also assume an unpolarized electron beam and an unpolarized proton target. Equation 12.9 gives for the differential cross section, summed over all hadronic final states:

$$d\sigma = \frac{(4\pi\alpha)^2}{2m_p 2E} \frac{d^3 p'}{2E'(2\pi)^3} \frac{L^e_{\mu\nu}}{q^4} \sum_n L^P_{\mu\nu}(2\pi)^4 \delta^4(p_f - p_i). \quad (12.12)$$

$L^e_{\mu\nu}$ is the square of the lepton current averaged over initial spins and summed over final spins. It is the same tensor that we calculated in electron-muon scattering in Chapter 3. Ignoring the electron mass at high energies:

$$L^e_{\mu\nu} = \tfrac{1}{2} \text{Tr}[\not{p}'\gamma_\mu \not{p} \gamma_\nu] = 2[p'_\mu p_\nu + p_\mu p'_\nu - p \cdot p' g_{\mu\nu}].$$

We leave the cross section differential in the electron scattering variables, which will be the independent variables, and sum only over the hadronic variables.

To arrive at an inclusive cross section, we sum the tensor involving the square of the proton electromagnetic current matrix element to a given final state over all final states that satisfy 4-momentum conservation, and average over the proton spin. Thus, $L^P_{\mu\nu}$ is $\langle n|J_\mu|P\rangle^* \langle n|J_\nu|P\rangle$ averaged over the proton spin; it depends on $|n\rangle$. $L^P_{\mu\nu}$ can then be summed over states $|n\rangle$ in the expression

$$\sum_n L^P_{\mu\nu}(2\pi)^4 \delta^4(p_f - p_i), \quad (12.13)$$

giving a tensor independent of $|n\rangle$. This tensor depends in general only on the vectors and invariants that describe the process. The sum over n includes the sum over the invariant phase space for each explicit set of final particles. From the delta function expression, $p_n = P + q$ fixes the 4-momentum of the final hadronic system, which is the same for all hadronic states that contribute.

We will write the expression in Eq. 12.13 as

$$4\pi m_p W_{\mu\nu} = \sum_n L^P_{\mu\nu}(2\pi)^4 \delta^4(p_f - p_i). \quad (12.14)$$

The historical constant $4\pi m_p$ is included to simplify the expressions that occur later. The tensor $W_{\mu\nu}$ satisfies the following conditions:

1. It is gauge invariant. Therefore $q_\mu W_{\mu\nu} = q_\nu W_{\mu\nu} = 0$.
2. It can be taken to be symmetric, since $L^e_{\mu\nu}$ is.
3. It is constrained by parity conservation of the electromagnetic interaction to transform under parity as a proper tensor of rank two.

4. It depends only on the two invariants q^2 and ν, on the vectors P and q, and the tensor $g_{\mu\nu}$.

The above constraints result in the following unique general form for $W_{\mu\nu}$:

$$W_{\mu\nu} = \left[-g_{\mu\nu} + \frac{q_\mu q_\nu}{q^2} \right] W_1(q^2, \nu)$$
$$+ \left(P_\mu - \frac{P \cdot q}{q^2} q_\mu \right) \left(P_\nu - \frac{P \cdot q}{q^2} q_\nu \right) \frac{W_2(q^2, \nu)}{m_p^2}. \quad (12.15)$$

W_1 and W_2 are Lorentz-scalar functions of the available invariants. The physics can therefore always be expressed in terms of these two functions.

At high energy we have, in the proton rest frame, $Q^2 = 4EE' \sin^2(\theta/2)$, where θ is the electron scattering angle. Multiplying together the following two tensors gives, in this frame,

$$L^e_{\mu\nu} W_{\mu\nu} = 4EE' \left(\cos^2 \frac{\theta}{2} W_2(q^2, \nu) + 2 \sin^2 \frac{\theta}{2} W_1(q^2, \nu) \right). \quad (12.16)$$

This calculation is facilitated by noting that the terms proportional to q_μ or q_ν in $W_{\mu\nu}$ do not contribute because of the gauge invariance of $L^e_{\mu\nu}$.

Using $d^3p' = p'E'dE'd\Omega \simeq E'^2 dE' d\Omega$, Eqs. 12.12, 12.14, and 12.16, the cross section is

$$\frac{d\sigma}{dE'd\Omega} = \frac{\alpha^2}{4E^2 \sin^4 \frac{\theta}{2}} \left(W_2(q^2, \nu) \cos^2 \frac{\theta}{2} + 2 W_1(q^2, \nu) \sin^2 \frac{\theta}{2} \right). \quad (12.17)$$

Measuring the electron scattering cross section at different initial and final energies and scattering angles allows a measurement of W_1 and W_2. These functions are called structure functions. The measured structure functions are not small as Q^2 gets large; this is expected for point-like constituents in the proton, as we show in the next section. The data indicate a number of other regularities, which can be more clearly explained once we introduce the calculation in terms of constituents.

12.4.2 ■ Structure Functions and Constituents

Since Eq. 12.15 is entirely general, we expect that we can write the two structure functions in terms of the properties and distribution functions for the constituents. We assume constituents of charge e_i (in units of e) and spin $\frac{1}{2}$ as expected for quarks. We state the resulting structure function formula for spin 0; this spin hypothesis gives a different result from the spin $\frac{1}{2}$ case and is ruled out by the data. The proton constituents were initially called partons. The picture of the scattering process where the partons are identified as quarks is called the quark-parton model.

12.4 Deep Inelastic Lepton-Proton Scattering

Define $f_i(\xi)$ as the probability distribution to find constituent i, with charge e_i, carrying a fraction ξ of the proton momentum in a frame in which the proton is moving rapidly. We will calculate $W_{\mu\nu}$ in such a frame. For point-like constituents, the scattering is elastic and the tensor $L_{\mu\nu}^P$ is calculated directly in terms of the current operator for the constituents. It is similar to the tensor for the scattered lepton, with the constituent momenta replacing the lepton momenta. Thus, with these assumptions:

$$W_{\mu\nu} = \frac{1}{4\pi m_p} \int \sum_i e_i^2 f_i(\xi)\, d\xi \left(\frac{2E}{2E_c}\right) L_{\mu\nu}^P \frac{d^3 p'_c}{2E'_c(2\pi)^3} (2\pi)^4 \delta^4(p_f - p_i). \tag{12.18}$$

Here p'_c is the final constituent 4-momentum, p_c the 4-momentum for the initial state of the constituent. The rate is proportional to e_i^2, since the interaction is electromagnetic, and the sum includes the incoherent contribution of all constituents. For each constituent the sum over final states is given by the integral over the phase space for a single final particle.

Note that a factor $2E/2E_c$ appears. This results from the use of covariant fields, implying a flux of $2E$ particles instead of one particle. Thus, when using the covariant proton fields, the probability density for incident constituents is actually $2E f_i(\xi)$. Similarly, to describe a scattering with one constituent in both the initial and final state, we have to multiply by

$$\left(\frac{1}{2E_c}\right)\left(\frac{1}{2E'_c}\right),$$

with the second of the two already included in the invariant phase space. Thus we add only the factor of $1/2E_c$.

A convenient first step in simplifying Eq. 12.18 is to add an integral over energies by replacing

$$\frac{d^3 p'_c}{2E'_c} \quad \text{with} \quad d^4 p'_c \delta(p'_c \cdot p'_c - m_c^2).$$

This allows an integration over the four energy-momentum delta functions, eliminating them. This gives then:

$$W_{\mu\nu} = \frac{1}{2m_p} \int \frac{E}{E_c} \sum_i e_i^2 f_i(\xi)\, d\xi\, L_{\mu\nu}^P \delta(p'_c \cdot p'_c - m_c^2),$$

where $p'_c = \xi P + q$. Therefore, ignoring m_c^2 and m_p^2 relative to Q^2, we have

$$p'_c \cdot p'_c - m_c^2 \simeq 2\xi(P \cdot q) - Q^2.$$

Integrating the last delta function against $d\xi$ results in a factor of $2(P \cdot q)$ in the denominator and gives the constraint that

$$\xi = \frac{Q^2}{2P \cdot q} = x.$$

Note that this result is an output of the calculation, giving again the relation in Eq. 12.10. The result for $W_{\mu\nu}$ is then

$$W_{\mu\nu} = \frac{1}{2m_p} \sum_i e_i^2 x f_i(x) \frac{L^P_{\mu\nu}}{xQ^2}. \qquad (12.19)$$

Eq. 12.19 is so far quite general and would apply to spin 0 constituents with appropriate choice of $L^P_{\mu\nu}$.

For spin $\frac{1}{2}$ constituents:

$$L^P_{\mu\nu} = 2\left[p'_{c_\mu} p_{c_\nu} + p_{c_\mu} p'_{c_\nu} - (p_c \cdot p'_c) g_{\mu\nu}\right],$$

analogous to the lepton tensor. Using this expression, we are ready to identify corresponding terms in the expressions for $W_{\mu\nu}$ given by Eqs. 12.19 and 12.15. To make this easy, we can leave out all terms proportional to q_μ or q_ν, since they give no contribution to the rate. They are present to enforce a gauge invariant expression. Therefore:

$$W_{\mu\nu} = -g_{\mu\nu} W_1 + \frac{P_\mu P_\nu}{m_p^2} W_2 + \text{terms involving } q_\mu \text{ or } q_\nu.$$

Similarly,

$$L^P_{\mu\nu} = 4x^2 P_\mu P_\nu - Q^2 g_{\mu\nu} + \text{terms involving } q_\mu \text{ or } q_\nu.$$

Therefore, using Eq. 12.19:

$$W_1 = \frac{1}{2m_p x} \sum_i e_i^2 x f_i(x),$$

$$W_2 = \frac{1}{\nu} \sum_i e_i^2 x f_i(x). \qquad (12.20)$$

We see several regularities:

1. $\nu W_2 = 2m_p x W_1$. This equality, called the Callan–Gross relation, is actually a unique prediction for spin $\frac{1}{2}$. An analogous calculation for spin 0 constituents gives the same result for νW_2, but $W_1 = 0$. The data support the spin $\frac{1}{2}$ assignment.

2. Both νW_2 and W_1 are functions of x only. In principle they could be functions of Q^2 and ν separately; instead, they are functions of the ratio. This prediction by Bjorken is called Bjorken scaling and is the signature of a point-like,

12.4 Deep Inelastic Lepton-Proton Scattering

quasi-free substructure in the proton. We will see later in the chapter that this prediction is violated by slowly varying logarithmic terms. It turns out that the logarithmic corrections are calculable in QCD, allowing an excellent check of the theory. The scaling assumption is a good first approximation.

For reference, we list a few alternative expressions that are used for the structure functions and cross section. To use functions that scale, W_2 and W_1 are replaced by structure functions

$$F_2 = \nu W_2 = \sum_i e_i^2 x f_i(x),$$

$$F_1 = m_p W_1. \qquad (12.21)$$

The Callan–Gross relation is then $F_2 = 2xF_1$.

The differential cross section is often expressed in terms of other variables. As an example,

$$dE' \, d\Omega = \frac{\pi}{EE'} \, dQ^2 \, d\nu$$

can be used to write the differential cross section

$$\frac{d\sigma}{dQ^2 \, d\nu}.$$

Another choice involves using dimensionless variables, for instance, x. Another is:

$$y = \frac{P \cdot q}{P \cdot p}, \qquad (12.22)$$

which for a proton initially at rest is ν/E. Both x and y can range between 0 and 1. Using the Callan–Gross relation and these variables, the cross section has the nice form:

$$\frac{d\sigma}{dx \, dy} = \frac{2\pi\alpha^2}{Q^4} s \left[1 + (1-y)^2\right] F_2. \qquad (12.23)$$

Here, at high energy, s is approximately $2m_p E$ in the proton rest frame, and we have ignored terms proportional to m_p/E. A measurement of the differential cross section allows the extraction of the structure function F_2.

12.4.3 ■ The Quark Picture for the Structure Function

The discussion of the structure functions has so far not made the connection to the picture of the proton as a bound state of two u quarks and one d quark. We now build these ideas into a parameterization of the structure functions. We will typically discuss the quantities

$$F_2(x) \quad \text{or} \quad \frac{1}{x} F_2(x) = \sum_i e_i^2 f_i(x).$$

Although we have discussed only ep scattering, very important measurements, tests, and constraints come from en scattering (measured using deuterons or other nuclei that have proton and neutron contributions), neutrino scattering on nucleons, and the production of $e^+ e^-$ (or $\mu^+ \mu^-$) with accompanying hadrons in pp or pn scattering. These reactions have different couplings to the constituents and allow a separation of the individual constituent terms contributing to the structure functions. All the data provide a beautiful confirmation of the picture outlined below. For the present discussion we will study ep and en results only.

Define the distribution function for the u quarks in the proton, which correspond to the u quarks in the bound state picture, as $u_v^p(x)$. Here p tells us we are looking at a proton and v stands for a "valence quark." The valence quarks account for the internal quantum numbers of the proton. Therefore, we require: $\int u_v^p(x) \, dx = 2$; $\int x u_v^p(x) \, dx$ is the fraction of the proton's momentum carried by the two u valence quarks. Similarly, the valence d quark distribution in the proton is $d_v^p(x)$. The simplest expectation is that $d_v^p(x) = \frac{1}{2} u_v^p(x)$.

For a very weakly interacting system, the valence quarks would account for the entire structure function, in which case F_2^p would equal $x[\frac{4}{9} u_v^p + \frac{1}{9} d_v^p]$. However, QCD is a field theory allowing the creation of particles, and it is also strongly interacting on the scale of the proton radius. Thus we expect to find gluons, as well as valence quarks, if we look at the proton on a short-distance scale. The gluons do not interact electromagnetically and appear as a noninteracting neutral component that carries some fraction of the proton's momentum. However, the gluons can also create pairs, which are electrically charged. This component interacts electrically and provides constituents that are called sea quarks and antiquarks. These are denoted $u_s^p(x)$, $\bar{u}_s^p(x)$, $d_s^p(x)$, $\bar{d}_s^p(x)$, $s_s^p(x)$, and $\bar{s}_s^p(x)$, for u, d, and s quarks and antiquarks, respectively. They satisfy relations following from production in pairs, for example,

$$\int u_s^p(x) \, dx = \int \bar{u}_s^p(x) \, dx.$$

The simplest assumption is that

$$u_s^p(x) = \bar{u}_s^p(x) = d_s^p(x) = \bar{d}_s^p(x) = s_s^p(x) = \bar{s}_s^p(x);$$

although $SU(3)$ violation makes it likely that the strange quark and antiquark distribution are suppressed relative to the lighter constituents. Thus the simplest approximation involves one valence distribution function and one sea distribution function.

The structure functions extracted from the data, which we turn to shortly, indicate a few complexities that are nonperturbative in origin:

12.4 Deep Inelastic Lepton-Proton Scattering

1. In the proton, $d_v^p(x)$ does not equal $\frac{1}{2}u_v^p(x)$ as $x \to 1$. It turns out that only a u quark can be found at very large x, indicating that, for the remaining valence quarks, an isospin 0 ud state is preferred in this configuration. Apparently, the probability for a uu state accompanying a d quark that has x near 1 is suppressed.
2. The probability of finding u and d sea quarks are not equal. The integral of the function $\bar{d}_s^p(x)$ is larger than that for $\bar{u}_s^p(x)$. The valence and sea quarks are not entirely independent of each other.
3. $SU(3)$ violation suppresses strange quarks.

We use a few simple forms to fit the data in terms of the valence and sea distribution functions. For example, for each type of valence quark a function of the form $A(1-x)^n$ is a good first approximation, with a more complicated polynomial used for higher-precision fitting. For a sea quark distribution,

$$\frac{B}{x}(1-x)^m$$

is a good first approximation, where again a polynomial can be used for more accuracy. Since the sea quarks arise through a bremsstrahlung and pair production mechanism, they have on average much smaller x values than the valence quarks. Hence we expect, and in fact find, that $m \gg n$. As $x \to 0$ the sea quarks dominate $F_2(x)$ and lead to a nonzero value, as seen in the data.

Figure 12.11 shows F_2^p versus x for a Q^2 of approximately 5 GeV^2. The cross section data are taken using muon beams, instead of electron beams, which should

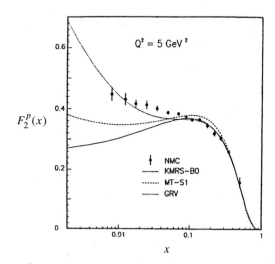

FIGURE 12.11 $F_2^p(x)$ determined in muon-proton scattering. Curves are predictions of several models that were formulated prior to the data collection. The logarithmic scale in x allows better visualization of the region at small x. [From P. Amaudruz et al., *Phys. Lett.* B295, 159 (1992).]

give the same structure functions. F_2^p is large for x values out to about 0.2 and then drops rapidly. A very large range of data for F_2^p are displayed in Figure 12.12. The data clearly indicate that scaling is only an approximation. The solid lines in this figure are the impressive fits to the Q^2 changes in F_2^p based on *QCD*, which we will discuss later in the chapter.

The use of neutrons as the scattering target provides a nice additional data set, where isospin symmetry provides useful extra constraints. In particular,

$$u_v^p(x) = d_v^n(x),$$
$$d_v^p(x) = u_v^n(x),$$

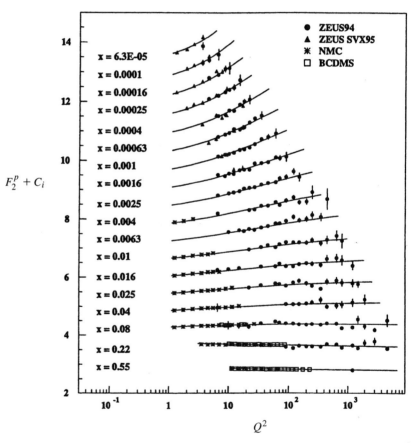

FIGURE 12.12 The proton structure function at fixed x for a very large range in Q^2. Curves are fits based on *QCD*. Data come from a number of experiments. For clarity, an amount $C_i = 13.6 - .06i$ is added to F_2^p, where $i = 1(18)$ for the lowest (highest) x-value, to allow each curve to be followed separately. [From J. Breitweg et al., *Eur. Phys. J.* C7, 609 (1999).]

12.4 Deep Inelastic Lepton-Proton Scattering

with analogous relations for the sea quarks. Figure 12.13 shows the ratio of F_2^n to F_2^p. We see that

$$\frac{F_2^n(x \to 0)}{F_2^p(x \to 0)} = 1,$$

as expected from an identical sea quark dominated region at very low x, and

$$\frac{F_2^n(x \to 1)}{F_2^p(x \to 1)} = \frac{1}{4}.$$

This ratio corresponds to the ratio of the d quark to the u quark charge squared, a result expected if the region $x \to 1$ can only have a u quark in the proton, leading to only a d quark in the neutron by isospin symmetry.

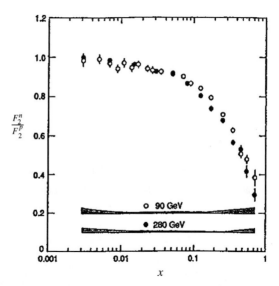

FIGURE 12.13 Ratio of F_2^n/F_2^p versus x. Data were taken with muon beams of 90 GeV and 280 GeV incident on proton or deuterium targets. Bands in the figure indicate the systematic error versus x for each beam energy. [From P. Amaudruz et al., *Nucl. Phys.* B371, 3 (1992).]

Using the quark charges and assuming an identical sea quark distribution for the proton and neutron, we can use isospin symmetry to obtain

$$\left[F_2^p(x) - F_2^n(x)\right] = \frac{x}{3}\left(u_v^p(x) - d_v^p(x)\right). \tag{12.24}$$

The data for F_2^p minus F_2^n are shown in Figure 12.14. The difference peaks around 0.3, as expected naively for three quarks that share the nucleon momentum. In the

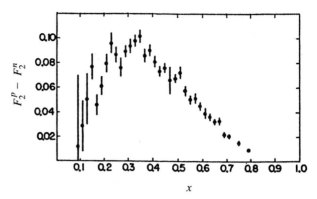

FIGURE 12.14 Difference of $F_2^p - F_2^n$ versus x. Data were taken using electrons on proton or deuterium targets. [From A. Bodek et al., *Phys. Rev. Lett.* 30, 1087 (1973).]

approximation where $d_v^p(x) = \frac{1}{2}u_v^p(x)$, and ignoring sea quarks, we expect from Eq. 12.24 that $3[F_2^p(x) - F_2^n(x)] = F_2^p(x)$. This is roughly true for $0.25 \leq x$, indicating that in this region we find valence quarks predominantly. Relaxing the assumption about identical sea quark distributions in the proton and neutron, we can derive the interesting result:

$$\int_0^1 \left(F_2^p(x) - F_2^n(x)\right) \frac{dx}{x} = \frac{1}{3} - \frac{2}{3}\int_0^1 \left(\bar{d}_s^p(x) - \bar{u}_s^p(x)\right) dx.$$

The integral over the measured structure functions is 0.24 ± 0.03, indicating that the light-quark sea distributions are not equal for d and u.

Finally, if we ignore the strange sea quarks, we can use isospin symmetry and the data to derive the momentum fraction carried by gluons. Defining the fractional momentum carried by the sum of quarks and antiquarks of a given flavor:

$$f_{u,\bar{u}}^p = \int dx\, x \left[u_v^p(x) + u_s^p(x) + \bar{u}_s^p(x)\right],$$

with $f_{d,\bar{d}}^p$ the analogous quantity for d and $\bar{d}$ quarks, can calculate:

$$\int dx\, F_2^p(x) = \frac{4}{9}f_{u,\bar{u}}^p + \frac{1}{9}f_{d,\bar{d}}^p,$$

$$\int dx\, F_2^n(x) = \frac{1}{9}f_{u,\bar{u}}^p + \frac{4}{9}f_{d,\bar{d}}^p. \qquad (12.25)$$

The measured values of these two integrals are 0.18 and 0.12, respectively. We can then solve for each fraction, getting the result:

$$f_{u,\bar{u}}^p = 0.36, \quad f_{d,\bar{d}}^p = 0.18.$$

Assuming that gluons carry most of the rest of the momentum gives a fraction:

$$f_g^p = 0.46.$$

This is a large number; gluons carry nearly half the momentum at Q^2 values $\simeq 5\ \text{GeV}^2$, which are typical values for the data used.

12.5 ■ SCALING VIOLATIONS

In the scaling picture of the structure functions, the scattering of either valence or sea quarks by the virtual photon results in a final state that consists of the scattered quark and the proton remnant. Both of these are colored and result in jets of hadrons after hadronization. The final state structure therefore consists of two jets.

The *QCD* interactions can, however, provide additional dynamic processes. As an example, the quark that absorbs the virtual photon can emit a gluon leading to a three-jet final state. This is an order α_s correction to the scaling cross section, analogous to the α_s correction to the formula for R in e^+e^- annihilation. It can, however, be included within the structure function framework given by Eq. 12.15, which was completely general. The question is, however, how does it modify the very nice picture of these functions in terms of constituent distribution functions, for example, as given in Eq. 12.21, which led to the scaling behavior?

The diagrams for the lowest-order *QCD* corrections to the simple quark constituent picture are shown in Figure 12.15. Note that the diagrams involving an initial state gluon allow a measurement of the gluon distribution function, which does not contribute directly at lowest order. For sufficiently large Q^2, the short-distance behavior of *QCD* makes a perturbative calculation of these diagrams a good approximation. Using perturbative *QCD*, we can predict violations of the simple scaling behavior as well as the distribution and rate for multi-jet events. Then we can compare these predictions to experimental data and validate the dynamics of *QCD* at short distances. The data also provide a good measurement of $\alpha_s(Q^2)$. This turns out to give one of the most accurate measurements of this parameter. It agrees nicely with the other determinations, such as that obtained from the value of R discussed earlier in the chapter.

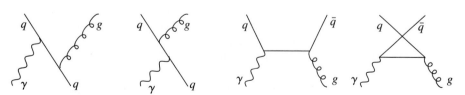

FIGURE 12.15 Order α_s corrections to simple constituent picture. The lepton that emits the virtual photon is not shown.

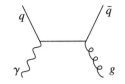

FIGURE 12.16 Contribution of gluons to quark scattering by the photon.

In order for the higher-order QCD processes to make a significant contribution of order α_s, the kinematics of the various initial constituent particles are constrained to be quasi-one-dimensional in a frame where the proton moves at very large momentum. Thus, the higher-order processes share some features with large cross section processes, particularly the pair-production process discussed in Chapter 11. To see how this works, we take as an example the gluon initiated diagram, where the constituent that absorbs the virtual photon is a light quark. This is shown in Figure 12.16.

For the gluon 4-momentum in the diagram, we take a value that is y times the proton 4-momentum. We specify the fraction of the gluon's momentum carried by the quark as z. The antiquark carries a fraction $1 - z$. The quark therefore carries a fraction zy of the proton's momentum. We keep the notation that x is the fractional momentum for the quark that absorbs the virtual photon; therefore $x = zy$. We assume that the various energies and longitudinal momenta are large compared to the constituent particle masses and the transverse momentum generated in the QCD part of the process. These can then be ignored to first approximation, allowing the constituent momenta to be defined in terms of fractions of the parent particle momentum. It also allows decoupling, to good approximation, of the longitudinal and transverse degrees of freedom in integrals over phase space. These features are similar to approximations used in the pair-production calculation in Chapter 11.

Next, we calculate the cross section in a coordinate system in which the proton momentum, taken to be very large, and the momentum carried by the virtual photon are collinear, defining the longitudinal direction. In this frame, the constituents have the following 4-momenta, where we now include the transverse momentum generated through the diagram of Figure 12.16 as a small correction:

$$p_g = (E_g, E_g, 0),$$

$$p_q = \left(zE_g, zE_g - \frac{|\vec{p}_t^2|}{2zE_g}, \vec{p}_t\right),$$

$$p_{\bar{q}} = \left((1-z)E_g, (1-z)E_g - \frac{|\vec{p}_t^2|}{2(1-z)E_g}, -\vec{p}_t\right).$$

Here we ignore the quark and antiquark masses. We will use these expressions in the following calculation of the contribution to the rate for this diagram. In calculating a rate, the amplitude corresponding to the Feynman diagram for Figure 12.16 has to be squared, with appropriate averages and sums over spins and colors taken.

This calculation involves four elements that we consider in turn.

QCD vertex. For this term in the calculation, we want the contribution from the $g \rightarrow q\bar{q}$ part of the diagram given in Figure 12.16. A calculation of this vertex factor gives a function of z and $|\vec{p}_t|^2$. If we consider the situation of small $\vec{p}_t$, we

12.5 Scaling Violations

can expand this function in terms of a Taylor series around $\vec{p}_t = 0$. Because of gauge invariance, which requires that only the two transverse gluon polarizations contribute, this expansion gives a term of the form $|\vec{p}_t|^2$ times a function of z. This result is true for all of the *QCD* vertices in the diagrams in Figure 12.15.

To calculate the contribution to the rate, first square the quark-antiquark current and sum over spins. The result is the same as that given in Section 3.3.2:

$$4(p_\mu{}^q p_\nu{}^{\bar{q}} + p_\mu{}^{\bar{q}} p_\nu{}^q - g_{\mu\nu} p^q \cdot p^{\bar{q}}).$$

Next, dot this result with the tensor resulting from averaging over gluon polarizations, which is

$$\tfrac{1}{2}(\hat{e}_y \hat{e}_y + \hat{e}_z \hat{e}_z),$$

where the transverse directions are chosen to be the y and z directions. After the dot product, this gives

$$2|\vec{p}_t|^2 \left(\frac{1-z}{z} + \frac{z}{1-z} \right).$$

This expression must be multiplied by the strong coupling constant squared, $4\pi\alpha_s$, and averaged over initial and summed over final colors. We return to the color average at the end of the calculation.

The propagator. The next piece of the diagram is the quark propagator. We look only at the denominator in the propagator and assume that the spinor terms arising at each vertex are lumped with the given vertex, as we did with the *QCD* vertex. The denominator in the propagator is given by:

$$(p_g - p_{\bar{q}})^2 - m_q^2 = -2 p_g \cdot p_{\bar{q}}.$$

In the frame where the x axis is defined by the gluon direction, the dot product is:

$$2E_g(E^{\bar{q}} - p_x^{\bar{q}}) \simeq \frac{E_g}{E^{\bar{q}}} \left(|\vec{p}_t|^2 + m_q^2 \right) = \frac{1}{(1-z)} \left(|\vec{p}_t|^2 + m_q^2 \right).$$

This has to be squared, yielding a factor in the overall rate calculation (ignoring m_q^2) of

$$\frac{(1-z)^2}{|\vec{p}_t|^4}.$$

Combine this with the factor of $|\vec{p}_t|^2$ from the *QCD* vertex to get an overall factor of $1/|\vec{p}_t|^2$. This factor results in a cross section that favors low transverse momentum.

Sum over states for the constituent that doesn't scatter. To calculate the rate, we must sum over the antiquark final states. This is given by

$$\frac{d^3 p^{\bar{q}}}{(2\pi)^3 2 E^{\bar{q}}},$$

which is

$$\frac{dz \, d\, |\vec{p}_t|^2}{16\pi^2 (1-z)}$$

after integrating over the azimuthal angle.

In addition, we have a factor of

$$\frac{2E^q}{2E_g} = z$$

from the state normalization, when we go from an incident quark to an incident gluon. This allows the final quark scattering term to be normalized as if the quark were the incident particle.

The electromagnetic vertex where the quark scatters. If we ignore the transverse momentum, compared to the large longitudinal momentum, as we did earlier with the transverse momentum stemming from the proton wave-function, then this vertex involves the interaction of a quark of fractional momentum zy with the virtual photon. This interaction gives a contribution to F_2 at a value of x given by zy. In this approximation, the contribution at the electromagnetic vertex is the same as the lowest-order quark contribution at the given x, outlined in Section 12.4.2. The contribution to the structure function of this diagram therefore satisfies the Callan–Gross relation, since this relation works for any x value. Therefore, the diagram in Figure 12.16 provides a modification of the density function for finding a quark in the proton, since it contributes the same way as the initial density function. This key idea allows the higher-order corrections to be included in the distribution functions, although it results in slow Q^2 violations of scaling. In fact, these considerations are independent of which probe scatters the quark, and work for other kinds of short-distance interactions.

A few additional comments concern the above structuring of the calculation. We considered only one of the two diagrams initiated by the $g \to q\bar{q}$ process. Treating the two contributions, where the q or $\bar{q}$ absorb the virtual photon, as separate with separate rates, is justified provided Q^2 is large, since the two processes then do not overlap in the phase-space for the final state. Thus the contribution of the diagram where the virtual-photon is absorbed by the quark can be added to the quark contribution to the structure function; the diagram where the antiquark absorbs the photon can be added to the antiquark contribution to the structure function.

Our calculation assumes the production of light quarks. The same diagrams can, however, be used to calculate the rate for heavy quark and antiquark production explicitly, in terms of the gluon structure function. Such a calculation gives a good prediction of these small contributions to the cross section. Finally, we could keep the transverse momentum in the calculation of the matrix element

12.5 Scaling Violations

at the virtual-photon quark vertex. In this case, the resulting cross section yields structure functions that violate the Callan–Gross relation. These small violations are in fact found in the data. They also result from the transverse momentum stemming from the proton wave-function, which we also ignored.

We next multiply the four factors in the quasi-one-dimensional approximation for the order α_s calculation. If we write the photon-quark cross section as $\sigma_0(x)$, the contribution of the diagram in Figure 12.16, for a fixed y value, can be written as

$$\frac{d\sigma}{dz\, d|\vec{p}_t|^2} = \sigma_0(x) \frac{\alpha_s}{2\pi} \frac{1}{|\vec{p}_t|^2} f(z). \tag{12.26}$$

The strong coupling constant is factored out explicitly and $f(z)$ is the product of the various z-dependent terms. The fractional momentum $x = zy$, and the cross section still needs to be integrated over the distribution function for various y values. Including the color averaging, $f(z)$ is

$$f(z) = \tfrac{1}{2}\left[z^2 + (1-z)^2\right]. \tag{12.27}$$

The factor of $\tfrac{1}{2}$ in Eq. 12.27 arises because of color and is called a color factor. Each diagram in Figure 12.15 has such a factor arising from averaging over initial, and summing over final, colors. These factors can be calculated from the color interaction discussed in Section 4.5, in particular Eq. 4.7. Thus for given colors, the $g \to q\bar{q}$ interaction is proportional to T^α_{nm}, where α specifies the gluon color and n, m the quark and antiquark colors. When the matrix element is squared and then summed over final colors, we get a factor:

$$\sum_{n,m} T^\alpha_{nm} T^\alpha_{mn}.$$

Here we use the fact that the generators are Hermitian (the index α, specifying the gluon type, is not summed over). The sum above is just

$$\mathrm{Tr}\left[T_\alpha^2\right].$$

Using the general formula,

$$\mathrm{Tr}\left[T_\alpha T_\beta\right] = \frac{\delta_{\alpha\beta}}{2},$$

gives a factor of $\tfrac{1}{2}$. Since this is true for each gluon, averaging over gluon colors gives the same result. Contrast this result with what we would get for a transition of the gluon to a colorless single particle state, for example, a single vector meson. In this case the colors interfere in the amplitude and the amplitude is proportional to

$$\frac{1}{\sqrt{3}} \sum_n T^\alpha_{nn} = 0,$$

since the generators are traceless. The color factors weight the different contributions in Figure 12.15 differently and are explicit predictions of the theory.

In order to complete the cross section calculation, we return to Eq. 12.26 and integrate over $|\vec{p}_t|^2$ first. The upper limit in this integral occurs for 90° scattering for which $|\vec{p}_t|^2 = \hat{s}/4$, where $\hat{s}$ is the square of the center of mass energy. We can calculate $\hat{s}$ by squaring the initial 4-momentum, which is $yP + q$. Ignoring the proton mass, this gives a maximum value of

$$|\vec{p}_t|^2 = \frac{Q^2}{4}\left(\frac{1}{z} - 1\right).$$

Integrating over $|\vec{p}_t|^2$ gives then a factor

$$\log\left[\frac{Q^2}{\mu^2(z)}\right]. \tag{12.28}$$

The lower cutoff of the integral includes the z-dependent factors and depends on physics at the scale $\sim \lambda_{QCD}$. This includes, for a more exact calculation, the particle masses that were omitted in the earlier equations as well as transverse momentum in the bound state. Since these nonperturbative features are not completely understood, it seems difficult to complete the calculation and obtain the cross section contribution of the diagram in Figure 12.16. Note, however, that we do not know a priori what the initial constituent distributions are, so it is difficult to attribute a given fraction of the cross section to Figure 12.16. Instead, we combine the data with the theory in order to calculate the change in the distribution with Q^2 as opposed to its exact value. The result in Eq. 12.28 tells us how this diagram contributes as we go from a given Q^2 to a slightly higher Q^2. This depends only on the Q^2 part of the logarithm with dependence on $\mu^2(z)$ dropping out. Thus what we actually calculate is the evolution of the structure functions with Q^2, which is a logarithmic effect. This effect provides the violation of the scaling behavior, with structure functions no longer dependent only on x, as seen in Figure 12.12.

12.5.1 ■ Evolution Equations for Structure Functions

We will rephrase the results above from the point of view of the consituent distribution functions. As opposed to writing the α_s contribution as a separate cross section, we use it to calculate the Q^2 evolution of the constituent distribution functions. These distributions are expressed as functions of x and Q^2, so we need first to sum over all y and z that contribute to a given x value in Eq. 12.26. This sum is weighted by the probability of finding a gluon with a given y value. Taking $g(y, Q^2)$ as the gluon distribution function, this will give, for the cross section, the integral

$$\sigma(x, Q^2) = \int_0^1 dy \int_0^1 dz\, g(y, Q^2)\, \delta(x - zy)\sigma_c(z, Q^2),$$

12.5 Scaling Violations

where $\sigma_c(z, Q^2)$ is the cross section in Eq. 12.26, integrated over the transverse momentum. Integrating over the delta function gives the result:

$$\sigma(x, Q^2) = \int_x^1 \frac{dy}{y} g(y, Q^2) \sigma_c\left(\frac{x}{y}, Q^2\right). \tag{12.29}$$

Only y values between x and 1 can contribute to a given x.

Using Eq. 12.29, we can calculate the change with Q^2 of the cross section. Rewriting this in terms of the constituent distribution functions and including the contribution from all the diagrams in Figure 12.15 for the quarks, in addition to the result for $g \to gg$ (which modifies the gluon distribution), results in a set of differential equations:

$$\frac{dq_i(x, Q^2)}{d(\log Q^2)} = \frac{\alpha_s(Q^2)}{2\pi} \int_x^1 \frac{dy}{y} \left[P_{qq}\left(\frac{x}{y}\right) q_i(y, Q^2) + P_{qg}\left(\frac{x}{y}\right) g(y, Q^2) \right],$$

$$\frac{dg(x, Q^2)}{d(\log Q^2)} = \frac{\alpha_s(Q^2)}{2\pi} \int_x^1 \frac{dy}{y}$$

$$\times \left[\sum_j P_{gq}\left(\frac{x}{y}\right) q_j(y, Q^2) + P_{gg}\left(\frac{x}{y}\right) g(y, Q^2) \right]. \tag{12.30}$$

These are called the DGLAP (Dokshitzer, Gribov, Lipatov, Altarelli, Parisi) equations and provide the evolution equations for the fractional momentum distributions for the light quarks or antiquarks (both indicated as $q_i(x, Q^2)$, so that i runs from 1 to 6 for three light flavors) or gluons (indicated as $g(x, Q^2)$). Both valence and sea contributions to a given flavor are lumped together in the $q_i(x, Q^2)$. The functions P_{qq}, P_{qg}, P_{gq}, and P_{gg} are called splitting functions and are just the generalization for the various diagrams of the function $f(z)$ (which is $P_{qg}(z)$ in Eq. 12.27). The subscripts indicate both the initial constituent and the final constituent, which can then scatter. The splitting functions can be calculated from the various Feynman diagrams. They satisfy a number of relations that follow from momentum conservation at the vertices

$$P_{qq}(z) = P_{gq}(1-z),$$
$$P_{qg}(z) = P_{\bar{q}g}(1-z) = P_{qg}(1-z),$$
$$P_{gg}(z) = P_{gg}(1-z).$$

The gluon splitting through the term P_{qg} leads to a significant increase of $F_2(x, Q^2)$ at small x, as Q^2 increases. At large x, $F_2(x, Q^2)$ decreases with Q^2 due to gluon radiation, the P_{qq} term. At intermediate x, $0.1 \le x \le 0.3$, the various terms compensate each other approximately and $F_2(x, Q^2)$ is a very weak function of Q^2 for $5 \le Q^2 \le 500$ GeV^2.

In practice, the constituent distribution functions have to be fit to data at some modest Q^2, using a model such as the quark-parton model. The evolution with Q^2 is then given by the DGLAP equations. The constituent distribution functions, now including the logarithmic scale variation resulting from Eq. 12.30, can then be used in all hard scattering processes, for example, in scattering of constituents for pp or $\bar{p}p$ interactions, in addition to providing excellent fits to the deep inelastic data.

12.6 ■ RESULTS FOR LARGE TRANSVERSE MOMENTUM SCATTERING IN $\bar{p}p$ REACTIONS

We now return to the production of high transverse momentum jets in hadron scattering processes. We look explicitly at the case of $\bar{p}p$ interactions at a center of mass energy of 1.8 TeV, where a substantial amount of data exists. The constituent cross sections are given in Table 12.1 and we extract the constituent distribution functions from deep inelastic scattering data, as described in the previous section. The data for the differential jet cross section with transverse momentum are shown in Figure 12.17. Since the detector measures energy, the energy is used to approximate the momentum, so the variable is called the transverse energy. The cross section is integrated over an angular region specified by the rapidity range $0.1 \leq |y| \leq 0.7$.

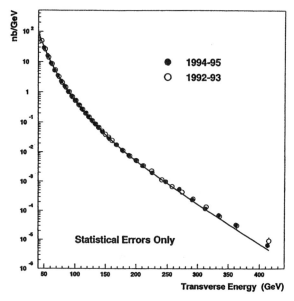

FIGURE 12.17 Inclusive jet cross section in 1.8 TeV $p\bar{p}$ scattering compared to QCD expectations. Data come from two different data-taking periods, as indicated in the figure. [From T. Affolder et al., *Phys. Rev.* D64, 032001 (2001).]

12.7 The Next Frontier

The data in Figure 12.17 cover a range of transverse momentum from about 50 GeV to 400 GeV. The cross section falls by seven orders of magnitude over this range and is nicely described by the *QCD* prediction. The measured jets come from a variety of constituent scattering processes. Figure 12.18 shows the expected parentage of the jets as a function of transverse momentum. The figure shows that gg scattering dominates for jets where the transverse momentum is significantly smaller than 5% of the beam energy. For jets where the transverse momentum is larger than 30% of the beam energy, the valence quarks dominate the rate.

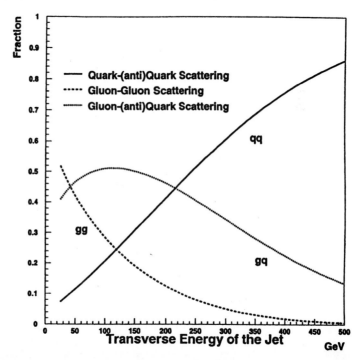

FIGURE 12.18 Fractional contribution of each subprocess versus transverse jet energy for the jet cross section in Figure 12.17. [From T. Affolder et al., *Phys. Rev.* D64, 032001 (2001).]

12.7 ■ THE NEXT FRONTIER

The data presented in this chapter, and methods to interpret the data, represent several decades of intensive work. The results are now comfortably part of the Standard Model. The continued study of physics at higher-mass scales and shorter distances will come with the next accelerator, which promises a center of mass energy of 14 TeV using *pp* collisions. This accelerator should, in particular, allow a large step forward in our search to understand electroweak symmetry breaking,

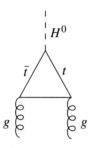

FIGURE 12.19 Gluon fusion diagram produces an H^0 via a $t\bar{t}$ loop.

for example, through discovery of the H^0. This follows from our earlier statement that information about the electroweak symmetry breaking mechanism must lie in the mass range from 100 GeV to 1 TeV. The processes we discussed, particularly QCD processes yielding high transverse momentum jets, will become backgrounds hindering the search for new phenomena. It is therefore quite important to understand these reactions.

Processes that can produce new phenomena, such as the H^0, will be analogs of Figures 12.5 and 12.6, where quarks and antiquarks replace the leptons, since the pp collider is being used as a quark and gluon collider. In the case of the H^0, the coupling to $t\bar{t}$ is particularly large because of the large value of m_t. This allows an additional diagram specific to a hadron collider, shown in Figure 12.19. The contribution to the cross section from this diagram is particularly large for a light H^0, where gluons with fractional momentum $\sim 1\%$ of the parent proton are involved. The combination of these processes should allow discovery of the H^0, if it exists. In fact, the extensive studies of how the H^0 would be produced, and how the signal for it would be isolated, have determined the energy and luminosity for the collider.

CHAPTER 12 HOMEWORK

12.1. Compare the fermionic cross sections for annihilation to a photon or a gluon at high energies. We will ignore all particle masses below.

(a) Consider the annihilation reaction $e^+e^- \to \mu^+\mu^-$ via a virtual photon, as discussed in Chapter 3. Show that you can write the differential cross section as

$$\frac{d\sigma}{dt} = \frac{2\pi\alpha^2}{s^2}\left[\frac{t^2+u^2}{s^2}\right]$$

(b) Consider the annihilation of quark and antiquark flavors $q_1\bar{q}_1 \to \mu^+\mu^-$ via a virtual photon. Assume that each q_1 and $\bar{q}_1$ come in an incoherent mix of the three colors. Show that the differential cross section is

$$\frac{d\sigma}{dt} = \frac{2\pi\alpha^2}{3s^2}\left[\frac{t^2+u^2}{s^2}\right]e_i^2.$$

The quark charge is e_i in units of e.

(c) Consider the annihilation of a quark and antiquark flavor pair to a different flavor pair $q_1\bar{q}_1 \to q_2\bar{q}_2$ via a virtual gluon. Show that the color part of the amplitude is given by the expression

$$\sum_\alpha T_\alpha^{kl} T_\alpha^{mn},$$

where k, l and m, n are given by the colors of the final and initial quark-antiquark pairs, respectively. The variable α runs over the eight gluons.

(d) Averaging over initial colors and summing over final colors, show that the differential cross section for the process in (c) is:

$$\frac{d\sigma}{dt} = \frac{4\pi\alpha_s^2}{9s^2}\left[\frac{t^2+u^2}{s^2}\right].$$

12.2. Suppose the proton were made of charged constituents of charge e_i and spin zero. Assume that the constituent distribution function for fractional momentum x is given by $f_i(x)$. Find the expression for the two structure functions in Eq. 12.15 in this case, assuming that scaling is a good assumption. Note that the electromagnetic current matrix element for spin zero is the sum of the initial and final constituent 4-momenta.

12.3. Verify the formula in the text:

$$\int \frac{1}{x}\left[F_2^p(x) - F_2^n(x)\right]dx = \frac{1}{3} - \frac{2}{3}\int\left[\bar{d}^p(x) - \bar{u}^p(x)\right]dx.$$

12.4. One of the *QCD* corrections to scaling involves the vertex $q \to q + g$. Show that the color factor for this vertex is $\frac{4}{3}$.

12.5. Consider the deep inelastic processes $\nu_\mu p \to \mu^- + X$ and $\bar{\nu}_\mu p \to \mu^+ + X$. The final state X will be summed over so that we are looking at an inclusive measurement. Ignoring Cabibbo suppressed transitions and strange quarks in the proton, which constituent scattering processes contribute to each of the two cross sections to lowest order?

12.6. Consider a quark doublet pair $\binom{U}{D}$, as well as the corresponding antiquark pair. Assuming that the quark current matrix elements are identical to those for the lepton doublets, show that the lowest-order differential cross sections for the following processes at very high energy are:

(a) $\nu_\mu D \to \mu^- U$: $\dfrac{d\sigma}{dy} = \dfrac{G_F^2 s}{\pi}$

(b) $\nu_\mu \bar{U} \to \mu^- \bar{D}$: $\dfrac{d\sigma}{dy} = \dfrac{G_F^2 s}{\pi}(1-y)^2$

(c) $\bar{\nu}_\mu U \to \mu^+ D$: $\dfrac{d\sigma}{dy} = \dfrac{G_F^2 s}{\pi}(1-y)^2$

(d) $\bar{\nu}_\mu \bar{D} \to \mu^+ \bar{U}$: $\dfrac{d\sigma}{dy} = \dfrac{G_F^2 s}{\pi}.$

The variable y is the invariant defined in the text and varies from 0 to 1; s is the center of mass energy squared for the neutrino-quark system. Can you figure out a general reason why some of the cross sections go to zero at $y = 1$ and why some don't?

12.7. Problem 12.6 is the starting point to derive the high energy neutrino-nucleon inclusive scattering formulas:

$$\frac{d\sigma}{dx\,dy}(\nu_\mu N \to \mu^- + X) = \frac{G_F^2 xs}{\pi}\left[D(x) + \bar{U}(x)(1-y)^2\right],$$

$$\frac{d\sigma}{dx\,dy}(\bar{\nu}_\mu N \to \mu^+ + X) = \frac{G_F^2 xs}{\pi}\left[\bar{D}(x) + U(x)(1-y)^2\right].$$

N can be a proton or a neutron and the quark and antiquark distribution functions for fractional momentum are those for the given nucleon. The invariant Mandelstam variable s refers to the neutrino-nucleon system.

(a) Antineutrino scattering with y near 1 is a good way to measure the sea quark distribution function. Why?

(b) Consider a target with equal numbers of protons and neutrons (called an isoscalar target). Show that if we can ignore antiquarks in the nucleon, then the cross section for ν_μ is three times the cross section for $\bar{\nu}_\mu$.

(c) Consider ν_μ scattering with large x and y. Which quark fragmentation functions can we measure if we look at hadrons that carry a large fraction of the momentum of the final state scattered quark?

CHAPTER 13
Physics at Higher Energies

The previous chapters have presented the wide range of data explained by the Standard Model of particles and their interactions. The electromagnetic and color interactions have the elegance and simplicity one might expect to find in a fundamental theory; however, the origin of the electromagnetic interaction through electroweak symmetry breaking means that at least the electromagnetic theory is only a low-energy approximation. Such a low-energy theory is called an "effective field theory." In fact the electroweak theory raises many additional questions.

1. Are the $SU(2)$ and $U(1)$ pieces of the theory related in some way? For example, can we relate the two coupling constants g and g'? Can we understand the pattern of weak isospin and hypercharge quantum numbers of the various quarks and leptons?
2. Is the difference between left- and right-handed interactions a low energy phenomenon only, so that further interactions favoring right-handed quarks and leptons remain to be discovered?
3. Why are there three flavors for quarks and leptons? What determines the masses and mixing matrices? No good suggestions now exist regarding how to solve these flavor questions.
4. What is the physics of the Higgs sector of the theory? Hopefully, we can make significant progress on this question in the near future when experimentation at the TeV scale begins.

If we exclude gravity, the Standard Model interactions can account for all processes observed in nature, as well as the pattern of internal quantum number changes seen in these processes, up to mass scales of a few hundred GeV. Using the data, extra interactions at these mass scales can be excluded provided their couplings are not extremely small. For processes completely forbidden in the Standard Model, limits on extra interactions are especially restrictive. As an example, proton decay, which requires the transition of quarks to lighter particles (say, leptons) has not been seen, with limits for partial lifetimes to explicit final states typically greater than 10^{32} years. In models discussed below, these limits correspond to masses greater than about 10^{15} GeV for hypothesized bosons that could mediate such decays. If such interactions exist, they have no effect on the present universe. Their impact would have been mainly on the evolution of the very early universe, when available energies were comparable to the mass scale of the phenomenon in question. This evolution would therefore include the freezing out of entire interactions, should such interactions exist.

The one realm in nature not well understood from the quantum field point of view is gravity, and the physics dominated by the gravitational interaction. The large scale gravitational interaction is described successfully by general relativity, but the microscopic foundation for this is not well understood. Important questions that can be addressed using general relativity include the geometry of the universe on a large scale and the evolution of this geometry. These features relate cosmology to the energy and momentum content of the universe. Exploring this connection is a very promising direction for the future, since experiments have been developed to measure the space-time geometry. The present data indicate that the universe began approximately 14 billion years ago from an expanding, incredibly dense, high-energy initial state. High-energy interactions are crucial for understanding the physics of this state during its early time history. A parameter allowed by general relativity is a cosmological constant. This parameter seems to be nonzero and its cause is called dark energy. Unlike ordinary matter, which tends to slow the expansion of the universe, dark energy is now causing the expansion of the universe to accelerate. The ultimate nature, or cause, of the dark energy is another mystery.

Gravity also presents us with a dimensional constant, Newton's constant G. From G we can form a mass or distance scale. Choosing a mass scale, $(1/G)^{1/2}$ is equal to 1.2×10^{19} GeV. It is called the Planck mass. Its inverse is the distance scale at which gravity cannot be ignored for the interactions of a pair of particles. The presence of this scale raises several questions.

1. The Planck scale is very large; does it perhaps reflect a lower mass scale and complex physics underlying gravity? This can occur in models with extra dimensions of space that are compactified at distances that are large compared to the Planck length scale. Or is perhaps the Planck mass a fundamental mass scale for physics?
2. With how many fundamental mass scales does nature provide us? Is the weak scale in any way related to the Planck scale?
3. Why are particle masses so small compared to the Planck scale?

The search for a quantum theory of gravity has led to the invention of string theory, in which the fundamental objects are quantum strings in a space with many more dimensions than the four we know. The extra dimensions are compactified and have a small extension. Other nontrivial geometrical structures can also exist in this space. The implications of these theories are not well understood, but it can be shown that they yield Einstein's gravity in the classical limit. But they do not explain the dark energy phenomenon. A nontrivial prediction of string theory is the existence of supersymmetry. This is a symmetry between fermions and bosons, which requires the existence of a boson partner for every fermion and vice versa. These partners have the same internal quantum numbers for their

Standard Model interactions. If the symmetry were exact, these particle partners would be degenerate. Since we do not see such partners (for example, there is no bosonic partner of the electron), the symmetry must be broken, if it exists. Supersymmetry typically has a new multiplicative quantum number, whose sign is determined by the number of supersymmetric particles present. This results in the stability of the lightest supersymmetric particle. These particles may well provide the dark matter that is visible gravitationally.

The linkage of supersymmetry with the Higgs boson is an attractive possibility, since it can help solve a problem in the Higgs sector. The mass renormalization of the Higgs particle diverges quadratically with the momentum scale up to which we consider quantum fluctuations. This contrasts with the electron mass renormalization, which diverges logarithmically and does not change the mass very much, provided new physics cuts off the logarithm at a mass scale that is at most the Planck mass. The quadratic mass divergence for scalars yields a mass shift that dwarfs the weak mass scale if we integrate up to the Planck mass. This problem needs to be understood, since the Higgs mass can be at most around 1 TeV. With supersymmetry, many extra particles exist that contribute to the mass renormalization. These contributions, as a consequence of the symmetry, cancel after summing over a supersymmetric group of particles, at mass scales large compared to the masses of the supersymmetric particles involved. For this mechanism to work for limiting the Higgs mass shift, the missing supersymmetric particles should have masses in the 1 TeV mass region. A disadvantage of this solution to the large Higgs mass shift is that it replaces the issue of electroweak symmetry breaking with supersymmetry breaking, so we still need additional physics that is not understood.

One important prediction of supersymmetry is that at least two doublets of Higgs scalars exist. This yields, after electroweak symmetry breaking, at least five physical scalar particles, three neutral and a charged pair. These particles should be found in high energy experiments, in addition to the many supersymmetric partners of the known existing particles.

The possibility of new physics at very high mass scales, which may be required for gravity, could also be relevant for the Standard Model interactions. A particularly appealing possibility is that the various gauge theories unify into one larger simple gauge theory at very high energies. Such theories are called grand unified theories. A grand unified theory would automatically relate the unusual pattern of internal quantum numbers of the quarks and leptons that lie in a common multiplet. This unification hypothesis may be surprising, since it requires unifying the color interaction with the other, much weaker, interactions. Since a non-Abelian gauge theory such as $SU(n)$ has one coupling constant, the disparity in couplings between the low energy interactions is a problem. However, the coupling constants run with mass scale and the strong interaction becomes quite weak at very high mass scales. If we attempt to discern whether the $SU(3)$ color and $SU(2) \times U(1)$ weak interaction couplings might intersect at one unification scale, we find that this energy scale is very large. Thus the physics over this entire range must be included, since all virtual objects whose mass is small compared to the grand unification scale contribute to the running of the coupling constants. If we include only the Standard Model physics, the couplings do not all intersect. If we also include supersymmetric particles with masses in the TeV region, the couplings may intersect at a scale $\sim 10^{15}$-10^{16} GeV.

The grand unified theories require symmetry breaking at two separate scales. At the very large scale, the symmetry is initially broken, leaving the $SU(3)$ color and $SU(2) \times U(1)$ weak interactions as remaining symmetries. A number of interactions are frozen out by the symmetry breaking. In particular, interactions that create transitions between quarks and leptons are frozen out, since the exchange of gauge bosons of mass $\sim 10^{15}$ GeV provide very low rates. A signature of such theories is proton decay, which is now not completely forbidden, although the decay rate in a given theory may be too small to measure.

At the lower symmetry breaking scale, which is the weak scale, the symmetry is further broken, leaving the $SU(3)$ color plus $U(1)$ of electromagnetism as unbroken gauge theories. Again, a number of interactions are frozen out, but now not as thoroughly, since the weak scale is much smaller than 10^{15} GeV. The frozen interactions, for example, the charged current weak interactions, can still be seen clearly because transitions occur, at a weak rate, which are forbidden in a theory containing only the remaining unbroken gauge theories.

Grand unified theories may play a role in explaining two other mysteries. The first is that the universe contains only matter and no antimatter, to good approximation. The second is that there is also very little matter! For example, the ratio of matter particles to photons is only about 10^{-10}. An appealing possibility is that the universe initially had equal amounts of matter and antimatter, which nearly all annihilated, leaving mostly photons. The small amount of matter left is then a consequence of some difference between particles and antiparticles. Such a scenario is in fact possible, provided we have several ingredients: very heavy particles that decay as the universe cools, as provided by the superheavy gauge bosons of a grand unified theory; CP violation, providing a difference in some decay processes for particles and antiparticles; and a lack of thermal equilibrium in the early universe. So far these ideas provide the ingredients for an explanation of the present cosmic abundances, but not a detailed explanation.

We have briefly discussed some of the interesting ideas presently being explored. How will they evolve? What new ideas will be suggested by the next generation of physics results? Will theorists invent further startling new ideas? The unfinished character of the Standard Model suggests that we have much still to be discovered.

APPENDIX

Conventions

A ■ UNITS

The conventions for units used throughout the text are $\hbar = c = 1$, with energies measured in eV or multiples thereof. Occuring most frequently are MeV or GeV, where $1 \text{ MeV} = 10^6$ eV and $1 \text{ GeV} = 10^9$ eV. With $c = 1$, the units for mass and momentum are the same as those for energy. Also, the units for length and time are $[\text{energy}]^{-1}$. We will also use the more conventional units of sec and cm, or multiples thereof. Useful choices for distances are: 1 Angstrom $= 10^{-8}$ cm and 1 fermi $= 10^{-13}$ cm. These units are convenient for describing atoms and nuclei, respectively. Cross sections are measured in units of cm^2, with a convenient multiple defined by 1 barn $= 10^{-24}$ cm^2.

To translate between units, we can use

$$\hbar = 6.582 \times 10^{-22} \text{ MeV sec},$$

$$\hbar c = 197.3 \text{ MeV fermi}.$$

For example, the relationship between lifetime and width is

$$\tau = \frac{1}{\Gamma} \quad \text{for } \hbar = 1.$$

Here the width is in energy units. To use more conventional units, the general procedure is to insert a number of factors of $\hbar$ and c to provide the desired units. Thus, the lifetime in units of seconds is:

$$\tau = \frac{\hbar}{\Gamma}.$$

A width of 1 MeV then corresponds to $\tau = 6.582 \times 10^{-22}$ sec. To go between units for cross sections, the correspondence is:

$$(1 \text{ GeV})^{-2} = 3.894 \times 10^{-28} \text{ cm}^2 = 38.94 \text{ millibarns}.$$

In addition to the foregoing, we must specify the units for the electric charge. We choose for the Coulomb potential of the proton:

$$V = \frac{e}{4\pi r}.$$

This leads to a dimensionless coupling constant, measured to be

$$\frac{e^2}{4\pi} = \frac{1}{137.036}.$$

In more conventional units, the dimensionless coupling constant is

$$\frac{e^2}{4\pi\hbar c}.$$

For other interactions, the dimensionless coupling constants all have a 4π in the denominator, as in the case of electrodynamics.

B ■ USE OF LORENTZ INDICES

The discussion in the text is 4-dimensional (however, only from the point of view of special relativity). Thus quantities are carefully specified by their transformation properties under Lorentz transformations. For example, x_μ (space-time coordinates), p_μ (4-momentum), and A_μ (4-vector potential) are Lorentz vectors; $F_{\mu\nu}$ (the electromagnetic field) is a second-rank tensor. We use a number of conventions regarding these, listed below.

 1. We do not distinguish between upper and lower indices (that is, we don't introduce contravariant and covariant versions of the same vector). Thus x_μ and x^μ are the same 4-vector with the same components: (t, x, y, z). This allows us the freedom to adjust the notation in cases with several indices, some of which may be Lorentz indices and some related to internal group transformations. When specifying components, we take the time component to be the zeroth component, the space components to be specified by indices 1, 2, and 3. The use of contravariant and covariant vectors is now the convention most common in field theory texts. However, to use a variety of texts and to read a variety of papers, the student will need to be comfortable with the various conventions that can be chosen.

 2. The dot product of two 4-vectors yields a Lorentz scalar. To simplify the notation, we use the convention that a repeated Lorentz index implies that a dot product is to be taken. Thus $x \cdot x$, $x_\mu x_\mu$, $x_\mu x^\mu$, and $x^\mu x^\mu$ all are the quantity $t^2 - x^2 - y^2 - z^2$ (note $c = 1$ here). Thus the definition of $x_\mu x_\mu$ is the same as $g^{\mu\nu} x_\mu x_\nu$, had covariant and contravariant vectors been used. As in the choice of the summation convention, the dot product convention chosen in the text provides an economy in the notation. The minus sign in front of the product of space components is part of the convention. As another example, the quantity $F_{\mu\nu} F_{\mu\nu}$ is also a Lorentz scalar, after doing the double sum. For $\mu, \nu = 0, 0$ or two space indices, the product enters the double sum with a plus sign; for μ or $\nu = 0$ with the other a space index, the product enters with a minus sign. For $F_{\mu\nu}$, the electromagnetic field tensor defined in the text:

$$F_{\mu\nu} F_{\mu\nu} = 2(B^2 - E^2).$$

B Use of Lorentz Indices

Two tensors also occur that are the same in all reference frames. These are $g_{\mu\nu}$ and $\varepsilon_{\mu\nu\sigma\rho}$. Here $g_{00} = 1$, $g_{11} = g_{22} = g_{33} = -1$, with all other terms $= 0$. The fully antisymmetric tensor $\varepsilon_{\mu\nu\sigma\rho} = 0$ if any index is repeated and $= +1$ or -1 if $\mu\nu\sigma\rho$ is an even or odd permutation of $0, 1, 2, 3$. Thus, for example, $\varepsilon_{0,1,2,3} = 1$, etc. A second rank tensor can be obtained by multiplying $g_{\mu\nu}$ by a scalar or dotting two 4-vectors into $\varepsilon_{\mu\nu\sigma\rho}$.

Finally, we note that we will also look at quantities that transform under internal transformations that are not space-time transformations. These will typically be unitary transformations in the space of quantum states. The transformation groups will typically be $SU(2)$ or $SU(3)$. We can define operators that are scalars, vectors, etc., under these transformations. We will use the summation convention for repeated group indices when making scalars out of two vector operators. In this case, when calculating the dot product, all terms enter with a plus sign in the sum.

Bibliography

GENERAL REFERENCES

Experimental Techniques

Ferbel, T. J. (ed.), *Experimental Techniques in High Energy Physics,* Addison-Wesley, Menlo Park, CA, 1987.

Kleinknecht, K., *Detectors for Particle Radiation* (2nd edition), Cambridge University Press, Cambridge, UK, 1998.

Green, D., *The Physics of Particle Detectors*, Cambridge University Press, Cambridge, UK, 2000.

Williams, H. H., "Design principles of detectors at colliding beams," *Ann. Rev. Nucl. Part. Sci.* 36 (1986), 361.

Particle Accelerators

Livingston, M. S. and Blewett, J. P., *Particle Accelerators,* McGraw-Hill, New York, 1962.

Humphries, S. Jr., *Principles of Charged Particle Acceleration*, Wiley-Interscience, New York, 1986.

Kohaupt, R. D. and Voss, G. A., "Progress and performance of e^+e^- storage rings," *Ann. Rev. Nucl. Part. Sci.* 33 (1983), 67.

Lederman, L. M., "The Tevatron," *Sci. Am.* 264 (1991), 26.

Myers, S. and Picasso, E., "The LEP collider," *Sci. Am.* 263 (1990), 34.

Particle Physics and Field Theory

Cahn, R. N. and Goldhaber, G., *The Experimental Foundations of Particle Physics,* Cambridge University Press, Cambridge, UK, 1989.

Perkins, D. H., *Introduction to High Energy Physics* (4th edition), Cambridge University Press, Cambridge, UK, 2000.

Halzen, F. and Martin, A. D., *Quarks and Leptons: An Introductory Course in Modern Particle Physics*, Wiley, New York, 1984.

Leader, E. and Predazzi, E., *An Introduction to Gauge Theories and the "New Physics,"* Cambridge University Press, Cambridge, UK, 1982.

Cheng, T.-P. and Li, L.-F., *Gauge Theory of Elementary Particle Physics,* Oxford University Press, Oxford, UK, 1984.

Georgi, H., *Weak Interactions and Modern Particle Theory,* Benjamin/Cummings, Menlo Park, CA, 1984.

Peskin, M. E. and Schroeder, D. V., *An Introduction to Quantum Field Theory,* Addison-Wesley, Reading, MA, 1995.

Itzykson, C. and Zuber, J.-B., *Quantum Field Theory,* McGraw-Hill, New York, 1980.

Ramond, P., *Field Theory: A Modern Primer* (2nd edition), Addison-Wesley, Redwood City, CA, 1989.

Brown, L. S., *Quantum Field Theory,* Cambridge University Press, Cambridge, UK, 1992.

CHAPTER REFERENCES

Chapters 1, 2, and 3

Brown, L. M., Dresden, M., and Hoddeson, L., *Pions to Quarks: A Collection of Articles on the Development of Particle Physics between 1947 and 1963,* Cambridge University Press, Cambridge, UK, 1989.

Hoddeson, L., Brown, M., Riordan, M. and Dresden, M., *The Rise of the Standard Model: Articles on Particle Physics in the 1960s and 1970s,* Cambridge University Press, Cambridge, UK, 1997.

Riordan, M., *The Hunting of the Quark: A True Story of Modern Physics,* Simon and Schuster, 1987.

Okun, L. B., *Particle Physics: the Quest for the Substance of Substance,* Harwood Academic Publishers, Chur, Switzerland, 1985.

Salam, A., "Gauge unification of fundamental forces," *Science* 210 (1980), 723.

't Hooft, G., "Gauge theories of the forces between elementary particles," *Sci. Am.* 243 (1980), 90.

Weinberg, S., *The First Three Minutes,* Fontana, London, 1983.

Wilczek, F., "Cosmic asymmetry between matter and antimatter," *Sci. Am.* 278 (1998), 30.

Feynman, R. P., *The Feynman Lectures on Physics,* Volume 3, Addison-Wesley, Reading, MA, 1963.

Feynman, R. P., *QED: The Strange Theory of Light and Matter,* Princeton University Press, Princeton, NJ, 1985.

Mandl, F. and Shaw, G., *Quantum Field Theory,* Wiley, Chichester, UK, 1984.

Sakurai, J. J., *Advanced Quantum Mechanics,* Addison-Wesley, Reading, MA, 1967.

Schweber, S. S., *QED and the Men Who Made It: Dyson, Feynman, Schwinger, and Tomonaga,* Princeton University Press, Princeton, NJ, 1994.

For a discussion of helicity amplitudes, see:

Jacob, M. and Wick, G. C., *Ann. Phys.* 7 (1959), 404.

Bibliography

Chapter 4

Lipkin, H., *Lie Groups for Pedestrians*, North-Holland, Amsterdam, Holland, 1966.

Cahn, R. N., *Semi-Simple Lie Algebras and Their Representations*, Benjamin/Cummings, Menlo Park, CA, 1984.

Georgi, H., "Lie algebras in particle physics," *Front. Phys.* 54 (1982), 1.

Sakurai, J. J., *Invariance Principles and Elementary Particles,* Princeton University Press, Princeton, NJ, 1964.

Glashow, S. L., "Quarks with color and flavor," *Sci. Am.* 233 (1975), 38.

Bartelt, J. and Shukla, S., "Charmed meson spectroscopy," *Ann. Rev. Nucl. Part. Sci.* 45 (1995), 133.

Berkelmann, K., "Upsilon spectroscopy at CESR," *Phys. Rep.* 98 (1983), 145.

Bloom, E. and Feldman, G., "Quarkonium," *Sci. Am.* 246 (1982), 42.

Lee-Franzini, J., "Physics of quarkonia," *Nuclear Physics B (Proc. Suppl.)* (1988), 139–178.

Chapter 5

Gasiorowicz, S., *Elementary Particle Physics*, Wiley, New York, 1966.

Johnson, K. A., "The bag model of quark confinement," *Sci. Am.* 241 (1979), 100.

Chapter 6

Kroll, N. M., Lee, T. D., and Zumino, B., "Neutral vector mesons and the hadronic electromagnetic current," *Phys. Rev.* 157, (1967), 1376.

De Rújula, A., Georgi, H., and Glashow, S. L., "Hadron masses in a gauge theory," *Phys. Rev.* D12 (1975), 147.

Chapter 7

Aitchison, I. J. and Hey, A. J., *Gauge Theories in Particle Physics*, Adam Hilger, Bristol, UK, 1981.

Wilczek, F. "Quantum chromodynamics: The modern theory of the strong interaction," *Ann. Rev. Nucl. Particle Sci.* (1982), 177.

Peskin, M. E., "Chiral symmetry and chiral symmetry breaking," SLAC-PUB-3021, published in Les Houches Sum. School (1982), 217.

Hosaka, A. and Toki, H., *Quarks, Baryons and Chiral Symmetry*, World Scientific, Singapore, 2001.

Rothe, H. J., "Lattice gauge theories: An introduction," *World Scientific Lecture Notes in Physics* Vol. 59 (1997).

DeTar, C. and Gottlieb, G., "Lattice quantum chromodynamics comes of age," *Phys. Today* 57N2 (2004), 45.

Chapter 8

Pais, A., *Inward Bound—A Historical Account of the Early Days of Weak Interactions*, Clarendon Press, Oxford, UK, 1982.

Perl, M., "The tau lepton," *Rep. Prog. Phys.* 55 (1992), 653.

Weinstein, A. J. and Stroynowski, R., "The tau lepton and its neutrino," *Annu. Rev. Nucl. Part. Sci.*, 43 (1993), 457.

Commins, E. D. and Bucksbaum, P. H., *Weak Interactions of Leptons and Quarks*, Cambridge University Press, Cambridge, UK, 1983.

Adler, S. and Dashen, R., *Current Algebras*, W. A. Benjamin, New York, 1968.

Berkelmann, K. and Stone, S. L., "Decays of B mesons," *Ann. Rev. Nucl. Part. Sci.* 41 (1991), 1.

Isgur, N. and Wise, M. B., "Weak transition form-factors between heavy mesons," *Phys. Lett.* B237 (1990), 527.

Isgur, N. and Wise, M. B., "Spectroscopy with heavy quark symmetry," *Phys. Rev. Lett.* 66 (1991), 1130.

Chapter 9

Dolgov, A. D., Morozov, A. Yu., Okun, L. B., and Schepkin, M. G., "Do muons oscillate?," *Nuclear Physics* B502 (1997), 3.

Nauenberg, M., "Correlated wave packet treatment of neutrino and neutral meson oscillations," *Phys. Lett.* B447 (1999), 23.

Wu, T. T., and Yang, C.-N., "Phenomenological analysis of violation of CP invariance in decay of K0 and anti-K0," *Phys. Rev. Lett.* 13 (1964), 380.

Bilenky, M., Pascoli, S., and Petcov, S. T., "Majorana neutrinos, neutrino mass spectrum, CP violation and neutrinoless doble beta decay. 1. The three neutrino mixing case," *Phys. Rev.* D64 (2001), 053010.

Bilenky, S. M., Giunti, C., Grifols, J. A., and Masso, E., "Absolute values of neutrino masses: Status and prospects," *Phys. Rept.* 379 (2003), 69.

Kuo, T. K. and Pantaleone, J., "Neutrino oscillations in matter," *Rev. of Modern Phys.* 61 (1989), 937.

Bahcall, J. N. and Ulrich, R. K., "Solar models, neutrino experiments and helioseismology," *Rev. of Modern Phys.* 60 (1988), 297.

Burguet-Castell, J., Gavela, M. B., Gomez-Cadenas, J. J., Hernandez, P., and Mena, O., "Superbeams plus neutrino factory: The golden path to leptonic CP violation," *Nucl. Phys.* B646 (2002), 301.

Chapter 10

Watkins, P., *The Story of the W and Z*, Cambridge University Press, Cambridge, UK, 1986.

Weinberg, S., "Unified theories of elementary particle interactions," *Sci. Am.* 231 (1974), 50.

Weinberg, S., "Conceptual foundations of the unified theory of weak and electromagnetic interactions," *Rev. Mod. Phys.* 52 (1980), 515.

Veltman, M., "The Higgs boson," *Sci. Am.* 255 (1986), 88.

Chapter 11

Collins, P. D. B. and Martin, A. D., *Hadron Interactions*, Adam Hilger, Bristol, UK, 1984.

Perl, M., *High Energy Hadron Physics*, Wiley-Interscience, New York, 1974.

Donnachie, A., Dosch, G., Landshoff, P. and Nachtmann, O., *Pomeron Physics and QCD*, Cambridge University Press, Cambridge, UK, 2002.

Rossi, B., *High-Energy Particles*, Prentice-Hall, Englewood Cliffs, NJ, 1952.

Amaldi, U., "Fluctuations in calorimetry measurements," *Physica Scripta* Vol. 23 (1981), 409.

Chapter 12

Feynman, R. P., *Photon-Hadron Interactions*, W. A. Benjamin, New York, 1972.

Altarelli, G., "Partons in quantum chromodynamics," *Phys. Rep.* 81 (1982), 1.

Fisk, H. and Sciulli, F., "Charged current neutrino interactions," *Ann. Rev. Nucl. Part. Sci.* 32 (1982), 499.

Bethke, S. and Pilcher, J. E., "Tests of perturbative *QCD* at LEP," *Ann. Rev. Nucl. Part. Sci.* 42 (1992), 251.

Chapter 13

Langacker, P., "Grand unified theories and proton decay," *Phys. Rep.* 72 (1981), 185.

Polchinski, J., *String Theory*, Volumes 1 and 2, Cambridge University Press, Cambridge, UK, 1998.

Amaldi, U., de Boer, W., Frampton, P. H., Furstenau, H., and Liu, J., "Consistency checks of grand unified theories," *Phys. Lett.* B281 (1992), 374.

Arkani-Hamed, N., Dimopoulous, S., and Dvali, G. R., "Large extra dimensions: A new arena for particle physics," *Phys. Today* 55N2 (2002), 35.

Haber, H., Kani, I., Kane, G., and Quiros, M., "Is nature supersymmetric," *Sci. Am.* 254 (1986), 42.

Index

Abelian transformations, 231
Accelerators
    accelerator-produced neutrinos, 337, 345
    Large Hadron Collider, 404
Accidental symmetry, 146
Amplitudes, calculations, 15–70
    action (S), 19–20
    as basic principle of quantum mechanics, 2–3
    charged scalar fields, 63–64
    current density (J_μ), 17–18
    electromagnetic interactions, 43–45
    Euler–Lagrange equation, 20
    fermion fields, 64–68
    first-order interaction vertices, 41–42, 43
    free particles, 15–16
    incoming (absorbed) particles, 37–38, 42
    Klein–Gordon equation, 16–18
    Lagrangian ($\mathcal{L}$) density, 18–20
    Lorentz covariant description, 3
    matrix elements (M_{fi}) (S matrix), 41–50
    momentum-space propagator (f), 47, 50–52
    outgoing (emitted) particles, 37–38, 42
    particle spin [$\chi(p_\mu)$] dependence, 15–16, 23–24
    scalar interactions, 45–46
    second-order interaction terms, 43–44
    spin $\frac{1}{2}$ particles, 27–37
        bilinear covariants, 35–36
        Dirac equation, 28–30, 37

        Lagrangian and symmetries, 30–32
        minimal substitution, 31, 32
        plane wave solutions, 32–34
        spin $\frac{1}{2}$ Hermitian operators, 35–36
        spin operator, 27–28, 31, 67
        spinors (helicities), 27–28, 30, 33–34
    spin 0 particles, 16–18
    spin 1 particles, 23–24
    spin 1 photons, 24–27
    time-dependent perturbation theory, 38–43
    time evolution and particle exchange, 43–50
    virtual scalar exchange, 45–46
    Yukawa potential, 17, 20
    See also cross sections (σ); decay rates, calculations; Quantum Field Theory; Yukawa potential
Antiparticles
    asymmetry between numbers of particles and antiparticles, 8, 12, 357–358
    forward-backward asymmetry, 358
    scalar particle charge operator (Q), 63–64, 65, 66

B^0, $\bar{B}^0$ and B_s^0, $\bar{B}_s^0$ systems, 329–333
    amplitude, 332
    CKM factors, 330–331, 332, 333
    CP violating parameter (β), 332–333

    CP violation, 331–333
    diagrams and terms for mass mixing, 329–330
    time-dependent asymmetry, 332
Baryon magnetic moments, 193–198
    anomalous moment interaction (Dirac–Pauli representation), 194
    constituent quark masses (m_u and m_s), 196, 202
    first-order magnetic dipole transition (M_{fi}), 197
    gauge invariant interactions of protons, 193
    Gordon decomposition, 194
    Hamiltonian density ($\mathcal{H}$) calculations, 194–195
    interaction Hamiltonian (H) for spin, 195
    magnetic dipole transition, 197–198
    measurement, 193
    predicted and measured values, 196
    See also magnetic moments
Baryons, 180–186
    baryon magnetic parameter, 207–208
    baryon mass formula, 205
    baryon number conservation, 149
    Clebsch–Gordan coefficients, 182–183
    decay of baryons, 184–186
    decuplet states with $J = \frac{3}{2}$, 181, 184

Baryons (cont.)
Δ baryon, 181, 202, 204
isospin shift operators (U_+, U_-, V_+, V_-), 184
Λ^0, 196, 197
masses, 184
predicted and measured baryon mass values, 206
proton spin state, 182
proton stability, 185
semileptonic decay of strange baryons, 185
Σ^+, 184, 196
Σ^-, 184, 196
Σ^0
  decay width (Γ), 197–198
  electromagnetic decay to $\Lambda^0 + \gamma$, 197–198
  magnetic moment, 196
  mass and states, 184
  spin rearrangement, 197
  spin states, 197
spin interaction effect on masses, 201–202, 204–205
spin (J), 180–181
spin states of baryons, 183, 184
strangeness values, 181
with three heavy quarks (ccc or bbb), 128–129
wave functions of baryons, 183
weak interactions leading to semileptonic decay, 185, 186
Ξ^-, 184, 196
Ξ^0, 184, 196
See also mass spectrum of baryons and mesons; neutrons; protons; quarks (spin $\frac{1}{2}$ particles)
Base states, 2, 3, 24
See also eigenstates
Bethe–Bloch energy loss formula, 388
Bhabha scattering, 92–93
Bjorken scaling variable (x), 418, 422
Bohr radius, 140, 384
Born approximation, 57
Bose symmetry, 174, 288

Bosons. See fundamental particles and interactions; gauge bosons; gluons; photons; spin 1 particles
Breit–Wigner propagators, 155–159, 160, 188, 267
Bremsstrahlung, 389, 425

C eigenstates, 131, 132, 177
C matrix, 177
Cabibbo angle (θ_c), 270, 337
Cabibbo suppression
  bottom (b) quark decay, 306, 307
  charm quark decay, 302, 303, 305
  D mesons, 328–329
  definition, 270
Cabibbo–Kobayashi–Maskawa (CKM) matrix, 268–272
  See also charged weak currents for quarks
Callan–Gross relation, 422, 423, 432, 433
Casimir operators
  conservation and additive properties, 115–116, 118
  $SU(2)$ and $SU(3)$ generators, 123–126
Charge conjugation symmetry, 110, 113, 143, 155, 229
Charged pion (π meson) decay, 274–280
  axial vector, 275
  branching ratios, 274
  charge operators, 277–278
  conserved vector current (CVC), 277–279
  current matrix element M_{fi}, 275, 277
  decay width, 276–277
  electromagnetic corrections, 276–277
  helicity suppression, 274
  leptonic decays, 274
  matrix element M_{fi}, 275
  π^- lifetime, 274
  π decay rate, 275
  rate for semileptonic decay, 280

vector current matrix elements, 278–279
See also mesons; weak interactions of fermions
Charged weak currents for quarks, 267–274
Cabibbo angle (θ_c), 270
Cabibbo–Kobayashi–Maskawa (CKM) matrix, 268–272
V_{UD} matrix, 268–269
See also CP violation; quarks (spin $\frac{1}{2}$ particles)
Cherenkov detectors, 335–336, 338–340, 344
Chiral symmetry, 240–244
  chiral generators, 245
  chiral symmetry rotations, 244–245
  conserved currents, 241
  conserved vector and axial vector currents in neutron decay, 242–243
  Goldberger–Treiman relation, 241–244
  left- and right-handed components of fermion fields, 241
  left- and right-handed components of $SU(3)$ flavor symmetry, 241
  left- and right-handed fermion weak interactions, 24–25
  parity operations, 245
  QCD Lagrangian several massless quark flavors, 241
  $SU(3)_L \times SU(3)_R$ flavor symmetry, 241
  See also $SU(3)$
CKM (Cabibbo–Kobayashi–Maskawa) matrix, 268–272
  See also charged weak currents for quarks
Clebsch–Gordan coefficients
  baryons, 182–183, 184
  K meson decays, 286, 289, 298
  mesons, 171, 173

Color gauge theory. *See* chiral symmetry; quantum chromodynamics (QCD); running coupling constant $\alpha(q^2)$; spontaneously broken symmetry
Color interactions, 126–130
  amplitude for quark-quark scattering via gluon exchange, 127
  Casimir operators, 128
  confinement, 11, 128–129, 135, 404–405
  Feynman diagram for quark-quark scattering via gluon exchange, 127
  generators, 127
  interaction energy between well-separated singlets, $V(R)$, 128–129
  interaction energy (U), 127
  quadratic Casimir operator for $SU(3)$ multiplets (U), 128
  quark transition with gluon emission, 126
  $SU(3)$ operator, 128
  $SU(3)$ symmetry due to three quarks in each singlet, 129–130
  T_α values for various multiplets, 128
  See also hadrons; quarks (spin $\frac{1}{2}$ particles); strong interactions
Color potential, 130–136
  C eigenstate, 131
  charge conjugation, 131
  Cornell potential, 134
  excited states for $c\bar{c}$ and $b\bar{b}$ systems, 132–135
  Feynman diagram of positronium decay, 133
  mass of $c\bar{c}$ and $b\bar{b}$ systems, 131, 133
  momentum space wave function, 131
  positronium (e^+e^-) production of $c\bar{c}$ and $b\bar{b}$ resonances, 130–132

  positronium lifetimes, 132
  potential $V(R)$ for $c\bar{c}$ and $b\bar{b}$ bound states, 134–135
  Richardson potential, 135
  spin and parity states for $c\bar{c}$ and $b\bar{b}$ systems, 133
  unstable heavy-quark bound states, 130–131
  See also quarks (spin $\frac{1}{2}$ particles)
Compton scattering ($\gamma + e^- \to \gamma + e^-$), 97–103
  annihilator currents, 97–98
  anticommutation relations, γ matrices, 99, 101
  cross section, 101–103, 212
  Feynman diagrams, 100
  M_{fi} expression in momentum-space, 99
  momentum-space electron propagator, 99
  momentum-space propagator, 98
  photon matrix elements, 97–98
  photon spin sum, 100
  S matrix expressions, 97–99
  trace calculation, 102
  See also processes with two leptons and two photons
Constituent Quark Model, 193
Cosmological constant, 442
Coulomb scattering, 380–389
  average momentum transfer, 382
  cross section (σ) calculation, 380–382, 386
  effective ionization energy, 386–387
  ion pair production, 388
  kinematics, 380–382
  momentum transfer (q^2), 385–388
  multiple Coulomb scattering, definition, 382
  multiple scattering angle, 382–384
  by muons, 389

  pion form factor [$F_\pi(q^2)$], 216
  radiation length (X_0), 383–385
  See also electromagnetic interactions
Coupling constants (g or e)
  effect on Yukawa potential, 17, 58
  in Hamiltonian density ($\mathcal{H}$), 41
  universality in $SU(n)$ non-Abelian gauge theory, 254
CP violation, 272–274
  in the $B^0, \bar{B}^0$ system, 331–333
  CP violating parameter (β), 332–333
  in kaon system, 321–321
  lowering operators, 272, 273–274
  in neutrino mixing, 344–345
  raising operators, 272–273
  weak decay of pseudoscalar mesons, 295
  See also charged weak currents for quarks
CPT theorem and CPT symmetry, 108–109, 273–274, 294
Cross sections (σ)
  calculations, 54–58
  definitions, 54–55
  large cross section processes, description, 378–380
  See also amplitudes, calculations

Decay diagrams, 169–177
  explanation, 171
  K^{*+} decay diagrams, 172
  ϕ decay diagrams, 169, 172
  ρ^+ decay diagrams, 173
  ρ^0 decay diagrams, 174, 175–176
  vector meson decay diagram to e^+e^- final state, 136
Decay rates, calculations, 52–55
  branching ratio, 54, 210
  decay lifetime (τ), 5, 54
  decay width (Γ), 5, 54, 84
  lifetime (τ), 5, 54

Decay rates (*cont.*)
  phase space factor for final state, 53, 83–85, 197, 298
  transition rate (decays per second), 16, 53–54
  See also amplitudes, calculations
Deep inelastic lepton-proton scattering, 417–429
  Bjorken scaling variable (x), 418, 422
  Callan–Gross relation, 422, 423, 432, 433
  differential cross section, 419, 423
  electron scattering cross section, 420
  electron scattering kinematics, 418
  Feynman diagrams, 417
  final momentum of the constituent, 418
  inclusive cross section (e^- scattering), 419–420
  muon-proton scattering, 425
  $W_{\mu\nu}$ calculations, 420–422
  See also large momentum transfer processes; protons; structure functions
Degrees of freedom
  amplitude linearity in spin degrees of freedom, 4
  charge (Q), effect on degrees of freedom of particle, 23
  color degrees of freedom, 268, 378, 396
  Dirac equation, 22–23, 31–32
  gauge freedom, 364–366
  spin degrees of freedom for a quark/antiquark state, 148
  spin degrees of freedom of an electron, 33
  in the spontaneously broken theory, 369
  in three-particle decay, 164
Deuteron, isospin states, 121–122
DGLAP (Dokshitzer, Gribov, Lipatov, Altarelli, Parisi) equations, 435

Dirac equation
  amplitude calculations for spin $\frac{1}{2}$ particles, 28–30, 37
  degrees of freedom, 22–23, 31–32
  Dirac neutrinos, 375
  Dirac spinor fields, 255–257
  Dirac–Pauli representation, 29, 35, 194
  transformed Dirac equation, 110
Distance scales in quantum mechanics, 1, 3, 418
Drell–Yan process, 413

e^+e^- annihilation. *See* positron-electron (e^+e^-) annihilation to antimuon-muon ($\mu^+\mu^-$)
e^+e^- pair production. *See* pair production
e^+e^- production of e^+e^-. *See* positron-electron (e^+e^-) production of positron-electron (e^+e^-)
e^-e^- scattering. *See* electron-electron (e^-e^-) scattering
Eigenstates
  description of, 3
  helicity, 4
  momentum eigenstates in scattering experiments, 4
  multiplets, 119
  See also base states; mass eigenstates
Electrodynamics. *See* quantum electrodynamics (QED)
Electromagnetic interactions
  cross section, 379–380
  as the only unbroken gauge symmetry, 253, 350, 444
  spin-independent electromagnetic contributions to meson masses, 206, 207
  spin rearrangements, 197, 199
  See also Coulomb scattering; electroweak interactions; fundamental particles and interactions; photon coupling to vector mesons; photons; radiative transitions between pseudoscalar and vector mesons
Electromagnetic showers
  conversion of photons into e^+e^- pairs, 389
  generation by electrons, 336, 389–390
  generation by photons, 390–391, 395
Electron-electron (e^-e^-)
  scattering, 89–90
  differential cross section in the center of mass ($d\sigma/d\Omega$), 90
  Feynman diagrams, 89
  matrix element (M_{fi}), 90
  possible transitions and time-orderings, 89
  trace calculation, 90
Electron-muon ($e^-\mu^-$) scattering, 74–86
  amplitude calculations, 43–50
  cross section, 85–86
  decay rate, 84
  differential cross section ($d\sigma/dt$), 87
  Feynman diagrams for $e^-\mu^-$ scattering after space-time integration, 75
  Feynman diagrams for $e^-\mu^-$ scattering via scalar exchange, 47–48
  4-momentum (q) transferred, 47, 50
  helicity (spinor) changes, 56
  higher-order $e^-\mu^-$ scattering, 105
  matrix element for single photon exchange (M_{fi}), 78–79, 87
  momentum-space propagator (f), 47, 50–52, 76

Index

momentum-space propagators, 76, 77
photon propagators, 77–79
rates (from $|M_{fi}|^2$), 79–80
repulsive Coulomb potential, 79
scattering cross section, 55–57
second-order calculation for electromagnetic interaction, 46–47, 75
space-time behavior of scattering, 46–50
space-time propagator (D_F), 46, 51, 75–76
space-time propagators, 76
spin-dependent expressions, 81
spin vectors ($\hat{e}_x, \hat{e}_y, \hat{e}_z$), 76–77
summing of spinors over helicities, 80–81
summing of spinors over helicities for spin $\frac{1}{2}$ particles, 80–81
tensors $g(p)$ and $g(-p)$, 76–77
time orderings in scalar exchanges, 48–49
trace calculations and formulas, 81–83
two-body phase space factor, 83–85
Yukawa potential energy, 58
See also electrons (e^-); muons (μ^-)
Electrons (e^-)
classical radius, 384
free field Lagrangian ($\mathcal{L}_{\text{Free}}$), 256–257
helicity state calculations, 258–259
interaction vertices, 7
mass, 7
muon decay products, 258
spin degrees of freedom of an electron, 33
See also electron-muon ($e^-\mu^-$) scattering; leptons
Electroweak interactions, 9–10, 348–376
charge operator (Q^Y), 349
commuting transformation of isospin doublets, 349–350

corrections to W and Z masses, 370–373
interactions among the gauge bosons, 358–360
mass eigenstate fields, after mixing, 350
mixing for B^0 and W^0 bosons, 350–352
parameters of the renormalized theory, 370
quarks, 10–12
W pair and Z pair production from e^+e^-, 359–360
weak hypercharge (Y), 349, 351, 354
weak interaction Hamiltonian density, 348
weak isospin (T, T_3), 349, 351, 354
weak mixing angle, 350
See also electromagnetic interactions; gauge bosons; neutral current mixing; $SU(2) \times U(1)$ gauge symmetry; symmetry breaking; weak interactions; weak mixing phenomena

Fermi, Enrico
density effect on energy loss relativistic particles, 388
Fermi four-fermion coupling (G_F), 256
Thomas–Fermi model for the atom, 384, 390
Fermions (spin $\frac{1}{2}$ particles)
fermion fields, 64–68, 241, 253–255, 349
fermion-scalar coupling, 373–375
free field equation for spin $\frac{1}{2}$ particles, 118
gauge symmetry breaking as a source of mass, 253, 373–375
left- and right-handed fermion weak interactions, 24, 36, 253, 255

left-right asymmetry (A_{LR}), 357–358
spin, 7
$SU(2)$ requirement for massless fermions, 253, 257
See also fundamental particles and interactions; leptons; quarks; weak interactions
Feynman rules
conservation of electric charge, 113
factors requires to calculate $-M_{fi}$ for any Feynman diagram, 105–106
higher-order terms in the perturbation expansion, 103–105
loop diagrams, 97, 103, 371, 372
Flavors (quarks), 10, 146–149, 241–246
flavor symmetry, 146–147, 241–242, 244, 246
flavor transformations predicted by the Standard Model assumptions, 309
quark flavor changes by weak interactions, 287
$SU(3)$ flavor multiplets, matrices and transformations, 147–149
See also quarks (spin $\frac{1}{2}$ particles); SU(3)
Free field equations, 16, 28, 59, 118, 194
Fundamental particles and interactions, 6–13
See also electromagnetic interactions; fermions; leptons; quarks; Standard Model; strong interactions; weak interactions

G parity, 177–178, 264
γe^- scattering. *See* Compton scattering ($\gamma + e^- \to \gamma + e^-$)

460 Index

Gauge bosons (W^+, W^-, and Z^0)
  corrections to W and Z masses, 370–373
  electroweak interactions, 9–10
  emission during transitions within lepton doublets, 8
  gauge boson and Higgs particle loops, 371–372
  gauge symmetry breaking as a source of mass, 253, 254
  interaction vertices, 9
  interactions among the gauge bosons, 358–360
  mass, 9
  mass-generating reaction, $SU(2)$ symmetry breaking, 9, 253, 254, 309, 365–370
  neutrino scattering processes via W or Z^0 exchange, 9
  $SU(2)$ symmetry, 9, 253
  symmetry breaking, 9
  as virtual particles, 8
  W^0, lack of eigenstate, 254
  W exchange, 9, 48, 255, 271, 287
  W pair and Z pair production from e^+e^-, 359–360
  Z^0 branching ratios, 355
  Z^0 coupling to fermions, 352–354
  Z^0 cross section, 355–356
  Z^0 decay width, 354–355
  Z^0 eigenstate, 254
  Z^0 mass, 352, 368
  Z^0 mass eigenstate, 254
  Z^0 propagator, 355, 371
  See also electroweak interactions; fundamental particles and interactions; weak decay of pseudoscalar mesons; weak decays of the heavy quarks; weak interactions; weak interactions of fermions; weak mixing phenomena
General relativity, 106, 442
GIM (Glashow–Iliopoulos–Maiani) mechanism, 300, 319, 328

Gluons
  color, 11
  confinement, 11–12, 396
  corrections due to loops, 238
  cross section factors for short distance scattering, 414–415
  exchange during meson annihilation, 150–151, 154, 162, 169
  exchange of eight colored gluons in $SU(3)$ color interaction, 126
  Feynman diagram for quark-quark scattering via gluon exchange, 127
  gluon annihilation diagrams, 150, 154
  gluon exchange during meson annihilation, 150–151, 154, 162, 169
  gluon-gluon scattering, 413, 414, 437
  interactions with gluons, 238
  interactions with quarks, 11, 238
  neutral color singlets in meson decay, 170–171
  neutral (single) gluon exchanges described by Cornell potential, 134
  penguin diagram, 287–288
  $q_1\bar{q}_1$ annihilation through a gluon or through a photon, 413
  quark-gluon scattering, 414
  quark transition with gluon emission, 126
  single (neutral) gluon exchanges described by Cornell potential, 134
  strength of gluon interactions in K meson decay, 287
  $SU(3)$ symmetry, 11
  use of jets in studying parent quarks and gluons, 11–12
  See also strong interactions
Goldberger–Treiman relation, 241–244
  $g_{\pi pn}$ constant (residue at the pion pole), 243–244

massless pion requirement, 203, 243–244
neutron decay, axial vector current, 242, 243
neutron decay, isospin raising operator, 242
partially conserved axial current (PCAC), 244
pion exchange amplitudes, 242–243
pion exchange current matrix element, 243
See also chiral symmetry; strong interactions
Gordon decomposition, 194
Grand unified theories, 443–444
Gravity
  description by general relativity, 442
  Newton's constant G, 442
  quantum theory of gravity, 12, 442–443
  spin, 112, 248

Hadronic scattering
  average number of particles produced in an event, 397
  collisions of hadrons, momenta of final state particles, 378–380
  cross sections, 378–379, 396
  diffraction dissociation, 396–397
  fractional energy, 397, 407
  hadron structure and short distance scattering, 412–415
  hadronic jets from e^+e^- annihilation, 6, 179–180
  high transverse momentum jets in hadronic scattering, 436–437
  leading particles, 397–398
  limited transverse momentum, 397
  pp or $p\bar{p}$ collisions, 413, 436–437
  scaling, 398
  See also inclusive distributions; jets; large momentum transfer processes

Index

Hadrons
  charge structure of strong interactions, 112–113
  collisions, measured cross sections, 378–379
  collisions, momenta of final state particles, 378–380
  constituent quark masses (m_u and m_s), 196
  hadronic decays, 286–292
  internal symmetries, 117–122
    constraints for fermions or bosons, 120–121
    creation operators for obtaining particle and antiparticle states, 118
    definition, 117
    free field equation for spin $\frac{1}{2}$ particles, 118
    general unitary matrix (U_{ij}) with unit determinant, 118
    multiplets, 118–119
    proton-neutron doublet $SU(2)$ isospin symmetry, 121–122
    $SU(n)$ symmetries, 118, 122–126
    symmetry transformation for three quark types, 118
    unitary transformations, 118–119
  quantum numbers and quantum rules, 113–117
  spin, 12
  spin-dependent interactions, 146
  See also baryons; mesons; quarks
Hamiltonian density ($\mathcal{H}$)
  coupling constants (g or e), 41
  definition, 19
  first-order interaction vertices, 41–42, 43
  $\mathcal{H}$ scalar $\rho\phi$, 20, 38
  second-order interaction terms, 43–44
  spin $\frac{1}{2}$ particles, 30
  See also Time history of states

Hamiltonian (H) operators
  calculation of interaction Hamiltonian, 38–40, 71
  charged scalar fields, 63, 65–66
  H_F time independence, 62
  with Lagrangian ($\mathcal{L}$) functions, 19–20
  Quantum Field Theory, 61–62
  with spin $\frac{1}{2}$ particles, 30
  See also Time history of states
Helicity (λ), 4, 23–25
Higgs mechanism, 360–365
  field equation for gauge field, 362
  gauge field [$A_\mu(x)$], 361–362, 364, 368
  Goldstone boson, 363–363
  Lagrangian ($\mathcal{L}$), 362
  mass generation, 362, 366–370
  nonvanishing vacuum expectation value, 360, 362
  potential energy term $V(\phi)$, 362, 365
  renormalizable terms in the Lagrangian, 360–361, 364–365
  scalar field [$\phi(x)$], 361, 364, 368
  spin 0 particles, 360
  symmetry breaking of the weak gauge symmetry, 246
  See also symmetry breaking
Higgs particle (H)
  energy scale, 412
  gauge boson and Higgs particle loops, 371–372
  interactions, 364–365, 369
  linkage with supersymmetry, 443
  mass, 364, 411
  production via e^+e^- annihilation, 410–412
High-energy interactions, 441–444
Higher-order terms in the perturbation expansion, 103–105
  electron propagation, 104
  electron vertex, 104
  higher-order $e^-\mu^-$ scattering, 105

loop correction to photon propagator, 97, 103, 104, 234
  verification by magnetic moment measurements, 104–105
  See also Feynman rules; S matrix
Hydrogen, resonance states, 5

Inclusive distributions, 395–401
  angle for produced particles, 399
  average number of particles produced in an event, 397
  fractional energy, 397
  fractional energy in the center of mass, 397
  number of particles, 399
  partitioning of energy, 398–399
  rapidity (y), 399–401, 406, 436
  scaling, 398
Isgur–Wise function, 309
Isospin
  baryon isospin violating mass splittings, 207–209
  conservation during meson decay, 155
  conservation during strong interactions, 289
  deuteron isospin states, 121–122
  equal masses of members of isospin multiplets, 148, 152
  isospin breaking, 155, 178, 212, 311, 313
  isospin conservation during meson decay, 155
  isospin multiplets in baryons, 181–185
  isospin rotations, 147–148, 178
  isospin symmetry of π mesons, 148
  meson isospin violating mass splittings, 205–207, 209–211
  proton-neutron doublet $SU(2)$ isospin symmetry, 121–122
  quantum numbers (I, I_3), 121, 147

Isosphere (*cont.*)
shift operators (U_+, U_-, V_+, V_-), 184
$SU(3)$ flavor multiplets, analog isospin states, 148
symmetry states of proton-neutron doublet, 121–122
$u \leftrightarrow d$ $SU(2)$ symmetry, 147

Jets
forward-backward asymmetry, 358
hadronic jets from e^+e^- annihilation, 6, 179–180, 406–410
high transverse momentum jets in hadronic scattering, 436–437
quark fragmentation functions, 406–410
use in studying parent quarks and gluons, 11–12, 358
See also hadronic scattering

Kaon decay (K meson), 280–300
annihilation of quarks in K^- decay, 281–282
branching ratios, 281, 282, 323
CP relations, 281
decay to 2π, 287–290, 318–319
decay to 2π, amplitudes, 295–297, 318
decay to 3π, 287–290, 318–319
decay to 3π, amplitudes, 297–298, 318
f_K, 282
hadronic current, 281
hadronic decays, 286–292
isospin changing operators (ΔI), 289, 291–292
isospin properties of hadronic K decay, 287–290
K^0, $\bar{K}^0$ system, 280–283, 318–327
amplitudes, 318, 323–324
asymmetry of intensity functions $[A(t)]$, 321–322
box diagrams involved in K^0, $\bar{K}^0$ mixing, 319
CP violation, 321–327
decay widths, 318–319
degree of mixing, 321
eigenstate notation, 318
GIM mechanism, 319
intensities for kaon components, 320, 321–322, 326, 327
mass difference (Δm_K), 319
mass eigenstate evolution, 326
oscillations, 320–321
propagation equation, 321
Re(ε), 322
time integrated mixing parameter (r), 321
$K^0 \to \gamma\gamma$ decays, 282, 298–300
$K^0 \to \mu^+\mu^-$ decays, 282, 298–300
$K^- \to \pi^0$, branching ratios, 284–285
$K^- \to \pi^0$, coupling constant (g_{WK*}), 284
$K^- \to \pi^0$, matrix elements, 286
K^+ and K^-, similar CP state and decay behavior, 280–281
K_L^0 (long lifetime) CP state, 280–281, 318–319, 323
K_L^0 (long lifetime) CP violations, 321
K_S^0 (short lifetime) CP state, 280–281, 318–319, 323
lifetimes of kaons, 281
mean flight distance, 281
model for 2π decays, 290–292
q^2 dependence of form factors, 284
scattering and annihilation diagrams, 287–288
semileptonic decays, 282–286
spectator diagrams, 287–288
strong interactions, 287
transitions in meson semileptonic decays, 283
vector dominance model for $K^- \to \pi^0$, 283–286
ways that exchanged W is absorbed for K^- and, 288–289
ways that exchanged W is absorbed for K^- and K^0 decay, 284–285, 287–288
weak effective Hamiltonian, 286–287, 289–290, 300
See also mesons; strange particles, interactions; weak interactions
Klein–Gordon equation, 16–18

Lagrangian ($\mathcal{L}$) functions, 18–23
action (S), 18–19
charge (Q), effect on degrees of freedom of particle, 23
conserved current (J_μ), 21–22
Euler–Lagrange equation, 19, 21–22
Hamiltonian ($\mathcal{H}$) density, 19–20, 21
Lagrangian density, 18–20
spontaneous symmetry breaking, 20, 245–246, 366–367
$SU(n)$ symmetries of $\mathcal{L}$, 118
symmetries and the Lagrangian, 20–23
λ_{QCD}
defined as mass unit, 135, 239
uncertainty of source for λ_{QCD}, 234, 239
See also quantum chromodynamics (QCD)
Large cross section processes. *See* electromagnetic showers; hadronic scattering; pair production
Large momentum transfer processes, 404–405
See also deep inelastic lepton-proton scattering; hadronic scattering; positron-electron (e^+e^-) annihilation to hadrons; short distance scattering

Index

Lepton currents, 71–74
  creation and annihilation operators, 72
  lepton current terms, 71–72
  matrix elements of electron current, 72–73
  processes contained in electron current, 73
  relations involving photons, 73–74
  transverse photons combined with current conservation, 74
Leptons (spin $\frac{1}{2}$ particles), 7–10
  charge, 7
  doublets (families), 8, 254
  gauge boson emission during transitions, 8
  interaction Hamiltonian, 71
  interaction vertices, 7
  interactions with photons, 7
  lepton numbers (L_e, L_μ, and L_τ), 113
  masses, 7
  spin, 7
  stability, 7–8
  weak decay processes, 8
  See also electron-electron (e^-e^-) scattering; electron-muon ($e^-\mu^-$) scattering; electrons (e^-); fundamental particles and interactions; muons (μ^-); neutrinos; positron-electron (e^+e^-) annihilation to antimuon-muon ($\mu^+\mu^-$); positron-electron (e^+e^-) production of positron-electron (e^+e^-); processes with two leptons and two photons; tau lepton (τ^-)
Lifetime (τ), 5, 54
Linearity of quantum mechanics, 2, 4, 40, 81, 136, 193
Local gauge symmetry, 231–234
  coupling constant (g), 234

current operators and terms, 232–233
free field equations for eight massless gluons, 234
generators (T_a) for $SU(3)$ color rotations, 231, 232
global gauge transformations, definition, 21, 231
gluon term, 233
infinitesimal transformation of quark fields, 232–233
Lagrangian, 231–234
local gauge transformations, definition, 25, 231
Lorentz covariance, 3, 4, 23, 106–108
Lorentz invariant quantities for two-body scattering
  differential cross section ($d\sigma/dt$), 87
  Mandelstam variables, 86, 414, 440
  matrix element M_{fi}, 86–87
  q^2 (used interchangeably with s, t, or u variables), 86
  s (square of the total energy in the center of mass), 86, 414
  t (momentum transfer squared), 86, 414
  u (momentum transfer squared), 86, 414
Lorentz transformations
  amplitude, 15
  in Lorentz field theory, 106–108
Lorentz invariant momentum space factor, 51
spin 1 photons, 24
unitary operator (U), 107–108

Magnetic moments, 193–200
  anomalous moment interaction (Dirac–Pauli representation), 194
  baryon magnetic moment values, 196
  decay width calculations, 198, 199–200

first-order magnetic dipole transition (M_{fi}), 197
gauge invariant interactions of protons, 193
Gordon decomposition, 194
Hamiltonian density ($\mathcal{H}$) calculations, 194–195
interaction Hamiltonian (H) for spin, 195
magnetic dipole transitions, 197–200
measurement, 193
ϕ meson decay to $\eta' + \gamma$, 198–200
photon wavelength and energy, 197, 199
Σ^0 decay to $\Lambda^0 + \gamma$, 197–198
Majorana representation of particle-antiparticle symmetry, 64, 375–376
Mandelstam variables, 86, 414, 440
Mass eigenstates
  D mesons, 303
  electroweak interactions, 350
  K^0, $\bar{K}^0$ system, 326
  mesons, propatation as mass eigenstates, 161–163, 313, 315–317
  neutral current mixing, 350
  neutrino propagation, 269, 313, 333
  quarks, 268, 287
  Z^0 mass eigenstate, 254
Mass-generating reaction, $SU(2)$ symmetry breaking, 9, 257
Mass spectrum of baryons and mesons, 200–211
  baryon isospin violating mass splittings, 207–208
  baryon magnetic parameter, 207–208
  baryon mass formula, 205
  λ_{QCD} contribution, 201, 202
  magnetic spin-spin interaction effect on meson masses, 206, 207
  meson isospin violating mass splittings, 205–207

Mass spectrum of baryons and
   mesons (*cont.*)
 meson mass values, 204
 predicted and measured baryon
   mass values, 206
 spin-independent
   electromagnetic
   contributions to meson
   masses, 206, 207
 spin interaction effect on baryon
   masses, 201–202, 204–205
 spin interaction effect on meson
   masses, 201, 202–204
 spin splitting, 135, 202
 u, d mass difference
   contributions to meson
   masses, 207
 u mass from QCD theory,
   202–203
 See also baryons; mesons
Masses, dynamic determination, 6,
   257
Maxwell's equations, 25–27
Measurement, 3–6
Meson decay. *See* mesons; weak
   interactions; specific
   particle types
Mesons
 annihilation of quark-antiquark
   ($q_i, \bar{q}_i$) pairs, 150–151,
   154
 B meson decay, 306–307, 308
 Clebsch–Gordan coefficients,
   171, 173
 Coulomb-like interaction
   energy, V (R), 128
 D mesons
   decay, 302–305, 308–309,
     327–329
   lack of mixing in $D^0, \bar{D}^0$
     system, 327–329
   mass eigenstates, 303
 eigenstates, 151, 152, 315
 η meson (eta) decay, 209–211
 g (constant), value and
   symmetry assumptions,
   173, 175
 general model for meson
   masses, 150–152
 gluon exchange during
   annihilation, 150–151,
   154, 162, 169
 ideal mixing and other mix
   states, 153–155, 162,
   217–220
 isospin conservation during
   decay, 155
 isospin violating mass
   splittings, 205–207
 K and K^* meson masses, 204
 K^{*+} meson decay, 172–173
 light mesons, 149–155, 309
 light vector meson decay widths
   to e^+e^-, 212
 magnetic spin-spin interaction
   effect on meson masses,
   206, 207
 mass eigenstates, 161–163
 mass splitting, 205–208, 313
 meson Hamiltonian (H_m), 150,
   151, 154, 162
 meson mass values, 204
 meson propagator, 157–158,
   161–163, 234–236
 meson states and quantum
   numbers, 149–150, 152,
   240
 octet and singlet states, 150,
   151
 Okubo–Zweig–Iizuka rule, 170
 ω meson decay to $\pi^+\pi^-$,
   217–220
 ω meson decay to $\pi^+\pi^-\pi^0$,
   164, 166–167, 168–169
 ϕ meson decay to $\eta' + \gamma$,
   198–200
 ϕ meson decay to K^+K^-,
   169–170, 171
 ϕ meson decay to $\pi^+\pi^-\pi^0$,
   169–170
 ϕ meson decay to $\rho\pi$, 170
 ϕ meson mass, 204
 π meson (pion)
   charged pion decay, 274–280
   mass, 204
   π^- lifetime, 274
   size, 142
   spin splitting, 202
 pion form factor [$F_\pi(q^2)$],
   214–216
 propagation as mass eigenstates,
   161–163, 313, 315–317
 pseudoscalar mesons
   constituent quark masses
     (m_P), 203
   decay rates summed over
     final helicities
     [$\Gamma(P \to V + \gamma)$], 221
   decays into two photons,
     224–226
   effects of spin interactions on
     masses, 201, 202, 203
   masses, 153, 203, 204
   matrix elements (M_{fi}) for
     radiative transitions, 221
   mixing, 153, 315–318
   octet and singlet states,
     152–153
   radiative decay widths, 225
 resonances
   amplitude for interaction
     [$\pi(q^2)$], 157–158, 162,
     234
   Breit–Wigner propagators,
     155–159, 160, 188, 267
   decay widths (Γ_R), 159, 160,
     162
   independent events, 160–161
   loop corrections due to
     interactions while
     propagating, 158, 234
   narrow resonances, 159–161
   ω^0 resonances, decay into
     pions, 5–6
   particle propagation and
     interactions, 157–158,
     161–163, 234–236, 315
   π mesons, interactions, 5
   production of vector meson
     resonances, 156–157
   ρ^0 resonances, decay into
     pions, 5–6
   S matrix element for
     resonance to go to final
     state, 158
 ρ^+ decay, 173
 ρ^0 decay, 173–174, 173–177

Index

ρ meson mass, 204
spin, 12
spin-independent
  electromagnetic
  contributions to meson
  masses, 206, 207
spin interaction effect on meson
  masses, 201, 202–204
spin splitting, 135, 202
strangeness, 149, 155
transformations between states,
  150
u, d mass difference
  contributions to meson
  masses, 207
vector mesons
  angular distribution, 163, 164
  constituent quark masses
    (m_V), 203
  Dalitz plot of decay
    $\omega^0 \to \pi^+ \pi^- \pi^0$,
    166–167
  decay rates summed over
    final helicities
    [$\Gamma(V \to P + \gamma)$, 221
  decay widths (Γ), 137, 140,
    156, 159, 163
  effects of spin interactions on
    masses, 201, 202, 203
  light vector meson decay
    widths to e^+e^-, 212
  masses, 154, 203–204
  matrix element (M_{fi}) for
    $\omega^0 \to \pi^+ \pi^- \pi^0$ decay,
    168–169
  matrix elements (M_{fi}) for
    radiative transitions, 221
  octet, 153–154
  production of vector meson
    resonances, 156–157
  ρ, ω mixing, 217–220
  strong decay to pseudoscalar
    mesons, 156, 163–164
  $SU(3)$ decay width
    expectations for vector
    mesons, 174–175
See also charged pion (π
  meson) decay; hadrons;
  mass spectrum of baryons

and mesons; photon
  coupling to vector mesons;
  quarks (spin $\frac{1}{2}$ particles);
  radiative transitions
  between pseudoscalar and
  vector mesons; weak decay
  of pseudoscalar mesons
Mixing phenomena
  CKM mixing, 272, 280
  ideal mixing, 153–154, 220,
    222, 266, 312
  isospin mixing, 154–155,
    209–211
  mixing angles, 153, 336–337,
    343, 349–350
  mixing in meson propagation,
    162, 315
  mixing in vector meson mass
    matrix, 188–189
  quark families, 10
  ρ, ω mixing, 217–220
  See also weak mixing
  phenomena
Multiplets
  Casimir operators, 119
  constraints for fermions or
    bosons, 120–121
  definition and rules, 119, 121
  eigenstates, 119
  pion exchange between proton
    and neutrons, 122
  proton-neutron doublet $SU(2)$
    isospin symmetry,
    121–122
  quark and antiquark multiplets
    (3 and $\bar{3}$), 125
  quark-antiquark bound states in
    mesons, 120
  state for multiplets made of two
    quarks, 125
  state of multiplets made of three
    quarks, 126
  state of multiplets with one
    colored quark and one
    colored antiquark, 125
  $SU(3)$ flavor multiplets,
    matrices and
    transformations,
    147–149

$SU(2)$ states formed from $|p\rangle$
  and $|n\rangle$, 123–124
unitary transformations,
  119–120
Muons (μ^-)
  charged current matrix element
    calculations, 258–260
  Coulomb scattering, 389
  decay width, 260–261, 261
  electromagnetic effects on
    muon decay, 261, 262
  Fermi four-fermion coupling
    G_F, 256, 370
  helicity state calculations,
    258–259
  interaction vertices, 7
  lifetime (τ_μ), 261
  lowest-order diagram for muon
    decay, 258
  mass, 7, 258
  neutrinos released, 258
  radiative processes, 389–391
  weak decay processes involving
    gauge bosons, 8–9,
    258–262
  See also electron-muon ($e^- \mu^-$)
    scattering; leptons

Neutral current mixing, 348–352
  charge operator (Q^Y), 349
  commuting transformation of
    isospin doublets, 349–350
  mass eigenstate fields, 350
  mixing for B^0 and W^0 bosons,
    350–352
  neutral current equation, 339
  Second-order weak neutral
    current process, 353
  $SU(2) \times U(1)$ gauge symmetry,
    348–349
  weak hypercharge (Y), 349,
    351
  weak interaction Hamiltonian
    density, 348
  weak mixing angle, 350
  weak neutral current coupling
    factors, 353
  Z^0 boson coupling to fermions,
    352–354

Neutral current mixing (*cont.*)
  Z^0 branching ratios, 355
  Z^0 coupling to fermions, 352–354
  Z^0 cross section, 355–356
  Z^0 decay width, 354–355
  Z^0 mass, 352
  Z^0 propagator, 355
  See also electroweak interactions; gauge bosons; weak mixing phenomena
Neutrino oscillations, 333–345
  CP violation in neutrino mixing, 344–345
  definition, 313
  intensity ($I_{\beta\alpha}$), 334–335, 337
  mass pattern, 313
  mass splitting, 313
  matter-induced oscillations in the sun, 341–344
  mixing angles, 336–337
  mixing of ν_μ and ν_τ, 335–336
  MSW (Mikheyev, Smirnov, and Wolfenstein) effect, 341
  oscillations of accelerator neutrinos, 337
  oscillations of atmospheric neutrinos, 335–336, 337
  oscillations of reactor neutrinos, 337–338
  oscillations of solar neutrinos, 339–341
  suppression of ν_e oscillations to ν_μ and ν_τ, 335, 336–337
  three-generation neutrino mixing, 336–338, 344
  two-generation neutrino mixing, 335, 337–338
  $U_{\alpha i}$ (neutrino analogue of the CKM matrix), 334
  See also neutrinos; weak mixing phenomena
Neutrinos
  accelerator-produced neutrinos, 337, 345
  atmospheric neutrinos, 335, 337
  charged current interactions, 333, 338–339, 342
  Cherenkov detectors, 335–336, 338–340, 344
  determination of neutrino direction, 338–339, 340
  elastic scattering on electrons, 339, 340
  lack of charge and interactions, 8
  left-handed neutrino production in weak interactions, 116
  lepton flavor eigenstates ($|\nu_\alpha\rangle$), 334
  Majorana neutrinos, 375–376
  mass eigenstates (ν_i) in neutrino propagation, 269, 313, 333
  neutral current interactions, 339, 342
  neutrino scattering processes via W or Z^0 exchange, 9
  nuclear transformations induced by ν_e, 340–341
  ν_e detection, 335–336, 338–341
  ν_μ detection, 335–336
  parity violations in weak interactions of neutrinos, 116
  π^+ decay to $\pi^+ \nu_\mu$, 314, 315
  production during neutron decay, 185
  reactor-produced antineutrinos, 337–338
  solar neutrino energy spectrum, 338, 339, 341
  solar neutrino production, 338
  $U_{\alpha i}$ (neutrino analogue of the CKM matrix), 334
  weak interactions involved in production and detection, 313
  See also leptons
Neutrons
  binding energy, 12
  free neutron decay (d quark decay), 12
  magnetic moment, 196
  neutron decay, current conservation by QCD chiral symmetry, 242–243
  neutron decay, neutrino production, 185
  proton-neutron doublet isospin symmetry, 121–122
Noether's theorem, 20
Nuclei
  binding energy per nucleon, 12

Okubo–Zweig–Iizuka rule, 170

Pair production, 212–213, 389–395
  cross section, 390, 391–394
  e^+e^- pair production diagrams, 213
  energy attenuation, 395
  minimum momentum transfer, 389–390
  radiation length, 394–395
  rate calculation, 391–395
  See also photon coupling to vector mesons
Parameters, 269–270, 271
Parity transformations, 25, 116–117, 168
Parity violations
  forward-backward asymmetry, 358
  left- and right-handed fermion weak interactions, 24–25, 253, 255
  left-right asymmetry (A_{LR}), 357–358
  weak interactions of neutrino, 116
Particle decay rates. *See* decay rates, calculations
Particle momentum, effects on amplitude, 15–16
Particle spectrum, measurement, 3
Particle-antiparticle interactions, 112–113
Penguin diagrams, 287–288, 298, 303, 305, 306
Perturbation expansion. *See* S matrix
Photon coupling to vector mesons, 211–220

Index

current matrix element, 211, 212
light vector meson decay widths
    to e^+e^-, 212
pair production, 212–213
values of $g_{\gamma V}$ from light vector
    meson decay widths, 212
vector meson dominance
    Feynman diagram for
        $K^- \to \pi^0$, 283
    Feynman diagram for virtual
        photons with $q^2 < 0$ and
        $q^2 > 0$, 216
    $K^- \to \pi^0$, branching ratios,
        284–285
    $K^- \to \pi^0$, coupling constant
        (g_{WK^*}), 284
    $K^- \to \pi^0$, decay width (Γ),
        284–285
    pion form factor $[F_\pi(q^2)]$,
        214–216
    q^2 dependence of form
        factors, 284
    ρ, ω mixing, 217–220
    ρ dominance in the π^+, π^-
        channel with $J^P + 1^-$,
        214–217
    ρ meson contribution for
        $\gamma + p \to$ hadrons,
        213–214
    vector meson dominance
        and radiative decays,
        225–227
vector meson width, 211, 212
See also electromagnetic
    interactions; mesons;
    photons
Photon propagators
    electron-muon $(e^-\mu^-)$
        scattering, 77–79
    loop correction to photon
        propagator, 97, 103, 104,
        234
    matrix element for single
        photon exchange (M_{fi}),
        78–79
Photons
    amplitude calculations, 24–27
    comparison to massive particles,
        24

electromagnetic current density
    (J_ν), 25
electroweak interactions, 9–10
gauge invariance, 7
helicity, 24–25
interactions with leptons, 7
lack of charge, 7
local gauge transformations,
    25–27
negative charge conjugation,
    112
photon field calculations, 25–27
polarization choices, 27
polarization vectors, 26–27
relations involving lepton
    currents, 73–74
spin, 7, 24–27
wavelength and energy, 197,
    199
Pions (π mesons). *See* mesons
Planck mass, 442, 443
Polarization vector, 23–24, 26–27,
    99, 117, 200
Positron-electron (e^+e^-)
    annihilation to
    antimuon-muon $(\mu^+\mu^-)$,
    86–88
    annihilation cross section (σ),
    88
    Lorentz invariant quantities,
    86–88
Positron-electron (e^+e^-)
    annihilation to hadrons,
    405–412
    cross section ratio R, 405–406
    energies below Z^0, 405–406
    energies in the electroweak
    regime, 410–412
    Higgs boson, 410–412
    quark fragmentation functions,
    406–410
    See also large momentum
    transfer processes
Positron-electron (e^+e^-)
    production of
    positron-electron (e^+e^-),
    90–94
Bhabha scattering, 92–93
bound states, 91–92

differential cross section in the
    center of mass $(d\sigma/d\Omega)$,
    92, 93
Feynman diagrams, 91
helicity conservation in the
    relativistic limit, 92–94
matrix element (M_{fi}), 91, 92
scattering current and
    annihilation current
    operators, 92–94
Positronium (electron-positron
    system), 91–92, 132, 133,
    227–228
See also positron-electron
    (e^+e^-) annihilation to
    antimuon-muon $(\mu^+\mu^-)$;
    positron-electron (e^+e^-)
    production of
    positron-electron (e^+e^-)
Potential, as a consequence of
    particle interactions, 3
Processes with two leptons and
    two photons, 94–103
    Feynman diagrams, 94, 95,
    96
    higher-order terms in the
        perturbation expansion
    higher-order $e^-\mu^-$
        scattering, 105
    loop correction to photon
        propagation, 97
    spinor expressions for $g(p)$ and
        $g(-p)$, 96–97
    virtual particle types, 94
    See also Compton scattering
        $(\gamma + e^- \to \gamma + e^-)$
Projection operators, 36, 68
Protons
    anomalous moment interaction
        (Dirac–Pauli
        representation), 194
    baryon number, 149
    gauge invariant interactions of
        protons, 193
    Hamiltonian density $(\mathcal{H})$
        calculations, 194–195
    interaction Hamiltonian (H) for
        spin in a magnetic field,
        195

468 Index

Protons (*cont.*)
  magnetic moment, 196
  momentum spectrum for
    constituents, 415–417
  muon-proton scattering, 425
  pp or $p\bar{p}$ collisions, 413,
    436–437
  proton decay, 441–442
  proton-neutron doublet isospin
    symmetry, 121–122
  size, 142, 418
  stability, 185
  structure functions, 423–429
  wave function, 183
  See also deep inelastic
    lepton-proton scattering;
    structure functions
Pseudoscalar mesons. *See*
  mesons
Pseudoscalar mesons, weak decay
  of. *See* weak decay of
  pseudoscalar mesons

q (4-momentum transferred), 47
QCD. *See* quantum
  chromodynamics (*QCD*)
QED. *See* quantum
  electrodynamics (*QED*)
Quantum chromodynamics (*QCD*)
  λ_{QCD}
    defined as mass unit, 135,
      239–240
    uncertainty of source for
      λ_{QCD}, 234, 239
  lattice gauge calculations, 248,
    282
  massless quark theory and the
    massless limit of *QCD*,
    202–203, 240–248
  meson propagator, 234–236
  pseudoscalar meson eigenstates
    from *QCD* Hamiltonian,
    209
  QCD Lagrangian, 233, 241,
    243, 246
  running coupling constant for
    QCD, $\alpha_s(|q^2|)$, 238–240
  u mass from *QCD* theory,
    202–203

  u mass relationship to λ_{QCD},
    202–203
  See also color interactions;
    color potential; running
    coupling constant $\alpha(q^2)$;
    structure functions;
    $SU(3)$
Quantum electrodynamics (*QED*)
  factors required to calculate
    $-M_{fi}$ for any Feynman
    diagram, 105–106
  as the first successful particle
    theory, 71
  relationship between q^2, m^2,
    and $\pi(q^2)$, 234–235
  renormalization, 236
  running coupling constant
    $\alpha(q^2)$ in electrodynamics,
    234–238
Quantum events, 1
Quantum Field Theory, 58–68
  annihilation operators, 59, 61,
    64, 67
  anti-commutation relations,
    64–66
  charged scalar fields, 63–64
  commutator, 59–60, 62
  creation operators, 59, 60, 61,
    64, 67
  eigenstates, 60
  fermion fields, 64–68
  Fock Space, 58–59, 63
  Hamiltonian operators, 61–62,
    63, 65–66
  H_F time independence, 62
  scalar particle charge operator,
    63–64, 65, 66
  spin operator, 67
  zero-point or vacuum energy of
    the field, 62
  See also amplitudes,
    calculations
Quantum mechanics, basic
  principles, 2–3
Quantum numbers and quantum
  rules, 113–117
  additive quantum numbers
    (energy, momentum),
    114–115

  angular momentum operators
    J_x, J_y, J_z, 115–116
  Casimir operators, 115–116
  charge conjugation operation,
    116
  conserved momentum, 114–115
  Feynman diagrams, 113
  Hermitian generator of the
    transformations, 114–115
  isospin quantum numbers
    (I, I_3), 121, 147
  lepton numbers (L_e, L_μ, and
    L_τ), 113
  meson states and quantum
    numbers, 149–150, 152,
    240
  multiplicative quantum
    numbers, 116–117
  parity rules and exceptions,
    116–117, 131
  positronium (electron-positron
    system), 92
  quantum conservation rules,
    113
  $SU(2) \times U(1)$ quantum
    numbers, 354
  vectorially additive quantum
    numbers, 115–116, 125
  Y (weak hypercharge quantum
    number), 349
Quarks (spin $\frac{1}{2}$ particles), 10–13,
  146–155, 193–227
  b (bottom) quark, mass, 130,
    136, 307
  b (bottom) quark, weak decay,
    10, 305–307
  baryon number conservation,
    149
  bottomonium ($b\bar{b}$), 131
  c (charm) quark, decay,
    302–305
  c (charm) quark, mass, 130,
    136, 307
  charged weak currents for
    quarks, 267–269
  charmonium ($c\bar{c}$), 131, 136,
    140
  colored quarks in colorless
    multi-quark states, 120

confinement, 11–12, 128–129, 135, 404–405
constituent quark masses (m_u and m_s), 195, 196, 202, 203–204
constituent quark (m_{eff}), 142
cross section factors for short distance scattering, 414–415
d (down) quark, weak decay, 11
DGLAP (Dokshitzer, Gribov, Lipatov, Altarelli, Parisi) equations, 435
doublet families, 10
e^+e^- production of $c\bar{c}$ and $b\bar{b}$ resonances, 130–131
e^+e^- production of hadronic jets, 179–180
electric charges, 10
electroweak interactions, 10–12
flavors (up, down, charm, strange, top, and bottom), 10
Heavy Quark Effective Theory, 307–309
interaction Hamiltonian (H_q) for spin in a magnetic field, 195
light quark flavors [u (up), d (down), and s (strange)], 146–149
light quark isospin quantum numbers, 147
light quark $SU(3)$ flavor symmetry, 146–147
masses, 10, 136
masses in the Lagrangian, 246–247
massless quark theory and the massless limit of QCD, 202–203, 240–248
mixing of states, 146
n_f description of "light" flavor, 135
particle exchange properties of $SU(n)$ generators, 123–124
$q_1\bar{q}_1$ annihilation through a gluon or through a photon, 413
quantum numbers of states, 131
quark and antiquark multiplets (3 and $\bar{3}$), 125
quark flavor changes by weak interactions, 287
quark fragmentation functions in jet structures, 406–410
quark-gluon scattering, 414
sea quarks, 425, 427, 428
size of bound states, 136–141
stability, 10–12
state for multiplets made of two quarks, 125
state of multiplets made of three quarks, 126
state of multiplets with one colored quark and one colored antiquark, 125
strangeness, 149
strong interactions, 11–12
structure functions, 423–429
$SU(3)$ shifts among three colored quarks, 124
symmetry transformation for three quark types, 118
t (top) quark decay, 130, 300–302
t (top) quark mass, 372
u, d mass difference contributions to meson masses, 207
$u \leftrightarrow d$ $SU(2)$ symmetry (isospin), 147
u mass from QCD theory, 202–203
valence distribution functions, 424
valence quarks, 423, 424, 425
vector particle Υ, size, 133, 137–141
See also baryons; charged weak currents for quarks; color interactions; color potential; fundamental particles and interactions; hadrons; mesons; multiplets; $SU(3)$; weak decays of the heavy quarks
Radiative transitions between pseudoscalar and vector mesons, 220–223
decay rates summed over final helicities, 221, 223
$f_{VP\gamma}$, 221–222
matrix elements (M_{fi}), 221
See also mesons
Rapidity (y), 399–401, 406, 436
Renormalizability, 248
Resonances
decay lifetime (τ), 5
decay width (Γ), 5
mass spectrum and pattern, 5
measurement, 5–6
at thermal equilibrium, 5
transitions from heavier to lighter states in accelerator experiments, 5
See also under mesons
Running coupling constant $\alpha(q^2)$, 234–240
in electrodynamics, 234–238
for large q^2, 238
meson propagator, 234–237
photon propagator, 234–235
for QCD, $\alpha_s(|q^2|)$, 238–240
for small q^2, 237
See also quantum chromodynamics (QCD)

S matrix
CPT theorem and CPT symmetry, 108–109
energy and momentum conservation (4-dimensional δ function), 41
higher-order terms in the perturbation expansion, 97, 103–105
loop correction to photon propagator, 97, 103, 104, 234
matrix elements (M_{fi}), amplitude calculations, 41–50
S_{fi}, equation, 41

S matrix (cont.)
  special symmetries of the S matrix, 108–109
  use in time history of states, 41–43
  verification by magnetic moment measurements, 104–105
s-wave
  energy states, 127, 131, 206
  kinetic energy, 142
  spin perturbation, 201–202
  symmetric s-wave final state, 211, 288, 297–298, 318
Scalar particle charge operator (Q), 63–64, 65, 66
Scattering. See Coulomb scattering; electron-electron (e^-e^-) scattering; electron-muon ($e^-\mu^-$) scattering; jets; positron-electron (e^+e^-) annihilation to antimuon-muon ($\mu^+\mu^-$); positron-electron (e^+e^-) production of positron-electron (e^+e^-); processes with two leptons and two photons
Semileptonic decay
  charged pion (π meson) decay rate, 280
  kaon decay (K meson), 282–286
  strange baryons, 185–186
  transitions in kaon decay (K meson), 283
  weak interactions leading to semileptonic decay of baryons, 185, 186
Short distance scattering, 412–417
  cross section factors for light quarks and gluons, 414–415
  differential cross section, 416
  Drell–Yan process, 413
  gluon-gluon scattering, 414, 437
  Mandelstam variables, 414
  pp or $p\bar{p}$ collisions, 413, 436–437
  $q_1\bar{q}_1$ annihilation through a gluon or through a photon, 413
  quark-gluon scattering, 414
  See also large momentum transfer processes
Sizes of bound states (quarks), 136–141
Space-time processes contained in the weak Hamiltonian, 272–273
Special relativity, 15, 54, 446
Spectator diagrams, 287–288, 303–305, 306
Spin 0 particles
  amplitude, calculations, 16–18
  antisymmetric spin state, 180
  pion, as a spin zero particle, 146, 246
  propagators, 76, 234, 248
  $W_{\mu\nu}$, 422
Spin 1 particles
  amplitude, calculations, 23–24
  momentum-space propagators, 76, 77
  photon propagators, 77–78
  space-time propagators, 76
  spin vectors ($\hat{e}_x, \hat{e}_y, \hat{e}_z$), 76–77
  tensors $g(p)$ and $g(-p)$, 76–77
  See also gauge bosons; gluons; photons
Spin dependence [$\chi(p_\mu)$]
  free particle amplitude (ϕ) calculations, 15–16
  polarization vector (e_μ), 23–24
  spin 1 particles, 23–24
Spin notation, 4-component, 67
Spin operators, spin $\frac{1}{2}$ particles, 27–28, 31, 67
Spin splitting, 135, 202
Spontaneously broken symmetry, 244–246, 366–367, 369
Standard Model
  flavor transformations predicted by the Standard Model assumptions, 309
  general concepts, 6–13
  interactions at high energy, 441–444
  predictions for e^+e^- annihilation to hadrons, 405, 411
  predictions for three neutrino species, 356
  requirements for a renormalizable theory, 248
  See also fundamental particles and interactions
Strange particles, interactions
  baryon strangeness values, 181
  charged weak currents for quarks, 267–269
  mesons, strangeness, 149, 155
  quarks, strangeness, 149
  semileptonic decay of strange baryons, 185–186
  tau lepton (τ^-), decay to strange quarks, 265–267
  See also kaon decay (K meson); quarks
Strangeness, 149
Strong interactions
  charge structure of strong interactions, 112–113
  hadronic collisions, 378–379
  in pion mass determination, 5
  in ρ^0 and ω^0 decay, 5
  See also baryons; fundamental particles and interactions; gluons; hadrons; mesons; multiplets
Structure functions, 420–429, 420–436
  antiquark final states, 431–432
  calculation, 420–423
  Callan–Gross relation, 422, 423, 432, 433
  constituent distribution functions, 429, 434–436
  contribution of gluons to quark scattering by the photon, 430
  DGLAP (Dokshitzer, Gribov, Lipatov, Altarelli, Parisi) equations, 435
  differential cross section, 423

electromagnetic vertex where
quark scatters,
432–433
lowest order corrections to
simple quark constituent
picture, 429–433
QCD interactions, 429–433
QCD vertex calculation,
430–431
quark picture for the structure
function, 423–429
quark propagator, 431
scaling violations, 429–436
sea quark distribution, 425, 427,
428
splitting functions, 435
valence quarks, 424, 425
$W_{\mu\nu}$ calculations, 420–422
See also deep inelastic
lepton-proton scattering;
protons; quantum
chromodynamics (QCD)
$SU(2)$
Casimir operators, 123
different W couplings to left-
and right-handed fermions,
253
equal masses of members of
isospin multiplets, 9, 148,
152
lowering operators, 272,
273–274
mass-generating reaction,
$SU(2)$ symmetry breaking,
9, 257
matrices ($T_i = \sigma_i/2$), 122,
124–125
matrix commutation relations,
122
particle exchange properties of
$SU(n)$ generators,
123–124
raising operators, 272–273
requirement for massless
fermions, 253, 257
space-time processes contained
in the weak Hamiltonian,
272–273
$SU(2)$ generators, 122, 253

two-particle states formed from
$|p\rangle$ and $|n\rangle$, 123–124
$u \leftrightarrow d$ $SU(2)$ symmetry
(isospin), 147
See also gauge bosons; weak
interactions
$SU(3)$
approximate $SU(3)$ flavor
symmetry for u, d, and s
quarks, 146–147, 241–242,
244, 246
Casimir operators, 124
Coulomb-like interaction
energy, $V(R)$, 128
diagonal operators, 123, 125
isospin raising operators, 184,
242
matrices ($T_i = \lambda_i/2$) for quark
states, 122–123, 124–125
matrix commutation relations,
123, 124
particle exchange properties of
$SU(n)$ generators,
123–124
prediction of equal messes in a
multiplet, 205
quark and antiquark multiplets
(3 and $\bar{3}$), 125
shifts among three colored
quarks, 124
$SU(3)$ decay width expectations
for vector mesons,
174–175
$SU(3)$ flavor multiplets, analog
isospin states, 148
$SU(3)$ flavor multiplets,
matrices and
transformations, 147–149
$SU(3)$ generators, 122–123
$SU(3)$ symmetry for quarks due
to quarks in each singlet,
129–130
$SU(3)$ vector addition, 125
$SU(2) \times U(1)$ gauge symmetry,
348–349
quantum numbers, 354
quantum numbers for
fermion-scalar coupling,
373

Weinberg and Salam theory,
365
See also electroweak
interactions; symmetry
breaking
Supersymmetry, 442–444
Symmetry breaking
corrections to W and Z masses,
370–373
mass of gauge bosons (W^+,
W^-, and Z^0), 9–10,
254
spontaneous symmetry
breaking, 20, 245–246,
366–367
in vector meson decays, 163
weak gauge theory, 253
See also electroweak
interactions; Higgs
mechanism

Tau lepton (τ^-)
amplitude for decay, 265–266
axial vector current, 264
branching ratios, 263–265
coupling constant g_W, 266
decay to hadrons, 263–267
decay to strange quarks,
265–267
decay to u and d quarks, 263
decay width, 266
effective weak Hamiltonian,
255–256, 264
electromagnetic current
operator, 266
G parity, 264
gluonic corrections to branching
ratio, 264
hadronic vector current, 264
interaction vertices, 7
lifetime (τ_τ), 262–263
mass, 7, 262
polarization (left- and
right-handed τ^-),
357–358
weak decay processes involving
gauge bosons, 8–9,
262–267
See also leptons

Three-body decay, 164–168
    Dalitz plot density calculations, 165–166
    Dalitz plot of decay $J/\psi \to K^+K^-\pi^0$, 166–168
    Dalitz plot of decay $J/\psi \to \pi^+\pi^-\pi^0$, 166–167
    Dalitz plot of decay $\omega^0 \to \pi^+\pi^-\pi^0$, 166–167
    Dalitz plot of event density versus E_1 and E_2, 165
    invariant masses related to the center of mass energies, 165
    variables describing configuration in the decay plane, 164–165
    variables describing dynamics, 164–165
Time history of states, 38–50
    electromagnetic interactions, 43–45
    Feynman diagrams for $e^-\mu^-$ scattering via scalar exchange, 47–48
    first-order interaction vertices, 41–42, 43
    momentum-space propagator (f), 47, 50–52
    S matrix elements, 41, 43
    scalar interactions, 45–46
    second-order interaction terms, 43–44
    space-time propagator (D_F), 46
    time-dependent perturbation theory, 38–43
    time evolution and particle exchange, 43–50
    time orderings in scalar exchanges, 48–49
    See also Hamiltonian density ($\mathcal{H}$); Hamiltonian (H) operators
Two-body scattering. See Lorentz invariant quantities for two-body scattering; specific particle types

Uncertainty principle, 313–314
Unitary matrices and quark masses, 268
Unitary operator (U), 39–41, 44, 107–108, 118
Universality, 254

Vacuum state, notation, 38
Vector meson dominance. See photon coupling to vector mesons
Vector mesons. See mesons
Virtual scalar exchange
    amplitude calculations, 43–50
    Feynman diagrams for $e^-\mu^-$ scattering via scalar exchange, 47–48
    4-momentum (q) transferred, 47, 50
    momentum-space propagator (f), 47, 50–52
    space-time behavior of scattering, 46–50
    space-time propagator (D_F), 46, 51
    time orderings in scalar exchanges, 48–49
V_{UD} matrix, 268–269
    See also Cabibbo–Kobayashi–Maskawa (CKM) matrix; charged weak currents for quarks

W bosons. See gauge bosons
Wave-particle-duality, 2
Weak decay of pseudoscalar mesons, 292–295
    amplitudes, 294–295
    CP violation, 295
    lifetime (Γ_P), 294
    mass matrix (Δm) for mixing, 316–317
    mixing phenomena, 315–318
    role of unitarity, 293
    strong phase δ_{ni}, 294
    weak phase ϕ_{ni}, 294
    weak scattering matrix, 293–294

width matrix (Γ) for mixing, 316–317
    See also mesons; weak interactions; weak mixing phenomena
Weak decays of the heavy quarks, 300–307
    B meson decay, 306–307, 308
    bottom particle lifetimes, 306
    bottom quark (b) decay, 10, 305–307
    charm quark (c) decay, 302–305
    charmed particle lifetimes, 302–303
    D meson decay, 303–305, 308–309
    decay widths, 301, 302
    Heavy Quark Effective Theory, 307–309
    Isgur–Wise function, 309
    top quark (t) decay, 300–302
    See also quarks (spin $\frac{1}{2}$ particles); weak interactions
Weak effective Hamiltonian, 271, 286–287, 289–290, 300, 353
Weak gauge group. See gauge bosons
Weak interactions
    comparison of weak and strong interaction theories, 253
    quark flavor changes by weak interactions, 287, 300
    space-time processes contained in the weak Hamiltonian, 272–273
    weak effective Hamiltonian, 271, 286–287, 289–290, 300, 353
    See also electroweak interactions; fundamental particles and interactions; gauge bosons; weak decay of pseudoscalar mesons; weak decays of the heavy quarks; weak interactions of fermions; weak mixing phenomena

Index

Weak interactions of fermions
  change in fermion charge, 253
  charged current interactions, 254–255
  charged current matrix elements (M_{fi}) for low q^2, 255–256
  charged weak currents for quarks, 267–274
  comparison of weak and strong interaction theories, 253
  different W couplings to left- and right-handed fermions, 253
  effective weak Hamiltonian, 255–256, 264
  Fermi four-fermion coupling G_F, 256, 262, 370
  fermion field doublets (ϕ_L), 253–254, 255
  free field Lagrangian ($\mathcal{L}_{Free}$), 256–257
  interaction Hamiltonian density ($\mathcal{H}$), 253–254, 255
  lepton (μ^- and τ^-) decay by emission of gauge boson, 8–9
  matrix elements (M_{fi}) for low q^2, 255–256
  muon (μ^-) decay, 8–9, 258–262
  neutral weak currents for quarks, 268, 348
  neutrino scattering processes via W or Z^0 exchange, 9
  parity violations by left- and right-handed fermions, 24–25, 253, 255
  second-order charged current process for low q^2, 255–256
  $SU(2)$ generators (T_k), 253
  tau lepton (τ^-) decays, 8–9, 262–267
  W propagator, 255
  weak eigenstate fields, 267
  weak Hamiltonian, 272–273
  weak interactions leading to semileptonic decay, 185, 186
  See also electroweak interactions; fundamental particles and interactions; gauge bosons; $SU(2)$

Weak mixing phenomena
  B^0, $\bar{B}^0$ and B_s^0, $\bar{B}_s^0$ meson systems, 329–333
  amplitude, 332
  CKM factors, 330–331, 332, 333
  CP violating parameter (β), 332–333
  CP violation, 331–333
  diagrams and terms for mass mixing, 329–330
  time-dependent asymmetry, 332
  weak neutral current couplings, 353
  D^0, $\bar{D}^0$ system, 327–329
  degree of mixing (χ), 327–328
  K^0, $\bar{K}^0$ system, 280–283, 318–327 (*See also* mixing phenomena; weak interactions)
  amplitudes, 318, 323–324
  asymmetry of intensity functions [$A(t)$], 321–322
  box diagrams involved in K^0, $\bar{K}^0$ mixing, 319
  CP violation, 321–327
  decay widths, 318–319
  degree of mixing (r), 321
  eigenstate notation, 318
  GIM mechanism, 319
  intensities for kaon components, 320, 321–322, 326, 327
  K_L^0 (long lifetime) CP state, 280–281, 318–319, 323
  K_L^0 (long lifetime) CP violations, 321
  K_S^0 (short lifetime) CP state, 280–281, 318–319, 323
  mass difference (Δm_K), 319
  mass eigenstate evolution, 326
  oscillations, 320–321
  propagation equation, 321
  Re(ε), 322
  mass matrix (Δm) for mixing, 316–317
  time integrated mixing parameter (r), 321, 327
  weak decay of pseudoscalar mesons, 315–318
  width matrix (Γ) for mixing, 316–317
  See also mixing phenomena; neutrino oscillations; weak decay of pseudoscalar mesons

Weinberg and Salam theory, 365–370
  gauge boson mass generation ($\mathcal{L}_{scalar}$), 365, 367–370
  parameters determining $\phi(x)$, 366
  weak isospin rotation, 366
  See also electroweak interactions; Higgs mechanism; $SU(2) \times U(1)$ gauge symmetry; symmetry breaking

Wigner–Eckart theorem, 286, 289

Yukawa potential
  amplitude calculations, 17, 20
  Born approximation, 57
  cutoff Coulomb potential, 384
  effect of coupling constants (g), 17, 58
  electron-muon ($e^-\mu^-$) scattering, 58
  interaction potential energy, 20
  particle-antiparticle interactions, 112
  spin zero particle amplitudes, 17, 20

Z^0 bosons. *See* gauge bosons